A Physicist's View of Matter and Mind

A Physicist's View of Matter and Mind

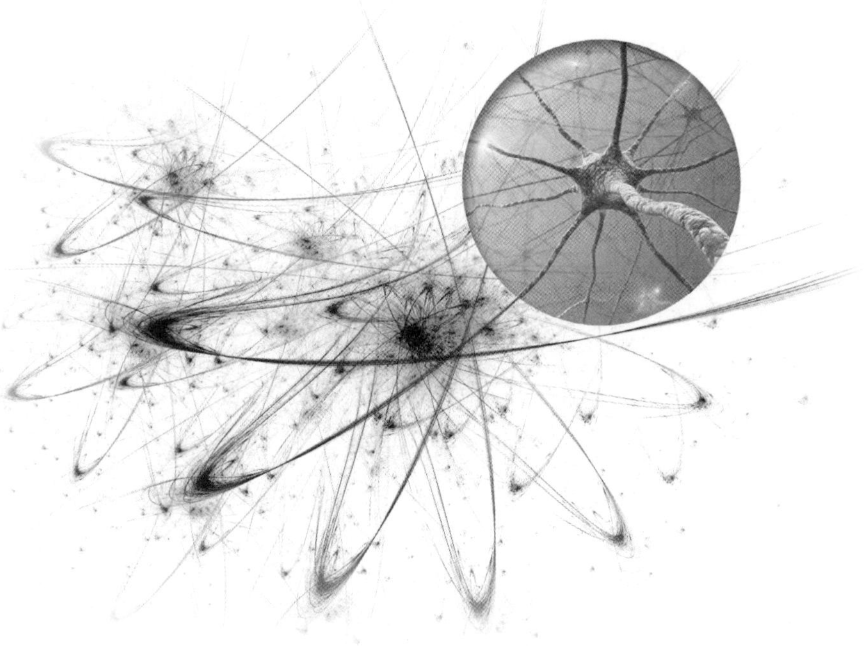

Chandre Dharma-wardana

National Research Council of Canada

University of Montreal, Canada

World Scientific

NEW JERSEY • LONDON • SINGAPORE • BEIJING • SHANGHAI • HONG KONG • TAIPEI • CHENNAI

Published by

World Scientific Publishing Co. Pte. Ltd.
5 Toh Tuck Link, Singapore 596224
USA office: 27 Warren Street, Suite 401-402, Hackensack, NJ 07601
UK office: 57 Shelton Street, Covent Garden, London WC2H 9HE

Library of Congress Cataloging-in-Publication Data
Dharma-wardana, Chandre.
A physicist's view of matter and mind / Chandre Dharma-wardana, National Research Council of Canada, University of Montreal, Canada.
pages cm
Includes bibliographical references and index.
ISBN 978-9814425414 (hardcover : alk. paper)
1. Physics--Philosophy. 2. Physics--Popular works. 3. Consciousness. 4. Dharma-wardana, Chandre--Philosophy. I. Title.
QC6.D474 2013
530.01--dc23

2013005327

British Library Cataloguing-in-Publication Data
A catalogue record for this book is available from the British Library.

Printed in Singapore by B & Jo Enterprise Pte Ltd

Trust in consilience is the foundation of the natural sciences. For the material world at least, the momentum is overwhelmingly toward conceptual unity. Disciplinary boundaries within the natural sciences are disappearing, to be replaced by shifting hybrid domains in which consilience is implicit. ... Nothing fundamental separates the course of human history from the course of physical history, whether in the stars or in organic diversity.

Edward O. Wilson, Consilience, 1998

Preface

This is a physics book that attempts to place physics in the context of a wider framework arching over biology, philosophy, and across C. P. Snow's two cultures. That is, it discusses things that other books, with a narrower vision, neglect to treat. We explore the unity and consilience of our knowledge about 'matter', and what we intuitively understand as 'mind', within the perspective of the 'working physicist' who assumes the existence of an independent external reality.

We attempt to demystify quantum physics as well as the study of consciousness. Of course, a small but influential number of physicists like Eugene Wigner — and *Wigner's friend* [228] — might insist that ultimately we need an observing mind to ensure 'reality'. Some popular-science writers prefer to present quantum physics as something utterly 'weird', leading to what van Kampen has called the 'scandal of physics'. We avoid such pitfalls by following not only Bohm's reinterpretation of quantum mechanics, but also the epistemological consequences of density functional theory. This offers the possibility of describing many-particle quantum systems purely as functionals of the one-particle density, without the need for a wavefunction. We regain the world of everyday reality at human length scales and time scales, without the usual conceptual difficulties of standard presentations. Living organisms and 'consciousness' are presented as emergent properties of physiology. It is fair to say that we fully understand the reductionist path of how to start with a very complex system (e.g., a living cell) and reduce it down to its physics. The path for the build-up from the basic physics to reach the complex system is clear in principle; but the technical steps for going beyond quantum chemistry and biochemistry are still being spelled out in laboratories grappling with nano-science, neuroscience, computational chemistry and related disciplines. However, integrating everything into a consilience including the social sciences may become the common task of physics, philosophy and culture.

We start from fundamentals and attempt to chart how to go to more complex things. However, fundamental things turn out to be surprisingly complicated, and we don't even know what is fundamental! Big things and small things seem to connect up in the most amazing way. Cosmologists who study the grand scale of the universe and particle physicists who study the smallest objects end up asking each other for help, while physicists or philosophers look like fools, and fools look like philosophers or physicists. It is not surprising that humanity has debated essentially the same basic questions about the cosmos, consciousness and life, in the language and idiom of each époque. In this book we address these topics, and discuss them at several levels of complexity and within their historical, philosophical and scientific perspectives.

Hence many topics and many names would be paraded rather quickly in the very first chapter. There we only have space and time to race through the topics, merely to indicate their consilience. We take up most of those topics in greater detail in the course of the book. Chapters can be skipped and returned to, if needed, using cross references. Hence, when the reader comes across a concept or notion that he is not familiar with, he may see its re-emergence further upstream, with further explanations. When a deeper explanation is called for, instead of footnotes or end-notes, we use small text to read on or skip out. The mathematical content (given in small print) will also increase as we enter more and more into the chapters dealing with quantum mechanics, and decrease as we go into biological topics. However, biology entails more demanding discussions of chemistry. Some readers may skip many of these tortuous paths and take their own short cuts, using the table of contents or the index to follow up any topic.

Finally, it is not just myself, but also the readers of this book, who would have cause to be thankful to several of my colleagues who read through the manuscript in full, and others who read it in parts, or in chapters, and indicated the need for corrections of various severity on the 'Richter scale', *a.k.a.* 'the writer's scale'. My colleagues have also suggested clarifications and modifications in regard to various aspects of the presentation of, e.g., Bohmian mechanics, or 'consciousness' etc. However, in the end the book reflects my own views, and I alone bear the responsibility in regard to all errors as well as excursions into what may seem speculative thinking!

Chandre Dharma-wardana
Ottawa, Canada
15 November 2012

Contents

Part I.
The Nature of Physical Law

Chapter 1

A Bird's Eye View of our Quest

This is an overview of the ideas covered in the book, parading many topics and their authors. This overview is not a direct summary, but serves to set the stage. Part I of the book looks at the hierarchical structure of science and physical systems arising from the existence of different energy scales and length scales. 'Scientific revolutions' occur when different energy and length scales are exposed by new tools. The themes unifying physical law at different length scales and at the most fundamental level are discussed. Part II looks at the emergence of collective phenomena, self-organizing processes, persistent structures, membranes, living cells, vision, memory, and consciousness as we consider increasingly complex systems at human length scales.

1.1 Duality, plurality, and reality

The idea of duality is often used in many traditions to imply that the world that we live in has a dual character — mind and matter, soul and soma, *yin* and *yan*, Adam (*Om*) and Eve (*Jeeva*), etc. The denial or defence of this duality constitutes the stuff of most philosophies and religions. Even the methodologies of analysis are categorized as reductionist or constructionist, bottom-up or top-down, and echo one's attitude to this duality. The *Jyana* tradition of Buddhism, renamed as 'Zen' in the hands of Japanese masters, claims to reject duality and embraces a transcendent holistic unity. Willard Quine, espousing an utterly different secular philosophical tradition would also end up in demanding a holistic approach [177].

Modern physics, equipped with mathematics, a language infinitely more subtle and yet more limited than the language of human transactions, includes not only opposing dualities, but also commuting and non-commuting realities. However, in the physics approach, we need to isolate the part of nature that is to be studied, thus setting up a boundary. These boundary conditions are usually imposed to define a simpler system. This process involves working within a larger system and reducing a part of it for study. Many problems (see Sec. 2.2) associated with 'holism, inductive knowledge', or the 'theory-laden' nature of information and such matters can be resolved if the boundary conditions of the reduced system could be clearly stated. Sometimes the selected boundary conditions may have unintended symmetries that allow for dualities (Fig. 1.1) or pluralities in the perceived reality. In practice these symmetries cannot be exactly imposed and a selection is made by the observer (see Sec. 12.3.1). For instance, perception via our senses (vision or other), involves an interpretation of the input data where our brains supply the necessary information for 'framing it', or 'breaking the symmetry'. This view plays a prominent role in Gestalt psychology [147].

If we isolate bits of matter from each other by setting up suitable boundaries, then we can study atoms and even smaller constituents contained in them. This is the *reductionist approach* of the experimental physicist. The basic building blocks of physical reality at this scale are 'excitations' of an under-lying energy field. These excitations may be waves (Fig. 1.2) or particles (Fig. 1.3), depending on whether you set up experiments designed to 'chase' them along (i.e., determine the momentum) or catch them in one spot (i.e., determine the position).

Localized excitations of nature at the lowest level of reduction available to us are known as 'elementary particles'. Frequently mentioned elementary particles are electrons, photons, gluons, neutrinos and quarks. These and other known building blocks of nature can be 'cataloged' in a sensible, comprehensible way using what is known as the 'standard model' of particle physics.

1.1.1 *Contextuality, non-locality and de-coherence*

The existence of a 'wave-particle duality' and its reconciliation are the essential features of the quantum nature of reality which underlies our experience. In the quantum world, actions are called operations, which themselves have

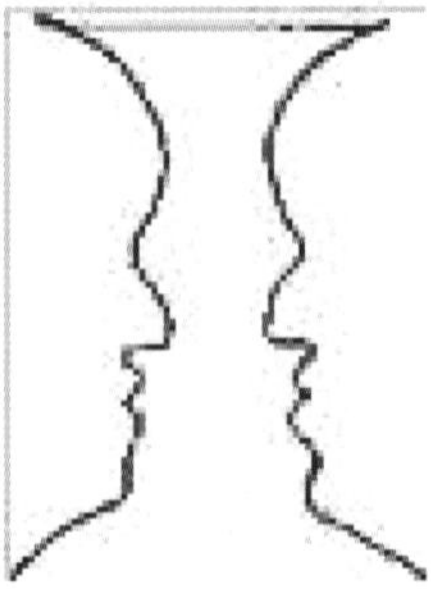

Fig. 1.1 Duality — the Greek vase which flips into two staring faces or reverts to a vase, as one keeps on looking at it. The frame, boundary conditions, or 'gestalt structures' needed to interpret the visual images are provided by the brain itself.

Fig. 1.2 An excitation of a sheet-like surface (an 'energy field') which is 'wave-like'. The wave is not specifically localized at any spot but pervades the whole space defined by the boundary.

mathematical analogues known as operators. Their compatibility or incompatibility is described by mathematical rules (known as commutation rules) somewhat similar to the rules of protocol and precedence that we are familiar with, from actions in everyday life. For instance, in everyday life the operation of crossing the road and then closing our eyes, could be very different from closing our eyes and then crossing the road. So the order of operations, and the context matters. Similarly, sometimes philosophers of science [178] talk of *contextuality* to mean a similar property analogous to those of the quantum theory.

Another property of quantum systems is their extreme inter-connectedness. If there are a thousand particles in the system, the behaviour as observed at one end depends on every part of the system, somewhat as in a hologram. The system is described by a wavefunction which samples every bit of phase space available to it. A type of correlations found in quantum systems, i.e., quantum *entanglement*, is found to involve correlations far exceeding anything that can be

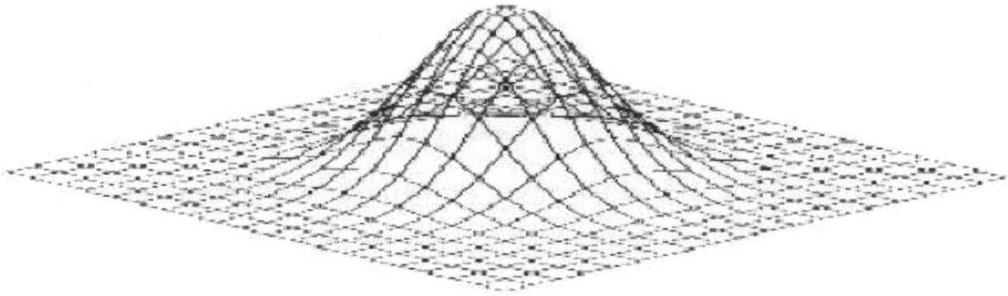

Fig. 1.3 A highly localized excitation of an 'energy field'. This is a 'particle-like' excitation. This is built up by suitably adding (i.e., superposing) a lot of simple waves. An 'anti-particle' can be thought of as made up of waves which form a depression. Then the 'vacuum state' is the state where the underlying 'energy field' is 'flat', in an average sense (i.e., there can be spontaneous fluctuations known as 'vacuum fluctuations' made up of an equal number of particles and anti-particles). The particle can be made as sharply localized as we please by using suitable combinations of waves. However, such wavepackets may not hold together, i.e., stay in coherence, for very long.

produced using classical inter-connectedness, e.g., as found in a jigsaw-puzzle or a rubber netting. This brings to one's mind some 'spooky action' connecting any part of the system to some part not local to it. This non-local property, or *non-locality* of quantum systems, is hard to maintain for long, or over long distances, as the boundaries set up to define the system have to be maintained against all external disturbances. Otherwise, quantum correlations undergo *decoherence* due to perturbations from outside the boundaries of the system.

The larger the system, the more difficult it is to 'protect it' from decoherence. If such decoherence effects are ignored, all sorts of paradoxical conclusions can be drawn. Thus, we do not expect 'spooky action at a distance' within our everyday concepts of 'localized reality'. Macroscopic quantum phenomena would indeed display 'spooky action at a distance'; but decoherence destroys any such possibility. Furthermore, when we consider many-particle systems. the inter-particle distance begins to set a length-scale for quantum processes and electrons begin to form localized 'chemical bonds' (see Sec. 7.3.1). Then normal chemistry takes over from basic quantum phenomena. Loosely speaking, the resulting decrease in long-range quantum correlations leads to the property of *near-sightedness* proposed by Walter Kohn. Furthermore, when everyday objects (quantum systems of many particles) are considered at room temperature rather than at absolute zero, additional considerations (that we discuss in chapter 8) contribute to restore the reality that is familiar to us. Nevertheless, behind that familiar world, there is an enriched quantum reality revealed by physics.

In this enriched quantum reality, we have moved from a duality to a multi-scale plurality in a higher-level 'unified' view of the world, harmonizing the reductionist, holistic and non-commutative content of our knowledge. This does not really mean that we know the answers to many questions. However, the rise of molecular biology and neuroscience has presented a clear method of dealing with some of these complex questions. Furthermore, the universe is believed to be made up of dark energy (DE, 73%) and dark matter (DM, 23%). At present we have no idea of the nature of DE and DM. It is the few per cent of visible matter that is directly relevant to the human scale of things! This book deals only with those.

1.2 The 'standard model' of matter according to physics

Ancient thinkers distrusted the senses and resorted to introspection and usually had an idealist view of the world, although there were some materialist thinkers who were more close to the practical technologies that were known as 'natural philosophy'. This tradition, combined with the interrogation of nature via experiments evolved into the disciplines of modern science. This was driven on the one hand by pure curiosity and theorizing about the world, and on the other by the activities of conquest, commerce, and social organization. Later in the book we discuss the nature of scientific knowledge and how successive theories have arrived at the contemporary views of our world. The most successful view of the world is the one codified by scientists who have investigated the 'external world' as a system made up of matter and its interacting fields. Here we use the word 'scientist' loosely to include chemists, biologists, geologists, physicians, engineers and others even in the social sciences who use the methods of 'natural philosophy'.

The *standard model* of physics attempts to describe all phenomena with a few dozen 'fundamental particles', and energy fields that mediate the interactions among these particles. If we include fields, particles as well as their anti-particles, we need some five dozen fundamental objects in the 'standard model'. Obviously this is 'too many' for the economical mindset of the theoretical physicist, or that of the mystic who wants every thing to come out of a single sacred word like *Om*.

These fundamental objects may well be the varied manifestations of some other, even more fundamental field. However, the laws associated with these elementary objects, viz., the laws of physics, are amazingly simple and fascinating. Chemistry, biology, and the more complex stuff of the world are all regarded as consequences of the laws of physics. A popular, easily readable presentation of this point of view (which is also largely shared by this book), has been given by Murray Gell-Mann in *The Quark and the Jaguar* [83].

For our purposes, the standard model is a sufficiently fundamental starting point. Hence in this section we give a brief account of the 'standard model' of matter, as formulated in the latter half of the 20th century. Many physicists contributed to this model [4] which eventually took formal shape in the hands of Abdus Salam, Steven Weinberg, and Sheldon Glashow.

Physics reveals that matter is made up of atoms which are mostly empty — even more empty than the solar system where there are asteroids and debris in the vast spaces between planets. Everything that matters to humans depends mainly on what goes on at the very edge of these atoms. Chemistry happens at such edges. Biology happens at an even larger length scale, at the edges of molecules. Nevertheless, the inner structure of matter, as seen in the very core of atoms, and in the sub-nuclear world, turns out to be directly connected with questions of cosmology and the large-scale structure of the world in regard to space and time. Hence, the nature of the world at very general length scales, from the smallest to the largest, is of great interest. We examine selected aspects of physical theory in detail in later chapters of the book. Hence, the following section may be skipped or merely skimmed by the readers whose interests reside in larger length scales.

1.2.1 *Atoms, nuclei, quarks and quantum fields*

Matter appears to exist in various phases, i.e., solids, liquids, gases etc. Many idealist philosophies of the East and the West have attacked this, by claiming that all reality is 'mental', or that it is an illusion — a *Maya*. The ordinary person's reaction to such philosophical doubt is typified by that of Dr. Samuel Johnson to Bishop Berkeley's public statements. Bishop Berkeley (1685–1753) was a philosopher whose public claims irked Dr. Samuel Johnson, a literary luminary of the period and compiler of the first English dictionary. Johnson is supposed to have 'refuted' such views by proving 'the existence' of matter by 'kicking a stone'. We mention this because the physicist's approach to understanding matter is indeed based on 'kicking' matter in various ways. The physicist probes matter with projectiles, heat, intense light, or chemicals that are just other forms of matter. One approach would be to use increasingly energetic probes to penetrate or 'bombard' matter, and then examine the debris of the collision to understand the composition of the targeted matter. Today such experiments are done using 'particle accelerators', e.g., at CERN, that provide well controlled particles of known energy, for use as missiles, to be directed at atoms, nuclei or other objects under study.

Initially, natural fast particles were used as the energetic probes. Radioactive atoms emit energetic 'α-particles', later identified to be Helium nuclei. Rutherford, Geiger and Marsden (1909) studied how α-particles are scattered from atoms, and showed that atoms are not hard diminutive 'marbles'. They are mostly empty space containing a positively charged, heavy, point-like central core (nucleus) and very light negative charges (electrons) distributed sparsely around it.

Similar scattering experiments targeting nuclei revealed that nuclei are also *not point particles*, but are made up of even smaller particles named protons and neutrons. The nucleus appeared point-like only for the rough probes (less energetic particles) that could only resolve the larger length scales of atoms and molecules. Protons and neutrons are collectively called nucleons. They are also called *baryons* as they are heavy compared to electrons which are classified as *leptons*. Leptons have a spin angular momentum of (1/2) in units of $\hbar$, the quantum-unit of angular momentum. The photon is a zero-mass particle but not included with leptons as it has an angular momentum of unity. Subsequently, intermediate-mass particles were identified, especially in Cosmic ray showers. They are known as *mesons*. Mesons decay into electrons, positrons, neutrinos and photons. They are particles with integral values of spin (in units of $\hbar$). Baryons and mesons are collectively called *hadrons* as they display short-ranged, 'hard' interactions.

Thus, taking the proton mass as the unit, an Aluminum nucleus with a mass of ~27 consist 13 protons and 14 neutrons interacting among them and holding together. The Al atom is mostly empty space, containing a nucleus of 27 nucleons at the tiny core, and 13 electrons distributed at length scales which are many thousands of times larger than the size of the nucleus. The 'radius' of a nucleon is of the order of a femtometer (10^{-15}m), while the 'radius' of an atom (10^{-10}m) is some 10^5 times larger. A human hair is some ten million times thicker.

The strong repulsion among positively charged protons is balanced by the attractive forces among the nucleons. The proton, p, and the neutron, n, have almost the same mass, and since mass and energy are equivalent ($E = mc^2$), Heisenberg in 1932 suggested that the p and n were a manifestation of the same particle in two nearly identical energy states. This basic particle is assigned a property called 'isospin' defining two states (which may be termed 'up, down', or p, n). If you are in space where 'up' and 'down' do not matter, there should be no distinction between these two particles. This deep symmetry is mathematically denoted by the symbol SU(2). That there is such a distinction between p and n shows that the 'isospin symmetry is broken'. The isospin concept, which involves a non-Abelian gauge symmetry (see Sec. 3.5), enabled nuclear physicists to classify the internal energy states of nuclei, and propose that there is a *weak interaction* that 'splits' the exact degeneracy between the proton and the neutron, thus making their ground-state energies (i.e., rest masses) slightly different in energy. The isospin concept explained the existence of other heavier relatives of the proton and the neutron, like the Λ (Lambda), Σ (Sigma), and the Ξ (Xi) particle. Meanwhile, Dirac's theory (Sec. 6.10.2) of electrons and their anti-particles led naturally to the concept of anti-particles being associated with every type of particle.

The establishment that quantum particles have wave properties, and *vice versa*, meant that particles have fields associated with them, and fields have particles associated with them. The photon is the field-particle of the electromagnetic field.

Nucleons were postulated to interact via 'pions' (π). However, although π^-, π^0, π^+ particles were discovered, we see below that nucleons and pions are themselves composite structures whose interactions occur at a deeper level involving quarks and gluons.

Are p, n and other baryons point-like particles? In a sense, the question is meaningless as 'size' is not an attribute of a particle. A particle can have mass, charge, spin, isospin, helicity and so on, but its size only appears in terms of the scattering cross-section when probed with other projectiles. When composite particles are probed, they can split into smaller fragments. Free neutrons, and indeed all baryons are unstable. The neutron decays into a proton emitting an electron, and an anti-neutrino denoted by the symbol $\overline{\nu}_e$. The latter is a particle with nearly zero mass and postulated theoretically by Wolfgang Pauli in 1930. It was observed only in 1956. Hence neutrons were already suspected to be composite particles. Protons seemed to be stable elementary particles — but today we know better.

Scattering experiments on nucleons, and results from cosmic ray studies where unknown intermediate mass particles, i.e., mesons (μ) were found, suggested a deeper set of particles at yet higher

Table 1.1 The elementary particles, i.e., the building blocks of Nature used in the Standard Model.

Particles			Fields		
quarks	charge	leptons	fields	interactions	particle
up	2/3	e	Electro-magnetic	e.g., electrons-positron	photons
down	−1/3	ν_e	QED		
charm	2/3	μ	Chromo-dynamics	e.g., inter-quark	gluons
strange	−1/3	ν_μ	QCD		
top	2/3	τ	Electro-weak	muon decay	W and Z bosons
bottom	−1/3	ν_τ			
			Higgs field	sets masses	Higgs particle
			gravitation	interacting masses	gravitons

energy scales and controlled by even deeper symmetries. In 1964 Murray Gell-Mann and George Zweig proposed that baryons were made up of 'quarks', i.e., unusual particles of fractional charge. The quark concept was proposed by Gell-Mann and Ne'emann in 1961 harnessing the symmetry known mathematically as SU(3), i.e., a generalization of the two-fold symmetry contained in Heisenberg's SU(2), to present a fascinating eight-fold symmetry based on three fractionally charged particles known as quarks. This was a non-Abelian gauge symmetry (see Sec. 3.5) hidden deep inside the sub-nuclear structure of matter. This theory had revolutionary features, like proposing fractional charges for the 'quarks', a name borrowed by Gell-Mann from James Joyce's *Finnegan's Wake*. Additional features of the theory meant that the quarks could not exist freely, but could only be found 'enslaved' as components of nuclei. Thus, according to the quark theory, the eight baryons (neutron, proton, $\Sigma^{\pm}$, Σ^0, Ξ^0, Ξ^-, and Λ) all form an 'Octet' made up of various combinations of the u, d and s quarks.

The theory had to face a decade of skepticism. However, the theory systematized the known facts, predicted new particles and their experimental masses. The year 1974 was very eventful for particle physics as experimental evidence tumbled in from the Stanford Linear Accelerator (SLAC) and Brookhaven, confirming the new theories. These experimental groups discovered, independently, states of a type of quark (known as the charm quark q_c) bound to its anti-particle $\overline{q}_c$. Such bound states are analogous to the electron-positron bound-pair, and are known as 'charmonium'. The lowest energy state of charmonium is known as the J/ψ particle. Other predictions of the theory were also verified in the heady years following 1974.

We briefly summarize the quark model of elementary particles (see Table 1.1). The earliest known family consisted of two quarks, called the up-quark (q_u) and the down-quark (q_d), and two leptons, called the electron (e^-) and electron-neutrino (ν_e). The names 'up', 'down' are simple labels and nothing more (i.e., names, like Paul and Peter). The neutrino was a name given by Fermi, for a neutral particle which probably has nearly zero mass and postulated by Pauli.

Experiments, as well as theory using symmetry and conservation laws revealed two other quark families. Thus we have the 'muon' family which contains the 'strange' quark and the 'charmed' quark (q_s, q_c), and the associated leptons called the muon and the μ-neutrino denoted by (μ^-, ν_μ). Finally, the 'tau-family', contains the top quark and the bottom quark (q_t, q_b), and the leptons (τ^-, ν_τ).

The Higgs particle, proposed by Peter Higgs and others in 1964 is included in Table 1.1, and had eluded confirmation up to the first decade of the 21st century. The standard model does not explain why the particles have their observed rest masses, but assigns them in terms of interactions with the Higgs field. To simplify matters drastically, the Higgs field may be conceived as providing resistance to longitudinal excitations (thus providing mass to them). This does not affect photons and other transverse excitations. The Fermi Lab physicists had dubbed it the 'god-damn' particle because of its elusiveness. However, Leon Lederman, the director of the Lab had to change the nickname to the 'God particle' in writing a popular book about the Higgs, and other particles. Experiments at the Large Hadron Collider (LHC) at CERN in July 2012 have revealed signals in the $\sim$125 GeV region, providing strong support for the Higgs mechanism.

We have included gravitons in the Table, being the bosons that mediate gravity in a quantum field. The relevant theory of 'quantum gravity' is not yet known.

Quantum chromdynamics (QCD) is a quantum filed theory elaborating on quantum electrodynamics (QED) where the excitations are called quarks and gluons. Quarks can have three varieties of charge, known as 'color'. The 'colors' must always neutralize each other to give no color (color singlets). Thus we have the three colors 'red, green' and 'blue'. The antiquarks carry the 'anti-colors', just as a positron carries the charge opposite to an electron. Unlike QED which obeys Abelian Gauge symmetry (Sec. 3.5), QCD involves a non-Abelian gauge symmetry. The 'color' here is just a label; the spelling of 'color' is that used by Gell-Mann, a founding father of quark theory and an accomplished linguist.

What is the nature of a quark? Why is the electron charge exactly equal to that of a proton? If quarks cannot be seen in a free state, but only as bound states of one another, can all of them be excitations of something even more fundamental? Would such a theory bring in the graviton also as one of the excitations that arise naturally from the theory, thus doing away with General Relativity as a separate theory for gravitation which stands outside the quantum theory? *String theory* is one such attempt to provide a more compact grand unified theory (GUT), a name introduced by Schrödinger already in the 1940s. In string theory, the elementary particles and fields are conceived as excitations of one-dimensional objects called strings. This theory is at present only a research program without new predictions in the energy and length scales currently accessible to experimentalists. The LHC may provide the experimental results at higher-energy scales and shorter length scales where the Standard Model needs modification. However, only a small sub set of the standard model is needed for understanding matters at the human scale of things. Hence new experiments from CERN etc., are not expected to change biochemistry and physiology that work at very low energy scales.

1.3 Scales of energy and length

Consider the 'nascent universe' — or at least, the part accessible to us. It begins in some highly energized metastable initial state sometimes called a 'false vacuum' state. We don't know much about this state but attempt to deduce whatever we can about it. Unlike ancient philosophers, theoretical physicists are guided by mathematics as well as the laws they have uncovered. By 'vacuum' we mean an initial state with no excitations in it. We know that there are fluctuations in any vacuum, i.e., zero (or 'emptiness') is after all made up of -1 and $+1$, or -2 and $+2$, as they add to zero when averaged. Thus we may think of vacuum fluctuations in terms of lots of -1s and $+1$s distributed randomly. That is, the vacuum is empty on the average, and yet it is full of fluctuations. These fluctuations or excitations of the vacuum are what physicists call 'elementary particles', like electrons and positrons, gluons, quarks etc. One could think of a very large energy fluctuation putting the system into a metastable 'false vacuum'.

As we move from the very high energy scales of elementary particles towards smaller energy scales, interactions (among the particles known as gluons and quarks) give rise to more complex 'particles' like protons and neutrons. These are made up of quarks, where the inter-quark interactions are more or less balanced. The forces between nuclei are the weak remnants of the inter-quark forces left over after the formation of nucleons. The lightest nucleus is the proton. The proton is the nucleus of a hydrogen atom. If it were a sphere, its radius would be about 1×10^{-15} m, i.e., a femtometer. Nuclei and electrons interact via the exchange of photons (they are the 'elves' of the electromagnetic field). Electrons and nuclei form atoms that are many orders of magnitude bigger than nuclei. A proton and an electron bind to become a hydrogen atom. The 'radius' of the hydrogen atom (the Bohr radius) is about 5×10^{-11} m. It sets the length scale of atoms and molecules and is known as an *atomic unit* of length which is about a million times bigger than the scale of nuclei. Thus at each stage, the parameters, or factors which determine the reality at the lower scale, get replaced by new, more extended or coarser types of parameters or collective variables. However, as the length scales get bigger and bigger, the energy scales get smaller and smaller. That is, energies associated with nuclear forces are enormous, compared to the energies associated with electronic states of atoms which are counted in electron-Volts (eV). Molecular-formation energies are even more fine grained, and are counted in milli-electron-volts (meV), i.e., a thousandth part of an eV, or even in calories.

- Molecular Energies, e.g., body heat at $\sim 37°$C is in the $\sim$25 meV. range. This corresponds to 'infrared' light.
- Energy levels in atoms are in the eV range; e.g., the hydrogen atom has an

ionization energy of $\sim$13 eV, similar to the thermal energy of an object heated to 151,000 K. The visible, ultra-violet and X-Ray range are reached with heavier atoms. Atoms and small molecules are nanoscale (10^{-9} meters = 1 nm) objects.
- Nuclear energies are given in MeV (10^6 eV). The temperatures inside the sun are in this, γ-ray regime. Nuclei are many orders of magnitude smaller than atoms.
- The rest mass m_0 of an 'elementary particle' corresponds to an energy m_0c^2, where c is the velocity of light. These energies range from MeV to hundreds of GeV and up. The CERN laboratory attempts to reach such high energies.
- The primary patch of spacetime that underwent exponential expansion (inflation) during the big-bang is about 10^{-36} meters in size (i.e., at least a hundred-billion-billion times smaller than a proton). The 'Planck energy scale' that prevailed in the earliest universe is believed to be $\sim 10^{19}$ GeV.

This sequence is also dictated by the four fundamental forces of nature that emerge as the universe cools down from its initial intense heat (See Fig. 1.4).

- Strong interactions prevail between quarks, which are the components of heavy nuclear particles (e.g., protons, neutrons), known as baryons. The gluons are the excitations that mediate the interaction among the quarks. The quarks have 'flavors' and 'colors' — these are just descriptive name tags. The excitations of this force field, known as the chromodynamic field, are quarks and gluons. The strong interactions have a short range comparable to the size of a nucleus, i.e., about a femtometer (10^{-15} m).
- 'Weak interactions' (WI) are a million times weaker than the strong interactions. These are very short-ranged interactions, dying off within a billion-billionth of a meter (10^{-18} m). WI are capable of changing the flavors of quarks. This force is mediated by two particles known as the W-boson and the Z-boson. The W-boson is charged, while the Z-boson is not. These two particles are similar to the photon, except that the photon has no mass. Also, the energy of these particles is associated with a 'shoving backwards and forwards' motion (longitudinal oscillations), unlike the photon which carries its energy with an up-down motion (transverse oscillation) relative to the direction of propagation. The WI can change a proton to a neutron and *vice versa*. The weak interaction and the electromagnetic interaction, mediated by the three bosons (the photon, W boson and Z boson) are grouped together under the name 'the electro-weak' interaction.
- The electro-magnetic interaction. This interaction governs our entire everyday world of physics, chemistry, cyberspace, life and neural activity. The ratio of the strength of the electro-magnetic interaction to the strength of the strong interaction is called the 'fine-structure constant'. It is equal to 1/137. When you sit on a chair, it is the electrostatic repulsion between the electrons in your body, and the electrons in your chair, which you 'feel' as a hard surface. Samuel Johnson is said (by Boswell, the biographer of Johnson) to have used the hardness of a kick in refuting Bishop Berkeley's idealism, i.e., just the Coulomb repulsion of like charges! It is also the energy in thunder, lightening and laughter. The photon carries the energy of this interaction. Unlike the strong and weak interactions, it has an infinite range of action. That is, electrostatic forces decrease as $1/r^2$ (the

Coulomb law), and the electrostatic potential decreases as $1/r$.

- Finally, there is the weakest of all forces, namely, gravity. Since gravity also obeys Newton's inverse square law of force, $1/r^2$, gravity has an infinite range. It becomes weaker and weaker, but ever so slowly.

Just as the six sides of a dice are found to be manifestations of the same single object (i.e., the dice), one would like to understand how to obtain a 'unified' view of all these fundamental forces.

In Fig. 1.4 the universe is assumed to begin in some unknown manner, from a very hot, dense, initial state ('metastable vacuum'). This leads to an exponentially expanding *inflation* phase. The inflationary theory of the early universe (i.e., the first 10^{-35} sec) is driven by ideas from the quantum Theory (QT). This theory then partly 'hands over' the discussion to the theory of General Relativity (GT) which also predicts the existence of space-time singularities (i.e., mathematically, situations where the solutions seem to require dividing by zero, or lead to some such mathematically uncontrolled situation). Examples of such singularities are black holes where spacetime collapses, or big bangs where spacetime blows up.

We are concerned with *The* big bang (BB) explosion associated with a singularity in the field equations describing our part of the world, restricted by our 'event horizon'. Of course, there can be many such BBs going on in the 'multiverse', just as there can be many bubbles in a boiling kettle. But we are trapped in our BB bubble, so our knowledge is restricted to our BB. General Relativity becomes increasingly inapplicable at the short lengths associated with the singularity. In fact, here we are thinking of lengths perhaps as small as what is known as *the Planck length* which is about 10^{-35} meters. Thus the initial universe must be discussed using some theory other than GR. However, the cosmology that developed in the middle of the 20th century, in the hands of Gamow and others [14], predicts the big bang using the physics of general relativity. In the 1948 paper by George Gamow *et al*., where the author names (Ralph Alpher, Hans Bethe, George Gamow) are arranged to sound like the first letters of the Greek alphabet (α, β, γ), they showed how the relative abundances of elements found in stars, as determined by spectroscopy, could be used to deduce the physical conditions which existed in the early universe. Thus the 'inputs' to the Big Bang theory included well-established spectroscopic information, nuclear physics and relativity.

Inflation theory discusses an even earlier period of the universe, before the formation of nuclei or elementary particles. The rise of string theory in the 1980s, and the more recent *m*-brane theory (see Sec. 5.9.1) inspired the ideas [134] behind the 'inflationary model' of the universe. Inflation involves the exponential stretching of the original patch of instability to an enormous extent in an incredibly short time. The simple symmetries that existed in the early universe evolved into new

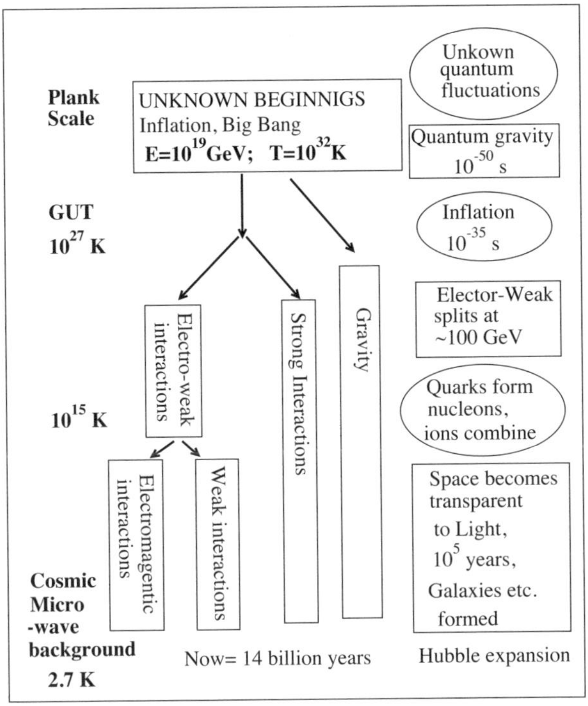

Fig. 1.4 As the early universe cools, the high-symmetry of the grand-unified theory (GUT) spawns the lower symmetries of the Standard model in $\sim 10^{-10}$ s. Four types of interactions, viz., gravity, electromagnetic, weak, and strong interactions emerge. The high energy scales and short length scales get replaced by smaller energy scales and bigger length scales. These allow many energy states to exist in a narrow range of energies, allowing the emergence of complex systems typical of biology.

symmetries, and the four fundamental interactions shown in Fig. 1.4 emerged. As the universe cooled, elementary particles combined to form complex composites. Large scale structures like galaxies formed — i.e., the very short Planck scale of length (see Secs. 5.9, 5.8) has now got replaced by much bigger length scales.

It is this fine-graining of the energy, associated with the coarse-graining of the length scale that brings about complexity from the simplicity of the elementary interactions. At the molecular level, the energy scales are much smaller than at the atomic-scale. Thousands of molecular energy configurations are possible within an energy range where only one or two atomic states are possible. Atoms and molecules give rise to more fragile biological structures (cells, plants and animals) having more coarse-grained (i.e., bigger) length-scales pertaining to the biological world. As we move from the atomic world to the biological world, the boundary conditions defining the way we isolate the system for study become different to those in the quantum world. The higher-level variables usually contain 'collective' effects as well as 'damping effects' or 'decoherence factors' which determine the 'lifetime' of the new, higher-level variables.

In technical jargon one would say that the 'propagators (Green's functions) describing the excitations no longer have sharp, discrete poles — instead the poles have become complex, with branch-cuts'. These new modes represent 'new physics' appropriate to the new characteristic energy scales.

Herbert Simon [197] argued in 1962, in an essay entitled *The architecture of Complexity*, that complex systems are organized in a hierarchical fashion. Thus sub-atomic particles build atoms, atoms make molecules, cells, organisms, organisms make tribes, societies and nations. Here Simon too was discussing energy scales and length scales. On the other hand, Landau and Ginzburg had formulated a very beautiful and comprehensive quantitative theory of order mainly for physical systems, but equally applicable to any complex system. When dealing with physical systems, the energy scales and length scales are globally useful, convenient measures of hierarchical position. The history of science itself can be understood (Sec. 2.4.1) as an unraveling of new energy and length scales as our tools gets sharper and more powerful. New theories are needed to encompass these new energy scales and length scales.

In this book we argue that the appearance of collective modes, moderated by damping and decoherence, and involving complex systems with many nearly equivalent energy minima introduces irreversibility and indeterminism into our physical theory, paving the way for a theory of the mind where consciousness and free will, deductive reason and intuitive insight, become comprehensible within a relatively standard interpretation of physics and chemistry.

1.3.1 *Hierarchical structure in logic, language and life*

This hierarchical structure that emerges as we move from elementary particles (e.g., electrons, photons, quarks etc.) to more complex objects, as well as the indeterminism arising from the damping effects at each stage is also seen in other systems as we go from simplicity to complexity. Very simple formal-language systems, e.g., those containing a few symbols and a few strings (sentences) derived using a basic set of rules, are such that all statements in such a language are logically related to each other. Such a language, or a 'propositional calculus', may be said to be complete, or closed. Russell and Whitehead, in their *Principia Mathematica*, attempted to show that mathematics is such a complete, consistent system derivable from the axioms of logic. Such sentences would be called 'analytic sentences' by the Logical Empiricists who were very influential in the early half of the 20th century. Russell and Whitehead hoped to reduce all arithmetic to a system of logical relations among 'sets' or 'groups'. A 'set' is simply a collection of elements that obey specified rules of transformation, e.g., addition, association, commutation etc. However, this effort unearthed various logical paradoxes, e.g., Russell's paradox of sets, which suggested that self-referential statements in a language have to be made using a higher-level 'meta' language.

The set of tables is itself not a table. Thus we have sets which are not members of themselves. The set of ideas is itself an idea, and hence should be a member of itself. Now, is the set of all sets which are not members of themselves, contained in itself or not?

The meta-language will have its own meta-paradoxes. A landmark result from the foundations of mathematics is the now famous Gödel's theorem, which showed that formal systems, if they are sufficiently complex, contain valid statements not deducible within the formal system. A similar result in computer science is known as Turing's *halting theorem*, and asks if a computer can decide by itself that it has solved a given problem, and come to a halt on its own.

In this book we examine the hierarchical structure moving from elementary particles to atoms, molecules, cells and other complex adaptive systems including human beings. A human is both a biological organism and a person. Is s/he a 'conscious machine'? On your table you have a computer, an able robot executing many complex algorithmic operations for you. It may even use algorithms based on 'neural-network models' (see Sec. 2.2.5). It will do this as long as the computer is maintained at the voltages and power levels specified by the manufacturer, and as long as the conditions of humidity and temperature remain within factory specifications. You can open several 'windows' and the computer behaves as several virtual computers doing several tasks, as long as you obey rules about creating such windows. The computer is a system that inputs energy, outputs wasted energy (entropy), and performs tasks. It interacts with you in a predictable way.

Across the table Bob is seated opposite you. In fact, there are two individuals, you and him — two personalities. The inter-personal interaction occurs via

Fig. 1.5 Bertrand Russell, Kurt Gödel, and Alan Turing (courtesy APS archives).

language, gestures, and within social conventions of protocol and precedence — commutation rules — that exist between you. These are related to the variables that are effective at this level of coarse-graining of the physical world. On the other hand, he can be seen as a 'thing', i.e., an *it*. That is, a conglomeration of cells held by biochemical forces, and showing a bio-chemical characteristic known as 'being alive' (existence). Existence depends on the input of energy and nutrients, the presence of thousands of other living bacteria in the gut and other parts of the body. It is really many living organisms working collectively, with no specific part identifiable as the 'individual' named Bob. He is in many ways a biological computer. The architecture for his/its construction is specified in Bob's DNA. This DNA embodies the specifics of Bob that encodes Bob sufficiently for him to be uniquely identified. In fact this code is used incessantly to repair and replenish body components (cells) which need to be replaced due to wear and tear. The copying process itself introduces errors, wear and tear, leading to a gradual aging of Bob. That is, just as my desk-top computer is constantly updated and repaired (*via* my on-line connection to the manufacturer who knows the basic architecture of my computer) against wear and tear etc., Bob's system is also updated, by a self-contained system of protein synthesis and cell repair.

Knowing Bob for many years, his behaviour is reasonably 'predictable'. Also, just like the computer, he has several windows or personalities. If his body temperature, oxygen intake, blood-sugar level, or serotonin level in the brain were modified by even a few per cent, he too, just like a computer which is deviating from specifications, begins to 'behave' strangely and in unpredictable ways. Who then is Bob? Is he a 'bio-chemical computer'? Or is he the 'person' that you have known for many years? Clearly, they are two different ways of looking at the same object. Figure 1.1 displays the dual image of a Greek Vase which also looks like two human faces, depending on the glancing stance used by the perceiver. The relationship between the image of Bob as an organism, or as a person is clearly more subtle. Many philosophers or religious thinkers would argue that the move from Bob the organism, to Bob the living, conscious individual who is, say, a lawyer, a husband, a father, a colleague and a friend, involves some 'esoteric principle' or an *élan vitale* beyond analysis or 'reductionism'. The electronic computer needs some one to give it commands. Is it too much to envision a 'self-learning' computer which constructs its own commands and generates its own responses, based on the inputs it receives from the outside world? Isn't that Bob?

In the 19th century, it was believed that certain chemical substances ('organic' compounds, e.g., acetic acid or vinegar) found in living matter could not be made in the laboratory. Only living matter could produce them. Similar 'impossibility' barriers, from sugar to chlorophyll and DNA, were suggested from time to time, and yet they have all been surpassed. The complete analyses of the coding in the DNA of the primitive worm *Caenorhabditis Elegans* and other benchmark organisms, and finally the human Genome itself have been completed. The role of each gene in the neural wiring and behaviour of *C. Elegans* and similar small organisms are now well understood. Synthetic organic chemistry can use the very techniques of nature to create novel proteins that never existed in nature. The cloning of large animals, developments in computer science and neuroscience have generated new optimism, especially among life scientists, that the gap between Bob the mechanistic organism, and Bob the person may finally be bridged, without claiming mysticism. It is true that 'higher-level' variables can control lower level variables (as happens when you, a higher-level complex-adaptive system, decide to move a knight in a chess game). However, the effect of the higher-level variables is to set the boundary conditions that constrain the physical laws of the lower-level. They do not change the validity of the microscopic description.

A number of physicists have invoked the quantum theory to claim that a 'conscious observer' is an essential feature of reality. This view, starting from von Neumann and Wigner, is presented in, e.g., *Mind, Matter and Quantum Mechanics* by Henry Stapp [201], in Hameroff and Penrose (see Ch. 13) or in *Quantum Enigma: Physics encounters consciousness* by Rosenblum and Kuttner [182]. While Eugene Wigner and others began this discussion in the context of what is known as the 'measurement problem' in the quantum theory, some writers have spun a highly metaphysical, even sensational picture of reality. Thus, the running of computers, lasers, atomic clocks and all the physical underpinnings of the modern world depend on some almost magical 'act of observation' giving reality to the world. This reality, generated by the observer may be a mere momentary selection from an infinity of many worlds *a la* Everett [56].

We discuss the measurement problem and the 'quantum paradoxes' in Ch. 7. Quantum Enigmas and the 'role of the observer' get clarified when the 'boundary conditions' (BC) associated with the preparation of the system are included in the discussion. This is neatly done in David Bohm's approach where the effect of the BC enters into the Bohmian 'quantum potentials'. Specifying the boundary conditions to sufficient accuracy brings in all the issues of sensitivity to the input information, as well as Bayesian or other estimates of the probable validity of such information. Not surprisingly, some writers (e.g., Caves *et al.* [44, 77]) have gone back to review the meaning of probability to elucidate quantum mechanics.

When dealing with many-particle systems at human length scales, decoherence and finite-temperature (finite-T) effects have to be taken into account in a systematic way, so that popular metaphysical views associated with 'Schrödinger's cat' etc. (see Ch. 7) do not cause confusion.

Density-functional theory (DFT) states that a quantum systems can be completely described by a functional of the density distribution, Hence it *dispenses with the many-body wavefunction*. This resolves many issues in understanding 'quantum reality'. DFT can conveniently treat finite-T systems, without heavy formulations like thermo-field dynamics etc., and seamlessly treat classical and quantum 'hybrid systems'. Thus, in our view, the quantum theory, discussed in Part I of this book, presents a very comprehensible non-mysterious resolution of dualities, subjectivities and paradoxes that plague many discussions.

The early part of the book will look at the nature of physical law, scientific theories, and the critiques of some philosophers of science who have developed a new genre of 'science studies' where they claim that science is reducible to a relativism of belief — a socio-political belief system. Having dealt with such epistemological questions, we go on to study the basic laws of physics, emphasizing how the quantum theory of elementary particles, relativity and the underlying laws of symmetry and simplicity pave the way for complexity. A remarkable epistemological consequence of the simplicity of the laws of physics is the emergence of a form of teleology in the principle of least action (Sec. 3.6). A few symmetry rules, least action, and the equation of continuity seem to be the essential underpinning of the laws of classical physics as well as modern physics.

The world of atoms, molecules and ions involves longer length scales and smaller energy scales. However, it is still governed by the same basic laws. However, the deterministic interactions among charged particles at low-energy scales lead to collective modes like plasmons which have acquired a degree of indeterminism (reflected in the complex-pole structure of their propagators) by decoherence and by the loss of information across the boundary of the system. Other classes of collective modes are charge-density waves, magnetic structures, and crystalline order that self-assemble in materials systems.

In Part II of this book we examine the path of astrochemistry and pre-biotic chemistry of the early earth. The formation of incubators of living organisms in various type of 'Darwin's warm little ponds', and in hydrothermal vents are discussed. Froth bubbles, lipid bilayers, and biological structures (Sec. 10.4.2) are also self-assembled complex molecular quasi-equilibrium systems. Quasi-equilibrium stationary states of such material systems, formed within naturally formed lipid bilayers may involve feed-back loops that allow for adaptive modifications of the structure of such systems to conform to the requirements of the ambient environment. Such systems can form classes of complex adaptive systems which utilize (i.e., 'feed on') lower-scale systems for maintaining their stationary state. Such systems which can replicate themselves are essentially living organisms. Darwin's principle of the survival of the fittest applies to such systems.

Complex systems obey the basic laws of physics, but also spawn new, derived laws that hold among the new collective modes.

In this context, Darwinian selection (see Sec. 1.5) may be thought of as a quasi-equilibrium complex-adaptive version of dynamical energy optimization, i.e., *efficiency* demanded by *a higher-level principle of least action*.

The pressure of evolution favours primitive cells that form multi-cellular systems, with some cells doing specialized tasks. We discuss vision, neurons, memory and the emergence of consciousness, intuition, feeling and sensation usually associate with the mind of an intelligent, alert being. Here we have inputs from neuroscience, artificial intelligence, as well as the physics of complex systems. The energetics of complex proteins, their essentially non-computable conformations, and the indeterminacy and decoherence arising from the ambient environment conspire to replace the seemingly rigid determinism of small dynamical systems by the very opposite of determinism (see Sec. 8.6). That is, we begin to understand the existence of conscious machines which are largely biological automatons, but having a margin of free will and a capacity for strategy and action based on their perceptions. However, this 'direct' physico-chemical approach of the hard-headed scientist is assailed from many angles, as discussed in the next section.

1.4 The direct approach and its modern critics

If we leave the mystics and obscurantists to themselves, we have more subtle critics who question the very 'method' and 'validity' of the search for understanding. In fact, 'science studies', or inquiries into the nature of sciences, have become a popular component of the non-science part of academia today. When Feyerabend [72] concludes that there is no 'discernible method' in science, and that 'anything goes', he began as a critic of the efforts of philosophers of science to describe and define the scientific method. But the criticism is in fact leveled against any type of claim to understanding, be it in the sciences or the humanities.

Scholars armed with case studies from the history of science would present multi-faceted arguments claiming that science is itself a result of social paradigms or personal dispositions, with little or no basis in 'reality', and no better than the folk lore which worked very well for the needs of a previous era. And yet, even the very air we breath is 'air-conditioned' by the technological fruits of science. These same scholars would not hesitate to use airplanes, computers, or other products of this 'very doubtful enterprise'.

Instead of such strongly 'epistemological' attacks, others would prefer to accept a 'rational approach' tempered with some form of mysticism perceived as 'intelligence' or 'design' in natural processes. The mysticism can be pushed to a view of the world where everything has happened just to produce human beings. We need to answer why we should not reuse such very comforting 'anthropic' world views that are claimed to have worked 'very well' in earlier times.

1.4.1 *The 'reductionist' approach and its alternatives*

The laws of physics dictate how nuclei, atoms and molecules are formed. The rules for atoms and molecules, i.e., the laws of chemistry, follow from the laws of quantum mechanics. Since biochemical processes are determined completely by chemistry, biology gets assimilated into chemistry. At this point a murmur of dissent against this 'reductionist' point of view is possible, although the application to inanimate matter, plants and perhaps animals is uncontested today. If we extend the claim to include human beings and all aspects of human activity, then the initial rumbling of disagreement can grow into a tidal wave of opposition. Even those who are sympathetic to this line of reasoning begin to express doubts about 'excessive' or 'simplistic' reductionism and 'physicalism'.

On the other hand, physicists of the grand tradition, from Archimedes to Weinberg [220] have always accepted such a process. Biologists like Edward Wilson (II-[104]), or Neuroscientists like Eric Kandel have found this 'reductionist approach' to be the most fruitful (II-[51]).

A primary objective of 'reduction' is to ultimately achieve the construction [7] or 'synthesis', once 'the parts' and the 'whole' are understood. Some anti-reductionists may insist that there is 'meaning and purpose' — teleology — in everything. Any such subtle features are allegedly lost in the reductive process, just as the aroma in the wine gets lost when the components are taken apart. The mild anti-reductionist will claim that 'the whole is more than its parts'. S/he will tell you that the water molecule, i.e., H_2O, is not just two atoms of hydrogen and one atom of oxygen. Water has new properties not found in the components.

Philosophers of all ages and in all cultures have claimed that there are emergent properties, which make the 'whole more complex than the parts'. Even the Logical Empiricists who began with a clear-cut program of 'analytic sentences' and 'synthetic sentences', within a reductionist approach to language and logic, ended up with, e.g., the out-and-out *Holism* of Willard Quine [177]. He claimed that one needs to 'face the tribunal of sense experience ··· as a corporate body'. If philosophers think that elements of language or sense-experiences need to be treated holistically, then more traditional personal entities like consciousness, 'qualia' etc., would be claimed to be even more beyond reductionism. The appeal to holism is not at all new, and goes back to antiquity.

> The author of the *Suthra-kritange*, a Sanskrit text going back to Upanishadic times, perhaps five or six centuries before Christ, explains consciousness as an emergent property 'which manifests itself when the elements come together in the body, like the intoxicating power when the ingredients are mixed' [116].

An alternative view is that the whole is never more than its parts inclusive of the interactions and boundary conditions defining the system! Given the 'parts', the actual physical realization of the whole is the bottom-up process of building a complex system. This is usually a *selective* unification of one possible combination of the parts, assuming that we even know how to enumerate the 'component parts'. In fact, when we study the quantum theory of entangled systems, we learn that knowledge of a part can dictate the very nature of the whole, and *vice versa*. In effect, given the interdependence of the 'components' and the 'whole' implied by reductionists and anti-reductionists alike, it should not matter what picture we use unless the specification is incomplete! In our view, the error of much past thinking lies in how the parts (which make up the whole) are counted and boundary conditions are listed. Can we even list the components?

If we consider H_2O, it obviously has H, O, and H atoms, further specified as nuclei and electrons. We need not go beyond that level of reduction if the energy range is specified to be those of molecules. What is often forgotten is that there are interacting fields among these particles, as well as boundary conditions defining the extent of the system. Bohmian mechanics or density functional theory incorporate these in the 'quantum potentials', Kohn-Sham potentials etc., acting on the particles. Interactions are mediated by 'field particles' which are, in this case, Coulomb interactions (mediated by 'longitudinal photons'). It is only when these are included with the atoms that 'all the parts' of the initial system could be considered to have been counted — if they could even be counted! Even then the story is not complete. The specification of the parts has to include a description of the initial state of the component parts. The water molecules, i.e., the 'whole', are a specific selection from the many possibilities that one may construct from the initial parts (other possibilities are, e.g., an ionized plasma, a metastable H^+, OH^- system, etc., depending on the total energy). All the properties of the water molecule can be predicted to the precision needed in chemistry or physics, and both the reduction process, as well as the reassembly process can be executed within quantum chemistry, up to the required level of molecular complexity.

However, is there a non-physics approach with better predictive power? Even if some properties of water are deemed 'emergent', they are fully accounted for in every detail. What about the 'wetness' of water? This is a property of the joint system consisting of the sample of water and the observer's tactile organs taken together. Can we then extend this reasoning to cover the observer? What about properties like consciousness, thought, volition and life itself? Already in 1890 Thomas Huxley [112] asked the same questions and gave the correct answers:

> 'When hydrogen and oxygen are mixed in a certain proportion, and an electric spark is passed through them, they disappear and a quantity of water ... appears in their place. ... we do not hesitate to believe that ... (the properties of water) result from the properties of the component elements of the water. We do not assume that a something called 'aqueosity' entered into ... the oxide of hydrogen ... On the contrary, we live with the hope that ... we shall by and by be able to see our way clearly from the constituents of water to the properties of water., as we are now able to deduce the operations of a watch from its parts and the manner ... they are put together. Is the case in any way changed when carbonic acid, water and ammonium disappear, and in their place, under the influence of a pre-existing living protoplasm, an equivalent weight of the matter of life makes its appearance? ... What better ... status has 'vitality' than 'aqueosity'? ...'

It is precisely this 'determinism' of all properties, *vis vitalis* or not, that is thought to be implicit in this 'reductionist' scheme that many dislike. If physics is to 'govern everything', and if we can predict everything with machine precision, are we all 'clock-work mechanisms' contained in a universe, which is itself dictated by the same set of unrelenting physical laws? Are we claiming that the physics of the *Big bang* completely determined the actions of hijackers who rammed a jet plane into the World Trade Center on 9/11? Is there, in effect no fundamental difference between an intelligent human being, a brainwashed human being, and a properly programmed supercomputer?

The aversion to a complete 'laws-of-physics description' (LPD) of the physical, biological and human realms arises not only in the context of human determinism and free will, but already in the application of LPD to 'organic, living, conscious' creatures and personal pets. We hope to show that the possibilities contained in the laws of physics stretch beyond our best imagination. It turns out that the properties of any real system have indeterminacies built into them, whatever the length scale or time scale, be it in the quantum limit or in the classical limit. This does not essentially depend on Heisenberg's uncertainty principle, quantum measurement theory, chaos theory and so forth, although all these contribute to the escape from 'rigid physical determinism' at each length scale. In our view, rigid determinism is a special property of *finite few-particle dynamical systems*. The many-body problem of the real world has just the richness and *complexity* we need to resolve this issue (e.g., see Ch. 8 or Ch. 14).

Let us now return to the great program of 'reductionism' initiated by the ancients. The Greek atomists had already understood the importance of 'movement' and reached a reductionist, if poetic, world view which may have achieved dominance if not for the rise of Christianity wrapped in Aristotelianism. When Galileo published his *Discourse on Two New Sciences* [78] in 1638, after his earlier work on *Two world systems*, he returned to the study of motion within a conceptual framework rooted in the tradition of Democritus, Aristarchus and Archimedes. In contrast, the Aristotelian approach was to assimilate inanimate motion into those of living things having an end purpose, *a teleology*. Thus the application of a reductionist program for physics, initiated by Galileo in the 17th century was indeed the resurrection of a bold step.

William Harvey, a graduate of the University of Padua in 1602, worked in England on the circulation of blood and cleared up much metaphysics in his work *De Motu Cordis*. In this he resorted to studies on animals, snails and even fish, implicitly granting the unity of the 'tree of life'. A more extreme step would be to reduce the physiology of the brain to a science which is not based on a doctrine of the soul. In 1664 Thomas Willis [229] published his *Anatomy of the Brain*, and followed it up with *Cerebral Pathology and Two Discourses Concerning the Soul of Brutes*, where he discussed psychological disorders. Unlike Galileo's celebrated work, well enshrined in subsequent philosophical and scientific thought, Willis's work is not well known outside the scientific community. Nevertheless, modern-day attempts to describe human consciousness, brain function and physiology in chemical terms have to be viewed as a continuation of the 'reductionist' program of Willis, an associate of Robert Boyle who was also at Oxford.

1.5 Evolution, and 'human' questions

In this section we glance at a number of topics that begin in biology and fall within several humanistic disciplines. They achieve a *consilience* within the Darwinian evolutionary picture. This Darwinian picture is not universally appreciated, and hence we review it briefly, noting that some of the salient topics would be taken up again in Part-II of the book.

1.5.1 *Evolution as a physical process*

Darwin's principle of natural selection [53] is usually viewed as an additional 'non-physics' principle, which is needed to describe the evolution of complex systems (see, discussions in, e.g., Dennett [55]). It is the great unifying principle of the life sciences. However, the usual statements of the 'principle of natural selection' appear subjective, creating intense debate even among Darwinists.

What is natural selection? Pittendrigh calls it 'Darwin's demon' [164]. The critique of 'hard adaptationist' Darwinism by Gould and Lewontin in their well known 'Spandrels of St. Marco' article [90], and the opposing points of view expressed by Dawkins [54], Dennett [55] and others serve to illustrate the problem of 'pinning down' what 'natural selection' is about. In effect, if natural selection is summarized by, for example, the 'survival of the fittest' dictum, we cannot define the 'fittest' as 'those animals that survive'. In effect, as Ernst Mayr [140] had noted, many adaptationist discussions can sound like glib historical narratives, 'just-so' stories, or Freudian accounts of dream interpretation. Sometimes it may be written in the fancier jargon of 'fitness landscapes' and other elaborations. This emphasizes the need for a better analytical and physical discussion. In fact, some researchers like Stuart Kauffman have looked for ways of rephrasing Darwinian evolution in terms of self-organization and off-equilibrium dynamics [120].

It is quite likely that the 'principle of natural selection' can be economically and usefully expressed in terms of other fundamental variational principles, which apply to all physical systems. The principle that the whole system evolves towards a state of lower free energy includes the possibility that some parts of the system decrease their entropy while other parts increase much more. When a system is driven, we have the principle that all processes occur so as to minimize the amount of action consistent with the applied set of constraints and energy inputs. This is the principle of least action. The evolutionary process may be viewed as a Monte-Carlo type simulation of a system of movers in a landscape, and held at some temperature and other environmental parameters that change slowly. Here the action principle comes in via a move-selection scheme, as in Verlet's algorithm or that of Metropolis and Teller (see Sec. 2.4.7) [145]. In fact, modern quantitative biology has already adopted many of these ideas from condensed-matter simulations into the domain of 'soft condensed matter' and biology. Within such simulations of driven systems, it is possible to show that complex organisms evolve from random initial states, as proposed in Darwinian evolution.

A crystal formation is an organized complex, but it is an 'equilibrium' structure. Its capacity to replicate is instructive but uninteresting. The various possible crystal phases are described by the phase diagram of the material. This phase diagram becomes extremely rich and complex when we consider 'non-equilibrium' situations and multi-component dissipative phases. Living organisms belong to this extended, dissipative part of the non-equilibrium phase diagram. A soap bubble is a rather simple non-equilibrium dissipative system which is in quasi-equilibrium for the short 'life-time' τ_b of the bubble. It forms using the energy of the air-water interface; it lasts for τ_b seconds and then decays as the ambient parameters change to a non-equilibrium regime. Living organisms are such dissipative systems, though more complex and also adaptive. They too obey variational principles giving a direction to the process of evolution. This gets us out of the paradox where evolution is generally thought of as a completely unguided, 'blind' process, while optimizing the 'fitness' of whatever is evolving. We may even picture the unfolding of the universe from the big bang as a Cosmic Simulation guided by a weight function which does the appropriate 'natural' selection at each step. The weight function ensures that an effective free energy and an effective action are minimized, subject to various boundary conditions. 'Adaptation' is the biologist's term for what the physicist calls "the response of the system to the external perturbation"; this response provides the accommodation of the variational principles within the energy and time scales of the problem. Furthermore, the stability of certain evolutionary species for certain lengths of time requires the concurrence of a suitable set of time scales (damping times) in the characteristsics (normal modes) of the total system, as discussed in Sec. 2.3.6.

Looking at evolution in this way helps to 'explain' why nature seems to be so wasteful. Why does nature have to sow millions of seed to germinate only a few grains? Why produce millions of sperm only to hatch a few Salmon, or churn out millions of galaxies and stars for no apparent purpose? They are all necessary configurations required to evolve towards the 'answer' in a giant simulation. In fact, the bigger the number of configurations sampled, the more 'accurate' is the answer, in the sense of better satisfying variational principles!

1.5.2 *Existential and social questions*

This brings us to existentialist questions that we deal with in the last chapter of the book. Why are we here, what is our purpose and destiny etc.? One might even dismiss such questions as meaningless, following the modern positivists or the early Buddhists of the 6th century B.C. According to H. H. Price [173], we have to reach Cambridge of the 1920's with its positivist philosophers to encounter such a point of view. Positivists dismiss experimentally untestable questions as meaningless — just as questions about the 'sex' of the 'triangle ABC' have no meaning, even though syntactically correct. However, unlike in ancient times, many questions about our cosmological origins, consciousness and future destiny can be examined within the laws of physics.

We mention the early Buddhists or Gnostics to emphasize that human destiny, morality, human suffering and salvation are traditionally the terrain of religious and philosophical thought. While many science-oriented readers may dismiss such religious and philosophical thought as antiquated, their enormous cultural presence cannot be ignored. They are an empirical social fact. In every civilization, be it Chinese, Indian, or European, there were thinkers who were primarily scientific, and others who were primarily religious. A figure like Plato straddled both camps, and owed much to the religion of Bacchus and the tradition of Orpheus. Plato was the source of later developments that came to be embodied in Christian theology. Hence we have not hesitated to reach into the insights and points of view that may come from such sources.

Hellanic rationalism and Christian mysticism could be reconciled in the idea that the world is a product of intelligent design. William Paley [153], writing in 1802, was one of the clear proponents of intelligent design (ID). Paley discussed ordinary mechanical clocks that are utterly simple in comparison to those produced by the evolutionary process. In Paley's *Natural Theology: Or Evidence of the Existence and Attributes of the Deity* he argued that the elaborate complexity of nature is clear evidence for an intelligent creator. Darwin tacitly accepted this point of view before he set off on the voyage of discovery on the *Beagle*. Paley, in the introduction to his book brought up the analogy of the watch. Suppose one happened to find a watch on the ground. Would not its intricate mechanism imply 'that the watch must have had a maker ... who comprehends its construction, and designed its use'? The intricate working of the natural world was vastly more impressive than any watch. Paley could not see it to be anything but the product of a supernatural being acting with conscious intent. This is the argument based on 'Intelligent Design' that has been resurrected in more recent times by the religious apologists. It has been embellished with new clothing by creationists as a counter to the teaching of evolution in American schools [20]. However, it should be remarked that 'Intelligent Design' and 'natural theology' were a great advance over the thinking based on purely capricious divine wrath that preceded them. Even mountain formations were regarded as punishment dealt to earth by a creator who was angered by the misbehavior of the humans he had created. It was Sir Charles Lyell, the geologist who preceded Charles Darwin, who reformed such medieval beliefs to conform to the concept of *a rational* creator.

William Paley did not know about 'circadian clocks' that are products of evolution and well suited to the needs of living systems. Circadian clocks are the biological time keepers of metabolic and behavioral activity which relate to the cycle of day and night, seasons [164] and latitude. They are temperature compensated biological clocks found in mammals, flies, plants fungi and even cyano-bacterial colonies! These clocks depend on bio-chemical feedback loops associated with cellular transcription regulators. Every protein activity seen in the feedback loops used in the fruit fly is also present in the mammalian clocks [51], testifying to their evolutionary kinship. The daily flux of photons, made up of visible light as well as infra-red heat provided an additional window for organisms that learned to time their chemical activity in an advantageous manner, by evolving an internal clock.

A modern form of 'Intelligent Design' contends that the universe exists for producing the Human species, and that there are strong signatures of this revealed in 'otherwise unexplainable coincidences' in physical theory. Brandon Carter in 1974 [43] called this the *Anthropic principle*, a type of 'argument for design' [15].

In essence, this is a theory without predictive power, but has a capacity to say 'all this is too much of a coincidence, it must have been designed to be so'. If the electron did not have a spin, the periodic table of elements would not be as we know it, *ergo*, there would not be any carbon, and carbon-based covalent bonds are not possible, and we wouldn't be here. If carbon did not have a nuclear energy state at about 7 MeV, then Carbon would not have been produced in the stellar nuclear reactions that produce solar energy. Then there would be no carbon — *ergo*, we wouldn't be here. All these surely indicate that the great designer thought of us and made sure that electrons are half-spin particles, and arranged that carbon had the right energy states! This type of argument is a modern version of Rev. William Paley's old arguments (1802) about the marvelous mechanism of nature. The creationists [192, 76] seek to find unexplainable, awe-inspiring mysteries in a world where *Man* is the central fact. They unfortunately ignore the vast amount of Dark Energy and Dark Matter that seem to be irrelevant to the 'anthropic principle'.

Weinberg [218] has proposed a simple possibility where theology has been replaced by the usual picture of the uneconomical working of Nature that sows millions of seeds to sprout a few progeny. Our part of the universe is regarded as a tiny speck in a super-universe containing all possible types of universes. Life as we know it appears in some rare parts of the super-universe where the cosmological constant and all other physical parameters are in the right range. We happen to be in it and the anthropic principle becomes an empty statement.

Is there a concept of 'good' and 'bad' in the Universe? Clearly, nature stated in physics terms is *value neutral*, as further discussed in Sec. 2.1.1. This point of view is, however, not universal, even among physicists.

Christian rationalists, Buddhist or Jain epistemologists started from very different assumptions. Early Buddhists claimed that the external world is a sequence of mental and physical events (*Nama, Rupa*) concatenated by 'laws of causation'. Cosmological or mind-body questions were deemed 'metaphysical', and such inquiry was discouraged (see Sec. 2.1). However, they postulated 'transcendental laws' that were distinct from physical law. One such transcendental law admitted by these ancient thinkers was a moral law, based on 'fate', or the more complex concept of *Karma*. Fate was also a fundamental tenet of Greek belief. Similarly, the rational theologies of Judeo-Christian traditions also invoke a moral law, although founded in an act of mystic faith flowing from a personally addressable God, some sacrificial act or eschatological event. However, in traditional Christianity, every one, even a new-born infant, is considered wicked. Individuals are 'elected' to the city of God only by the grace of God — not because they lead a good and virtuous life. Anthony Burgess's novel *A clockwork Orange* and Stanley Kubrick's film feature the errant adolescent Alex; they explore the conflicting views of Saint Augustine and Pelagus on the notion of Original Sin and the freedom to chose good or evil [159]. The view that living virtuously by our own moral choice would get us to heaven is known as the *Pelagian heresy*. The Council of Orange and other subsequent authorities crushed this heresy.

In modern times, the noted cosmologist George Ellis has written *On the moral nature of the Universe* [151]. Similarly, various 'dialectical' types of reasoning, be they in the hands of Hegelians, materialists or 'humanist thinkers', sometimes imply an agenda based on 'moral' objectives. Unfortunately, there does not seem to be any justification by way of physical law, or empirical evidence, for any such 'moral laws' or punitive fate permeating the processes that occur in our world. In fact, the pain, misery and patent injustice ('evil') prevalent in the world could equally well be used as a basis for a belief in a malevolent creator. This point of view, held by the adherents of Catharism, and persecuted by the Inquisition as a heresy, has found a modern idiom in popular movies like the *Matrix*, where the world is just a dream sequence controlled by an evil computer.

1.5.3 *Exorcising the soul and inducting new notions*

We noted how some physicists, intrigued by the enormous number of steps needed to create life from the basic particles, and the low probability of many such steps, gave up the intellectual endeavor in favour of a quick solution. This is couched in 'anthropic principles' and metaphysical explanations. The 'soul' is of course the most famous example of giving a distinctive 'spiritual' character to matter. The Sanskrit word *jeeva* (life spirit) was probably the etymological source of 'Eve', while *om*, or *āthman* (soul) in Sanskrit, may have been the progenitor of the name 'Adam'. It has survived in modern German as 'atmen' (breath).

During Buddhist times (6th century BCE), Indian materialists ('Lokayathas') like Payasi had actually attempted to detect the existence of a soul by weighing a man immediately before and after his death. Confusingly, the corpse could be heavier than the living body!

While the idea of a persisting soul was fundamental to Hebrew, Hindu and Orphic belief systems, some ancient thinkers like Confucius had no use for the concept of a soul. The Buddhists (and some Jains) had developed detailed theories of the world as being made up of impermanent streams of mental and physical events, or *nāma* and *roopa*. The Buddha in the 6th century BCE explicitly rejected the doctrine of an enduring soul (aathma), and proposed the doctrine of *anatta* or *anathma*, i.e., non existence of an enduring 'soul' or 'self'. Nevertheless, most Indian thinkers re-introduced a 'consciousness' which goes from one life to another in a cycle of rebirths, although ever changing, and also carrying long individualized concatenations of 'Karma'. Remarkably, most thinkers, be they of the East or West, have ignored the dual parentage of every child, merging two identities of mother and father in the off-spring. Interestingly, there has been no religious teaching where the child's soul is proposed to be a hybrid of those of the two parents.

This old battle regarding the soul could be the new frontier of contention between science and religion in the 21st century, just as geocentrism was to the renaissance, and Darwinian evolution was to the 19th and 20th centuries. As neuroscience advances, comprehending human personality falls within the realm of physics and chemistry. The 'ghost in the machine' of Gilbert Ryle [185] is found to be analyzable and explainable in terms of brain function. This first became clear in the field of motor control of perception [118]; but today, brain processes controlling personality are regarded as cardinal aspects of personhood. So then, why do we need 'a ghost in the machine' at all? Although such views have been very common among scientists, it is also gaining ground among philosophers who have begun to look at consciousness in tandem with experimental research [146].

Does that mean that there is no 'identity' of a person? The first property that defines the 'identity' of a person is the encoding of information in his/her DNA — genomics. This is complemented by a knowledge of folded protein-structure, cells, neurons and the map of acquired traits whose physio-chemical description is the work of current research in neuroscience. The effective length of the DNA in terms of its binary bit length (see Secs. 2.3.5, 12.6.2) is a rough measure of the *complexity* of a living organism viewed as a *complex adaptive system*. The DNA determines the proteins and they in turn form the organism. However, the detailed three-dimensional structure of a protein (the conformation) is subject to the ambient force fields. Given that we are currently unable to even specify the full conformational details of a single protein molecule at room temperature, we are clearly unable to predict its time evolution at the level of detailed conformations. We will never be able to produce two identical brains!

This inability to synthesize or predict is not a refutation of the physical model, but an affirmation of the physics of complex systems and chaotic dynamics that applies under suitable circumstances even to hard-sphere billiard balls (see Sec. 8.6)! Thus new notions, coming from information science and the theory of complex systems may provide scientifically sound alternatives to old metaphysical concepts, and also define the limits to our knowledge.

Chapter 2

An Epistemic Hunt for Scientific Truth

In this chapter we examine the 'need' to know, i.e., epistemic hunger, as well as theory-laden, or morality-laden approaches to understanding, We look at model construction in science, reductionism, holism, complexity, inductive knowledge, neural nets, falsifiability and scientific method. The existence of a hierarchy of energy and length scales is emphasized. Scientific revolutions occur when new energy scales and length scales are exposed using new tools that probe new regimes of energy and length. The scientific method is viewed as the working of a weighted stochastic molecular-dynamics-like simulation. Different theories are limited perspectives of a more complete 'higher-dimensional' whole. The limits of knowledge are discussed.

2.1 The 'fundamental' questions

The fundamental questions debated today are no different from those debated by the thinkers of classical Greece or their contemporaries in India and China, recorded in the Dialogs of Socrates, the Discourses of the Buddha, or the Analects of Confucius. The living conditions of those times, the misery and pain, hunger, hardship and sickness, famine and plague, and the incredible cruelty of despotic rulers and their agents — these are usually not part of our experience. Not surprisingly, a large part of the effort in those times was to understand how to escape from this world into a better world. The Gospel announced the imminent reality of the kingdom of God where the good are selected to live in a heaven, an idealized model of Imperial Rome governed by God, the divine equivalent of Caesar.

Even in such a milieu where only a few could lead a life of contemplation, the basic questions regarding the origin of the universe, the nature of mind and matter, life and death, and the destiny of man were the classic questions that troubled the thinkers of the time. That is, there was a need to know. There was an 'epistemic hunger' that was as important as physical hunger, for those who had the means to overcome the daily needs of their bodies. The epistemic hunger of Greek thinkers is too well known to need any comment here. Similar thinking existed in other parts of the ancient world. A Sanskrit author of the 6th century CE, Kamalasila wrote [203]: 'It is natural, on the part of a normal human being who is engaged in the pursuit of his daily aims to inquire about the existence or non-existence of everything — not to do it would be abnormal.'

From the 6th century we come to the 20th century, when John Archibald Wheeler [225] wrote: 'Among all the mysteries that still confront us in our probing of nature, none present more challenge than these,

- How come existence?
- How come the quantum?
- How come one world 'out of the registrations of many observer-participants'?

The difference between ancient thinkers and moderns like Wheeler is important. The search for the so-called 'detached' knowledge was recognized as a valid post-renaissance preoccupation. Today this faces opposition from the so-called 'strong program' of some sociologists and partisan (e.g., feminist) philosophers who claim that everything is driven by class or group interest. Similarly, governments greedy for profits no longer support 'disinterested' research.

A famous Buddhist discourse involves a Brahmin philosopher-monk named Vachha-Gotta, and the monk Malunkyaputta who raised questions regarding the origin of the cosmos, and other 'fundamental questions'. Ten such questions are mentioned in the Pali texts [138], while four more are in later Sanskrit texts. Hence perhaps fourteen questions were put to the Buddha. Having at first remained silent, the Buddha subsequently explained his position that there are unanswerable, futile questions which are irrelevant to his fundamentally pragmatic effort to diminish suffering in the world via a moral and psychological approach.

The first two questions asked if the world is eternal, or not eternal, while the third and fourth asked if the world was spatially finite or not. The fifth and sixth questions were about the identity or duality of matter and mind. The questions 7–10 were about the 'after-death' state of a morally perfect being. The additional questions in the Sanskrit texts were elaborations on the previous ten questions.

One version of the story (the parable of the arrow) involves Malunkyaputta, a monk torn by deep epistemic hunger. The Buddha maintained a Zen-like silence to the questions that were put to him by the monk. Finally, by way of clarification the Buddha brought up the parable of a man shot by an arrow and writhing in pain.

'It is as if a man, stricken with a poisoned arrow would insist on knowing if the arrow was shot by a warrior, priest, merchant or worker, whether he was tall or short etc., whether the bow was a cross-bow or a long bow; whether the bow-string was fiber, bamboo thread, hemp, etc.; whether the shaft was of cultivated or wild wood; whether the feathers of the shaft were ... etc., before removing the arrow and taking remedial action. The man would die of the poison, without ever knowing the answers to all these inopportune questions'

It is as if all fundamental questions were beside the point for them, when faced with the utter suffering and pain endured by all living creatures. The immediate needs of compassionate action and universal love are given priority over understanding nature.

The understanding of reality is to be achieved by a meditative process of reflection, taught even today in Zen and other Buddhist traditions [152]. Nevertheless, in a number of remarkable discourses (e.g., the Kalama Sutta, Gnana Sutta and several other discourses (suttas), the Buddha sets down, circa 6th century BCE., a theory of knowledge which rejects the authority of sacred texts, rejects appeal to tradition etc., and insists on personal experience and the value of verification, 'as one would test for gold or base metal on a touch stone'.

This concern about suffering and existence was also found in the Orphic tradition of the Greeks, although there was less emphasis on altruism and compassion. Some modern evolutionary biologists have also attempted to invoke love, altruism and also selfishness as evolutionary results that confer adaptive advantages on those organisms or genes having such traits [170]. David Haig in his article entitled *The social Gene* [94], and Dawkins (*selfish gene*, Sec. 14.4) have in fact used words borrowed from human characteristics to describe the behaviour of genes. Bryce Dewitt, a well known cosmologist suggested [57] that the historical emergence of early Christianity was derived by its message of love. It is noteworthy that DeWitt does not bring in the older Christian conceptions of nature as 'God's Book' which is on par with the scriptures — as was claimed during the renaissance, and that the study of nature and the scriptures had the power of moral upliftment and ennoblement. In this view, truth is some how 'morality-laden'.

The early myths and belief systems actually fulfilled essentially similar functions in regard to what we have called 'epistemic hunger'. As François Jacob has asserted, myths and science 'both provide human beings with a representation of the world and of the forces that are supposed to govern it. They both fix limits of what is considered possible' (part II-[49]).

Although God-fearing religions are nowadays presented as religions of love, the historical facts may indeed be different. Even the religious books make no secret of it. Thus the Bible has 'good Samaritan' stories, as well as horribly cruel stories of bigotry and fanaticism. God, like Caesar, was feared and not loved. Belief systems turn to humanistic values when they do not wield authority, as in the early days of Christianity, or in modern times. The more recent change began in the 19th century, when a more humane face of God as 'love' became increasingly fashionable, due to the evolution of humane values from the renaissance onwards. In *The End of Faith*, Sam Harris [98] has given an excellent account of how religions have institutionalized themselves and linked up with terror and oppression. On the other hand, these religions offered a simple formula of eternal salvation and heavenly bliss through faith in a personal God to whom one might pray in adversity — in effect, a God of hope. Even today, while traditional religion is waning, fundamentalist and evangelist versions of religions are growing in many parts of the world where 'hope is the only hope'. In effect, Karl Marx's curt characterization of religion as the 'opium of the masses' was not pure Victorian venom.

2.1.1 *The morality debate — science as a vocation*

Modern day scientists and philosophers have engaged in 'science studies' that have led to 'science wars'. The so-called 'Strong positions' in sociology contend that scientific knowledge is a subjective construction, motivated by the self interest of a dominant social group, or perhaps even worse [127]. Hence it is interesting to return to the beginning of the 20th century when more moderate points of view prevailed, and retrace our steps back to contemporary times.

In a famous discourse delivered at the University of Munich, Max Weber [217] presented a modern framework for a discussion that had already been raised by the Greeks and the ancient Indian writers (Sec. 2.1). Is the search for knowledge a 'higher calling', associated with a moral purpose? Or is it another 'job', where the scientist works on a 'project', and delivers the result for a fee or salary? Or is it a bit of both? The same questions can be asked regarding music and art. It was normal for European aristocrats to 'commission' works for their use. Even Johann Sebastian Bach and Da Vinci had to 'produce', irrespective of whether it was a 'calling' or a 'job'. Modern engineering science as well as commercial art embodies this aspect of science as a trade. More recently, Steven Shapin [193] has addressed similar questions, taking into account the fact that today scientific research has become an occupation with a large body of practitioners, working in industry, government and academic laboratories. Many governments that believe in 'less-government' have followed policies where science is funded only to the extent that it leads to narrowly focused short-term commercial innovations.

Max Weber alludes to how Tolstoy regarded science and knowledge as being 'irrelevant or meaningless', as they do not provide him an answer to the 'meaning of life and death, purpose and end'. Indeed, David Hume had already remarked, in the 18th century, that rational philosophy cannot pass from a set of 'is' sentences, which affirm the nature of things (e.g., *the blue whale is endangered*), to 'ought' sentences (e.g., *we ought to protect the blue whale*), indicating what we *should* do. G. E. Moore [150] had put Hume's arguments in a more general and systematic footing in his *Principia Ethica*. Arguing from 'natural' premises to 'ethical' premises is known as the 'naturalistic fallacy' in Moore's work. Clearly, these arguments, to the extent they are valid, apply not only to ethical concepts like good and evil, but also to aesthetic concepts and other sentiments which were called passions by David Hume, and extolled by Jean Jacques Rousseau. This apparent gulf between scientific fact and ethical ends was aptly summed by Richard Feynman who said that it 'was a matter between you and your Rabbi'. Bertrand Russell, and indeed G. E. Moore held that we perceive good, and make moral judgments in the same way that we perceive a colour ('yellow'), or develop love or affection. Just as you can be colour blind, you can be morally blind. Modern neuroscience (see Ch. 14) seems to come to similar conclusions, where we find that most of our actions are pre-determined by the adaptive unconscious. Thus, within this point of view, our morality begins in our DNA and manifests as a part of the neural wiring of our mental equipment that gets modified by up-bringing.

Just as a child can be taught to appreciate the arts, music etc., by a suitable up-bringing, one may perhaps train a person to be 'good'. Tania Singer, a Swiss neuroscientist found that brain areas involved in pain perception also respond when people observe the suffering of others [198]. Small children as young as 18 months, and even chimpanzees were shown to have altruistic tendencies [215]. Thus a 'Center for Compassion and Altruism Research and Education' (CCARE) has been launched by a group of neuroscientists and associates of the Dalai Lama.

2.1.2 *Belief, probability and doubt*

Strong beliefs grounded in faith erode into doubt when confronted with experience that persistently goes against such beliefs. Experience is what the hunter, the farmer or the architect gathers as 'know-how'. These get codified as a series of 'is' statements about the world and how to control our environment. These 'is' statements are propositions of the form 'the sky is blue', 'the grass is green', 'lead is heavier than copper' etc. Science deals with the world of 'is' statements codified in the theories that seem to work in the range of length scales and energy scales used by the investigators. However, these 'is' statements are not articles of faith, but subject to doubt. Science provides a controlled *doubt system* — i.e., a systematized program for doubt. Constructive doubt, as opposed to empty total skepticism is possible only when we have something firm to stand upon. That is, the machinery of doubt deployed in science rests on having established a *terra firma* of trust which the practitioners of science have largely already accepted. This *terra firma* is the set of theories, models, protocols and procedures that constitute the established scientific body of knowledge that has largely become engineering.

An engineer working on a project assigns measures of probability to events, safety and failure based on his understanding of the properties of matter. Knowing that a cube has been machined accurately to have equal sides in a piece of uniform matter, s/he would assign that any side would come up with a probability p of 1/6. One can verify this using statistical trials, and confirm that p is also a measure of the frequency of a given side turning up. This is the 'frequentist' view of probability where the objective aspect of probability is emphasized. Furthermore, the engineer would hold the belief that *all* dice would behave in that manner. Of course, he has no way of testing this statement. The engineer is ready to assert that not only would 'properly manufactured' dice work in the 'expected way', but so would wheels, jet engines, vacuum pumps, thermometers, spectroscopes, clocks, and other instruments, manufactured with the well-tested knowledge that belongs to the energy and length scales where the engineer operates.

In classical probability theory (e.g., Fermat, Poisson, Laplace and others), it was believed that there was a 'law of large numbers' whereby objective statistical conclusions are asymptotically recovered from many observations. This view was attacked by von Mises and others who established an 'objective', frequentist definition of probability. The contemporary approach, due to Kolmogorov (1930s) is axiomatic, with 'probability' left as an undefined element whose algebra is specified by the mathematical theory which concerns itself with the sample space S, (e.g., two for tossing a coin), the field of events F, and the probabilities P associated with the events. In practice, experimentalists may have intuitive, non-rigorous ideas of probability; but the final results get codified in the working manuals of engineers and technicians where they *claim* objective probability to hold.

The scientist, armed with these reliable tools, pushes his experiments into new, unexplored scales of length and energy. This is done by mapping the more refined, inaccessible signals into those which are magnified into the realms of our instruments. Thus the microscope converts the small length scales of the amoeba to a magnified scale where our instruments can study it. A bubble chamber converts the trajectories of subatomic particles to visible paths that can be measured.

Given *new* energy and length scales, we are no longer on *terra ferma*. The physicist moves with caution since the old theories may not hold in this new, refined length scale. This is the borderline between knowledge and belief, trusted practice and groping doubt. Since the 'new experimental territory' is unknown, assigning probabilities cannot be fully objective or 'frequentist'. The assignment of a probability p to an event is mostly a measure of the degree of belief held by the researcher. This approach to probability theory was pioneered by Thomas Bayes (1702–1761) and the great mathematician Pierre-Simon Laplace. It can be shown that the standard rules of probability follow from the Bayesian tenets of probability which can be written as conditional probabilities capsuling some prior information that goes into the 'belief' [119]. It has to depend on prior assignment of probabilities based on extrapolating from one's interpretation of how one thinks 'it is going to be'. Thus we have a more subjectivist approach to the meaning of probability as *a measure of belief*, rather than as a measure of our knowledge of the relative weights of alternative 'objective' possibilities.

As experiments get firmed up, more and more researchers would express beliefs on the predicted outcomes, and then one's beliefs become less personal, and consolidates as the mainstream views in the subject. In this manner, subjective belief becomes gradually transferred into a corpus of commonly held beliefs that may have undergone radical revision during the assimilation process. By then, the various competing models at the finer length scales would have become well established in new theories and models. Then the Bayesian probability assessment gets replaced by an objective (e.g., frequentist) description by common consent. Once again, science has become codified and absorbed into engineering practice.

In the following sections we look at the nature of scientific knowledge, model construction etc., and fill in some of the ideas expressed so far.

2.2 Scientific knowledge and model construction

The ancients distrusted their senses, and tried to guide themselves by thought and introspection. The resulting ideas reflected the cultural paradigms of their social milieux. Most scientists, but not necessarily philosophers, would argue that this introspective process needed the touchstone of experiment to ensure that fact was sorted out from fantasy. Today many would grant that we possess a well tested, i.e., empirically acquired, body of public knowledge known as scientific knowledge. We discuss later in the book that this knowledge applies not only to phenomena on the earth, but even to events taking place in distant galaxies (i.e., at large length and time scales), or inside atomic nuclei (i.e., at very short length and time scales). Unlike 'subjective, personal, post-modern' or 'mystic' awareness, this body of knowledge is not private. This knowledge is organized into models (i.e., theories) that provide a structure to what is known to us. These models have a predictive capacity extrapolating beyond our empirical database.

For example, although the sun has risen every morning, it does not follow from this set of observations that we can infer that the sun would duly rise tomorrow morning. Just because something happened ninety-nine times, it does not follow that it would happen the hundredth time. This is the *problem of inductive knowledge* or *induction* not leading to universal knowledge. However, we include these observations of the sun in a mathematical model of the solar system which describes the motion of the planets, the rotation of the earth on its axis, and so forth. If this model is a good one, it embraces *all the data* in a self-consistent, holistic manner as explained later in our discussion. The model allows us to predict what we should see in the sky tomorrow, or on any other assigned date falling within the time scale associated with the estimated validity of the model. The existence of a model does not guarantee that we have converted empirical data into universal knowledge; however, it provides the basis for Bayesian belief in a model of the solar system where the sun rises every morning during its life span. When we need to distinguish this, limited type of induction that holds within the length scales and time scales of a model, we will call it *bounded induction*.

2.2.1 *Induction*

Let us look at *induction* and the construction of theories based on observations. Such observations may have been prompted by previous theoretical considerations; that is in fact the norm in the physical sciences. However, in some of the more descriptive sciences, data collection precedes model construction or theory.

Many writers use the word 'induction', or words derived from it rather loosely and even construct their negation. Feyerabend uses words like 'counterinduction'([72], Ch. 2, 4 etc.). The latter is described as follows: 'It is an essence of empiricism. The 'counterrule' corresponding to it advises us to introduce and elaborate hypotheses which are inconsistent with well-established theories and/or well-established facts. It advises us to proceed *counterinductively*'. We have discussed Feyerabend's views in some detail in Sec. 3.4.1.

We do not use the word 'induction' in that manner in this book. The concept of 'induction' used in science is what we called 'bounded induction'; i.e., it applies within a domain of energy scales, length and time scales where the theoretical model is expected to hold. Usually the theory comes *before* the data as a model imposing various rules of symmetry, invariance principles etc., that 'should' hold. Thus the Greeks insisted on looking for perfect spheres and other geometric symmetries. Some of the pre-imposed theoretical constraints used by modern theorists are discussed in Ch. 3. Thus scientific theories are not based on induction, or just looking at patterns or correlations in sets of data. Hence one might even claim that the assumption of such symmetry principles and invariance principles constitutes the metaphysics of science. However, there are cosmological arguments based on the isotropy of the initial state of the universe prior to the big bang, as well as internal consistency arguments that show that they need not be considered as metaphysical.

The process of bounded induction may be illustrated as a problem of extrapolating or interpolating a set of data to domains not directly covered by the input data. This extension of the data is done by a theory, or model, which establishes a functional form or co-relation between the input data and the observational data. Thus, for instance, the observational data are the n data points $y_i, i = 1, 2, \ldots, n$ relating a variable y to n data points of another variable, say time measurements, at $t_i, i = 1, 2, \ldots, n$, defined in the time interval t_1 to t_n. The problem of induction is equivalent, as far as we are concerned, to the problem of constructing the function $y = f(t)$, valid in an extended domain t_a to t_b, where, e.g, $t_b > t_n$, and/or $t_a < t_1$. The data points may have measures of confidence or weights w_i, margins of error $\pm\delta y_i$ etc. The only sense of a 'counter-inductive' use of a data point would be to exclude a data point by giving it a zero weight. It is not meaningful to give it a negative weight. The physical theory enables us to construct the function $f(t)$. Its output values $f(t_i)$ at the input values t_i should agree as closely as possible with the observed set y_i in an overall sense, and also give predictions which can be examined by experiments in the extended domain. Even if the domain is not extended, as is often the case, the theory enables us to *interpolate between* observations.

In Ch. 3 we suggest that Feyerabend is mistaken, or misrepresenting matters (Sec. 3.4.1) to attack Galileo, using 'counterinduction' as one of his missiles. One may select data from a given sample and use the selection to justify a different theory $g(t)$; but this is not counterinduction. It is still induction.

It should also be noted that we do not use the mathematician's definition of induction. In mathematics, the name 'induction' denotes a method of proof. We establish that if the n-th example of a statement is true, then the $n+1$-th example is necessarily true. Then we show that the result is true for the case $n = 1$. That proves the validity of the preposition for all n. For example, the statement 'the sum of the series of numbers, $1, 2, 3, \ldots, L$ is $L(L+1)/2$' can be easily shown to be true using mathematical induction. Goldbach's conjecture states that every even number greater than 2 is the sum of two prime numbers. This can be checked manually for simple cases. Thus 4 (=1+3), 6 (=1+5), etc. However, even though it is true in each case that we may test, we need to prove that if $2n$ is the sum of two primes, then the same property holds for $2(n+1)$. Unfortunately, no one has demonstrated that, and the proposition is labeled a 'conjecture'.

In philosophical discussions statements like, 'all ravens are black', are used to construct paradoxes irrelevant to the way induction is used in science. The word 'all' is an extremely dangerous word where boundaries of validity have been extended arbitrarily. Many of the difficulties in algebraic analysis were resolved already in the 19th century by considering limit processes where 'all', i.e., infinities and reciprocals of zeros were carefully dealt with. The universal 'all' must be used with great care, as we learn from Cantor's development of the mathematics of infinite sets. The statement, 'all ravens are black', when used by bird watchers, contains the word *all* in the weaker sense of a Bayasian statement of belief. The bird watcher is ready to bet in full confidence that the next observed raven is going to be black — but he is not talking of a full Cantor set. One can attempt to rebuff the bird watcher by presenting him a pathological construction.

For example, we can claim (purely semantically) that at some future time (i.e., for all $t > t_F$ where t_F is a future time), ravens are blue, and that for all locations inside a black-hole, all ravens are purple. We could add to it the possibility that all non-observed ravens are blue in all time domains and purple in all space domains. Note that this is *just a theory* about ravens and nothing more. In fact Goodman [88] introduces such pathological cases (limited only to the time domain and not to space) by defining a class of emeralds which are said to be 'grue'. Grue is the color of those emeralds which are, as usual, green for time $t < 2100$ CE, and blue for $t \geq 2100$ CE. Thus, all observed instances of green emeralds validate the sentence 'all emeralds are grue'. As noted by Alex Rosenberg [180] the 'grue emerald problem' is often presented as a 'new challenge' to induction. However, it is exactly the old problem of n observations (occurring during the period $t < 2100$ CE) not validating a new observation at $t \geq 2100$ CE. Goodman could have equally well said, 'anything could happen in 2100 CE'. If humans are to store nuclear waste for millennia they need to make long-term predictions. What bets would Goodman place on the change of colour of emeralds in 2100 CE?

Seemingly 'improbable' events may occur when these are least expected. This phenomenon has been popularized by Nicholas Taleb who calls them 'black swans' (cf., blue emeralds) that unexpectedly appear when everyone believes that there are only white swans [205]. If the problem is one of representation of data using curve fitting, we see that the curve which 'fits' the *green emeralds* in the time domain $t < 2100$ CE will equally well fit the *grue emeralds* up to $t < 2100$ CE and deviate from then onwards (the quantity fitted is the wavelength of the light reflected from a test emerald, as a function of time). If indeed there are 'grue' emeralds that become blue abruptly at $t \geq 2100$ CE, we have a function that bifurcates, with a new branch for the emeralds that became blue. That is, Goodman has a new theory about emeralds! The 'green branch' of the curve becomes the 'unphysical curve' after 2100 CE for Goodman.

The setting of the sun is an inseparable part of the theory of the solar system; the latter is based on bounded induction of diverse astrophysical data. The colour of emeralds is also a part of the physics of the electronic structure of atoms in crystals. Emeralds are 'expected' to be green not only because of past observations of emeralds, but also because the quantum theory requires it. The quantum theory is nearly a theory of everything, and satisfies various invariance principles and symmetries (see Ch. 3) that go beyond the inductive observation of the color of emeralds.

Chromium (Cr) or Vanadium (V) impurity atoms embedded in a matrix of the Beryllium-alumino-silicate, ($Ba_3Al_2(SiO_3)_6$), known as Beryl, acquire an electronic structure where the *d*-electrons of the impurity atom absorb part of the incident light, leaving the green part of the spectrum unaffected. The process is completely understood and emeralds can be grown in the laboratory or industrially. Indeed, the color of an emerald can be changed by applying compression or an external potential which can affect the crystal field acting on the Cr^{3+} ion embedded in the Beryl. Thus, emeralds could become blue if the conditions on planet earth changed abruptly, say due to a meteor impact. The necessary conditions for having grue emeralds can be calculated using quantum mechanics. A philosopher can claim that the quantum theory also would become 'unquantum' in the year 2100 CE etc., and invent a large number of pathologies that take place on 2100 CE, or even claim the end of the world. The Bayesian credence that one would attach to such constructions would diminish in proportion to the need to adumbrate the pathologies. Thus there is nothing new in the 'grue-emerald' problem.

One does not need bizarre constructions like 'grue emeralds' to establish difficulties in representing a set of points using a model (i.e., a fit function representing a theoretical model). Even if we remain within the domain of the data, follow the rules conservatively and make no pathological constructions, a model may yet predict wild oscillations between the given data points while representing the data very well. Such a theory is unphysical and contains some pathological features. Good scientists expect nature to obey certain symmetries and continuities. The intuition of great scientists like Galileo enables them to zero in on good models, even with fragmentary data, whereas lesser scientists would not see the emerging picture, even with masses of data. As we learn from neuroscience and the theory of neural nets (see Ch. 12), the brain is a special type of computer which is good at pattern recognition as well as model construction. Undoubtedly, some brains are better at it than others. The human brain has evolved the capacity to analyze data and construct models which it uses for immediate action. Such models are valid within the length scales and time scales associated with the immediate environment of the evolving species. Such models are tools for prediction within a limited range, and a re-tooling is needed for timescales longer than the relaxation times (characteristic times) and energies of the given system.

In effect, the model is only good for the length scales and time scales associated with the input data. If these scales are changed, the model may fail, and an extended theory is needed. This is like the failure of Newtonian physics when applied to particles moving at large velocities. The fitting is done under the assumption that if the number of observational data were increased, then the error in the model would decrease very rapidly, as in the bell-shaped 'error curve' of Gauss (see Fig. 6.1). This asserts that the probability of observing very large deviations $\Delta y = y - y_0$ from the best fit (given by the model) y_0 drops rapidly, as $\exp\{-\sigma(y-y_0)^2\}$. Errors arise because the tools of observation are not perfect, and the 'boundary conditions' used to isolate the observed system from the wider world are not perfect. It is the boundary conditions which enable us to replace a complex system by a 'reduced system' containing just a few variables that are under the control of the observer. The Gaussian assumption is economical as it involves a probability theory based on specifying just two parameters (mean value y_0 and σ, the standard deviation) to characterize the quality of the data.

This assumption is in fact not valid even for all physical systems, let alone for complex, open systems controlled by many variables, and evolving in a chaotic (essentially non-predictable) manner. Simple physical systems near phase transitions show deviations (fluctuations) which obey *power law* behaviour instead of Gaussian behaviour. Classical collisions of billiard balls rapidly become unpredictable as they become very sensitive to the boundary conditions, specification of initial states etc. In Sec. 8.6 we discuss that we are indeed unable to predict the actual conformation (detailed 3-D structure) of most proteins so fundamental to life. Hence, simplistic attempts to apply these methods to very complex living systems, or sometimes even to the social sciences, should be viewed with due caution.

However, the short-comings of the standard methods, and the new possibilities are understood by the computing community (see Sec. 2.2.5). Caution has also been expressed by, e.g., Frederich Hayek [101] even in his 1974 Nobel address where he labeled economic theory as 'Pretense of Knowledge'; yet he too fell into the same trap supporting Milton Friedman's questionable models. More recently, Nicholas Taleb [205] justly points out that 'black-swans' do occur, totally unexpectedly!

We may summarize this section by noting that puzzles of induction are no longer puzzles when they are embedded in the relevant physical model and regarded as problems in the mathematical representation of a theory. A theory has a domain of validity, and contains symmetry principles and invariance laws that are not included in a purely inductive approach to the data. The sun is expected to rise only within the expected lifetime of the solar system. Thus the induction is bounded to the domain invoked by the model. The expectations of validity are Bayesian in character. Furthermore, a theory is a very different object, quite unlike the 'data points' that were included in it, just as 'mankind' is something very different to an individual 'man' that is included in mankind.

2.2.2 *Using data to model, verify or refute a theory*

Let us now consider a set observations, say of an astral object seen in the horizon. It has been observed for about 10 hours. We may assume that this observation was prompted by previous theories or reports of such phenomena. However, to be brief, we make a first attempt to understand the data by a quantitative theory. The astral object may be a moving body, or a satellite which should show periodic motion. We construct three theories to model the data, and examine how the theories explain the data, and how they could be verified, modified or falsified.

Given a set of data y_i for observations t_i, the fit function $y = f(t)$ is not meant to be purely a best-fit numerical representation using a large number of arbitrary parameters. Instead, $f(t)$ is dictated by the theory, or theories, which are under the scrutiny of the researcher. Having the *least number* of fit parameters — simplicity (Sec. 2.3.1) — is favourable to a theory.

Suppose we are given a data set $\{y_i\}$ of nine points, e.g., 0, 3, 7, 6, 4, 2, 1, 0.5, 0, i.e., positions in the sky (in some units), associated with the observations of an object seen in the sky at time values $\{t_i\}$, i.e., 0, 1, 2, 3, 4, 5, 6, 7, 8, hours. If this is an object moving with an initial velocity u and an acceleration a, and if the damping due to winds etc., be deemed negligible, Newtonian theory asserts that $f(t) = ut + (1/2)at^2$, i.e., a polynomial form involving two unknown constants u and a having the physical meanings of 'velocity' and 'acceleration'. Their extraction from the observed data is thus 'theory laden'. In our view this is not a negative feature, but a positive feature of model construction used in science. Theory is the tool which reveals the parameters u and a. However, the data are not found to be a good fit to such a simple polynomial. One may choose to examine the data (see Fig. 2.1) using a polynomial $y = c + ut + (1/2)at^2 + bt^3$ involving four unknown constants u, a, b, c defining the model to now include damping or perturbation effects; or one may choose a fit based on a Fourier expansion in time, viz., $y = a + b\cos(ct + d)$ where the constants now have other physical meanings. Thus c is a cycle frequency for a periodic occurrence of the observed phenomenon, d is a phase shift, and b is an amplitude. Alternatively, one may choose a Padé form $y = (ut + (1/2)at^2)/(c + dt^2)$ where the four unknown constants reveal more complex interactions. One theory may give a monotonic fit, while another would give an oscillatory fit. Which one do we choose as reflecting reality? At this point it is only a belief that one of these models may work.

2.2.3 *Verification, assimilation and falsification*

The observed data have been used to form three theories (the polynomial, Padé and Fourier models) which seem quite promising within the window of the available data. Either we need more data to distinguish between the theories, or theoretical

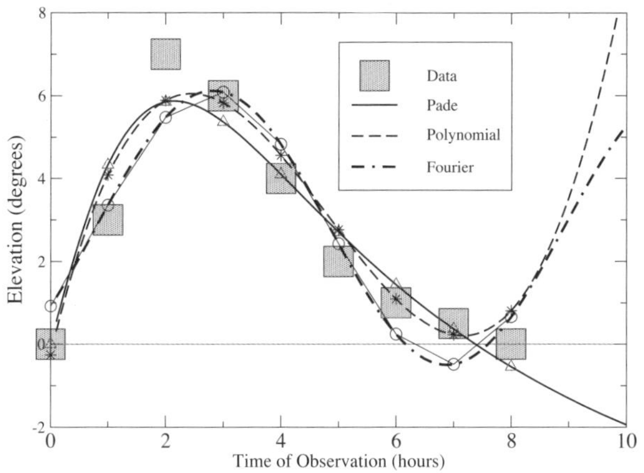

Fig. 2.1 A set of observations is fitted to three theories. These are represented by models yielding a polynomial, a Padé form and a Fourier expansion with four parameters. The size of the gray boxes represent the expected uncertainty in the data. The Third data point is poorly fitted by all theories, and needs re-measurement to confirm it or rule it out.

insights which rule out the polynomial expansion and accept the Fourier expansion or the Padé form. For example, if the observed phenomenon is *periodic* in time, the Fourier form *makes sense*. We note how the different models differ markedly in time domains outside the fitted domain. It is this feature which gives one the capacity to rule out models by prediction and *falsification*. However, two theories may be too close to each other in the accessible domain for direct falsification. Then we need to go to *new scales of time or length*. The polynomial model, although agreeing with the periodic model initially, could represent a process which readily distinguishes itself from a periodic process. Thus *weighted amounts* of verification, falsification and statistical methods (e.g., least-square deviations) are all included in the scientific testing of theories by model construction (i.e, in this type of bounded inductive scientific process). We note that the third data point stands in disagreement with the three theories represented by the Polynomial, Padé, and Fourier forms. If this observation is reconfirmed by additional data near the $t = 2$ observation, then clearly, all three theories need *modification*. One examines if perturbation terms (physically meaningful within the models) can be included to obtain better agreement. It is when such additional efforts fail, that the theories begin to be falsified. If such efforts succeed, the theory is improved, and also its basic tenets are verified, and the theory is assimilated into the body of accepted practice which goes as ‘engineering’. We can use the term *assimilation* to denote the minor modification of theories and their final absorption into engineering.

Thus, verification and falsification are not sharp *yes/no* processes but weighted inputs. The deviation of an observation from theory has to be viewed in regard to its *statistical weight*, and also within the context of simple modifications of the theory. There is no ‘critical mass’ of deviations, as conjectured by Kuhn.

Some theories, for example the theory of electricity and magnetism, known as electrodynamics, have had no exceptions. Electrodynamics is discussed in Ch. 4. In such a theory, even a single deviation, by even a tiny amount would be totally unacceptable. Even a single difference in the last valid decimal of a measured value and the predicted value is critical for such a theory. The quantum theory is another theory with no known exceptions and hence even the slightest observed anomaly would create a great sensation. On the other hand, predictions where both General Relativity and Quantum Theory are used together to get an answer may be suspect as they are still not 'unified' into a coherent whole. More extreme situations are found in some 'sciences' like Economics. Here predictions are virtually never found to be correct. It is not uncommon for the best economists to proclaim the continued stability of the financial system, but only to announce a major market crash within a few days. Thus, the significance of deviations from predictions depends on the quality of the science as a whole.

Suppose that additional observations (e.g., at other observatories) confirm that what is observed is a periodic motion of a mini-planet or satellite. Thus the Fourier form is provisionally selected as explaining our data. Let us write our conclusion in the form *planet sightings fit a Fourier form*. Now we recall a favorite hobby horse of philosophers. The sentence may be compared with *all ravens are black*. Here, 'planet sightings' play the role of 'ravens', and 'Fourier form' plays the role of 'blackness'. Now consider the inversion *all non-black objects are not ravens*. In effect, if the set of all objects is A, and if the set of all black objects is B, this inversion is a true statement about the set $A - B$ that includes, say, goldfish. Now, many philosophers claim that every time one encounters a member of this set, e.g., a goldfish, that is an inductively significant observation which supports the claim that 'all ravens are black'. If we return to our planetary example, the inversion reads *All non-Fourier forms are not planet sightings*. The content of this assertion would refer to any point not lying on the Fourier fit curve which has a thickness equal to the error bars in the data. In fact, the theory has *already* included, i.e., taken account of all those points by *not* having them on the curve. If a data point is far away from the curve, that point becomes less significant. Thus there is no additional inductive content in such data falling outside the error bars. The usual philosophic discussions do not include the consideration of such *weight factors* that in effect limit the theory to a limited domain — this is bounded induction.

2.2.4 *Isaac Newton and the Copernican revolution*

In Sec. 2.2.2 we discussed how the observations of an astral object were used to construct a model — a toy model of a theory putting together a set of observations, and also giving us the capacity to make new predictions. It was precisely such observations that gradually led to the most famous scientific revolution of our era, even with mathematics and theoretical physics still in their embryonic stage. This was the development of astronomy, and mathematics *via* Ptolemy, the medieval astronomers, and then Copernicus, Kepler, Galileo, Descartes and others that culminated into a towering intellectual achievement in Newton's *Philosophiae Naturalis Principia Mathematica*, i.e., The Mathematical Principles of Natural Philosophy. It is usually known as Newton's *Principia*, published in July 1687.

Newtonian astronomy is sometimes quoted to claim that science is 'organized common sense' that reaches further. In fact, Newtonian astronomy is astonishingly *contrary* to the common sense of that era. Eugene Wigner [227] writes:

> Philosophically, the law of gravitation as formulated by Newton was repugnant to his time and to himself. Empirically, it was based on very scanty observations. The mathematical language in which it was formulated contained the concept of a second derivative and those of us who have tried to draw an osculating circle to a curve know that the second derivative is not a very immediate concept. The law of gravity which Newton reluctantly established and which he could verify with an accuracy of about 4% has proved to be accurate to less than a ten thousandth of a per cent and became so closely associated with the idea of absolute accuracy that only recently did physicists become again bold enough to inquire into the limitations of its accuracy.

The initial models of geography and astronomy were indeed founded on naive observations and the 'common sense' of the age. The simplest 'inductive' model of the earth is that it is flat! Insightful observation of how the hull of a ship disappears at the horizon before the mast disappears had to be given great weight before the flat-earth inductive theory could be overturned. The motions of astral objects observed in the heavens were taken to actually occur, with the observer at rest, at the center of this theater of observations, but in terms of motions guided by the crystal spheres of the heavens. There was no 'empty space'. This results in geocentric astronomy — the astronomy of Ptolemy. The astronomy of ancient India and China was similar but 'empty space' was not denied. Geocentric astronomy provided an adequate account of celestial phenomena, and it is used even today in drawing up the farmers' ephemeris and for similar purposes.

However, the geocentric model contains arbitrary recipes that 'made no sense' but get the 'right answers'. The planets moved in circles, as well as some additional circular motions known as 'epicylces'. Planets seen in the sky display retrogressional motion through these epicycles. Although the sun was just one of the seven planets going round the earth, the sun is implicated in working out the motion of all the other six, while these other six played no such role. Mercury and Venus had to be always associated with sunrise or sunset, but never in the true night sky. To get this correct, the epicycles of Mercury and Venus have to always lie between the earth and the sun. Mars, Jupiter and Saturn go through their retrogressions when they are away from the sun, and the epicycles have to be arranged in relation to the sun. The model did not indicate any order or arrangement to the planets, their sizes, orbital times etc. It was Copernicus who understood how to clean up Ptolemy's 'system' to create a simple accurate model by discarding the common-sense view of a fixed earth. He made the earth and all the planets move around the sun, but remained wedded to the Greek idea of circular orbits. Kepler gave up the concept of the crystal spheres and began to think in terms of an empty space where the planets were 'free to move just like birds', guided by the sun and the other planets. Thus Kepler discarded 'perfect spheres' and followed mathematics when he introduced elliptical orbits for the planets, and came out with the now famous Kepler's laws of planetary motion. Newton would show that these are simple consequences of the ordinary terrestrial laws of motion.

The claim that the earth is spinning and moving at tremendous speed through the skies would have been shocking to the medieval thinker. It was Galileo's job to explain how it 'made sense'. Concepts shrouded in 'common sense' have to be stripped nude and re-shrouded into new notions — this is the process of creating *theory-laden concepts* that are more meaningful in the extended length- and time scales of the new world view. Initially, when Galileo was leading the Copernican revolution, it was the elegance of the mathematical theory that enthralled its protagonists. Unfortunately, mathematical elegance is appreciated only by a few. The vast majority believed that the 'commonsense' world of a fixed earth, perfectly circular orbits and epicycles were more beautiful and 'logical' than ellipses placed round the sun, with the earth moving at break-neck speed. However, the arrival of the telescope extended the length scales of observation, and the pendulum clock gave better chronometry, favoured the new theory, and damned the old. Galileo's observation of the phases of Venus was totally incompatible with the Ptolemaic system, while it was fully compatible with the new heliocentric system.

This was the scientific legacy of Isaac Newton who was born in 1642, the year of Galileo's death. Newton inherited the heliocentric model while the geocentric model was by no means dead. He also inherited Galileo's new ideas on motion and mechanics (worked out while under house arrest). However, these studies had not entered the curriculum of the University which remained true to Aristotle. Classical mathematics was a part of the university curriculum. Mathematics was held in high esteem by these learned men — and indeed there were very few women — who studied the heavens. Voltaire's lady friend was the exceptional woman who translated the *Principia* into French. At that time the most important form of mathematics was geometry. The most important paradigm for mathematics was the style of demonstrations found in Euclid's books. According to Galileo [80],

> 'Philosophy is written in this grand book, the universe, ... It is written in the language of mathematics, and its characters are triangles, circles, and other geometric figures without which it is humanly impossible to understand a single word of it ...'

Newton inherited other medieval traditions of learning, e.g., theology and alchemy. Although these are often neglected in discussing Newton, they played an important role in his intellectual life, and a deeper study of Newton shows that he approached these topics in a scientific manner [222]. Newton held many unorthodox theological views, but knew that it was best to keep them to himself.

The belief that 'God had laid out the design of his Universe in the scriptures in an occult language understood only by the pious', is a tenet held even to this day by the religious. Even non-theistic belief systems (e.g., Buddhism) claim that meditative states of the mind, *jyana* or 'Zen' states, are needed to reveal some truths which are not comprehensible by other means. Thus Newton inherited two traditions, viz., the occult-mystical, and Archimedean-observational, with mathematics as the over-arching rational guiding hand.

Newton became an expert grinder of lenses and built new telescopes and optical instruments. The results of these experiments on optics and light, re-tested in the true style of experimental physics, were duly reported to the *Royal Society*. He also did experiments with pendula; dropped different bodies from different heights, repeating Galileo's results. The Royal society was the forum for peer review of the discoveries in 'natural philosophy' at that time, where discoveries had to be enacted in public, departing from the secrecy of the alchemist. In medieval India, secrecy of discoveries was maintained as *guru-mushti*, i.e., a secret held by savants and revealed only to the favorite pupil. That is, no mechanism for open critical discussion was invented. Today, Royal societies and Academies of science play no significant public role, while the openness of science has been threatened by commercial concepts of 'intellectual property'.

In addition to experimental physics and mathematics, Newton carried out detailed studies of theological texts and alchemy. These too were recorded meticulously in notebooks. Newton checked and rechecked his conclusions, but he was not satisfied with the 'reading of the Universe by the occult clues left by the designer'. These dominated Newton's attention from the 1660s, till about the 1680s. He never reported his theological studies to the Royal Society as they were inconclusive. There is no reason to fault Newton's rationality on the basis of his theological studies, as has been done by some historians of science. In addition to all this, Newton led an active political life, and represented Cambridge in parliament; he was the Master of the Mint, and a public figure with all the usual traits of ambitious and dominating individuals. He was a conformist public figure and not a 'dangerous' social revolutionary like Galileo. The Newtonian revolution was a result of the landmines laid by Galileo and others.

Newton's first achievements in the university as an undergraduate were his rapid mastery of the corpus of mathematics available in his day, and his own creation of new mathematical techniques. He rapidly went beyond the clumsy geometrical approach of the day and invented the very powerful differential calculus — *the method of fluxions*. This is the standard mathematical tool of the quantitative sciences, even today. It is reported that Newton was brought back to astronomical calculations by the some what younger Edmund Halley who had asked Newton to provide him the trajectory of a body orbiting another and attracted to it by a force that varied inversely as the square of the distance between them. This is a classic problem in 'dynamics'. It was posed by Halley before the science of dynamics had been formulated. Kepler's ideas on planetary motion, and Galileo's ideas on moving bodies and acceleration had not been systematized. This was the impulse that set Newton in motion. The scientific genius shown in Newton's experimental studies, and the mathematical genius in Newton combined to produce one of the greatest codifications of science in Newton's *Principia*. Newton reduced celestial mechanics to terrestrial mechanics, and introduced the concept of gravitation that unified our understanding of the external world in a manner that had no precedent. One might argue that if Newton had not done this, perhaps somebody else — may be a Laplace — would have done it. After all, Halley was already talking of inverse square laws of attraction in 1684, while Robert Hooke also knew of the inverse square law. However, only Newton had the mathematical ability to use this knowledge. Hence each epoch produces its own genius, and it was Newton's turn to create what Thomas Kuhn called *the paradigm which modern science in its various dimensions has been attempting to emulate ever since*.

The *Principia* followed the Euclidean style of geometric proofs rather than modern methods based on the differential calculus that Newton had invented and used in his research. This makes the *Principia* hard reading for modern physicists. Book I of the *Principia* formulates the dynamics of point masses moving in space subject to various force fields. Book II deals with motion in resistive media, and discusses the work of Descartes, and shows that the Cartesian theory that planetary motion was a result of vortices in space is unable to yield Kepler's three laws. Having demolished Descartes, Newton in Book III picks up the dynamics of Book I and discusses observed phenomena, both on earth and in the heavens.

Newton uses the dynamics of Book I to discusses the planets, their moons, and all of astronomy then known using only the three laws of motion and the hypothesis of gravity. The inverse square law of force that held the solar system together, and made stones or apples fall to the ground, was identified by Newton as the force of gravity — *gravitas*. Thus heaviness is not a property of a body, but the result of an interaction between bodies, each having *mass*. The origin of gravity was a puzzle to Newton. That awaited Einstein, some three centuries later.

The rest of Book III deals with phenomena that were not clarified in the earlier parts. Newton uses his three laws of motion and the law of gravitation to quantitatively *explain* hitherto ill- understood phenomena. Pendula at the equator that had an oscillation time of one second have to be shorter than those in the northern (or southern) latitudes, and this was explained by the oblate shape of the earth, as required by the dynamics of spinning bodies. Newton invented and used 'perturbation theory' to explain the origin of the tides as caused by the pull of the moon and the sun. He explained the precession of the equinoxes — the day and time when the sun crosses the equator — that had been empirically observed since ancient times. Book III treats comets, that traditional harbinger of divine messages, entirely in the non-mystical language applicable to the other planets, and successfully describes the observed locations of the comet of 1680.

The *Principia* reduced all dynamics, terrestrial or celestial, to a matter of three dynamical laws and the law of gravitation, with such precise prediction that there was no room for further doubt of the model. It was only in the latter half of the 19th century that questions arose, in regard to extremely tiny errors in the predicted orbit of mercury when passing close to the sun.

The challenge laid before all physics by Newton was to reduce chemistry, biology and other phenomena, still enshrouded in alchemical and animistic mysticism, into a similar rigorous model as the *Principia*. This is the paradigm mentioned by Thomas Kuhn. Newton's three laws, and the law of gravitation are still taught in our physics courses even today, in essentially the form enunciated by Newton. They have not been discarded or dethroned, as claimed by some philosophers of science who think that science jettisons one view for another and lurches back and forth with no convergence towards an improved vision (see Sec. 2.4.2). The *Principia* holds valid within the length scales and time scales of Newtonian theory. Hence they are *assimilated into engineering*, and in the construction of trusted tools. These are used in pushing research into other length scales and time scales of unknown regimes where doubt prevails. There we find that the new models based on relativity and the quantum theory become necessary. These new theories have the characteristics of aiming to explain all of physics, chemistry and biology, using the basic laws of physics as well as ingenious new insights needed at each length-scale, as emphasized by P. W. Anderson [7]. They have given rise to highly precise predictions for which no exceptions have been found — so far.

2.2.5 *Neural-networks for modeling natural phenomena*

The approach to the construction of models or theories presented in the previous sections is useful for physical systems that depend on a small number of parameters. The astronomy of the Copernican revolution and the physics of particles fit into this essentially Newtonian paradigm. However, in the real world of complex systems, many-agent problems with many interacting variables come into play.

Scientists confronted with analyzing the power spectra of cosmic microwave background radiation ('after-glow of the big bang') have to deal with vast amounts of data using multi-dimensional cosmological models. Modeling such complex data sets is also typical of applications in the social sciences. In fact, patterns of traffic flow, the growth of the the Internet, *pattern recognition* and voice articulation are also phenomena of the manifest world. Given sets of symbols corresponding to the words of a language, how do we learn it? The inputs to the 'theory' are the alphabet, and the output or results of the 'theory' should allow us to predict the articulated sounds. In dealing with a very simple language, a simple look-up table would suffice. Such approaches were in fact the early starting point of AI (artificial intelligence) models of the 1960s, and such studies are ripening even now [99]. Pattern recognition and language translation proved to be hard nuts within those early approaches. This was particularly evident in early attempts at language translation and voice articulation. A naturally occurring language is complex, contextual and has to be 'learnt'. Further, the algorithm must be capable of handling or 'vocalizing' words which are new to it but belong to the language.

A new line of attack came from the structure of neural networks.

Biologists like Hodgkin and Huxley were pioneers who studied the electrical signals from the nerve cells of the squid and other animals. The sea slug *Aplysia californica*, a large creature with a simple brain was another rich goldmine for neuroscience, exploited by Eric Kandel and others. Such studies paved the way for our modern understanding of neuroscience. The brain of a sentient creature is constantly modeling the complex situations encountered by the creature, constructing stratagems (or theories) which enable it to make an immediate prediction for action. Scientific theories are also stratagems for prediction. Hence theory construction can be done by emulating the behaviour of neural networks.

A neural network (NN) in a brain consists of cells known as neurons and their connections (see Sec. 12.4). These connections occur through junctions known as synapses. These allow signals to pass mainly in one direction, as in diode-rectifiers. The synapse is approximately like two membranes separated by a space or gap. A chemical (e.g., glutamate) is spewed across the gap to transmit the signal across the synapse. It is mainly the synapses which get modified chemically

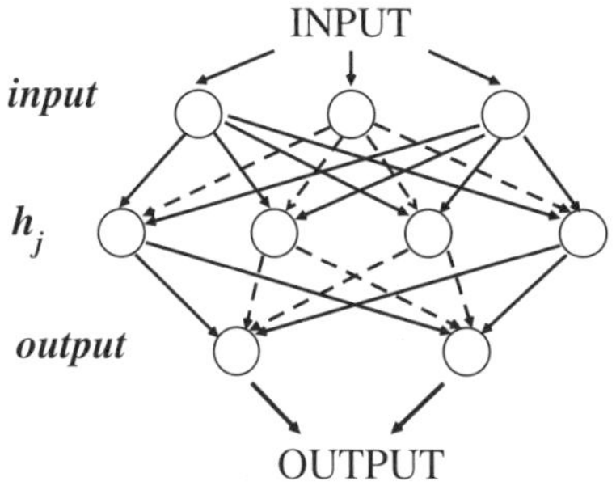

Fig. 2.2 An artificial neural net with one hidden layer (h_j). It processes information (input) and outputs a signal. This may be a prediction of a future state based on information on present conditions. Neural networks can adapt or 'learn' to accommodate new structure in the data.

during learning or adaptation. For example, the work of Kandel (II-[51]) showed that we learn by strengthening or building new synapses, or connections, between them. This capacity to adapt is known as 'synaptic plasticity' (see Sec. 12.4). The neurons can assume several states associated with the membrane potential of the cell. Neurons send out (or 'fire') spike-shaped electrical signals, known as *action potentials*, to neighbouring neurons when changing from one state to another. Whether a neuron would send out a spike or not depends on certain biochemical thresholds. In artificial NNs, this is described by some mathematical rule. Each neuron is sensitive to the firing sequences of its neighbours, and when the same neurons fire together frequently, their synapses get strengthened and correlations grow. This type of process is known as *Hebbian learning*. Thus, given a set of inputs, the neuron network can generate a set of outputs which can 'improve' with more and more data inputs (or learning opportunities).

The mathematical algorithms, or their physical realizations known as artificial neural networks [8], are highly simplified caricatures of the biological neural networks (the bit capacity of the brain is discussed in, e.g., Sec. 12.6.2). In Fig. 2.2 we show a schematic neural network, consisting of nodes and connections. The outer nodes are sensors which receive input signals. For instance, in a language-vocalization network, the inputs may be electrical pulses corresponding to keyboard symbols. These inputs travel into the inner nodes of the NN. Each node is assigned a number of states and they can be excited or made inactive (i.e., 'moved up and down' such states via the interactions). The signals get modified depending on the states of the node, the strengths of of the connections (e.g., the weights assigned to each node) etc. They also combine with signals coming via different routes. The signals finally emerge in the output nodes. At the start, in some fully stochastic models, the states of the inner nodes and the strengths of the synapses are selected at random. However, as signals (information) flow through the network, they gradually begin to take specific values. The output is compared with the target output, and the error or deviation can be used as feed-back or feed-forward inputs. The NN is designed to minimize the error or deviation between the output signals and the target signals expected for the given data base.

The *perceptron*, introduced by Rosenblatt in 1958 [181, 216] is the 'model neuron' used in some of the simplest neural networks. It maps an input vector $\vec{x} = x_1, x_2, \cdots, x_k$ to a scalar output $f(\vec{x}, \vec{w}, \beta)$. The quantities $\vec{w}^{(l)}, \beta^l$ are the weights and 'bias' of a perceptron in the layer l. Multilayer perceptron neural networks are a type of feed-forward network composed of a number of ordered layers of perceptron neurons that pass scalar messages from one layer to the next. Each layer outputs a weighted linear combination of its inputs. We use the letter l for the layer index and write

$$f^l(\vec{x}, \vec{w}, \beta) = \sum_k w_i^{(l)} x_k + \beta^l \tag{2.1}$$

In Fig. 2.2 we show a 3-layer NN. It consist of an input layer, a hidden layer and an output layer. 'Activation functions' $g^{(l)}$ are associated with each layer l. In such a network, the outputs of the nodes in each layer combine the signals coming from the previous layer. The index j is used to count the nodes (neurons) in the hidden layer, and i to count nodes in the output layer: Thus the signal from the nodes in the output (o_i) layer is:

$$o_i = g^{(2)} f_i^{(2)}; \text{ where } f_i^{(2)} = \sum_j w_{ij}^{(2)} h_j + \beta_i^{(2)} \tag{2.2}$$

The functions $g^{(l)}$ are chosen to be bounded, smooth and monotonic, but some of them are non-linear. At the training stage a model-response signal is used to obtain the least-square error (LSE) between output and the model response. The weights w and biases β are the unknowns (denoted by the vector $\vec{u}$) that are determined by minimizing the LSE. As these parameters vary, wide ranging non-linear mappings between the inputs and outputs are possible. According to a 'universal approximation theorem' [132], a standard multilayer feed-forward network with a locally bounded piece-wise continuous activation function can approximate any continuous function to any degree of accuracy if and only if the network's activation function is not a polynomial. This result applies when activation functions are chosen *a priori* and held fixed as $\vec{u}$ varies. The accuracy increases with more nodes and more hidden layers. The above theorem implies that adding sufficient hidden nodes produces higher accuracy.

The output signals are also electrical signals, and can vibrate the diaphragm of a speaker and make sounds. The experimentalist or programmer does not 'teach', modify, or optimize the neuronal states or their connections. S/he only supplies the inputs and collects the outputs. A properly designed NN 'learns by itself'. Such a system, when provided with a simple data base of words and sentences, begins to 'babble' like a child, and then, as it acquires more experience, learns to articulate the language more fluently as it is subject to more data-input sessions (training). Such neural nets, (e.g., NetTalk etc.) became common by the 1990s [216]. If some of the nodes in the NN are removed (simulating a small 'brain injury'), only a minor impairment in functionality results, especially if the NN contains many nodes. If the NN algorithm has been successful, the language capability extends considerably beyond the input data base. This feature is similar to a principle of induction where the domain of information, as established by the database, is extrapolated to a wider domain of applicability. In effect, an artificial NN is an algorithm which takes input data and makes predictions in a wider domain, as expected from a scientific theory. However, an artificial NN remains a bare calculational algorithm until the conceptual meaning of the initially unknown parameters ($\vec{u}$) defining the physics of the net are elucidated. If the input data was about a process very similar to what the neural network is doing, then the node structure may turn out to be an optimal map of that process itself. For instance, if we simulated interactions between three layers of an executive hierarchy in a firm, the best fit NN may actually be a 3-layer NN which is a close depiction of the actual structure of the firm. Similarly, if neural networks could be constructed for some complex natural system with an economy of nodes and with an uncluttered optimality, such NNs may be much more than mere calculational algorithms. (see Ch. 6 of Ref. II-[88] for a more detailed discussion).

2.2.6 *Searle's Chinese room and child's play*

An interesting thought experiment presented by John Searle, and known as the *Chinese Room* problem has attracted much attention in the context of AI and cognitive comprehension [190]. Searle considers a monolingual English speaker (the 'Agent') located in a room where s/he receives inputs $\{I\}$ of messages written in unintelligible squiggles. These are, unknown to the Agent, questions written in Chinese. A new batch of inputs written in squiggles arrives, together with instructions $\{J\}$ telling the Agent how to match the new set to the first set, and construct an output $\{O\}$. The net effect of this process is that, unbeknown to the Agent, s/he is providing intelligent answers in Chinese to the questions posed in Chinese. An outside observer may think that the Agent is proficient in Chinese.

Searle claims that although the overall system (the Agent combined with the instructions $\{J\}$) seems to understand Chinese, there is no real understanding. The monolingual agent is 'just doing syntax matching'. The way to demystify this is to slightly modify the Chinese room. We can choose graded questions and answers, as well as our Agent. So we replace the Agent by a child, input questions $\{I\}$ as simple Chinese child talk $\{C\}$. The set $\{J\}$ is replaced by graded answer-sounds $\{A\}$ of another child in play. At first, the child would babble out unintelligible sounds which may mean nothing. But after sometime, she would begin to speak good Chinese, and make $\{O\}$ given $\{I\}$ effortlessly! Learning languages is a play activity for children. Thus the solution to Searle's problem is 'just child's play'!

The machine resolution of this paradox is to replace the Agent with a suitable neural net, feed it with the inputs $\{I\}$, and adjust the weights of the nodes of the neural net iteratively, so that the outputs begin to increasingly agree with the target output, $\{O\}$. The neural network, just like the child, would babble incoherently at first, but the learning process would certainly take place, and *without* the need for producing instruction sets $\{J\}$ for each input set $\{I\}$. The Chinese room can begin operations only after an apprentice period has gone by. This is true even within the older AI approach, where the algorithms which produced the instruction sets $\{J\}$ have to be coded and tested in advance.

Searle's problem is constructed on the *assumption* that the AI algorithm is set in stone and not capable of learning. The agent is implicitly assumed by Searle to be like a mindless robot that matches syntax, without the slightest curiosity. When this assumption is removed, the whole argument falls apart, and the Chinese room becomes a Chinese school.

2.3 Description, complexity and quantifiability

The example discussed in a previous section (see Sec. 2.2.2), viz., the observations of the celestial position of an astral body and fitting those observations to a mathematical form is like a standard 'lab exercise' in any of the 'hard' sciences. It is sometimes argued that such a 'quantitative approach' is well neigh impossible in many of the 'soft' sciences. It may be claimed that a given 'social science' is merely classificatory, or it may deal with 'qualitative' features not amenable to quantification. One may argue that 'simple' systems can be easily quantified, while 'complex systems' need other methods like neural networks as already discussed. In normal parlance we use two concepts, simplicity and complexity, although they are just two extremes of a description of complexity.

In describing a specific characteristic, we need to convey a message in the most efficient coded form possible. The description of a complex system would include, say, information on all the items (e.g, species of trees, insects, etc., in describing a forest ecology), the interactions among the various items, as well as their time evolution. A number of authors call this description the parameter space, or algorithmic information content (AIC) of the system being described. Clearly, such descriptions becomes interminable unless we specify *boundary conditions* that limit the problem suitably. Such boundary conditions may involve *coarse graining*, where we deal with the effect of all smaller length scales by simply taking an average over them. For instance, when we discuss a fluid using the language of fluid mechanics, we do not worry about individual atoms and molecules that form the fluid. Instead, we think of the fluid as a continuum. We have averaged over the atoms and molecules. Structure at fine-grained length scales (associated with sub-micron lengths) do not exist in fluid mechanics. Another aspect of setting boundary conditions involves, for example, limiting the fluid mechanical considerations to a specific basin of water, and excluding other basins or lakes

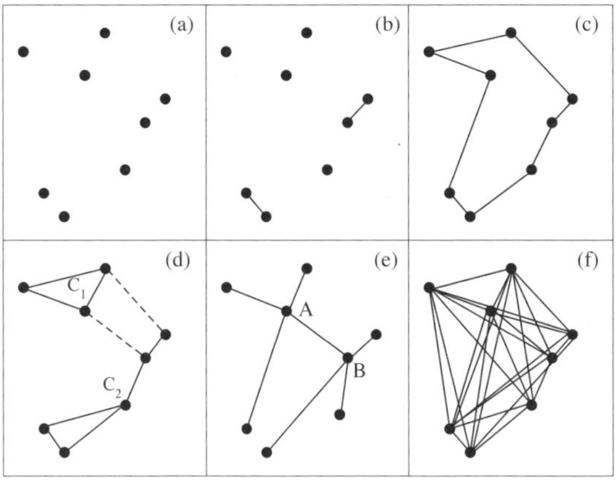

Fig. 2.3 Description of a system with un-connected objects is simple, as in (a), or when fully connected, as in (f). The intermediate cases are more complex. Panel (d) shows clusters with strong (solid lines) and weak interactions, (dashed links). *A* and *B* shown in (e) are highly connected. Complexity is characterized by a long description; simple systems have a concise description, i.e., 'fewer parameters'.

which are only weakly connected with the system under study. Thus *connectivity* is another important aspect of the identity of a *whole system*, often referred to as *holism*, especially outside scientific writings. The lack of use of the word 'holism' in scientific writings is because other more precise words (based on order parameters, distribution functions etc.,) are available in statistical physics. The common usage of the word 'holism' is imprecise. Here we attempt to use the word, within some tightened definitions.

In Fig. 2.3 we explore the concepts of connectivity, complexity and holism using a simple example of eight objects which have been selected by excluding the rest of the world, by setting a boundary. Thus the system involves just these eight objects, and their interactions. The interactions determine the connectivity, as well as the complexity of the system. In Fig. 2.3(a), all the objects are non-interacting, or basically isolated from each other. This is a very simple system. Its behaviour is just like that of any individual object in it. In panel (b) we have a situation where neighbours which are sufficiently close interact. Thus we have *two species* in (b), viz., monomers and dimers, or couples and celibates, since the distance between two objects need not be spatial, but it may be sexual or cultural. That is, what is relevant is *the distance in phase space*. These paired objects, or dimers have a 'holism' of their own. The system (b) is more complex than that in panel (a). In any case, the interaction between two individuals i, j can be specified by a description $v_{ij}(r)$ where r is some measure of the phase-space distance between the two individuals. In panel (c) only the individuals in the periphery are allowed to interact, while in panel (d), two types of interactions are allowed. Weaker interactions (or kinships) are shown by dashed lines. Thus the system breaks up into two clusters, C_1 and C_2. In panel (e) only the internal objects A, B interact with the peripheral objects. Thus A and B have high connectivities compared to the other objects. In panel (f), once again, all objects are on an equal footing, and they are all connected to each other, i.e., interacting with each other. This object has to be regarded as a single 'holistic object', just as the object in panel (c), which is again a different 'holism'. These connected systems have 'emergent properties' *not* found in any individual member. Connectivity can confer the properties of non-locality and contextuality, as seen in quantum systems (see Ch. 6).

The system in panel (a), and the system in panel (f), i.e., the least connected and the most connected, are the simplest to describe, while the intermediate panels are examples of more complex systems. Sets of objects, agents, or nodes only become a holism when they are sufficiently interconnected. General theories of networked systems can be developed without specifying what the 'nodes' are. They can be protein molecules in a cell, power stations, or computers linked to the internet. As the interactions are multiplied, a finite fraction of the nodes gets interconnected, and the character of the interconnected cluster begins to differ from the unconnected regions. The overall symmetries found in the whole system are not found in the clusters and they usually show *broken symmetries*. The transition from lack of connectivity to connectivity is a *phase transition* analogous to a gas of molecules condensing to form droplets of the liquid phase. Such connectivity transitions are known as *percolation transitions*, and the system or part of it that has undergone such a transition is characterized by various order parameters (aspects of its holism). The less connected system, existing in states prior to the percolation transition, does not have those characteristics of long-range or short-range order. The concept of 'holism' would be discussed further in Sec. 2.3.2.

The number of links to a node is known as its degree k. Different nodes may have different values of k, and so we have a distribution of degrees $p(k)$. Numerous networks, including metabolic networks in cells, social-media networks, the internet, etc., are found to have approximate power-law distributions for $p(k)$, i.e., they have the form $p(k) \sim \Sigma_i a_i |k - k_i|^{n_i}$, rather than Gaussian distributions (or bell-curve forms), $\sim \Sigma_i a_i \exp\{-(k-k_i)^2\}$. This type of 'scale-free' behaviour has vast implications on, e.g., how an epidemic starting at one infected node, or a political viewpoint, would propagate in the network. Such rapidly propagating modes are examples of the existence of collective modes and dynamical instabilities emerging out of the holistic structure. A discussion of such modes of a connected system, emergence of new order etc., will be given in Secs. 2.3.3 and 9.3.2.

As in the quantum theory of fields, the simplest models dealing with such systems use representations involving a large collection of coupled oscillators. Similar models are used for arrays of Josephson junctions (see Sec. 7.7.4.1). These simpler models need to be further elaborated to understand cascading processes that may lead to large-scale breakdowns in complex systems, as in extensive power outages that could darken a whole continent. Such situations are intimately connected with extreme sensitivity to initial conditions typical of chaotic dynamics.

2.3.1 *Simplicity as a desirable attribute of a theory*

In Chapter 3 we learn about the need for scientific theories to satisfy various inherent symmetries, topological properties, continuity etc. Such symmetries are very important in deciding how to deal with observations which always contain scatter. The *simplicity of a theory* is often related to it satisfying various symmetries which are holistic in character. Simpler theories utilize models which have a fewer number of structure constants or parameters. Thus their description is 'simple' and concise. For example, the three theories that we used in Fig. 2.1 all had four such constants, and hence are simpler than models with, say, five constants. Thus observations should be obtained from nature using a reductionist preparation of the holism that preserves its innate symmetries and simplicities.

The nature of the inflationary process, and the big bang which generated our universe provide explicit reasons for the existence of symmetries and simplicities in nature (Sec. 5.9). The universe most probably started as a very dense, highly connected homogeneous object. Contrary to the belief that there are many theoretical models that fit our observations of the physical world, the requirements of symmetry and simplicity severely restrict the viable candidates.

As the universe evolved, different energy and length scales began to be formed, and coarse graining averaged out some interactions, while others remained dominant. Thus new, complex structures [as in panels (b) to (e)] began to be formed. Atoms and molecules, non-equilibrium dissipative systems or complex adaptive systems which could gather and use information came into being (see Ch. 12). Our understanding of complex adaptive systems is certainly rather limited, but there is little doubt that we already know the fundamental laws, viz., the laws of physics, that have unleashed the process.

While scientists believe that nature is inherently simple, philosophers of science (e.g., Kuhn, Psillos [175] and others) seem to believe that nature is utterly complicated. They contend that, in the end, all paradigms will fail. In any case, even when paradigms succeed to some extent, Kuhn and many others would claim that successive paradigms do not converge to a final consilience (see Sec. 2.4.2). The opposite attitude of scientists is typified by Einstein who optimistically proclaimed that 'the most incomprehensible thing about nature is its comprehensibility'. The Kavali Institute of Theoretical Physics at the University of Santa Barbara, Calif., has chosen to inscribe this view in a thought-provoking plaque for all to see.

2.3.2 *Definition of holistic systems and complexity*

It is clearly important to define what we mean by a holism. Similarly, we need at least a working definition of complexity. Different disciplines (e.g., computer science, economics, genomics) may use different, suitably restricted definitions.

Some may ask 'is there is a single holism, or is there is a hard core together with a boundary etc., within the holism'. Under what conditions can one expose a characteristic feature of a system which is actually holistically connected? Answering such questions requires being a bit more precise, as we saw from the example in Fig. 2.3.

'Holism' implies that the elements of a system are interconnected and hence one cannot just take one element in isolation and study it. Scientists are extremely familiar with such systems. They are studied by 'many-body theory', after selecting a suitable reduced system. Here we discuss how this is done by keeping in mind a definite example. Thus a set of tuning forks connected together forms a holistic object as the tones of individual tuning forks are now replaced by 'combination tones' which are a property of the whole system. The 'Kuramoto model' of collective behavior in coupled networks is in fact a model of coupled oscillators, and has been used to discuss synchronization phenomena.

A holism consists of elements labeled by an index (or many indices), say i, and interactions among these elements. If they interact by pairs we need to specify the pair-interaction functions $V(i,j)$. They may also interact in threes, specified by a $V(i,j,k)$ or in n-tuples etc. Each individual element may also be subject to an external influence, say $U(i)$. The $U(i)$ is known as a 'one-body' term, while the $V(i,j)$ are known as two-body terms, and so on. Hence we may specify the complete holism by the function:

$$H = \Sigma_i U(i) + \Sigma_{ij} V(i,j) + \Sigma_{ijk} V(i,j,k) + \cdots \tag{2.3}$$

Plucking one of the tuning forks would be like applying a $U(i)$ to the i-th fork. Let the operation of exciting the i-th tuning fork be denoted by a_i^+, while the corresponding de-excitation which brings the fork to its quiescent state be a_i. Then it is customary to write the 'one-body term' associated with the external influence as:

$$H_1 = \Sigma_i U(i) a_i^+ a_i \tag{2.4}$$

But this excitation is connected to all the other components by the coupling terms $V(i,j,\cdots)$. Thus the actual observed (or heard) sound of the coupled system of tuning forks is a combination mode of the whole system. Many-body theorists know how to deal with such coupled systems and reformulate the model in terms of new combination modes that have emerged due to the interactions. Unlike the classical series expansion methods developed in celestial mechanics, modern many-body theory successfully sums infinite classes of terms in such expansions, and presents the summations in terms of integral equations which have to be solved using computers. Direct numerical simulations without approximations are also possible for holistic systems with a finite number of members.

Kuramoto's model of coupled oscillators, proposed in 1975, and other developments have been discussed in a book by Strogatz [183]. Here the two nodes i,j interact via the form $V_{ij} = K_{ij}\sin(\theta_i - \theta_j)$, coupling the phases $\theta_i - \theta_j$ of the oscillators. When the interaction is weak, each oscillator oscillates independently at its own frequency ω_i, while increased coupling tends to synchronize everyone. Thus, when a threshold is reached, *synchronization emerges spontaneously*; a phase transition occurs.

This type of theory has been used for idealized descriptions of phenomena like schooling fish, simultaneously emerging insects, bursting neurons etc. The synchronization is measured by an order parameter that characterizes the 'holism' that emerges because of the interactions.

In Sec. 1.4.1 we pointed out that if we are discussing a system of three atoms, H, O, and H, then it is not enough to consider it as consisting of just three parts. It is also necessary to include the 'longitudinal photons', i.e., the Coulomb interactions between all the charges. These are precisely what is contained in the coupling terms $V(i, j)$ for such systems.

2.3.3 *Normal modes of a holism*

Combination modes associated with the whole system are known as 'normal modes'. They are not at all like the original modes, but they are effectively independent of each other and can be studied, measured or manipulated for various technological uses. In the case of the set of coupled tuning forks, the normal modes can be heard by the ear while the sounds of the individual tuning forks cannot be discerned unless the $V(i, j)$ couplings happen to be negligible. The new modes do not sound like the old modes, and they will be differently damped (i.e., their durations or auditory lifetimes are different).

The 'normal modes' of a dynamical system are determined by a diagonalization of the Hamiltonian of the system. We can explore similar methods for any interacting system.

Whether the holism can be represented by a 'hard core' and a loose boundary would depend on the strengths of the coupling functions $V(i, j)$. This is also evident in neural nets where some of the nodes may be eliminated without impairing the total functionality. However, a decrease in the 'quality' or intensity of the output from the NN, for a given input may occur. This is equivalent to stating that the response function or susceptibility of the system is affected by removal of any nodes.

If a proposition is to be tested, it has be a part of a model. If the model is an actual language, as in the program of the Logical Empiricists, then we have a very complex, virtually insoluble holism. However, scientists are concerned with models dealing with nature, simplified and selected for study within carefully chosen boundaries (possibly in the form of suitable test samples). The model usually has a structure representable by a set of interacting modes. Then the interconnected modes can be transformed into a set of 'normal modes' which are independent of each other. For instance, if the theory can be represented by a Hermitian matrix of numbers, then it can be transformed into diagonal form, and each renormalized mode associated with the eigenvectors can be measured independently of the other modes. The diagonalization mixes the 'bare facts', or bare modes, and produces renormalized modes which depend on the theory which modeled the interactions among the bare modes.

Thus the bare modes become *theory laden* in a perfectly controlled manner in terms of the interactions $V(i, j)$ included in the theory.

2.3.4 *Reduction of a holism, and external interactions*

Consider the reductionist process within the holistic picture. In studying nature, i.e., a total holism or an ecology denoted by H_T, we construct a reduced system R, with a reduced holism H_R containing its internal interactions, and the rest of the world or 'environment' E, containing the holism H_E. The reduced system (or abstraction) interacts with the environment E. This interaction between R and E is denoted by H_{RE}. This H_{RE} depends on the boundary conditions used in constructing the reduction:

$$H_T = H_R + H_E + H_{RE} \tag{2.5}$$

In the simplest reductionist approximation we study only H_R with the interactions internal to it, neglecting the interaction with the environment. The idealized motion of a single particle moving in free space, and subject to a constant force would be the H_R studied in the equations of 'free' motion of Galileo or Newton. The effect of the environment, i.e., H_{RE} is ignored. However, when the moving object is placed on a plane surface, the effect of H_{RE} on the dynamics may be modeled by introducing the concept of *friction* or *damping* of the motion. In quantum mechanics, instead of friction, we talk of *decoherence* arising from the interaction with the environment.

The boundary conditions implicit in H_{RE} may also be used to specify whether the atoms in a sample of matter are isolated, or in contact with a thermostating bath holding the system at some temperature T. The boundary may also allow particles to move in and out of the reduction, thus setting the stage for a 'chemical potential' controlling the flow of matter, or even information.

2.3.5 *Measuring complexity*

We noted that complexity is smallest for fully disconnected systems and also for fully connected systems. A simple binary model of connectivity may use 0, or 1 to define bonding or non-bonding. Then a fully connected system has only one possible configuration. Similarly, a fully disconnected system has just the configuration of one of its independent components. At intermediate situations we have the possibility of many configurations. The (0,1) model can be generalized to *the probability* that a given node i (e.g., in Fig. 2.3) is connected to the j-th node of the community by the symbol P_{ij}. A measure of the configurational richness or *entropy* of the system can be written as

$$S = -\Sigma_{i>j} P_{ij} \log P_{ij} \tag{2.6}$$

If we have a set of unconnected (i.e., non-interacting) objects, or if everything is connected, the entropy is zero, with log 1 being zero.The function reaches a maximum magnitude when the probabilities become of the order of 0.5, with the likelihood of linking or not linking being about equal.

The probability of bonding is also affected by *the strength* of the interactions. In physical systems, this is simply the energy of bond formation. It is denoted by $v_{ij}(r)$. In general situations, we may not have a clear meaning of what this interaction is, and we need to allow the theory to give it a proper meaning, by a process of theory construction and comparison with experiment.

If i, j are fixed nodes, r_{ij} is fixed, and the effect of the interactions can now be summed as

$$E = \Sigma_{ij} P_{ij} v_{ij}(r) \tag{2.7}$$

The complexity of the system is determined by an interplay between the two terms E and S. If the system were in equilibrium, then a temperature T can be assigned where T is a parameter of the theory. If we are dealing with a physical system (e.g., a system of spins on a lattice), T is the usual thermodynamic temperature in energy units. If we are dealing with some other system (e.g., an ecology of interacting species), the relevant concept of temperature has to be deduced from the data and the modeling process. The free energy F of the system, i.e., the energy available for transforming the system or doing useful work turns out to be $E - TS$. In dealing with complex adaptive systems which are non-equilibrium dissipative systems, it is not possible to assign a unique thermodynamic temperature to the system. Instead, it may be meaningful to assign a parameter T_{ij} which is physically similar to temperature, but takes some non-equilibrium stationary value for each bond or interaction.

Within such a point of view, we define the complexity function F_C by

$$F_C = \Sigma_{ij} P_{ij} v_{ij}(r) - \Sigma_{ij} T_{ij} \{P_{ij} \log P_{ij}\} \tag{2.8}$$

This definition of complexity is a generalization of standard ideas in thermodynamics and statistical mechanics. A more commonly discussed definition is the binary bit length of the 'most concise description of a system' as its measure of complexity. This is discussed lucidly by Murray Gell-Mann [83]. Our definition, Eq. 2.8 can also be reduced to a multi-state spin description in terms of pseudo-spin states. Thus as noted earlier, P_{ij} can be (0,1) for 'bonding' or ' non-bonding' configurations. In fact, any bit-based description of a system can be reduced to a suitable statement of an 'algorithmic information content' (AIC).

2.3.6 *Energy scales, length scales, and complexity*

The emergence of complex systems out of simpler systems is closely related to existence of different scales of energy and length in nature. Thus, the detailed processes that occur at the scale of quarks and elementary particles need not concern the chemist who deals with atoms and their interactions. He has developed a vocabulary, based on the concept of valency, and the nature of the chemical bond, to describe chemical phenomena. Valency, as well as the dependence of such chemical properties on the periodic table of Mendeleev, can be understood quantitatively using the quantum theory. Whenever quantum calculations are carried out to the required accuracy, it is found that chemical phenomenology can be recovered. However, the chemist or mineralogist does not usually need quantum calculations. He can use the methods of physical chemistry, the rules of chemical bonding etc., to work entirely with a set of rules that require no reference to the Planck constant. Thus, a chemist can write down the reaction

$$2H_2 + O_2 = 2H_2O.$$

He is using the concept of valency (i.e, H atoms are univalent, while oxygen atoms are divalent), as well as the principle that the number of atoms remains constant under chemical transformations. All the objects under consideration, i.e., the atoms H, O, or the molecules H_2, O_2, and H_2O are our basic objects. The detailed interactions of the more primary constituents (nucleons, electrons etc.) have been averaged over, and it is only the residual effects of those primary interactions that persist in the more coarse-grained chemical description (Sec. 7.3.1).

An engineer studying the properties of water may know little about chemical bonding. He describes water in terms of the density, specific heat, viscosity, the melting- and boiling points. A more general description is provided by what is known as 'the equation of state' (EOS) of the substance. It relates the volume V of the substance to its temperature T and the pressure P, or it gives P, V for a given temperature T. A graph of P, V at various values of T will show the various phases of the substance (vapour, liquid, hexagonal ice, cubic ice etc.), revealing its *phase diagram*. Another description specifies the free energy F of the system containing a given amount of matter, as a function of P and T. Thus the engineer does not deal with atomic scales of length and energy. The engineer, working with steam uses an average description which is very accurate since he is dealing with millions of molecules of water. An individual molecule suffers many collisions with other molecules in a second, and hence its 'identity' is lost. Thus molecular dimensions get replaced by coarser length scales typical of the order of the wavelength of visco-elastic waves (e.g., sound waves). Sound waves, or plasma oscillations in the case of charged fluids, are the new 'modes' which characterize the new holism.

If the water in a lake is subject heat from the sun, the surface water begins to evaporate. The hot vapour rises up, and condenses into a cloud. The cloud is a non-equilibrium system held temporarily in a steady state. The cloud decays by dispersal of the water molecules to the surroundings due to winds, diffusion and condensation into rain drops. Thus the *lifetime* τ of the non-equilibrium dissipative structure is determined by a balance between processes which grow the cloud, and processes which evaporate the cloud.

It is the sun's energy acting at the molecular level that evaporates the water. The cloud is simply a larger-scale quasi-equilibrium structure that has arisen as a response to the sun's energy, to dissipate the energy via sub-processes. The formation of the cloud involves a reduction in the entropy of the collection of water molecules when compared to a set of free water molecules. But it is still a great increase in entropy when compared to water molecules in the liquid state. The cloud-like structure was a result of the interplay between the interaction term (Eq. 2.7) and the entropy term (Eq. 2.6) appearing in the expression for the complexity function (Eq. 2.8). If we also include all the corresponding processes in the environment of the region containing the cloud, all the processes associated with the absorption of sunlight etc., we see that the net effect is to convert the energy of the sun to degraded, randomized energy of the dispersed water molecules. This is entirely in keeping with *the second law of thermodynamics which states that all natural process lead to a degradation of available energy*.

Each of the sub processes, viz., evaporation at the molecular level, vapor accretion into the cloud, evaporation of the accreted molecules, diffusive processes, wind-driven processes, all have their characteristic *time scales* or *lifetimes* for each process to play its role. Let the time scale of the i-th process be denoted by τ_i, Then it turns out that some of the primary processes τ_m (at molecular scales) occur so fast that they average out during the time scales associated with the steady-state of the cloud. On the other hand, the ambient weather should remain steady while the cloud is being formed. The time scales associated with the changes in the ambient environment, sunshine, weather pattern etc., i.e., various τ_w have to be sufficiently slow to permit the cloud to reach a steady state.

What is true for the cloud holds for more complex systems like living organisms. We see that the formation of stationary dissipative structures requires the concurrence of a hierarchy of time scales $\tau_1 < \tau_2 < \cdots < \tau_n$. This is particularly important in evolutionary adaptation. The time scales associated with the changes in the habitat need to be slower than the times scales needed for the generation of mutations and their adaptation to the habitat. Those structures, or species whose time-scales of adaptation are too slow become extinct.

2.4 Paradigms and scientific revolutions

In Sec. 2.2 we discussed how observations are formulated into a theory or model of nature, incorporating interactions that depend on the interplay of various size scales and symmetries associated with the physical system.

Do we call such a 'theory' a model or some sort of a *paradigm*, at least in a narrow sense? We have avoided the word 'paradigm' as it has developed its own connotations in the philosophy of science, mainly as a consequence of Thomas Kuhn's influential book [122]. This book appeared in 1962, just when the Logical Empiricist program had made every one tired of arguments which seemed less and less relevant to science as such. Instead of working with logic and the structure of language as the Logical Empiricists were wont to do, Kuhn returned to a less technical approach based on a study of the history of science and how 'scientific paradigms' change. It was as if the castle gates were flung open; storytellers, poets, and even impostors were allowed into the arena which had, until then been dominated by a few acknowledged experts of highly esoteric martial arts.

Those martial artists were the *Logical empiricists*. They had been European members of the Vienna circle, or associated with Cambridge, England. Many had migrated to the USA fleeing Nazism in Europe. Rudolf Carnap, Carl Hempel, Herbert Feigel, and Willard Quine were well established figures when Kuhn published his book. Roughly speaking, the Logical-Empiricist program had consisted of attempting to justify inductive knowledge, accumulated as empirical data, by a study of logic and language or by an analysis of probability. The Logical Empiricists, sometimes also narrowed down as Logical Positivists, concerned themselves with analytic and synthetic sentences, the *nature of meaning*, and the *verifiability* of statements. The statement 'all bachelors are unmarried' is an analytic statement whose truth is analytically evident. Synthetic statements include external knowledge, and their nature has to be probed. They regarded these questions as being cardinal to an understanding of scientific truth. In the simplest form of the theory of meaning, a sentence which could not be verified (i.e., put to test) was deemed meaningless. Thus the sentence 'this triangle is masculine' may be deemed meaningless. However, such methods and protocols required embedding the sentences in a whole language [177]. Hence 'holism' came to dominate the partially successful but extremely limited approach using elementary prepositions. An important objective of the Logical Empiricist was to provide a deeper underpinning of science. This exercise attempted to define the scientific method itself.

Another school of science philosophy was centered around Karl Popper [166]. Popper claimed that verification involving mere empirical instances (induction) was irrelevant and that theories are evaluated on how they stood against attempts to falsify them. Thus *falsifiability* became the touchstone associated with Karl Popper. If a preposition, or more generally, a theory, was such that it is not possible to falsify it, then it is not science. Popper used his criterion as a means of separating 'good science' from 'pseudoscience'. It was on this basis that Marxism and Freudian psychology were regarded as pseudo-scientific doctrines. In such pseudosciences, the practitioners would re-interpret their theories, when challenged, in order to accommodate the challenge — perhaps as in Jesuit casuistry. While Popper is correct if we restrict ourselves to certain sub-fields of study or well-chosen scientific episodes, his methods largely fail when viewed more broadly. Popper's contention [167] that scientists proceed by making 'conjectures', developing the conjectures into theories, and subjecting them to 'refutations', applies more to the behavior of philosophers arguing at their seminars than to what actually happens in mainstream science. Here again, a largely theoretical sub-discipline in science may conform to Popper's model, but not in general.

The ideas of the Logical Empiricists, and the parallel programs of Popper, Kuhn and others developed in unexpected anarchic directions in the hands of Imre Lacatos, Larry Lauden and Paul Feyerabend [72]. Feyerabend concluded that there is really no scientific method, and that *anything goes*, in the sense that everything must be tried out. If establishment science 'prevents' a scientist from trying out a maverick idea, then the state should intervene! Thus Feyerabend, while supporting extreme freedom, also supported the political intervention of the Chinese government against western medicine and affirming folk medicine. And yet, he also seems to side with the Holy See regarding the Church sanctions against Galileo!

Scientists usually expect their theories to be obeyed without any exceptions whatsoever. This is true, as far as we know, for physics, chemistry, molecular biology and other subjects flowing directly from physics. This norm is not what is found in philosophical theories about science. The philosophical accounts are like case histories where some aspects of science are explained, while others are not. In fact, if we use the very criteria of Popper to asses his writings, or other philosophical writings, they risk being condemned as pseudoscience in the very sense used by Popper. In fact, Godfrey-Smith notes that there has been a century of Science studies [87] in the modern era, and much discussion about the nature of scientific truth while there has been less and less discussion about the discourse engaged by the philosophers themselves. It seems that when a given school of philosophy proposes some conjectures, their own students as well as others attempt to gain visibility by proposing refutations which support an opposite position. While Kuhn and his followers launched the theory of scientific revolutions, today we have Shapin [194] and others suggesting that there really was no revolution in 17th century science. The same kind of antithesis evolved from Popper to Feyerabend. Thus, in 'philosophical truth', as opposed to scientific truth, the *thesis* seems to generate an *antithesis*, or a branching out of many incompatible theses, each supported by scholarly but selective case studies. The criteria of truth proposed by these scholars are rarely re-applied to their own theories.

Alex Rosenberg [180], commenting on the work of Kuhn and others, states that: 'For all of Kuhn's insights into the history of science, most philosophers of science consider that something has gone seriously wrong in the development of the social studies of science since his time.'

Philosophical analysis which attempts to lay bare the nature of scientific truth is itself lacking a firm foundation. If there is only one kind of truth, then philosophy and science have to integrate seamlessly, as happens in certain areas of theoretical physics, cognitive science or evolutionary studies.

If science works by gathering 'empirical data', it faces the problem of induction that we already discussed. The observation of the sun rising every day, even for thousand days in succession, does not guarantee that it will rise on the 1001th day. The Logical Empiricist may look for a justification of induction in some formalism. The scientist is willing to forfeit universal knowledge of the sort aimed at by the philosopher, and place his bets on more modest forms of Bayesian belief. Furthermore, although many philosophers seem to think that the physical sciences have something to do with induction, there is little or no evidence for that in the hard sciences. Most scientists attempt to comprehend some 'key fact or insight' at a new length scale which may be totally at variance with 'accepted beliefs' at a familiar length scale. That is reflected in Einstein's optimism that 'the most incomprehensible thing about nature is its comprehensibility'.

If the sun rises regularly every morning, the scientist assumes that there is an underlying reality which causes this repetitive action. Thus a theory (i.e., a model) of the heavens is constructed to account for the observed motion of the sun. The earliest such theory was geocentric, with the sun moving exactly as seen in the sky. This theory successfully incorporated the motions of other heavenly bodies. Its most mature form in the middle ages was Ptolemy's model of the heavens. It fitted in with the anthropocentric theology of the Church, the Greek ideas of the perfection of simple geometric shapes, and what people saw with their own eyes. It was a good scientific theory with predictive power and a capacity to unify a lot of empirical data. It was both theory laden, and morality laden.

How was this model overthrown? Were there a heroic figure who went about trying to falsify this theory *a la* Popper? Was it a large number of reasons slowly coming together? Does Kuhn give us an answer to these questions? In fact, all the usual theories of scientific change seem to miss some very basic issues here.

According to Thomas Kuhn, the transition from the Ptolemaic world view to the Copernican world view was a 'scientific revolution'. Kuhn argued that science has periods of steady activity within a well accepted 'paradigm', followed by periods of revolution when the paradigm breaks down. Then a new paradigm begins to come into place. Kuhn claims that a 'crisis' arises when the practitioners of mainstream science begin to question the paradigm because a 'critical mass' of anomalies have cropped up, and because a rival paradigm has appeared on the horizon. Kuhn does not explain how the anomalies cropped up. Once the revolution is over, we are told that the new paradigm becomes mainstream science. The old paradigm is thrown out. The new paradigm speaks a new jargon and posits a world view which is said to be largely irreconcilable or *incommensurate* with the old world view. In fact, Kuhn seems to go even further; he claims that external reality itself actually changes, in some deeper ontological sense. When we discuss the theory of measurement in quantum mechanics, we come across a somewhat similar 'many worlds' ontology proposed by some philosophically inclined physicists. Everett and others claim that each time a quantum measurement is made, the act of observation creates a new world and a new reality, while only some (commuting parts of the) old world exists in a parallel universe [56, 58]. Of course, Kuhn has nothing as radical as that, but only a suggestion that paradigm changes are not just intellectual exercises, but are actual changes of the external world. Given the timing of those theories, Kuhn may even have been influenced by Everett, Wheeler, DeWitt [57] and others. Those who followed Kuhn's footsteps and took even stronger positions of 'relativism' of reality (e.g, Latour, Nelson Goodman and others) claim that reality (or nature) is a construct made by scientific theorizing. In this sense, the geocentric world created by Ptolemy and others really existed, and then Galileo and others *created* a new, heliocentric world. According to Feyerabend, this was done by trickery and propaganda by Galileo and others.

Although the Ptolemaic astronomy was more than adequate for explaining the observed data, Kepler, Galileo, and Newton went on to introduce a new astronomy based on flimsy data that was nevertheless what was needed by new mathematical models. They were following the very opposite of induction, and following insights that had arisen by looking at the available data in a new way, and using new tools (telescopes and clocks) reaching into new length scales.

Another famous scientific revolution, or 'paradigm change' is the replacement of Newtonian physics by modern physics. This began in the last decades of the nineteenth century, and extended into the first few decades of the twentieth century. Here too, when Einstein proposed a particle theory of light, based on a few observations of the photo-electric effect (and possibly Plank's work on black-body radiation), he was defying the principle of induction. After all, inductive evidence was at that time overwhelmingly for the wave theory of light.

The rise of modern physics involved two simultaneous paradigm changes. Kuhn seems to admit of only one dominant paradigm at any given time. Einstein's theories of relativity replaced Newtonian physics where the latter was found to be inadequate at large length scales and energy scales (see Sec. 1.3). Similarly, quantum mechanics replaced classical mechanics, because the latter was found to be inadequate at very small length scales as well as at energy scales where energy was found to come in discrete packets. Einstein was unhappy with quantum mechanics (QM), and tried hard to invent means of 'falsifying it', as Popper would have it. Meanwhile, attempts to deduce Einstein's theory of gravity (i.e., the general theory of relativity, GR) from quantum mechanics have so far not met with success (of course, in a different plane, Alan Sokal's tongue in cheek paper [199] on quantum gravity was very successful in the context of 'science wars').

Today we have two dominant paradigms, QM and GR, not posing difficulties to each other as they are associated with different scales of energy and length.

2.4.1 *Scientific revolutions and their genesis*

The failure to appreciate the importance of various characteristic energy, length and time scales to scientific theory has, in our view, strongly marred the attempts to understand the working of science. Scientific revolutions arise because tools of research, i.e., instruments, mathematical methods, calculating devices etc., slowly improve and allow scientists to change the scale and scope of their investigations. Then even just one striking observation that is contrary to the vast majority of available data would trigger a new vision in the fertile mind of some scientist.

Steven Shapin [194] has looked at some of these background developments associated with the introduction of a new tool, *the air pump*. The concept of a 'vacuum' did not exist in Aristotelian science.

In 'Leviathan and the air pump' [195] Steven Shapin and Simon Schaffer claim that Robert Boyle and his followers were creating a new way of using language, and literally *created* scientific reality. They view language somewhat as in Wittgenstein [231], using the concept of a 'sprache-spiel' (language-game), i.e., language as a culture which makes sense essentially within itself. For Wittgenstein, the social culture of language (i.e., the pattern of language maintained by society) is the essential 'meaning' of language. Armed with such a point of view, Shapin and Schaffer claim that instead of answering the arguments posed by Aristotelians about the vacuum, Boyle created a new language-game and reconstructed the questions, and introduced a modified way of using the word 'vacuum'. These authors claim that Boyle and other scientists *manufactured facts, or created a new reality*. Thus, in the end, Shapin and Shaffer also fall into a relativism of knowledge, although their position is clearly not as extreme as that of the strong sociologists like Latour and Woolgar [128].

An explorer who visits the polar region for the first time has to invent new terms to describe the new terrain. The invention of the air pump enabled scientists to reduce the density of matter available for study to new domains. A sample of air contains many millions of molecules, while the vacuua generated by Boyle reduced these numbers by orders of magnitude. The language used by Boyle and others to describe the new terrain is partly new, but dressed in the terminology of the old language. Thus the reconstruction of language and the questions by Boyle and others that Shapin and Shaffer refer to, can be understood without abandoning scientific realism. Indeed, the Aristotelians had no information about the new scales of low density that Boyle and others could produce. Hence Aristotelian questions and concepts about the vacuum were a construction without a substantive base, no different from those of theologians about the nature of heaven, or the wing structure of angels.

The constructions that Shapin and Sheffer refer to are actually based on new layers of experience which were not available prior to the invention of air pumps. The Royal Society gave Boyle and others a forum for establishing their observations as public facts, and not just personal experiences that cannot be shared by others. Here we have an excellent example of how re-tooling enables scientists to sharpen and deepen the length scales and energy scales available to the scientists, thereby calling for the revision of existing theories to take account of new phenomena, while the old-established phenomena also remain accounted for. The nature of scientific revolutions, and the essential role played by the hierarchy of energy and length scales that exist in nature, and their exposure using new tools, are once again displayed in Boyle's work.

The Copernican revolution was preceded, not by some sort of crisis, but by a steady but slow development of mathematics (e.g, introduction of the decimal systems, Arabic numerals), time-keeping equipment, improvement of observational tools and so forth. Given better techniques of measurement which probed beyond the limits of validity of the existing theories ('paradigms' if you wish), their shortcomings become obvious. Thus the telescope in Galileo's hands was an instrument that transcended terrestrial length scales and accessed astronomical length scales. Similarly, the pendulum clock pushed the accuracy of time measurement. With such tools, anomalous data can no longer be accommodated within existing theory by corrective adjustments. New length scales involve new hierarchies (Fig. 1.4). Everything was in place for a more fundamental revision of the old ideas of motion (mechanics), and Ptolemaic astronomy. That is, here again there were two distinct revolutions. Kuhn would have to call them the Copernican Paradigm (astronomy) and the Galilean paradigm (mechanics). However, this distinction is rarely made, although they corresponded neatly to two of Galileo's books [78, 79].

It was Newton who unified these two paradigms, i.e., astronomy and and mechanics into a unified whole by postulating an invisible, totally non-mechanistic force that he called gravity and could act mysteriously at a distance.

It is easy to see how new tools like spectroscopes, interferometers, galvanometers, as well as the development of differential equations and algebra paved the way for another major change in the length scales and time scales accessible to scientists. We noted how different scales correspond to different hierarchies of physics (see Fig. 1.4). Hence, when the tools become sharp enough to probe a new length scale, a new theory which accounts for the deeper hierarchy becomes necessary. Newtonian mechanics was thus replaced by quantum mechanics with leading contributions from Bohr, Schrödinger, Heisenberg, de Broglie, and others while Einstein and Planck had made early contributions. Talented experimentalists like Thompson, Stern, Gerlach, *et al.*, contributed to the program, using ever sharper tools. Meanwhile, a second front had opened where both particle physics and astronomy showed that Newtonian ideas of space and time had to be replaced by new ones from the special and general theories of relativity.

Whenever a theory is claimed to be 'overthrown', as a Kuhnian would have it, or as is popularly believed, it does *not* go out of circulation. A theory goes out only if it is wrong, as was the case with the Phlogiston theory of combustion. When a theory is found to be inadequate because of its scope of applicability, it still remains good within its own domain of energy, length and time scales. Even today, astronomical calculations for terrestrial length scales use a geocentric model. The astronomer who writes an ephemeris for a navigator or an amateur star gazer need not use general relativity or even a heliocentric model. The origin of coordinates is mathematically irrelevant, and it is merely a matter of numerical convenience whether we place the origin on earth, or on the sun.

Established science simply becomes engineering. Engineers are professionals who use a well accepted paradigm within its domain of length and energy scales. They are certainly not trying to falsify it. Thus, Newtonian mechanics may have been 'overthrown' by Niels Bohr and others in some sense, as asserted by some historians of science. But every engineer who designs a bridge or a dam uses Newtonian mechanics. Contrary to the claims of Kuhnian philosophers, Newtonian mechanics is alive and well within its own energy and length scales. The old established science, which works well within a given set of length and time scales provides the credibility base for new explorations. We alluded to this feature already when we stated that science provides a *terra firma* of established practice and also a *system of doubt*, while religion provides a system based on faith with no room for doubt. The system of doubt that occurs in Science applies to length scales and time scales that go beyond the already established theory which forms the *terra firma* of belief accepted by all practitioners. This is particularly evident, and even necessary in quantum mechanics, where every measurement (on submicroscopic quantum systems) has to be referred to *macroscopic* pointer reading of some apparatus which obeys classical mechanics that operates on *terra firma*.

Godfrey-Smith [87] has noted that the theory of 'punctuated equilibrium' in biology, proposed by Gould, Eldredge *et al.*, has some striking resemblances to Kuhn's theory of paradigm change. Eldredge and Gould [69] suggested that most species undergo little evolutionary change and exist in a state of 'stasis' for most of their geological history. Then, when evolutionary changes occur, they happen rapidly, by branching into two distinct species. Thus evolutionary change happens in bursts, and it is not a gradual transformation. Gould *et al.* hoped to explain gaps in fossil records using their theory of punctuated equilibrium. However, mass extinctions that create many vacant niches generate rapid evolutionary change. Most complex systems show such time variation which may involve discontinuities. We already noted how networks and systems with internal interactions could undergo phase transitions, and something of that sort could have been in the minds of Kuhn, or Gould and Eldredge. Hegelian-Marxist ideas of social revolutions may have played their part. However, what is offered are sketchy conjectures that do not fit well with modern DNA-based evolutionary studies.

Are different theories (or paradigms) incommensurate and in some sense so disparate that they do not converge towards an increasingly more coherent picture of the world? If so, our knowledge of the world lack consilience and cannot improve as one scientific revolution follows another. There can be no consilience. Kuhnian writings seem to suggest that there is no real progress or convergence to an improved understanding of nature. Other critics like Feyerabend, and many traditional existentialists also hold such views. Each world view for Kuhn is like a different dream which need not have a relation to a previous dream. And yet, Kuhn does admit that newer paradigms may be better at problem solving.

2.4.2 *Cross sections of reality*

If scientific theories merely jump from one 'paradigm' to another, without converging onto a definite consilience of views, we end up with a relativism where scientific research would be viewed as a futile exercise. In this book we argue that the denial of consilience by Khun and others is incorrect. Different theories are not disparate and incommensurate world views. They fit in very nicely within the hierarchic view of nature where *different types of physics* come into play at different energy and length scales. Their inter-relatedness becomes evident when looked at in terms of measures universal to all energy and length scales. The scale-

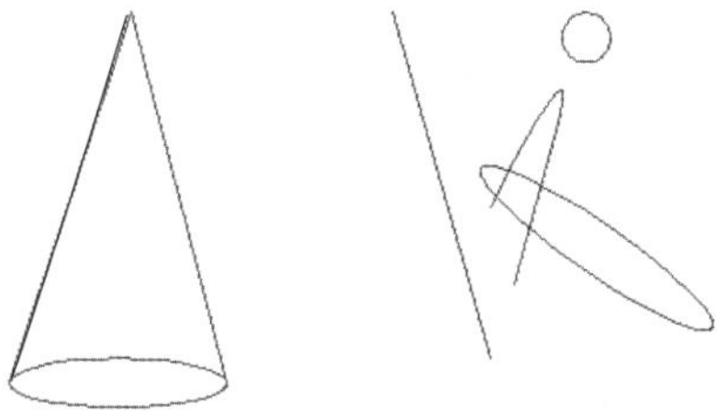

Fig. 2.4 Given an external, objective reality (e.g., a conical object), it can be cut into 'conic sections', which look very different, e.g., a straight line, circle, ellipse, parabola or hyperbola. Scientific theories are also partial descriptions of reality and may seem very different, even though they are approximating the same reality.

specific theories which may *appear* to be different paradigms are obtained using tools which project only a limited part of the full dimensionality of Nature. Here we use the word *physics* in the sense that all other sciences (chemistry, molecular biology, ecology, psychology etc.,) flow from the laws of physics as applied to increasingly complex systems associated with different scales.

Different theories are in essence different cuts or cross sections of reality. Newton's theories of motion and gravitation, as contrasted with Einstein's theories, are often presented as examples of the irreconcilability of the physical models, even though the earlier theory (i.e, Newtonian physics) may be obtained as an approximate form of the later theory (of Einstein). Einstein's results reduce to those of Newton if light has infinite speed. However, it is the seemingly different conceptualizations-geometry versus forces-that need to be explained.

Newton introduced the concept of gravity to explain the fact that all bodies fall towards the earth, and noted that the curved path of the moon around the earth could also be thought of as 'falling' towards the earth. Thus the concept of a gravitational force unified both terrestrial and celestial phenomena, not just qualitatively, but quantitatively. The curved motion of an object moving near another mass could be explained by assuming an inverse-square law of force between them. Einstein ascribed the curvature to an underlying fabric of space-time that gets distorted by the presence of masses. Newton did not explain gravity, but simply posed it as an unexplained hypothesis that unifies dynamics of celestial and terrestrial bodies. Einstein has given an 'explanation' of gravity that is completely compatible with Newton in the relevant length scales.

At a more fundamental conceptual level, the apparent dissimilarities in the physical-models underpinning two theories are usually a simple result of the 'projection' used in viewing reality. If we take a solid object in the form of a cone, and cut it in different directions (Fig. 2.4), we get various types of curves, known as 'conic sections'. If the cone is cut right at its tip, the conic section looks like a mere point. But it can be cut to give circles, ellipses, parabolas or hyperbolas. A hyperbola may look utterly different from a point or a circle. But they are all different cuts of the underlying reality (conical shape) which may be higher-dimensional (3-D) than any of the conic sections (2-D).

In such situations, advanced mathematical methods are needed to see how a progression of different theories (or approximate models) fits into a larger, more abstract (e.g., higher dimensional) reality. Indeed, when the theories of Newton and Einstein are looked at using such sophisticated tools, their inherent consilience becomes evident. Mathematical methods like 'fiber bundles' may be used to study a Newtonian space-time (a product space) whose topology is different at a global level. Such methods help to establish the consilience between seemingly disparate physical theories.

A much debated question is the nature of quantum reality, and whether the wavefunction ψ is a 'real physical entity'. Such doubts were raised regarding 'imaginary numbers', and when gauge potentials were introduced in electromagnetism. In Sec. 6.5 the reality (or not) of the ψ function, electron-spin etc., of quantum mechanics are discussed. We ascribe to them a reality in an abstract intrinsic physical space. Suitable experiments done in ordinary space-time can access that space, and we can only get statistical answers to our experiments by sampling that abstract space.

Another level of unity is found in all physical theories when it is found that all of them can be derived from a single principle of least action, coupled with a recognition of the existence of symmetries (conservation laws) in nature. In fact, with the rise of gauge theories in the middle of the 20th century, Utiyama [210] showed that the general theory of relativity and perhaps even the $1/r^2$ law could be derived by simply generalizing the global Lorentz invariance of special relativity to a local invariance. This added requirement was equivalent to imposing a gauge field which could be identified with the gravitational field. This may be taken as a further step in the consilience of our knowledge if we believe that gauge theory, rather than geometry, is going to be successful.

Thus it is clear that the interpretation of what gravity is, even at the non-quantum level remains an open question. Thus the existence of different perspectives (cross sections of reality) at this stage simply reflects the fact that relativity and the rest of physics have not been integrated seamlessly, as yet. These issues will be taken up in subsequent chapters of this book.

2.4.3 *Plato's cave, shadows and their explanations*

The previous discussion about cross sections of reality brings to our mind the discussion of knowledge contained in Plato's simile of the cave, where shadows of a higher reality are seen on the walls of a cave and mistaken for reality itself.

The world of the intellect is held by Plato to be distinct from the world of sense-perception. The intellect itself is divided into two parts, understanding and reason. Reason deals with 'pure truth' which is the higher form and it is dialectical. Understanding deals with mathematical thinking which can never assert what *is*, but only what *would be if* certain things were given; e.g., 'if AB and CD are parallel lines, then alternate angles are equal'. Plato seeks to clarify the difference between intellectual vision which is clear, and sense-perception which is confused and subject to error. The world of Platonic ideas is what the philosopher apprehends when the object is illuminated by the sun (the source of light, i.e., Goodness), while the transitory world of perception is a shadowy, twilight world.

The stage is set for the parable of the cave, where there are prisoners who can look only towards a wall. There is light from a fire behind them and empty space in front, towards the wall. The prisoners can only see shadows of themselves and of objects behind them. The prisoners regard these shadows as real. The prisoners have no investigative or experimental ability. They cannot even turn their heads. They see only a projection of reality. Plato uses a prisoner who escapes from the cave as a way of finding the truth about the world to 'explain' the shadows.

Given a set of circumstances X, does science explain why some Y happens? This is a popular question among philosophers [102] who posit that given a law of nature and a context as contained in X, science explains why some Y happens. Logical empiricists attempted to systematize *explanation* by providing a scheme where Y, the *explanandum* is the conclusion of the process, given the premise X, known as the *explanans*. Scientific theories do provide a relation $Y = f(X)$, where f embodies the theory or model. Philosophers who have written about 'explanation' consider that science 'explains' phenomena, whereas scientists tend to say that science *does not* explain the 'why', but provides the 'how'. That is $Y = f(X)$ is a relation or mapping, and it explains *how* X and Y are related within a set of boundary conditions. In 'one $\to$ one' maps the relation can even be inverted to yield $X = f^{-1}(Y)$. However, explanations look for 'causes', and we discuss the poverty and futility of such discourse in Sec. 14.3.

Some of the philosophical literature is dictated by the linguistic (i.e, prepositional) approach to problems, ignoring the structure-building nature of theory contained in $Y = f(X)$. These theoretical relations are not 'universal laws' in some timeless sense, but results established within a domain of observations and chosen boundary conditions. Let us consider a specific case of a much-discussed 'scientific explanation'. A well known debate over 'explanation', using objects and their shadows is found in the debates between Carl Hempel [102], Sylvain Bromberger [39] and many others. Some what like in Plato's cave, we have the sun shining on a flagpole, this constitutes the premise X. The shadow of the flagpole leads to the question 'why is there a shadow of length L?', denoted by $Y = L$. Philosophers contend that science gives a good explanation of why the shadow is of length L as one can deduce the value of L from the height h of the pole, altitude of the sun and a bit of geometrical optics. This mapping can be written as $L = f(X)$ where $X = (h, \theta)$ specifies the height h of the vertical pole and the angle θ made by the sun's rays with respect to the vertical. Then $f(X) = h\tan\theta$ gives the length of the shadow L. Much of the philosophical literature on 'explanation' accepts this as a good answer to *why* there is a shadow and *why* it is of length L.

If we observe only the shadow, but not the flagpole, can we equally well run the inverse argument, and conclude that there is a flagpole of height $h = L\cot\theta$ out there? However, it is contended that this does not 'explain' the height of the flagpole. That is, philosophers, especially those critical of the Logical Empiricist program, contend that although the explanation of the shadow and its length using science is good, the reverse process, i.e., explaining the flagpole and its height via the length of the shadow is not good. The shadow does not produce the flagpole. Thus explanations are claimed to show an asymmetry [39], running from 'causes to effects', but not from 'effects to causes'.

The question of causes and effects is not a standard part of standard scientific discourse (see Sec. 14.3). In fact, the flagpole does not cause the shadow, anymore than the presence of 'light causes the shadow'. One may still ask, *why* do flagpoles cast shadows? If we answer that light is scattered by the electrons in the flagpole, we again ask, why do electrons scatter light?, and so on (as in Aristotle's 'argument of the third man'), *ad infinitum*. That is, science does not answer 'why' questions, or assign causes. However, let us say that we accept that 'light is scattered by electrons in the pole' ($el - p$) as a scientific principle that we do not push any further. Thus our premise now involves not just rules about optics, direction of the sun and the height of the pole, but also $el - p$ which is short-hand for 'electrons are in the pole'. We now have as input $X = \{h, \theta, el - p\}$ and we can deduce that there is a shadow of length L, and also that this shadow is *not* made up of electrons. It has no material character because we can deduce that the shadow is merely a region with diminished fluence of photons. The inverse relation enables us to deduce the height of the flagpole from the length of the shadow, and also ascribe the ownership of electrons (i.e., material character) to the flagpole. The diminished fluence could have arisen from scattering of photons by other types of particles besides electrons. So we also need information about the environment which rules out other leptons. The explanation now runs correctly in both directions, while still respecting the asymmetry, if sufficient information is available. It is the availability of sufficient information that makes the mapping invertible in a unique manner.

Is such 'sufficient information' always available in principle? In quantum phenomena, we know that an observation (measurement) renders the discussion of previous histories very problematic. Approaches to 'non-demolition measurement' is an on-going topic of research. This ambiguity is not there at macroscopic length scales, but retrodiction or prediction of any non-linear process involves a choice among multiple solutions via the imposition of initial-state or final-state boundary conditions. Thus the philosophical discussions are limited by the lack of a framework where boundary conditions of differential equations could be considered. The use of cause and effect language, instead of relational differential equations introduces obvious pitfalls.

More loosely speaking, the general situation in science is close to that of the prisoners in Plato's cave, except that the prisoner-scientists have enough freedom to carry out experiments. The prisoners may not be able to turn their heads, but they have invented mirrors, microscopes, spectroscopes, and periscopes which can point in arbitrary directions. Now using the shadows, and the additional observables that the prisoners obtain through their new tools, they are able to construct a deduced reality, exactly as scientists do. They have even deduced the details of the first few seconds after the big bang, by observing the after-glow (cosmic background radiation) some 14 billion years later. So, contrary to Carl Hemple and others, we can indeed go backwards!

Caution is needed with words like 'observation, truth, reality', 'explanation', etc., as they are used differently by scientists and philosophers. Many philosophers think 'linguistically', i.e., within a language. Thus 'truth' is often regarded as some property of sentences or prepositions. Scientists tend to talk of 'validation' of data rather than about 'truth'. Similarly, 'observation' is associated by philosophers with direct 'sensations, sense data, phenomenal world, qualia' and so forth. Scientists include any recording of an event, e.g., a scintillation on a screen, a bubble-chamber track, or a pointer reading, an ultra-sound echo, etc., recorded by a camera or any other *suitable instruments*, as an observation. The issue of whether observation necessarily involves a conscious component will be taken up in the context of the quantum theory (Ch. 6). The *suitable instruments* work within length scales and energy scales already established by previous research. This is the *terra firma* that has become engineering. However, in a very young science some instruments may take some time before they are accepted , by the community. Thus Galileo's telescopic observations (Sec. 3.4.1) as applied to astronomy were not initially accepted by many of his colleagues. In addition, scientific observations are 'public' objects, in that they can be repeated by others.

One reason for setting up the 'Royal Society' was to provide a forum where savants could publicly witness a demonstration of a claimed observation. We will use the word *observables* to indicate these *validated* observations which may be obtained using any accepted instruments. The same word is used in the quantum theory, where taking an average over many observations is mandatory.

All observations should have passed 'verification' before assimilation into theory. Some sociologists like Hull [111] suggest that scientists no longer worry about *replication*, i.e., repeating the experiments for verification. Hull suggests that replication occurs mainly *via* the *need to use* the proposed theory. This claim is far from true in the physical sciences, and indeed in most sciences that deal with observables. The replication may not be an exact repetition, but it is usually done under more stringent conditions than in the original experiments.

A remarkable example of world-wide concerted attempts at replication occurred when two chemists, Fleischman and Pons, claimed in 1989 CE that they had achieved the fusion of Deuterium atoms at room temperature, i.e., *cold fusion*, by concentrating them in Palladium electrodes. Experiments attempting to reproduce these claims failed, refuting these claims within a short time [204]. This case provided students of the nature of science with a modern text-book example of how science works. Similarly, the fraudulent publications of Hendrick Schön, of Bell Lucent laboratories circa 2000 CE claiming the fabrication of molecular-scale electronic devices were also rapidly exposed by scientists who couldn't reproduce his results and were forced to re-examine his reported data in detail. By contrast, the social scientists did not have a mechanism for exposing the hoax carried out by Allan Sokal [199] who pretended to be a post modernist engaging in *science studies*. Finally Sokal himself had to reveal his subterfuge.

Observables, i.e., everything that can be validated, recorded, and publicly verified using acceptable instruments, extend our observations to regimes well beyond our *direct sensations*. Such regimes are referred to as the 'unobservable world' by some writers. In this book we regard such regimes as part of the 'observable world'. However, when we study relativity (Ch. 5) and the quantum theory (Ch. 6) we may refer to aspects of the external world which are unobservable even in principle. These are *unobservables*. Thus, according to current theory, no instrument however refined can find the simultaneous position and velocity of a subatomic particle. Similarly, no instrument can bring us information about parts of the universe expanding at speeds greater than the velocity of light.

Thus science constructs a 'deduced reality', based on all the observables. A complete theory, e.g., the grand unified theory (Fig. 1.4) would deduce all phenomena within the single framework of one model. Such theories may contain statements about unobservables. In fact we have two mathematically equivalent versions of the quantum theory (Copenhagen model, Bohm model) based on the Schrödinger equation while the interpretations are different. Here the two 'interpretations' make different assertions regarding the existence or non-existence of the trajectory of a quantum particle, its velocity etc., in their unobservable states. Such a trajectory is a non-observable. However, both theories give identical predictions for all the usual observables. The unobservables are theoretical intermediaries that leave no final trace in the result, although the philosophical implications may be quite different, see [32, 27] and Peat [158]. Similarly, there may be several artificial neural nets which may take a given set of inputs and produce acceptable outputs. In practice, one of them is likely to be simpler and easier to interpret further. If no such advantage is available, the preference of one theory over the other becomes a matter selecting the simpler one for visualization or calculation.

2.4.4 *Attempts to discuss scientific 'method'*

The attempts of philosophers of the positivist or empiricist school, as well as Karl Popper and his students to lay down 'methods' for scientists failed. Feyerabend and others used this very *failure to identify a scientific method* to claim that 'anything goes', and deny a place for scientific rationality. More nuanced positions have been taken by other writers. Unfortunately, these thinkers did not benefit from the developments in Artificial Intelligence (AI) and cognitive science (CS) that ran parallel to their philosophical lives. AI and CS showed that even to lay down rules for 'common sense', leaving aside 'scientific research', is a mind-boggling problem. Programming a robot to go to a room and bring out from that room an apple, as opposed to an orange, turns out to be an utterly non-trivial problem. This is an example of what is sometimes called a 'frame problem'. Stating the rules of the 'scientific method' in a complete sense would be equivalent to programming a robot to distinguish between a good scientific theory and a bad one. Scientists work with whole structures (theories) linking various parameters whose very meanings depend on the theory. Concepts like mass, iso-spin or helicity may have common English meanings. However, the theories usually specify how to manipulate them, and their 'meaning' is only implicit via their manipulations.

Feyerabend and others only dimly appreciated the 'feel for the subject', 'scientific method' and other tools unconsciously deployed by the brain of a trained researcher. Feyerabend and others search the history of science to construct anecdotal accounts to defend their views which sound like 'just so' stories. Horgan [109] branded Feyerabend the 'worst enemy of science', although Feyerabend was as much an enemy or supporter of Voodoo as he was of science!

Theories are also tools, like telescopes and cyclotrons used in research. The search for the right tools and the right theory is the business of good science. Thus the theory-laden nature of observations is a positive feature which ensures that the bare facts have been dressed in the appropriate way within a 'good theory', making it intelligible. The 'bare fact' is an unobservable *beable*, if we borrow a term from John Bell and reuse it in a wider context. If the theory is good, the 'theory-laden' nature of 'facts' is a desirable feature. However, Kuhn, Feyerabend et al. consider theory-ladenness a negative feature in all cases. Their concept of theory is not that of a tool which is under our control. For them theories distort reality and introduce bias. The strong-minded social scientists who embrace relativism would even assert that the 'theory-laden observations' are merely politically motivated results of a group safeguarding its own self-interest. In this context we cannot do better than to paraphrase Newton-Smith's 'pay as you go' argument [148]:

> ... if what we call *rational considerations* take their place on par with bribes, threats, and prizes, we have no reason to take the externalist seriously. Once in a discussion with two externalists I asked why we were not simply paid to agree. They answered that it had not proved effective to treat academics in this fashion!

2.4.5 *The scientific revolution in high-energy physics*

While Popper, Kuhn, Lakatos, Feyerabend *et al.* were debating their theories about scientific revolutions as found in the records of historians, a very important and instructive scientific revolution was occurring right under their noses.

Probing the atomic length scales and energies had ushered in the quantum theory, completely successful (i.e., no exceptions whatsoever) in all applications. Scientists had begun to probe nuclei and other innards of atoms. Nuclear physics had become well understood and absorbed into the *terra firma* of established knowledge, becoming the subject of nuclear engineering that deals with nuclear medicine, nuclear power stations, and nuclear weapons. This new engineering had also spawned new and more powerful tools for smashing atoms to reveal their sub-atomic contents. Thus new, more energetic and much shorter length scales had opened up for investigation. In addition, studies on cosmic rays had revealed a whole zoo of 'strange' particles known as mesons.

The second half of the twentieth century was the time when these events unfolded. The war, the weapons program, and the need for well-funded and expensive particle accelerators to study elementary particles, all contributed to the support of a community of highly gifted intellectual aristocrats who worked in distinguished universities and institutions. CERN, based in Switzerland and founded in 1952, Brookhaven and Argonne in the USA, were some of the centers involved in close international collaboration.

In the 1960s machines at CERN and Brookhaven could accelerate particles to energies in the 30 GeV range (30×10^9 eV). In the early part of the 21st century, machines at CERN and at the Fermilab had reached the TeV range (10^{12} eV), pushing the energy scales up by three orders of magnitude. The LHC (Large Hadron Collider) at CERN achieved a proton-proton collision event of nearly ten trillion eV in the year 2010.

Particle accelerators use electric and magnetic fields to accelerate elementary particles like electrons or protons. Galileo reputedly used the leaning tower of Pisa to accelerate particles in his study of mechanics. He dropped several objects of varying mass m, including a cannon ball, to show that all the objects reached impact at essentially the same moment, thus establishing that the heavier objects did not fall faster than lighter objects, but attained the same speed V. It was also noted that the particles fell vertically, and did not follow an arching curve that Aristotelians considered was a necessary result of any motion of the earth. Galileo's good grasp of experimental planning is seen in how he mitigated the effect of wind resistance etc., by using suitably heavy objects so that irrelevant issues did not fog the result. That is, his experimental set up followed a good reductionist approach.

The kinetic energy from this acceleration is $(1/2)mV^2$, and exactly equals the mgh where g is the acceleration due to gravity, and h is the vertical drop for the Tower of Pisa. Let the vertical drop be $h = 180$ meters, i.e., $\sim 2 \times 10^4$ cm., and $g \sim 1000$ cm.sec^{-2}. An electron (in the cannon ball used by Galileo) has a mass of about 9×10^{-28} grams. Hence the free-fall acceleration contributes an energy $mgh \simeq 2 \times 10^{-22}$ ergs, or 1.25×10^{-9} eV. That is, our capacity to accelerate particles selected under a well defined reductionist program, i.e., under well controlled conditions, has increased by a hundred-billion-billion fold (10^{20}) from the 17th to the end of the 20th century. Hence we are today able to reach length scales and energy scales way beyond what was even imaginable during the renaissance.

Heisenberg had noted the similarity of protons and neutrons, and proposed that they were like Tweedledum and Tweedledee — two forms of the same particle. They were understood as 'isospin' states (Sec. 1.2.1) obeying an $SU(2)$ gauge symmetry. Simple variants of this idea explained the existence of many of the particles known up to about 1950. However, as additional, unexplainable particles piled up, a new theory became necessary. As we outlined in Sec. 1.2.1, a very bold hypothesis involving fundamental units of fractional charge (quarks) was proposed in the early 1960s. It evolved, *via* $SU(3)$, i.e., the 'eight-fold path' of Gell-Mann and others, to the *standard model* of today.

Here we are dealing with a scientific revolution in the most mature of the sciences. Particle physics proceeds by formulating theory and setting up (usually very expensive) new tools for verifying, modifying (and if it so happens, falsifying) proposed theories. If the proton were made up of three quarks, viz., *u,u,d*, and if the neutron were made up of *d,d,u*, we see that they are indeed Heisenberg's Tweedledum and Tweedledee. The new theory does not discard the old theory. It

Fig. 2.5 Retooling to probe new length scales. The CERN bubble chamber, known as *Gargamelle*, was such a tool that has become a part of history. It was cooled by 10 tons of freon, and contained liquid hydrogen nearly at its boiling point. It was about ~ 5 m long and ~ 2 m in diameter. *(courtesy-CERN)*

gives substance to the old theory and enriches it. To test this, new tools that can probe inside protons became necessary.

An electron accelerator (named SLAC) was built at Stanford to generate energetic electrons. These could be sent into nuclei to see if they would be scattered by the point-like quarks of fractional charge. The theory had also predicted that neutrinos(ν_e) would impact a d-quark, convert itself into an electron, while the d-quark itself would change into a u-quark. Thus several neutrino-beam facilities were constructed at several laboratories, and specially at CERN. In addition, the French team from Saclay at CERN led the effort to build a special bubble chamber (Fig. 2.5) named *Gargamelle*, containing liquid hydrogen, and maintained just below its boiling point. Such super-critical liquids form tracks of gas bubbles when charge particles travel through them, leaving tracks much like the vapor trails of airplanes seen in the sky. Magnetic fields applied to the bubble chamber curl the tracks enabling experimentalists to identify the particles etc., by their magnetic, mass, and charge signatures. The discovery of the missing members of the predicted Baryon Octet, the meson nonet, and the decuplet of spin 3/2 particles which included the Ω^-, were all sagas which parallel the discoveries of new moons of Jupiter or the phases of Venus, some three centuries previously. Then came the discovery of the 'neutral currents' (i.e., processes where neutrinos interact with matter without changing their identity), charmonium (i.e., a charm quark which has paired off with its own anti-particle) and its J/ψ ground state, vindicating much of the theory and the experimental effort.

The last quarter of the 20th century saw the rise of the Standard Model (Sec. 1.2) as associated with Salam, Weinberg and Glashow, and Chromodynamics as associated with the names of Gross, Wilczek, and Politzer. The new theories did not discard Heisenberg's $SU(2)$ and other models as 'outdated', or inapplicable (as a Khunian might have it). Instead, an extended symmetry involving $SU(3) \times SU(2) \times U(1)$ which assimilated all the energy scales was achieved.

The scientific revolution of high-energy physics was closer to the verification/modification/assimilation approach to scientific revolutions than those discussed by Kuhn, Lacatos and others. It had some elements of the Popper approach, since some proposals were eliminated by experimental falsification.

2.4.6 *A mini-revolution in condensed-matter physics*

Another example of the impact of re-tooling took place in the early 1980s, and showed how new physics and conceptual surprises can arise even within a very well established theory, without the slightest change in the theory. The re-tooling of particle accelerators to push to higher energies and shorter length scales was possible due to parallel developments in electronics and materials physics. These had become possible due to our understanding of solids and their chemistry within the quantum theory. The Bell laboratories in New Jersey founded by Graham Bell, IBM and similar laboratories the world over were developing the capacity to engineer materials by controlled deposition of single layers of atoms (epitaxy).

Fig. 2.6 Retooling to probe new length scales. A molecular-beam epitaxy (MBE) machine is essentially a vacuum oven where vaporized atomic species are deposited in a controlled manner, to grow materials with the precision of nanometer length scales. *(courtesy-NRCC)*

Such a possibility arose with the development of 'molecular beam expitaxy' (MBE) machines (Fig. 2.6) and vapor deposition methods. Precision tools like scanning tunneling microscopes (STM), electron-microscopes, electron-beam lithography, and accurate chemical etching techniques that could be applied to MBE grown crystals also became available. These new tools, developed for making minute 'electronic chips', also gave scientists the possibility of studying how matter behaves in reduced dimensions. Electrons in a metal, or in a hot plasma, move in all three dimensions. However, using MBE technology, scientists could fabricate atomically thin sandwich structures of materials. For example, a layer of Gallium Arsenide (GaAs) can be sandwiched between two layers of Aluminum Arsenide (AlAs). The electrons get trapped in the GaAs layer where they only move in the plane of the sandwich (say, $x-y$ plane), and not across it (the z direction). Thus such electrons are confined to a two dimensional world. The physics of this confined world turns out to be quite surprising, even though they follow logically from the standard equations of quantum mechanics.

The study of such 2-dimensional electrons placed in strong magnetic fields produced a mini-revolution in the conceptual outlook of physicists who are dealing with devices and electronic gadgets. Experiments at the Max Plank institute were the first to show that 2-D electrons could flow without any resistance when subject to strong magnetic fields. Electrons in a magnetic field fan out to occupy energy levels known as 'Landau levels'. The phenomenon occurred when ever an integer number of Landau levels were filled by electrons. Thus Klaus von Klitzing became the discoverer of the 'Integer quantum Hall effect'. Soon after, experiments at Princeton seemed to suggest that when all the electrons are confined to just one Landau level, and if only an odd fraction (e.g., 1/3) of the Landau level were filled up by the available electrons, the system again behaved as if they were full Landau levels. Hence we have fully-occupied mini-Landau levels with mini-energies and corresponding length scales. This phenomenon came to be known as the 'fractional quantum Hall effect (FQHE)'. Robert Laughlin, then working at Lawrence Livermore Laboratories, came out with a surprising but very compelling theory that electrons are behaving like fractionally-charged particles (e.g., $|e| = 1/3, 2/3, 1/5, 1/7, \cdots$, etc.) in the FQHE.

While the particle physicists were discovering quarks with factional charges, condensed matter physicists were showing that even non-relativistic quantum mechanics was full of surprises. Complex systems could show the properties of simpler systems, via the emergence of composite objects that mimicked more elementary objects. Electrons in *fractionally occupied Landau levels* dressed themselves up with flux quanta from the magnetic field, and became *composite Fermions* that looked like ordinary fermions in a full Landau level!

The mini-revolution in physics associated with the integer and fractional quantum Hall effects could also be used as a text-book example of how scientific revolutions occur. They involved the opening up of new length scales or energy scales due to developments in tools and technique. Theories are constructed, verified, modified, and assimilated into the *terra firma* of accepted knowledge, while the false starts get refuted and die a natural death. For instance, when the FQHE was observed, an early theory of it was in terms of the formation of a charge-density-wave. Experiment and theory could both be adduced to refute that theory.

2.4.7 *Scientific method viewed as a simulation*

Extremely mature sciences like particle physics or condensed-matter physics may have scientific revolutions that follow the verification/modification/assimilation or refutation path, where judicious amounts of all these contribute to the process. However, we need to look at the more general canvas of creation of knowledge.

Feyerabend claims that the only valid method in science is the principle that 'anything goes'. This is an attempt to recognize that seemingly highly improbable, 'maverick' points of view may finally turn out to be correct. In effect, we need a process that judiciously weighs all possibilities, while even allowing for some rare (low-probability) moves. Such a process, or algorithm actually exists and it is frequently used by scientists in so-called 'molecular dynamics' simulations. These are merely stochastic simulations of a collection of movers subject to a 'landscape', (i.e., an external potential) and interacting among themselves according to some laws. The *Metropolis algorithm* (or modifications there of) is cardinal to such stochastic simulations. A modern form of this, used for moving 'particles' subject to Newton's equations of motion is known as Verlet's algorithm. In such a stochastic simulation all possible moves are allowed to occur, but with a weighted probability $\exp(-U/T)$ where the parameters U, T characterize the energy window open to the movers. That is, the local values of U, T depend on the instantaneous configuration of the collection of movers themselves, and T may be a 'temperature' or a measure of some environmental field.

The scientific method may also be viewed as the working of an algorithmic process where savants weigh (in their own minds, or using more formal methods) the possible moves, and carry out experiments based on some hypothesis which is as far as possible consistent with the general body of theory accepted in science, but may deviate in new energy and length scales. Usually what is obtained is partial verification rather than verification or falsification. This gives rise to a new configuration of movers (theories) for which a new set of moves becomes possible. All stochastic processes contain fluctuations including wild but rare fluctuations. The Metropolis algorithm and other such algorithms allow for such rare possibilities as well. One may picture research as a controlled simulation, inclusive of 'maverick moves', that finally anneals to the correct solution.

Hence scientific activity may be described as follows: when new length scales or energy scales open up due to new tools, new hypotheses (make new moves in the simulation) which deviate from established theory in the new regimes become possible, while remaining consistent within the established regimes of energy and length scales. When Einstein introduced the photon, a particle of light, to explain photo-emission, he was taking advantage of a new domain of experimental physics. However, it is not possible to simulate all the particles, i.e., electrons, ions, photons, that are relevant to the problem. Hence simple models which are consistent with the older scales of energy and length, but innovative in the new domains, are constructed and tested against experiment. In such processes we may come across a 'paradigm breaking' situation, or a more normal situation. There is no need for a 'critical mass' of deviations, as even a single deviation from standard theory is intolerable in a mature science.

New physics is emerging from nano-science, short-pulse laser experiments, molecular biology etc. Discoveries in black holes physics, quantum chromodynamics etc., are pushing us into ever more extremes. Although Horgan has claimed that the 'end of Science' has been reached [110], humans have yet to investigate dark matter and dark energy comprising $\sim$75% of the universe!

2.5 The cognitive conundrum

Let us imagine a major volcanic explosion. Let us say that *Mount Doom* has exploded. Molten lava — a hot structureless liquid gushes out of the mouth of the volcano. The lava is projected thousands of meters high into the sky and falls back, cooling and crystallizing as it falls. The heat from the volcano has also evaporated the water of a near-by lake and the falling lava crystals encounter a jet of steam and water which harbour a few thermophilic bacteria. Let us say that we have a particularly precocious relative of *Thermus Aquaticus* [35] turning up and colonizing one of the larger lava rocks.

Thermus Aquaticus is not just any ordinary beast. It is the first of the *Archaea*, the ancient lineage of micro organisms that may have played a key role in the 'tree of life itself'. It is a 'prokaryote' in that its cell does not even have a nucleus (see Ch. 12). But Carl Woese [232] who began to classify living creatures using ribosome DNA (or rDNA) would not call it a bacterium. In fact the 'tree of life' (Fig. 2.7) possibly branched into bacteria and archaea. Then the archaea split off the eukaryotes branch (which includes plants and animals) while accepting many genes from the bacteria, and keeping similarities to eukaryotes, thus muddying the neat branching of the tree of life by introducing cross links. The horizontal dashed lines connecting the branches indicate exchange of genetic material. Thermus Aquaticus has also given us an enzyme (i.e., an organic catalyst) known as the TAQ polymerase, the essential enzyme which facilitates the polymerase chain reaction (PCR). Unlike most biological materials, TAQ works at high temperatures and PCR can be carried out at a higher temperature where everything happens much faster than at room temperature. PCR is the tool for multiplying a bit of DNA billions of times in a few hours. Its applications in industry (e.g., to make sweeteners for the soda-pop industry), and in forensic science are now well known. A dramatic example was when PCR was used to amplify the bit of DNA [86] claimed to be from the Romanovs, the family of the Czar of Russia, murdered by the Bolsheviks. The amplified DNA provided conclusive positive identification. In 1993 Kary Mullis was awarded the Nobel prize for the PCR discoveries.

The lava rock finds a resting place at the edge of a hot volcanic spring. Our bug Thermus discovers the kind of mineral nutrients that it likes in the surroundings of the lava rock and soon begins to reproduce by cell division. Soon, a thriving colony of bugs populates the steaming waters around the lava rock. What if, by some miracle, this bug population becomes philosophic and begins to think of its destiny? The bugs begin to ask why they happen to live on the rock, the meaning of life, and other fundamental questions (see Sec. 2.1).

However, what the bugs can achieve is limited by their life cycle, the resources they have on the small hot spring near the rock, and the possibilities afforded by the location of the lava rock in a mountain stream. Unlike the proverbial frog in the well, the bug on the rock may observe a world beyond the borders of its home. However, can it ever find the answers to its riddles? The energy scales, the time scales and length scales, and the investigative tools available to the bug impose strong limitations on its possibilities. Even if this bug were inexorably

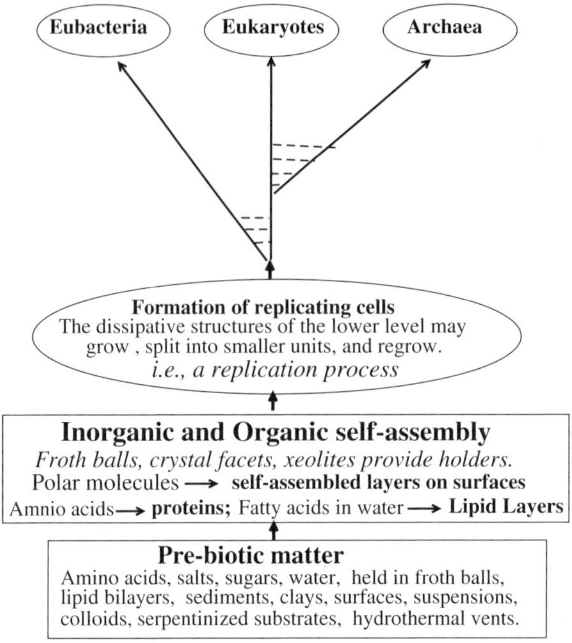

Fig. 2.7 The Tree of Life emerging from basic forms of matter *via* self-assembly of inorganic constituents. Froth balls, under-sea vent structures, crystals etc. can replicate by inorganic mechanisms. They provide templates as well as containers (holders) of prebiotic material that combine to form complex molecules capable of organic replication (See Chapter 10). Plants, animals, fungi are Eukaryotes. The early part of the 'tree' is muddied by cross-relations exposed from studies using many genes.

driven by what we called epistemic hunger, it has little possibility of understanding the world at large. This then is a cognitive conundrum that can apply not only to the bug, but to human thinkers as well. In effect, can we even in principle expect to know the answers to the sort of questions that we are asking?

The impossibility of knowledge has been a constant philosophic theme since antiquity. Greek skeptics like Protagoras as well as Indian skeptics such as Sanjaya (circa 6th century BCE) and Dhiganakha debated these issues. Many such views are discussed in the Buddhist *Dhiganakha sutta*. Modern skeptics [72, 127] claim a relativism of knowledge that borders on fantasy or solipsism.

We may follow the colony of bugs further along as it evolves. The colony eats up all its food supply as the well of hot water gradually dries up, leaving behind some organic waste on the Lava rock. But this does not mean that there is no life on the rock. Other organisms which can use up the waste left behind by the previous colony of bugs now begin to thrive. The new life forms lead their life cycles and become extinct, only to be replaced by another form of life which thrives on the waste of the previous life forms. Finally, another volcanic explosion propels the rock into a molten lava stream. The time scales of the volcanic eruptions and the rates of growth and decay of the bug populations are so different that there does not seem to be a possibility for the learned varieties of bugs to satisfy their epistemic hunger, or speculate on the possibility of a 'bugthropic' principle in the universe, analogous to the 'anthropic' principle!

2.6 Leaping over energy, length and time scales

We cannot answer any fundamental questions unless we can leap over many time scales, energy- and length scales. The place of humans in the Cosmos is at first sight not too different to that of the colony of bugs living on a lava rock produced by a volcanic eruption (the 'big bang' for the bugs). Observational Astronomy and modern theoretical physics provide strong evidence for believing that the universe came to existence some fourteen billion years ago in a big explosion, the so called big bang (see 1.4 and Sec. 7.5). Theory suggests that there are big bangs happening in the cosmos all the time. The theory of relativity forbids any communication of information beyond our *event horizon*. Our own big bang occurred in a tiny local patch of the cosmos. It inflates super-exponentially and produces a hot, structureless plasma which expands and begins to cool, just like the structureless lava of the volcano. The structureless plasma becomes grainy as it cools, forming the elementary particles (e.g., electrons, hadrons etc.) that are the basic constituents of matter. As the universe expands and cools even more, the elementary particles combine and the primordial gas undergoes 'phase transitions', just as water vapour condenses to form clouds, water and ice. The particles aggregate to produce clouds of nuclei, galaxies, stars, planets and so on. We humans are in many ways like the bugs on the volcanic lava rock.

Yet, unlike the bug on the lava rock, we speak as if we know what happened some 14 billion years ago! Indeed, amazingly, we can claim a precise understanding of what happened in just the first few moments after the big bang. This is because our present understanding of gravitation, particle physics, thermodynamics, and observational astronomy tie together and mesh so nicely that clear, unequivocal, testable statements may be made about the 'after-glow' of the big-bang. This after-glow is known as the 'cosmic microwave background' (CMB). The effect of the after-glow is to fill space with heat radiation at a temperature of about 2.275 ± 0.001 degrees Kelvin [139]. Balloon-borne and ground-based CMB experiments, and the space-borne Wilkinson Micro-wave Anisotropy Probe (WMAP) mission have revealed a wealth of experimental data about the slight anisotropy, the frequency spread and other details of the CMB which tightly constrain the dynamics, content and the possible shape of the universe (for details, see the bibliography under LAMBDA [125]). That is, we are not indulging in metaphysics when we talk about the early universe.

Two well known (if dated) popular books on the subject are by Steven Weinberg [219] and by Stephen Hawking [100]. The exponential 'inflationary' blow-up of the embryo-universe from a tiny uniform spot as small as 10^{-26} meters to a size of about a meter in about 10^{-35} seconds is truly mind-boggling. The initial tiny spot of 10^{-26} meters extent is about a hundred billion times smaller than a proton and is the space-time singularity (a 'white-hole') at which the laws of physics fail (or become undefined, as inside a black hole). Our universe is not just expanding because of the big bang and slowing down; in fact the expansion is accelerating — a runaway expansion? A popular account has been given by Alan Guth [93] who was a major contributor to the subject. The subsequent evolution of the universe, after the first few minutes is much more difficult to analyze. Given a set of initial conditions for any system, be it the Universe or anything else, what happens immediately afterwards is easy to predict. What happens later on is determined by subtle selections between many possible branching processes. Thus descriptions of the subsequent formation of galaxies, stars, and planets etc., become as difficult as predicting the exact details of eddies and swirls in a stream of water which initially gushed out in an orderly, streamline flow pattern. However, the broad outline is clear enough and we can indeed understand the formation of solar systems and Earth-like planets. For now, we are interested in the primitive human being who has appeared on earth.

The early human's effort was to find enough food to eat, and avoid being eaten before reproducing. Facing cyclic famine and drought, incessantly tortured by bugs and mosquitoes, braving internecine killings and savagery, groups of humans probably had to enslave others to sustain life styles capable of supporting some leisure and speculative thought. Such societies extended their observational capacities by building tools that enabled them to engineer the immediate world inhabited by them. The tools were a direct necessity for building shelters, making weapons and engaging in warfare. Tools for measurement grow out of such activity. These tools have enabled us to go beyond the limited length scales and energy scales set by our bodily biology. Unlike the bug in the volcano, or any other living species that we know of, humans and many primates (see Premark [172]) have a public description of the world they live in, given in terms of numbers — i.e., a digitized description. A few decades ago, some would have objected to such a statement on the ground that there are 'cultural' or 'emotive' entities that are not digitizable. However, most people today are familiar with digitized multimedia and it is generally recognized that all information, even the finest texture, far beyond those discernible by humans can be represented by strings of numbers.

Newton realized that the terrestrial laws of motion and gravity applied equally to extra-terrestrial objects like the planets. Newton analyzed light into the colours of the rainbow. When a substance, e.g., sodium, is heated in a flame it becomes yellow light, while copper gives a blue-green light. Spectroscopy analyzes the colour of light coming from stars and reveal their chemistry. Thus we find that the terrestrial laws of physics and chemistry are valid for the stars as well.

The power of spectroscopy is incredible in that it has opened up astronomical length scales as well as sub-atomic lengths scales. In astronomy, spectroscopes team up with telescopes. Bigger telescopes collect more light, enabling us to look at ever more distant stars and other stellar objects. We need not just look at the spectra of visible light. We can use radio waves, or any other convenient part of

Fig. 2.8 Henrietta Leavitt (1868–1921) and Edwin Hubble (1889–1953) were American astronomers whose work extended our knowledge of extra-galactic astronomy at length scales that go beyond parallax measurements.

the spectrum. We find that familiar spectral lines have shifted to the red end of the spectrum. That is, they have acquired a 'red Doppler shift' in their colours, telling us that the relative distances between us and the light sources are increasing. Every astronomical object is running away from us! But it is not expanding 'into some other empty space'. It is in this sense that the universe is said to be expanding.

The observed shift of spectral lines to lower frequencies (reddening of spectral lines) is called the *Hubble red shift*, after Edwin Hubble (see Fig. 2.8), although earlier observations by Vesto Slipher (Lowell observatory, Arizona) had already shown the reddening of spectral lines.

We come across various new stellar objects, e.g., quasars ('quasi-stellar objects') and black holes, which we cannot create in earth-bound laboratory experiments. But some of these strange objects have already been predicted by theory. We find that the laws of atomic physics and nuclear physics discovered on earth remain valid in the most distant reaches of the heavens, but some other laws (e.g., Newtonian gravitation) begin to require modification.

Let us take a more technical look at how we probe the deep Universe. Measuring distances, or finding out things about far off objects is difficult. The distance to stellar objects within 10–150 light years can be estimated using the method of 'parallax'. This is the 'wobble' in the position of the star arising from our motion around the sun. The distance between the earth and the sun is known as the astronomical unit (AU) of distance; here we note that the lower-case 'au.' stands for 'atomic units', at the very small length scales pertaining to atoms. The maximum parallax (angular shift) in the position of a star arises when the earth moves from one end of the diameter of the earth's orbit around the sun, to the other end. A parallax of one arc second (1/3600 of a degree) corresponds to a distance of 3.26 light years (ly), or one *parsec*, pc. This method of distance determination brings us to about the Hyades star cluster (48.4 pc), at a distance of $\sim$ 150 Ly.

Beyond 150Ly, a new method is needed. The brightness of an object diminishes with distance in a known manner. So we compare the star with a 'standard candle', i.e., a star whose absolute luminosity (brightness) is known. It was found in 1912, by Henrietta Leavitt (see Fig. 2.8) that the absolute luminosity of a type of star known as *Cepheid variables* could be determined from the periodic variation of their light. So Cepheid variables (CV) were recruited as 'standard candles' for use when the parallax becomes too small. The distances to the 'Magellanic Clouds' (105 Ly), and the Andromeda Galaxy (106 Ly) can be estimated using CV. Observations of CV revealed that their spectral lines (e.g., the yellow line emitted by sodium vapour) were shifted to the red. This indicates that all the stars are moving away from us (Doppler shift). Such observations showed that the universe is constantly expanding. This red shift of spectral lines of stars is known as the *Hubble red shift*.

To reach distances beyond the Andromeda galaxy and get to the Virgo cluster, the Coma cluster and the Hydra cluster (109 ly) we need the support of other methods. The most luminous Cepheids can be used to estimate distances as far as some 10^6 ly away. A type of star known as 'white dwarfs' containing carbon and oxygen tends to have a uniform composition and mass. The white dwarf in a binary system may have a companion of the type known as a 'red giant'. The white dwarf can pull matter from the red giant until the white dwarf (WD) reaches a precise mass when the WD can no longer support its own mass. Then the nuclei of the matter in the WD get squashed together so closely that the WD undergoes nuclear explosions (see Sec. 5.4.1). Here we are not talking of the quasi-stationary nuclear reactions that burn in stars like the sun. Such WDs generating nuclear energy explosively become extremely bright transient objects known as supernovae. The absolute intensity maxima of supernovae are well understood and this has lead to a revolution in cosmology. They have enabled us to probe in detail the energy content of the universe, leading to our understanding of dark energy and cold dark matter [95].

The star of Bethlehem and similar legendary sightings are claimed to be supernovae although the evidence is flimsy. Chinese records of a sighting circa 5 BCE, as well as 185 CE have been mentioned. However, of the ten historical supernovae, the first four (185 CE to 393 CE) are only given in Chinese records, while the last six (1006 CE to 1604 CE) are well authenticated [91].

Another method used to leap into large length scales is based on observing the light emitted by the tumbling motion of water molecules (water maser line) using an array of ten radio telescopes positioned around the world, from Hawaii to the Virgin Islands. A maser is like a laser, except that it works in the microwave region which corresponds to the energy associated with the tumbling of a water molecule. The Very Long Baseline Array (VLBA) acts as a giant TV dish thousands of miles wide, and can accurately measure the radio emission corresponding to the water maser lines in distant galaxies. This information can be processed to yield the distance of the galaxy from the earth. Thus supernovae and VLBA provide valuable standard candles which can be used up to a Hubble red shift of about one unit. The red shift z is the amount a spectral line has shifted to lower wavelengths (red side) due to the Doppler effect arising from the expansion of the universe. If we ignore relativistic effects, the redshift is approximately v/c where v is the speed of recession of the star, while c is the speed of light. Then the distance of the star d is given by $d \simeq zc/H$, where H is the Hubble constant which is $\sim 2 \times 10^{-15}$ per second. The further-away objects seem to expand more rapidly and have bigger red shifts. Then a relativistic formula for z has to be used. Ten units of the red shift (z=10), i.e., $v \sim c/2$ as estimated from a relativistic formula, correspond to deep distances where gamma ray bursts from core-collapsing supernovae can be used.

The deep symmetries found in the laws of electromagnetism provided the springboard for relativity and quantum mechanics. Just as spectroscopes probed the large-scale structure of the world we live in, the small-scale structure was also probed by spectroscopy. After all, the yellow light of sodium (Na) atoms excited by a gas flame, and the blue-green light of copper (Cu) atoms signified details of the atomic structure of Na and Cu. We learn more about this by breaking up atomic particles and examining the fragments. These experiments revealed a basic set of 'elementary particles' which all fell into place within the conceptual model which goes by the humble name, *the standard model* (see Sec.1.2.1). But this 'standard' is one of the highest standards of theory that we have so far! This theory is the cumulative product of the great minds of the twentieth century. The standard model based on quantum mechanics, taken together with general relativity, essentially encapsulates all of 20th century physics. Our working hypothesis is that all the other sciences result from these fundamentals, but with comprehensible (i.e., non-mystic) emergent properties at each new length scale.

The standard model describes all forces including electromagnetic and nuclear forces, but does not include gravity. Of course, in Fig. 1.4 we have shown gravity and the other three fundamental forces coming out of a Grand Unified Theory (GUT) that we don't still have. Gravity is described by a theory which has a completely different (geometric) flavour — the general theory of relativity. If matter is made up of elementary particles, and if this same matter is described by some geometric features of space-time, then there should be some means of putting everything together so that electromagnetism, nuclear forces (the 'weak' and 'strong' interactions), as well as the gravitational force all fall into some extended 'standard model', or a seamless GUT. The beginnings of such a theory are perhaps already coming up, in the form of string theory or *m*-brane theory. This seems to replace the geometric theory of Einstein by a super-quantum theory, which encompasses all the interactions including gravity.

Alternatively, it may turn out that all the presently known 'elementary' particles are really composites made up of other, even more fundamental objects involving incredibly large energy scales. That is, the simplicity and symmetry in nature that we discuss in a future chapter are simply a consequence of our being confined to the bottom of the energy pit, just like the bug on the lava rock that we discussed in Sec. 2.5. This would mean that if experiments probing really high energy scales become possible, then the world would not become simpler, but more non-linear and more complex. In effect, just when we thought that the onion has been peeled, it may be that we have to go on peeling even more [213]. The elaboration of such 'anti-GUT' theories, as well as their experimental tests will be the work of the future.

One the other hand, at human scales of energy and length, there is really no question whatsoever that chemical phenomena on earth are completely described by a subset of the standard model which is Schrödinger's quantum mechanics. In the following chapters leading up to part II of this book (and also Ch. 14) we attempt to show that the laws of physics are sufficiently rich to paint the canvas of all life, consciousness, personalities with free will, without invoking any non-physics principles or even any exotic interpretations of physics. The developments in molecular biology, genomics and neuroscience coupled with our increased understanding of statistical mechanics, the many-body problem, neural networks, artificial intelligence and computational algorithms, have made this perspective progressively easier, although still very incomplete.

Chapter 3

The Laws of Nature and the Supremacy of Symmetry

This chapter introduces the ideas of symmetry and how 'changes of perspective', i.e. coordinate transformations, impact on natural law. The laws of Galileo and Newton are the same for all perspectives and all observers. There are no privileged observers. Newton's laws of motion are the simplest laws compatible with Galilean relativity. This provides an opportunity to catch up with Galileo and his modern critics. Then we look at the principle of least action, and ask how a particle or a light pulse 'knows' to choose the shortest path as its trajectory. The local texture and the global texture form a duality that are in perfect harmony because of the simplicity of the laws of physics.

3.1 A grand-stand view of the laws of nature

Most of us remember having to learn dozens and dozens of laws in the science classes at school. Actually, all of them, including even the most arcane, seem to follow from a few basic ideas. These ideas are about the symmetry and isotropy of the 'arena' in which the laws of physics are played out.

The plethora of laws that we learnt at school was about levers, pulleys, floating bodies, laws of motion and so on. Others were about electricity and magnetism. Then there were others dealing with elasticity, acoustics, optics, thermodynamics, chemical valence, acids and bases etc. That wasn't the end of it — there were also laws of gravity, astronomy, and the quantum theory on the one hand, and Mendel's laws of heredity and all the nuanced laws of biology and natural selection.

Are all these laws as disconnected and as contrived as the laws that govern the local municipality? Are they mere products of the paradigms that shaped the social milieux of the scientists who created these laws — with Ptolemy replaced by Copernicus, Galileo and Newton, to be in turn replaced by Maxwell, Planck, Einstein, then Bohr, Schrödinger, Dirac, Heisenberg, and others, and then a cascade of newer names? The situation is in fact very different. The laws of science — indeed, all aspects of Nature — flow from just a few laws. The laws that we learn at school are a vast vegetative growth of innumerable branches that spread out from a 'single tree' of concepts. This single set of concepts has a universal applicability in that the laws found for one extreme have a direct bearing on the other extreme. Thus, theories of subatomic particles have a direct bearing on cosmology and large-scale phenomena. Hence it should not be surprising that they have a bearing on every scale of length and time — e.g., in the realm of human time scales and length scales.

Amazingly enough, today we are almost in a position to say that there is an underlying grand structure behind all laws of nature. This presents itself as symmetry principles that are intuitively easy to accept. That is, we first look at the principles that govern the laws of physics themselves. Some of these are so powerful that we can virtually 'derive' many of the laws of physics from them. In the following we present these ideas with little or no mathematics, and leaving out some of the usual qualifying clauses that make things look complicated. A few basic ideas are at the bottom of it all.

1. The principle that there is no privileged observer — all observers see the same laws of nature. This is a type of symmetry principle, in the sense that if you replace one observer by another who may be moving at a different speed, you still get the same laws. That is, it does not matter if you are on a rocket ship, on earth, or on a piece of lava rock hurtling off a volcano (cf., Sec. 2.5). In the final analysis, you would be able to conclude that the same laws of nature apply to you as well as to other observers. In the grand scheme of things, conservation principles, such as the conservation of energy, are consequences of symmetries.

2. Additional aspects of physical reality occur in certain inner spaces that are not directly accessible to us, as is typical of quantum phenomena. These spaces are known as Hilbert spaces. The laws of physics remain invariant under rotations of the coordinate systems needed to describe such aspects of reality. Invariance under such rotations leads to what is known as Gauge Symmetry.

Let a particular measurement of a physical system be taken from a position given by the vector $\vec{q}$. The general principle of covariance says that the laws of physics are the same for any origin or orientation of the vector $\vec{q}$. In abstract function spaces, orthogonal axes are replaced by an orthonormal basis set of functions. Here we consider a vector of functions, viz., $\psi(\vec{q}) = \phi_1(\vec{q}), \phi_2(\vec{q}), \ldots$, instead of $\vec{q}$. The laws of physics remain invariant under transformations where the norm $|\psi(\vec{q})|$ is conserved. This is equivalent to changing a wavefunction ψ to $\exp(i\theta)\psi$. The rotation of the vector ψ changes the phase factor $\exp(i\theta)$. If θ is fixed it is called a global gauge transformation, while if θ depends on local coordinates, we have a local gauge transformation.

3. Other yet to be specified symmetry principles and conservation principles.

The requirement that a property holds for a unique 'globally applicable' coordinate system (e.g., global gauge invariance), or that it applies to any coordinate system set up in any locality (e.g., local gauge invariance), links the 'bird's eye view' (BEV) of the world to the worm's eye view (WEV) of the world.

The attentive reader might note that the first principle can be absorbed into a generalization of the second since classical phenomena are included in the quantum description. However, it is useful to maintain this redundancy and keep the 'no privileged observer' rule and the 'no privileged coordinate system' as separate issues, since the first principle is what is used traditionally in the Theory of Relativity, while the second one is needed in the Quantum Theory. The 'observer' invoked in the theory of relativity is some one who receives information about the system without disturbing it. In contrast, quantum mechanics deals with subatomic objects like electrons that are disturbed by even the smallest possible disturbance (perturbation) which cannot be made arbitrarily small — the disturbances also come as quanta. Thus the 'observer' in quantum mechanics participates in setting up the system by the very act of observing the system. All this suggests that we retain, at least for the present, the distinction between the 'no privileged observer' and the 'no privileged coordinate system' in discussing symmetries.

It turns out that fundamental symmetry principles provide a common scaffolding to the laws of the quantum theory, and the theory of relativity [129, 4].

The way to make 'sense' of all the elementary particles is embodied in the 'standard model', associated with Steven Glashow, Abdus Salam and Steven Weinberg, and many others who contributed to 20th century particle physics. The standard model makes much use of symmetry, as well as symmetry breaking to give order to the particles and their fields (see Table 1.1). In the grand program of a unified theory of gravity and quantum mechanics, envisaged under supersymmetry, string theory, and perhaps involving new formulations of gravity, we may be able to present all our discussions using symmetries defined in suitable abstract structures that play the role of space-time.

However,without further ado we need to clarify what we mean by symmetry.

3.2 What do we mean by symmetry?

Perhaps some reader may feel that we are using the word 'symmetry' in a way different to common usage. We have a sense of symmetry that we use in art, architecture and sculpture. The concepts of symmetry used in science extend these intuitive ideas to a level of grandeur that cannot be fully appreciated without the aid of mathematics. Nevertheless, the basic ideas are the same.

We should not forget that the Greeks also used their ideas of symmetry in their Cosmology. Indian cosmology which predated the Greeks, invariably referred to the universe as *Charkravāla*, which implies a cyclic property as well as a circular cross section. The sphere and the circle were the perfect geometric figures, and Greeks assumed that the celestial bodies moved in circular orbits around the earth positioned at the center of the celestial sphere. Similarly, celestial objects like the moon had to be perfect spheres — as pointed out by Father Clavius in his arguments against Galileo. When Galileo claimed to see mountains and valleys on the moon, the fault must clearly be in Galileo's telescope!

If we consider a two-handled Greek vase with the two sides identical, we can say that the vase is symmetrical. Similarly, the vase in Fig. 1.1 is symmetrical as the left and right sides 'match'. In effect, if you turn the vase by 180 degrees so that the left is interchanged with the right, the object looks identical to what it was before this act of rotation. This is precisely the idea of symmetry used in mathematics and physics. The act of rotating the vase by 180° is called a 'symmetry operation'. Let us denote it by R. The vase has a two-fold symmetry. An equilateral triangle has three-fold rotational symmetry, while a regular hexagon has six-fold rotational symmetry. A symmetry operation is an action that we can imagine being carried out on an object, and at the end of the operation, the object is indistinguishable from its original form. Thus we may denote the symmetry operation for the vase by R_2, while that of the perfect hexagon by R_6. Another possible symmetry operation is reflection, where we determine if an object is identical with its mirror image. If the reflection is identical with itself the object is said to be invariant under the operation of reflection. The mirror image of the left hand looks like a right hand, and hence does not remain unchanged under reflection. Objects that are not invariant under reflection are said to have a handedness or *chirality*. Most naturally occurring sugar molecules have a right-handed chirality (hence we use names like 'dextrose'), while neutrinos, and the amino acids coded by DNA have a left-handed chirality. Since the neutrino is a very elusive, utterly small elementary particle, clearly we are already getting into deep abstraction. This emphasizes the need for a more systematic approach other than the simple act of looking into a mirror for testing reflection symmetry.

Once the mathematical operations for rotation, reflection etc., are identified, they can be used at any length scales or energy scales, be they for classical particles, Dirac electrons or neutrinos. Thus, while Shapin and others may claim within their sociological point of view that scientist are constructing facts [195] or inventing language when dealing with new phenomena, scientists are in fact using the over-arching language of mathematics which apply universally at arbitrary scales of length and energy.

Symmetry operations like reflection or rotation of an object ultimately leave the object in the same place, and are called 'point symmetry' operations, while the group of operations which requires moving (translating) the object is called the 'space group' of symmetry operations. Instead of a vase, we may have a brick (a polyhedral shape with dimensions a_1, a_2, and a_3 along the x, y, z directions), and ask if its replicas can be used to fill up 3-D space without leaving any gaps. This is possible and we simply move the brick by a_1, a_2, and a_3 along x, y, z directions at each step. Clearly, any arbitrary brick shape (crystal 'unit cell') will not fit without leaving gaps. In two-dimension this is the problem of tiling a floor, and clearly pentagonal tiles would not fit together.

A *tiling pattern* which repeats itself in a regular manner is a periodic pattern. In one-dimension, the tiling pattern *ABABABAB*, using two tiles *A* and *B* is periodic, since the pattern repeats itself. Similarly, $(ABB)(ABB)(ABB)\cdots(ABB)$ is a periodic pattern.

Many naturally occurring structures and processes are non-periodic. A famous non-periodic 1-dimensional tiling pattern is the Fibonacci scheme, $(A)(B)(AB)(BAB)(ABBAB)$ etc., where each new generation of tiling is the 'sum' of the two preceding generations taken in order: e.g., $(B)+(AB)\rightarrow BAB$. That is, the offspring is 'sum' of the two parents. This sequence originated from an attempt by Fibonacci (13th century) to establish the 'laws of breeding' of successive generations or rabbits. However, Goonatilake states that the sequence was known and used by Pingala, the author of a text on Sanskrit prosody in the second century BCE [89]. Kepler, having discovered the periodic regularities of planetary motion, was also intrigued by these regular but non-periodic sequences found in the arrangement of flower petals, branching of plants and sea-shell patterns. Molecular-beam epitaxy (MBE) of crystal-growth deposits individual atomic layers on surfaces. Thus Dharma-wardana *et al.* studied MBE-grown artificial 1-D Fibonacci crystals and their Raman scattering [62]. Non-periodic 2-dimensional tiling schemes were studied by Roger Penrose, and are known as Penrose tiles [92]).

On counting the number of possible symmetrical schemes available for packing particles into 3-D space, we find only 230 different 'regular' crystalline forms. Some examples are the cubic crystalline structure of rock salt, the 'diamond' structure of Carbon and Silicon crystals, and the 'hexagonal cubic' structure of ordinary ice. While we can actually draw and count the different crystal forms in three dimensions, we cannot even conceive of the possible crystal forms in, say, 4-dimensional space. That is why mathematics is called in when intuition becomes helpless. The theory of symmetry groups can be invoked to show, with some help from a computer, that there are 4895 possible crystalline hyper-solids in 4-dimensions. Even the most endowed artist might not conceive of them, or their undoubtedly intriguing projections on ordinary 3-dimensional spaces.

3.2.1 *Crystal defects and information — clay crystals*

A crystal is an ordered array of atoms, where the repeating unit is the unit cell. The formation of a crystal involves a reduction in symmetry, i.e., *a symmetry breaking* of isotropic space. Crystal growth may be thought of as a replication process, but occurring with greater simplicity than the replication of living organisms. However, a crystal, or any other non-living object at room temperature T attempts to reach a state where the free energy F is minimized. This free energy consists of its internal energy E, and a 'disorder energy' TS, where S is known as the entropy of the system. The increase of entropy decreases the free energy (see Chapter 9) which is of the form $F = E - TS$. The disorder is measured by the number of possible arrangements of its parts available to it. Thus a card pack with two kind of cards has more arrangements than a pack with just one kind of cards. Similarly, if there are defects (e.g, vacant lattice sites in the crystal) in addition to filled sites, there are many ways of arranging the vacancies among the filled sites, and hence increase entropy. Similarly, magnetic materials at very low temperatures have atoms with spins arranged in parallel, defining a direction of magnetization. However, at finite temperatures, these break up into domains, where some regions have reversed magnetization which costs internal energy E. However, if the decrease of internal energy E from the formation of defective regions is offset by the gain in entropy S, the free energy F can be reduced on forming point defects (e.g., vacant lattice sites) and/or domain structures (see Fig. 9.3).

Usually these processes do not reach equilibrium. Instead a metastable state is formed. A crystal at room temperature would move some of its defects or atoms to its surface and create vacancies inside the crystal, or on its surface, decreasing its free energy or reaching some metastable state. Natural crystals have many other types of defects, e.g., mismatched crystal planes, dislocations etc., arising from a combination of thermodynamics and growth kinetics. Such defect structures are often a 'fingerprint', or 'information imprint' of the initial growth conditions of the seed crystal. This information is in some sense similar to 'mutations' of the original crystal template. As the crystal grows, these defects also can be replicated, thus 'preserving the information' present in the initial crystal substrate, as seen even in molecular-beam epitaxy. In geological formations, a given type of crystal growth may influence the local environment sufficiently to promote its further growth. Clay crystals may form as sediment, and encourage further sedimentation. Thus we have the concept of inorganic selection processes occurring in nature, as part of geological structural growth.

Cairns-Smith [13], a chemist at the Glasgow University, noted the similarities of the above process to the steps needed for primitive cell replication, and proposed in 1968 the *clay theory* of the origin of life. Simple, inorganic selection processes are suggested to lead to a 'natural selection' for clay crystals (silicates) which trap certain types of prebiotic molecules to the clay surfaces. These surfaces act as catalysts for complex organic reactions which would otherwise have low probability of going to completion. If such processes lead to the formation of sufficient amounts of replicating and self-catalyzing molecules like RNA, then the clay scaffolding becomes redundant, and a 'Genetic Takeover' of the replication is envisaged to happen, thus launching further evolution into DNA and protein-based life as we know it. Theories of the formation of prebiotic matter and the evolution of life will be taken up in Chapter 10.

3.3 Coordinate transformations and Galilean symmetries

Just as we can carry out symmetry operations on objects, we can carry out various symmetry operations on the laws of physics. For example, if you rotate an object, would it still obey Newtons laws? The answer is yes.

Since the laws of physics deal with, say, the positions (e.g., x, y, z coordinates) of particles, these can be subject to various symmetry operations, while holding every thing else constant. Reflection puts the image on the side 'opposite' to you in the mirror, changing the depth d to $-d$. Thus x, y, z, reflected in a mirror in the $x-y$ plane will change to $x, y, -z$. Reflection symmetry is a form of 'point' symmetry. Translational 'symmetry' arises if space is the same everywhere. In that case, if we move the particle at x, y, z to some other point $x+a, y+b$, and $z+c$, it would seem as if nothing had happened. Then we say that the system is invariant under a 'position translation'. If we changed the time t to $t+d$, and if everything still seems to remain same, then the system is invariant under 'time translation'.

It can be shown that Newton's laws of mechanics, presented in *The Principia*, are invariant under space translation, space rotation and time translation. But there is something even more stunning.

Let us first discuss what is known as a *Galilean transformation*. This was presented by Galileo (Fig. 3.1) in a reply to Francesco Ingoli, the secretary of the 'congregation for the propagation of the Faith', who published [113] an attack on the Copernican system. Galileo's reply itself was never sent to Ingoli, but was indirectly addressed to the Pope, Urban VIII himself. We discuss this in a later section, where we examine why these ideas were deemed dangerous to the very foundations of 17th century society. After all, a modern American scientist working on super-symmetry would not address his scientific essay hoping that it would be read by who ever sits in the Oval Office, the modern equivalent of the classic power of the medieval church.

Fig. 3.1 The laws of motion discovered by Galileo were extended and formalized by Newton who was born on the same year that Galileo died. Galileo used his telescope to provide wide-ranging support to the theories of Copernicus and Kepler. Newton was able to unify the behaviour of terrestrial and celestial matter under one set of laws using his theory of gravitation.

We consider an 'observer' who is at the position (x, y, z), and another observer B who is moving at a constant speed v in the x-direction, such that his position becomes $x' = x - vt, y' = y$ and $z' = z$. Their clock times are the same so $t' = t$. This change from $x \to x', y \to y'$ and $z \to z'$ is called a 'Galilean transformation'. The laws of physics are found to be the same for these two observers who are moving at a constant speed relative to each other. This is known as *Galilean relativity*. It says that the laws of physics cannot depend on an absolute velocity.

A law of physics is a relation between input data and predicted results, that has always been found to agree with experiment over a given range of energy and length scales, and important enough to merit the name *law*. Given some input numbers (e.g., initial position and momentum of a particle) denoted by the x, p, let the predicted velocity at a given time be the set $y(t)$. Then the law may be written as $y(t) = f(x, p)$. When a symmetry operation (example, a rotation) is carried out, let x, p transform to x', p', and y transform to y'. If the law is invariant under the symmetry transformation we should have $y'(t) = f(x', p')$.

Now consider Newton's second law, which says that the acceleration a of a mass m is proportional to the applied force F. Suppose we did not know this law, but ask for the simplest form of the relation between a and F that would be the same for both observers A and B. Amazingly, it turns out that it is just Newton's law that does this. That is, *the principle that 'there is no privileged observer' under Galilean transformations leads to physical laws whose simplest forms recover Newtonian mechanics*. In contrast with Galileo and Newton, Aristotle would require a force F to produce a velocity v. That is instead of $F = ma$, where a is the acceleration, Aristotle's law of motion might be $F = kv$. This does not have the symmetries found in Newtonian mechanics.

The 'no special observer' point of view of the classical mechanics of Galileo and Newton may be contested by someone who might point out that if the observer B were himself accelerating, then Newton's laws do not seem to be the same when viewed by A and the accelerating observer B. The coordinate systems of observers where Newton's second law seems to fail are called 'non-inertial frames of reference'. If the observers are moving about with constant relative velocities (as in the Galilean transformation), then such observers are called *inertial observers*, and Newton's laws remain the same for such observers. We can bring the non-inertial observers back into the club by being a bit more artful. Consider a person in an elevator which is suddenly slowing down. He feels a force exerted by the floor of the elevator acting on him. Such forces are often called 'fictitious forces'. The 'centrifugal force' that one feels when riding on a merry-go round is also often called a fictitious force. However, if they are included in the analysis, then the applicability of Newton's laws for all observers, inertial or non-inertial, is restored. We must not forget that many concepts which are at first introduced as 'fictitious', turn out to be not so fictitious once we understand them. Negative numbers, irrational numbers and 'imaginary' numbers were introduced at various stages of the history of mathematics to maintain the familiar laws of arithmetic and to accommodate new types of solutions to arithmetical problems. We no longer doubt that imaginary numbers have as much reality as 'real' numbers!

However, it is interesting to note that the *strong program in Sociology* would probably claim that these concepts were a part of 'the social reconstruction of knowledge due to the changing social conditions in European society', just as statistical methods in mathematics were claimed to be 'a social instrument of the reformist English middle class' by MacKenzie [136]!

3.4 Before Galileo and Newton

The Galilean-Newtonian view of the world has become second nature to us. However, it is important to compare it with what was held to be true before these renaissance men of science shook up the intellectual firmament of their day. The revolution associated with Copernicus, Kepler and Galileo is perhaps far more important and cataclysmic than anything brought about by General Relativity (GR) or Quantum Mechanics (QM), at least at the human scale of things.

While QM deals with the subatomic world, relativity deals with large length scales and properties of bodies moving at speeds close to that of light. The Galilean revolution dealt with matter at the human scale of things. Galileo's first great work [79], entitled *Dialogue on two world systems*, (1632) was an attempt to popularize the Copernican system that he had confirmed using his telescopes. But the scientific basis of the Copernican system lay in the laws that govern the motion of ordinary objects. Galileo had to go on his knees in front of the Holy Congregation of the Inquisition and recant, saving himself from torture and death. However, his 'crime' was sufficiently serious that he had to remain essentially under house arrest (to the end of his life). He could continue to work in isolation as long as he did not attempt to advance his heretical position in any manner what so ever (In Latin: *ominio relinquat, nec cam de caetero quovis modo teneat, doceat out defandat, verbis aut acriptis*), as already decreed in the first trial of Galileo in 1616. However, Galileo's second great work [78], entitled *Discourse on two new sciences* (1638), written while under house arrest and printed in Holland, (i.e., outside Catholic Europe), dealt with laws of moving bodies and really set the intellectual foundation for the Copernican system. This however was not understood by the Church fathers, as the discussions about the motion of bodies on inclined planes, the parallelogram laws of composition of velocities and accelerations seemed far away from the polemics of the previous work of Galileo.

In reality, the study of the dynamical laws of matter was a politically even more subversive activity as it undermined the logical bias of the belief system of the day. This is why the Galilean revolution was an enormous cataclysmic event, affecting every level of social and cultural organization. General relativity and the Quantum theory, while involving far reaching conceptual as well as technological changes, do not have the same implications at the level of *cherished beliefs*, as was the case with Galileo and Darwin.

The 'standard' view point that most people adopt in their daily lives is that we have material objects (cars, trains, people, air molecules, nuclei) occupying space, and that these objects move around in this objective 'space'. As Heisenberg is supposed to have said, 'Space is blue and birds fly in it'. This space is perceived to be independent of each person; it is the arena where the 'external world exists'. Objects move freely in this space unless they are subject to forces which may accelerate or retard them, as stated by Newton's first and second laws — laws that are already fully discussed in Galileo's 'Discourse on two new sciences'. Newton believed that there was an 'absolute space', and hence he distinguished between relative motion and absolute motion with respect to this absolute space.

As already noted. the principle that the *laws of nature remain the same for all observers*, as long as they are moving at constant velocities, is known as *Galilean relativity*. It is consistent with the idea of absolute space. Galileo presented his analysis of Galilean relativity in his 'Reply to Francesco Ingoli', the Secretary for Propaganda in defense of the Faith. Ingoli's arguments were of three sorts: (i) astronomical, (ii) philosophical and (iii) theological. Galileo limited himself to replying only to the first two, and made ample salutations to the Faith. Nevertheless, his response was regarded as too ironic and irritated the Pope, Urban VIII. In criticizing the Copernican view, Ingoli had argued that when a stone is dropped from the top of a mast of a stationary ship, the stone falls vertically to the bottom of the mast. However, if the ship were moving, the stone should fall a bit behind the mast, due to the ship's motion while the stone fell. So, if the earth were moving, why would falling bodies fall vertically? Galileo points out that if you are inside a ship there is no way to know if the ship is moving or not, as long as the ship does no accelerate or come to an abrupt change of speed. In effect, the laws of motion are the same for an observer who is at rest, or moving at some constant speed, measured relative to absolute space.

Even the idea of 'space' was not a commonly accepted notion. The pre-renaissance view of reality, at least in the West, was dominated by the teachings of Aristotle as interpreted by the Church Fathers. Plato and Aristotle believed that 'extension' was the 'essence' of matter, and that there was matter everywhere — there was no void, or 'space', as we understand it. (This view is also found in Descartes and other renaissance thinkers who were not primarily scientists). This is why Robert Boyle's pump for producing a vacuum was of such great significance. And yet, Shapin and Schaffer [195] have aligned it to form the basis of a skeptical sociological commentary on the very process of science.

In contrast to the Aristotelian tradition, Indian philosophers who were contemporaries of Heraclites, had considered the existence of 'space' as a separate entity for holding the aggregates made up of the four elements, earth, water, fire and air. The Greek Atomists too had posited a 'space' where atoms moved.

Thus, if there were no void, there was no reason for matter to move, and the state of rest was considered to be the natural state of matter. Motion has to be initiated by a first cause — the Creator who is the unmoved mover. Things moved, or *things happened so as to satisfy a final purpose*. Non-material objects moved only to the extent that they partook some of the essence of living objects. The only natural movements allowed for inanimate matter were motion down towards the earth, and motion up towards the heavens. The governing principle of the world was *teleology*. Thus when Galileo explicitly stated a very different view (held secretly by many scientific men of the era) in his 'Dialogue on two world Systems', and claimed that the very earth moved, he uttered a heresy for which Giordano Bruno had already been burned at the stake. Bruno's 'Dialogues' had been published in 1584, where he had argued that heavenly bodies moved through an empty ether, under their own impetus. Bruno also held that Space and Time were infinite and there was no place for a creation or a last judgment.

In this context, it is interesting to note that the Greek Atomists (Leucippus and Democritus), and their followers like Archimedes and Aristarchus had anticipated the renaissance scientists in many ways. They believed in a void space where atoms moved on their own, mechanically, i.e., without an end purpose. Their approach went clearly against the usual traditions of Greek thinking — the search for geometric symmetry within an Orphic mysticism. They presented a description (reminding one of the kinetic theory of gases) of atomic matter, and even tried to construct a theory of turbulence and vortices to explain the origin of more complex material structures. Michel Serres [191] suggests that the Greek atomists invented a geometric form of the differential calculus (in their concept of the *clinamen*), and examined the question of how orderly motion could become chaotic or turbulent. It is of course well known that Aristarchus had proposed a complete Copernican theory of the Solar system, and that he too, two millennia before Galileo, was in danger of indictment for the impiety of making the earth move [184]. Why then did Galileo and Newton succeed, while the Atomists failed to sustain their science? The crucial difference arose from the slow accumulation of irrefutable empirical data. The gradual improvements in technology, acquisition of new experimental tools and developments in Mathematics (in the hands of Arab mathematicians), played a decisive role in the emergence and acceptance of new ideas. Greek astronomers like Aristarchus and Eratosthenes had no telescopes to support their claims. Unlike Kepler, Galileo had not realized the importance of

Tycho Brahe's observational work [84]. Galileo's own discovery of Jupiter's moons using his telescope, and the agreement in the variations of the brightness of Mars and Venus as they approach and recede from the earth, played a critical role in convincing him of the validity of the Copernican model.

The intellectual irony of it all is that, from a modern point of view, the debate over heliocentricity and geocentricity based purely on the observed *relative motion* becomes irrelevant. The equations of physics can be referred to any center, the earth, sun or moon, by simple coordinate transformations and the calculations remain just as valid. One may even reformulate a theory of epicycles *a la* Ptolemy as a method similar to a Fourier expansion of a time dependent process — a perfectly acceptable mathematical approach. However the simplest calculations are obtained for the heliocentric model. The choice of any other center of coordinates could give equally accurate results, but with more lengthy calculations. On the other hand, Galileo's optical observations (phases of Venus etc.) provided compelling evidence for the heliocentric model that was brushed away by those who refused to believe that the telescopes showed something real.

The two centuries that followed Newton led to an explosion of scientific and industrial activity, and spawned experimental developments in electricity and magnetism. This paved the way for three momentous contributions to human thought that came about in the 19th century. These are: the Theory of Evolution, Thermodynamics, and Maxwell's theory of Electromagnetism. Maxwell's theory is embedded with deep consequences that were to lead to the theory of special and General Relativity. Maxwell's equation's provided the model and paradigm for Quantum field theory and the current 'standard model' of particle physics. Maxwell's theory is the first theory of physics, which displays 'gauge symmetry', and a four-dimensional concept of the world. However, before we engage ourselves with the implications of 'gauge symmetry', we need to look into the concept of teleology, i.e., the theory which claims that physical processes are governed by a principle of final purpose. We do this in the following chapter, and close this discussion by a glance at one of the modern critics of Galileo.

3.4.1 *Modern critics of Galileo*

As an extreme example of a modern critic of Galileo, we examine Feyerabend's claim [72] that Galileo advanced his program by trickery and propaganda. Feyerabend contends (see his Ch. 13) that even today the original indictment of Galileo by the Church needs no revision. He also claims that Galileo had to proceed counter-inductively, i.e., use hypotheses which contradict well-confirmed facts.

In our view, Feyerabend perhaps willingly misrepresents the process of scientific discovery, and does not appreciate the role of newly revealed scales of length in the mechanics of discovery. We leave aside the polemical discussions in the text, and look at his Ch. 7 where Feyerabend offers to 'look at the situation from a more abstract point of view'. Here he asserts the existence of two paradigms regarding the nature of motion during Galileo's time.

- *Paradigm I* : This is the orthodox view of the 17th century for the motion of compact objects in stable surroundings. Thus the motion of a deer observed by a hunter is understood within this paradigm as absolute motion. Within this picture, the vertically falling stone *proves* that the earth is at rest, while if the earth were in motion, the stone would be expected to fall with an oblique motion.
- *Paradigm II* : This is the view that Galileo wishes to introduce to cover all phenomena. It forms the experiential basis for considering that *all motion is relative*. Thus a vertically falling stone *proves* that there is no relative motion between the starting point of the stone and the earth. However, even if the earth were in motion, no relative motion between the starting point and the stone exists.

Galileo realizes that both in paradigm I and in paradigm II, the earth is indeed at rest *with respect to the observer*. He realizes that the earth is in fact like a big ship, moving in space. This *vision* provides the basis for applying the *same model* to cover phenomena in both types of situations associated with the observed motion. The model construction (cf., Sec. 2.2), based on a fit $f(x)$ to the data points in the domain of paradigm II is now extended to all points in the domain of Paradigm I as well, with the same model, i.e., $f(x)$. It is the insight of a great scientist that does the model building, enabling him to see the connections between things which are seemingly contradictory or unconnected. There is no 'counterinductive' principle at play, as claimed by Feyerabend, and quite the contrary is evident. Feyerabend claims that 'an experience which partially *contradicts* the idea of the motion of the earth is turned into an experience that *confirms* it ...' (Feyerabend's italicization). The resolution of the alleged contradiction involves no trickery, and occurs transparently by simply applying the same $f(x)$ to all x.

It is interesting to examine Feyerabend's critique of Galileo's telescopic observations. Galileo had little experience in observational astronomy and so he had to learn. His observations of the moons of Jupiter etc., may have been initially somewhat problematic. However, the highly experienced Fathur Clavius, the Papal astronomer, easily confirmed Galileo's observations. Another observation of Galileo using his telescope was the change in the brightness of Mars and Venus as they approach and recede from the earth. The naked eye shows virtually no change, whereas the telescope (which collects more light on the surface of the lens which is bigger than the eye) showed a variation in agreement with the Copernican theory. Galileo accepted that the eye was highly unreliable in dealing with celestial objects. The moon looked very large and bright when near the horizon, but became small when it is at the zenith. A good scientist is some one who can select good data from a set of data which may contain bad data. The optical observations of Galileo were often poor, due to fringes and other problems of early telescopes. But he recognized the importance of the telescopic data for the brightness of Mars and Venus. The Galileo scholar Ludovico Geymonat [84] points out that although others had made telescopic observations of these planets, Galileo had the genius to recognize the stark importance of this confirmation of Copernican theory. Feyerabend (Ch. 10, p. 108) however asserts this to be a 'harmony between two interesting but refuted ideas which Galileo exploits in order to prevent the elimination of either'. While the orthodox Ptolemaic theory could not explain the nearly identical brightnesses of Mars and Venus as seen visually, it failed even more miserably with the new telescopic data, the newly observed phases of Venus etc. Of course, at first one could question the reliability of telescopes for non-terrestrial observations (as does Feyerabend).

Feyerabend claims that Galileo's mechanics and astronomy need the support of *ad hoc* hypotheses if we are to understand natural phenomena or experiments. Thus he points out ([72] Ch. 7, p. 77) that *friction* is introduced in a circular way to save the law of inertia, and not empirically, since experiments on friction were done only in the 18th century. Thus, when it suites him, Feyerabend demands that experiment precede theory, whereas there are ample examples of theory providing the cue for experiments. Feyerabend does not object to Aristotelians using theory without experiment. In Sec. 2.3.2 we already explained how the reductionist program naturally incorporates the treatment of such boundary conditions within a theory. They are not at all *ad hoc* steps, but simple, essentially universal assumptions that can be legitimately carried out in setting up a boundary between two regions in a holism. The mechanics of Galileo already contained the theory of forces and interactions, and friction is a straight-forward modeling of such interactions between a particle and a bounding surface. If such deductions augmented by modeling the interactions between the reduced system and the environment are not allowed, then Feyerabend is leaving no further room for investigating Nature.

3.5 Subatomic symmetry

In this section we revisit some ideas that have been alluded to and taken up in other sections of the book, within the perspective of this chapter. In Sec. 3.3 we learnt about the symmetries inherent in the world as understood by Galileo and Newton. In subsequent chapters we will see how the two principles: (i) the laws of Nature should be the same for any observer, and (ii) the constancy of the speed of light with respect to any observer, led to Einstein's ideas of relativity. A third principle which is known to be implicit in nature is the 'principle of least action' (Sec. 3.6), obeyed by light as well as by moving bodies.

In learning about the way light travels in space, we come across a very deep idea known as *gauge symmetry*. This is discussed in Chapter 4. This is associated with the fact that the observed phenomena are independent of the overall phase ('the phase factor') of a wave-like disturbance, e.g., a light wave. This is sometimes expressed by saying that the potentials of electromagnetic fields have a 'gauge' freedom. The observed phenomena are said to be 'gauge independent', and theories have to be 'gauge invariant'.

When observations are pushed into higher energies and smaller length scales typical of subatomic phenomena, descriptions of nature have to be done with quantum mechanics, where the use of a wavefunction, similar to the amplitude function of electromagnetic waves, becomes necessary. Such wavefunctions also carry phase factors. Thus, once again, ideas of 'gauge invariance', or gauge symmetries associated with the phase factors of wavefunctions come to the fore. The interactions of subatomic particles (Sec. 1.2.1) are also found to obey gauge symmetries. The phase factors behave not only as scalars (i.e., ordinary magnitudes), but also as operators. That is, they represent physical actions (like rotations, movements). Thus the outcomes of such actions *depend on the order in which the actions take place*. Such 'non-commuting' operations are very common even in everyday life. They are essential to the quantum theory. For many types of operations, say A, acting before B, gives an outcome different to B acting before A. If you have zero cash in the bank, withdrawing money (A) and depositing money(B) have different effects for A before B and B before A. Such processes where $AB \neq BA$ are known as 'non-Abelian'.

$$AB - BA = C \tag{3.1}$$

For Abelian operators, the commutator $C = 0$, and the two operators are said to *commute*. Alternatively, especially in the philosophical literature, the two operations A and B are known as 'non-contextual' operations.

The gauge symmetries associated with subatomic particles like quarks are found to obey rules of non-Abelian operators. As we discussed in Sec. 1.2.1, the non-Abelian two-fold symmetry known as SU(2) was used by Heisenberg to understand the existence of many nuclear states and short-lived mesons. Gell-Mann and Ne'emann used an eight-fold non-Abelian symmetry known as SU(3) to explain how a set of 8 baryons are all manifestations of such a symmetry class (Sec. 1.2). In effect, the Standard Model of particle physics, which describes the basic elements and interactions of nature reveals the deep simplicity and symmetry inherent in the world as found by experimental physics. A more detailed exposition of these ideas are found, for example, in Lederman and Hill [129]

3.6 The principle of least action and the simplicity of nature

Some twenty years after Galileo's death, Pierre de Fermat in Paris had discovered, in January 1662, an extremely interesting principle in Optics. Fermat attempted to explain the path followed by a ray of light when it traverses two different media — for example, a ray of light going through water, and then through glass. The Dutch physicist Willebrod Snell had already (1661) shown how the ray of light undergoes refraction at the interface, and travels at a steeper angle in the denser medium. However, Fermat's reformulation of the problem had an almost mystical, Aristotelian flavour. Fermat showed that the path traversed by the ray of light was in fact the shortest path, requiring the least time, given the different speeds of light in the two media. It seems as if the ray of light intended to get to its destination by *a principle of least time*. It was already known to the Greeks that the law of reflection of light obeyed a minimum principle. That is, if the incident and reflected angles were not equal, but were different, the distance and time of travel would not have the least possible value.

Two centuries later Hamilton showed that Newton's laws of mechanics may also be stated in terms of a *Principle of Least Action*, and provided the technical definition of action that has to be used in working out the motion of any object, within this seemingly *teleological* approach. Of course, these two centuries produced the less complete attempts by Maupertuis, Jacobi, and d'Alembert that preceded Hamilton's formulation of the Principle of Action. Fermat's principle was a particular case of this more general principle. At first sight, it seems that in this approach a place has been found for purpose and intent. Such interpretations have spawned much philosophical debate. It has been used in support of panpsychism, i.e., the belief in an intrinsic intelligence even at the most elementary level. Panpyschism could be regarded as a more sophisticated form of animism. Such thinking is partly a result of the inadequate exposition of the physical and mathematical concepts leading to the principle of least action.

3.7 How does a particle know the best path?

When an intelligent and thrifty individual, or a typically poor physicist, wishes to take a flight from one city to another, s/he would research the available options and *choose* the least expensive itinerary. It seems that mechanical motion in nature already incorporates such a principle of thrifty intelligence — a teleological principle. We explain this apparent mystery here.

The generalization of Fermat's principle to mechanics is usually known as Hamilton's principle of least action. Consider a particle (e.g., a ball) at some initial position A, and the ball is to finally arrive at the position B. There are many paths (trajectories) available for the ball. One may use a teleological language and say that the path 'chosen' by the ball is that requiring the least action — as if the ball evaluated the action for all the paths and selected the one requiring the least action! In fact, the principle of least action is not limited to Newtonian dynamics and light. The action principle, which selects an optimal trajectory, can be stated in such a way as to cover most physical processes including gravitational and electromagnetic phenomena. In 1942 Feynman [73] presented a generalization of the trajectory concept and the principle of least action to cover quantum mechanics (QM) as well (see Sec. 6.8). When dealing with the Lilliputian world of elementary particles (e.g., electrons), the particles also have to obey the wave-particle duality. The name 'particle' is used to designate a sharply localizable excitation of the quantum field. Hence particles have a definite, well-defined path, while a 'pure' wave pervades all space within the assigned boundary. This duality is resolved in QM by requiring that quantum particles can take all possible paths, including very unlikely ones. However, in QM the different paths have different probabilities for being realized. As the particle considered is made heavier and heavier, the path of highest probability for a quantum particle to go from A to B approaches the path of 'least action' used by a classical particle — e.g., a marble — moving from A to B.

Is Aristotle finally right, at least for classical particles? Is motion guided by final causes? Is there a modicum of intelligence even at the lowest level of matter, guiding it and gradually increasing as we go into more and more complex forms of matter? Amazingly, the answer is quite surprising and much more subtle than anything conjured up by the philosophers and mystics.

3.8 The local texture of the world and the simplicity of nature

We attempt a non-technical explanation of how least action turns out to be true without any intelligent choice by the particle 'selecting' its optimal trajectory. This is best done by considering an analogy. Let us think of an explorer who is lost at the summit of a mountain and night has fallen. He wishes to leave his present location at A and reach the cabin B at the bottom of the hill. He knows that the mountain is gently sloping and he wishes to get to the bottom of the mountain as soon as possible. So he simply feels the ground, and at each step follows the path of steepest descent. In effect, his path is dictated by the local characteristics of the ground at each step. This works out and he gets to the bottom cabin B. The next morning he surveys the mountain and realizes that the path he had taken in the night, without ever seeing the destination point B, is in fact the very special path that requires the least amount of action! That is, the local characteristics of the ground — the texture of the ground — at the starting point A determined the first step. And this determined the next step, and so on. There was never an actual evaluation of all possible paths and actively choosing the one giving the least action. Here we have a case of a local property of the initial point A (or any chosen point) giving rise to a global property, and this could arise in this case only because the mountain had a particularly simple, gently formed gradient. If the mountain slope were full of canyons and deep shafts, the explorer would never have made it beyond the nearest deep hole where he would have remained trapped. Furthermore, in real-life mountaineering, the path of least action may be a free fall from a cliff, a trajectory unsuitable even for a panpsychist.

The minimum principle embodied in the principle of least action is a global property that results from the local structure imposed by the laws of physics. It works only because the laws of physics have certain very simple forms. We had already noted that Newton's law of motion, which dictates that the acceleration is proportional to the applied force, is the simplest relation between force and acceleration that satisfies Galilean Relativity (Sec. 3.4) — i.e., that the law remains the same for any observer, as long as they are moving at constant velocities. It turns out that the principle of least action holds only for this simple form embodied in Newton's law. If the law had been something more complex (e.g., the acceleration is proportional to the cube of the applied force), then there would be no simple relation ensuring that the 'local properties' would lead to some global property similar to that of least action. Thus the existence of laws like the 'principle of least action' does not imply some mysterious teleology in nature related to intention or Aristotelian end purpose. It signifies that the laws of physics are the simplest possible laws that are consistent with the symmetry principles that we had already discussed in previous sections. This simple structure of the underlying laws implies that a purely local view ('worm's eye view', WEV) leads to globally minimizing the action ('bird's eye view', BEV) in carrying out the physical process. There is total consistency between the WEV and the BEV — another perfect duality! In addition to Newton's laws, we have the law of gravitation and electrostatics where the forces between bodies fall of as the inverse square of their distance of separation. The principle of least action arises from the local structure associated with these laws as well (see Sec. 5.5).

3.9 The worm's eye view — The Euler-Lagrange equations

Now we are ready for a slightly more technical account.

A moving particle, at some point x, has kinetic energy, i.e., its energy of motion. We denote this by $T(x)$. If the particle has gone up a hill in reaching the position x, then it has done work to get there and this potential energy is denoted by $V(x)$. The total energy $T(x)+V(x)$ is known as the 'Hamiltonian' of the system. The energy difference, i.e., $T(x)-V(x)$ is known as the 'Lagrangian' $L(x)$, after the famous French mathematician Pierre Lagrange. The 'action' in going along a given path is a cumulative value of this quantity, and differs from path to path. For instance, we can add up (i.e., integrate) the values of $L(x)$ as we step along a path p and estimate the total sum of $L(x)$, denote it by $S(p)$. This sum (or path integral, see Eq. 6.57) is known as the action for the path p. The optimal path is the one with the least value of this action (i.e., the principle of least action). However, this sum over the path can be replaced by a statement about $L(x)$ at the starting point, or at any other locality, without making any reference to the final end point. That is, the selection of the path by a minimum principle is mathematically completely equivalent to a purely local discussion in terms of what are known as the Euler-Lagrange equations. These equations contain various local gradients (with respect to time and space) of $L(x)$. They turn out to be a sophisticated re-expression of Newton's laws of motion. The simple mathematical structure of Newton's laws allows one to construct a *global property*, the minimum property of the action, out of the *local properties* dictated by the Euler-Lagrange equations.

3.9.1 *The inverse square laws*

Two of the most fundamental force laws that govern the world are known as 'inverse square' laws. The gravitational force due to a mass M decreases with the distance r from the heavy object, dropping off as M/r^2. The electrostatic force produced by a particle (e.g., an electron, a proton or any other charged object) of charge q also drops off as q/r^2. This is exactly what one would expect if one could imagine the gravitational field spreading out from a point if space were isotropic. That is, if space were the same in all directions, the 'force field' originating from the mass M or charge q would uniformly distribute itself in every direction. Since a spherical shell has a surface area of $4\pi r^2$, it is clear that the force would attenuate itself as the inverse square, i.e., $1/r^2$. These simple arguments suggest that two of the most basic forces of the world have a very simple structure. In fact, the force laws are often written in terms of potentials. Thus the electrostatic potential due to a charge q drops of as q/r, while the gravitational potential drops off as M/r. At the human scale of things, the only forces that we 'feel' are just these two forces, i.e., forces having an electromagnetic or gravitational origin. There is just no evidence for any 'mystical', paranormal or 'astral' forces.

The real force of this statement is that even hunger and pain, fatigue and other more complex biological effects all arise from electromagnetic forces, mediated via quantum chemical processes. All these complex effects arise from very simple laws whose structure ensures that the 'worm's eye view' of all processes, based only on a knowledge of the local forces, is sufficient to guarantee that the global property of least action associated with the beginning and end of a process is automatically satisfied. In fact, we should not attempt to understand electromagnetism or the interactions among quantum particles by appealing to the isotropy of ordinary space. We had already alluded to a more subtle principle known as the principle of local gauge invariance. It turns out that this gauge principle, (which in effect says that the laws of physics are independent of rotations in the local coordinate system in the 'Hilbert space' selected by any observer), necessarily dictates the form of the laws governing the forces among all the elementary particles of nature, (and possibly even gravitation, although this has not been established). A more detailed discussion of elementary particles, the gauge principle etc., has to be deferred till we discuss the quantum theory. However, the magic of the gauge principle is already embodied in Maxwell's vision of electromagnetism.

3.10 The 'unreasonable effectiveness' of mathematics

Many scientists and philosophers have remarked on the powerful role played by mathematics as the language of the exact sciences. The use of mathematics reveals unsuspected connections in nature and generates new discoveries and new concepts. The best scientists, from the days of the Greeks, and then to the leading figures of the Renaissance or the pioneers of modern physics talked with each other using mathematics. Today, investigations in the most advanced (i.e., well-codified) branches of science cannot be carried out without a very good command of abstract mathematics as well as computational mathematics.

The question prompted Eugene Wigner to write an article entitled 'The Unreasonable Effectiveness of Mathematics in the Natural Sciences' [227]. He wrote:

> ... The first point is that the enormous usefulness of mathematics in the natural sciences is something bordering on the mysterious and that there is no rational explanation for it. Second, it is just this uncanny usefulness of mathematical concepts that raises the question of the uniqueness of our physical theories. In order to establish the first point, that mathematics plays an unreasonably important role in physics, it will be useful to say a few words on the question, 'What is mathematics?', then, 'What is physics?', then, how mathematics enters physical theories, and last, why the success of mathematics in its role in physics appears so baffling ...

Wigner was a scientist and savant who formulated the abstract theory of symmetry in nature, made major contributions to particle physics as well as solid state physics. He justified most of his work in terms of aesthetic considerations. For him, mathematics was an exploration of beautiful abstract forms that challenged one's ingenuity. Discussing the nature of physics, Wigner points out that:

> "it is not at all natural that 'laws of nature' exist, much less that man is able to discover them. Schrödinger, in his book *What Is Life?* says that this second miracle may well be beyond human understanding ... All the laws of nature are conditional statements which permit a prediction of some future events on the basis of the knowledge of the present, ... the overwhelming majority of the determinants of the present state of the world, are irrelevant from the point of view of the prediction ..."

Thus we see that Wigner is simply stating that the laws of physics apply within the 'conditional statements' that we referred to as 'boundary conditions' of the reductionist physical model that we study. He notes that 'the overwhelming majority of determinants are irrelevant'. That is, the reduction is valuable in many circumstances and hence widely applicable.

In discussing the role of mathematics in physics, Wigner notes the secondary need of applying mathematics to do calculations, and proceeds to point out that mathematics plays a far more 'sovereign role', in that when the laws of physics are formulated in the language of mathematics, we seem to get 'more than we put in'. Wigner points out that Newton's formulation of planetary motion, based on very flimsy empirical evidence, yielded a rich harvest of new results and explanations that are demonstrated in Newton's *Principia* (see Sec. 2.2.4). Discussing a similar bounty in the quantum theory, Wigner points out how the application of mathematics to the problem produced a bumper harvest.

> ... Max Born noticed that some rules of computation, given by Heisenberg, were formally identical with the rules of computation with matrices, established a long time before by mathematicians. Born, Jordan, and Heisenberg then proposed to replace by matrices the position and momentum variables of the equations of classical mechanics. They applied the rules of matrix mechanics to a few highly idealized problems and the results were quite satisfactory. However, there was, at that time, no rational evidence that their matrix mechanics would prove correct under more realistic conditions. ... the first application of their mechanics to a realistic problem, that of the hydrogen atom, by Pauli ... gave results in agreement with experience. This was satisfactory but still understandable because Heisenberg's rules ... included the old theory of the hydrogen atom. The miracle occurred only when matrix mechanics, or a mathematically equivalent theory, was applied to problems for which Heisenberg's calculating rules were meaningless. Heisenberg's rules presupposed that the classical equations of motion had solutions with certain periodicity properties; and the equations of motion of the two electrons of the helium atom, or of the even greater number of electrons of heavier atoms, simply do not have these properties, so that Heisenberg's rules cannot be applied to these cases. Nevertheless, the calculation of the lowest energy level of helium, as carried out a few months ago by Kinoshita, at Cornell and by Bazley at the Bureau of Standards, agrees with the experimental data within the accuracy of the observations, which is one part in ten million. Surely in this case we 'got something out' of the equations that we did not put in ... physics as we know it today would not be possible without a constant recurrence of miracles similar to the one of the helium atom, ... The quantum theory of the Lamb shift, as conceived by Bethe and established by Schwinger, is a purely mathematical theory and the only direct contribution of experiment was to show the existence of a measurable effect. The agreement with calculation is better than one part in a thousand.

Thus we see that Wigner leaves the matter as a mystery, but a clarification is possible. We look at problems where the application of mathematics with the same vision as in physics has led to no progress. For example, Frederich von Hayek in his 1974 Nobel address, as well as in his earlier writings [101] pointed out that methods borrowed from the physical sciences, as applied to economics have led to a mere 'pretense of knowledge', where there is none. Attempts to understand financial markets using models borrowed from mathematical physics have not helped, as discussed in a popular work by Nicholas Taleb [205]. Leaving aside stock markets, more elementary complex systems (see Part-II of this book), e.g., simple protein molecules, have conformations that defy prediction even though such systems are governed by deterministic laws. The key to this problem was given by Poincaré in the latter part of the 19th century, when he showed that there are dynamical systems whose motion is essentially unpredictable since they follow chaotic dynamics. The trajectory is extremely sensitive to initial conditions (or boundary conditions) and no useful predictions can be made.

In contrast, in the physical sciences we choose to deal with simple reductionist models where the dynamics is well behaved. As discussed in this chapter (Ch. 3) and other chapters, most of the laws of physics can be extracted from simple symmetry principles and conservation laws. The conservation of particle number leads to the 'equation of continuity', while the need to make the laws of physics appear the same to all observers leads to Galilean or Einsteinian relativity and various 'gauge conditions'. When Paul Dirac attempted to derive a relativistic equation for the electron, he looked for a linear equation that satisfied continuity, relativity and energy conservation (see Sec. 6.10.2). Out of that effort, he not only obtained a theory of the electron, but also a multiple bonus of explaining the abstract up-spin and down-spin features of the electron, a prediction of the existence of the anti-particle of the electron (i.e., the positron), and a new vision of the vacuum state that led to the development of quantum field theory!

Thus the 'unreasonable effectiveness' of mathematics in its application to physics is probably anchored in the fact that nature at its most basic level turns out to be that of a homogeneous, space-time symmetric (isotropic) plasma with 'integrable' dynamics that obeys various symmetry principles and conservation laws. The metastable, highly compressed homogeneous initial state of the universe that gave rise to inflation and the big bang may be the 'reason' why such simplicity is inherent to basic physical phenomena.

Chapter 4

Maxwell's Magical Trinity — Electricity, Magnetism and Light

We discuss how the study of electricity and magnetism culminated in Maxwell's equations. These equations can be thought of as being a consequence of simple but very deep symmetries that underlie nature — viz., Gauge symmetry and Lorentz invariance. However, it was the experiment of Michelson-Morley, attempting to measure the speed of the earth relative to the aether that brought matters to the fore.

4.1 Electricity and magnetism

No lesser mortal than Richard Feynman is supposed to have stated that, in the far future, historians would judge that the most seminal event of the 19th century was the publication (in 1873) of Maxwell's Treatise on Electricity and Magnetism. Darwinists may not agree with Feynman on this. Also, the shine of statistical mechanics and thermodynamics born in the same era cannot be dimmed so easily. However, James Clerk Maxwell effectively planted the seed from which Relativity as well as modern gauge theories of elementary particles sprang up. Maxwell's theory stands valid even today, whereas Newton's theory underwent modification via Einstein's work applicable to relativistic energy and length scales. The quantum theory was born from attempts to explain how the energy of electromagnetic radiation depends on the frequency of the radiation. The theory of relativity was also born out of subtle questions regarding relative motion of magnets and coils in electromagnetism. Einstein's celebrated paper on Relativity published in 1905 in German was entitled 'Zur Elektrodynamik bewegter Körper', i.e., *Electrodynamics of moving bodies*, revealing the kinship of relativity to Maxwell's work.

Before Maxwell's work, there were several disparate laws of electricity and magnetism. Thus Gauss's Law related the electric field $\vec{E}$ to the charge density ρ. Ampere's law related the magnetic field $\vec{B}$ to the associated steady current $\vec{j}$, while Faraday's law related the rate of change of a magnetic field to the induced current. There was no hint of any connection with luminous light in all this, and the laws seemed more like a phenomenology than a complete science. Also, Maxwell noted that if the currents were moving electric charges, the existing equations did not satisfy local conservation of charge. Conservation of charge required that what flows into some volume element must flow out. We make a short digression to introduce the equation of continuity.

4.2 The equation of continuity

A more or less 'self-evident' principle of nature that crops up in many natural phenomena is associated with the concept that *what flows in must flow out*, unless there are some unknown sources and sinks. In effect, this is an invariance principle for the number of particles (or charges) in a given volume subject to inflow and outflow. It is called the 'equation of continuity'. If charges are present, and if they are moving, then these are 'currents', and their flow must satisfy the equation of continuity.

Maxwell noticed that there was a problem with Ampere's law, stated as 'curl of the magnetic field' is equal to the current. The curl of the magnetic field (i.e, $\nabla \times \vec{B}$) is a force which drives an electric current that has to be steady. This can only be true if the charge density ρ is constant in time. If ρ is changing with time, it can be denoted by $d\rho/dt$. Then this must be balanced by the corresponding net changes in the current components in the x, y, z directions. If we just take a one dimensional problem for simplicity, the change in the current at x is dj/dx. The *equation of continuity* says that the change in charge density at the location x gives us the charge current (if we assume that the charge us unity, this is also the equation for the particle current). Thus,

$$d\rho/dt + dj/dx = 0$$

The three dimensional form of this equation is

$$d\rho/dt + \vec{\nabla} \cdot \vec{j} = 0 \tag{4.1}$$

where $\nabla \cdot \vec{j}$ is called the 'divergence' of the current (another form of the equation of continuity is discussed in Eq. 4.7). Thus Maxwell modified Ampere's law:

$$\nabla \times \vec{B} = \vec{j} \text{ to } \nabla \times \vec{B} = \vec{j} + \partial \vec{E}/\partial t \tag{4.2}$$

and ensured that the equation of continuity is obeyed.

Maxwell's modification of Ampere's law ensures the *conservation of electric charge at each locality*. If only global charge conservation were valid, any net accumulation must be balance by a removal of charge some where else to compensate it. Maxwell's modification respected local charge conservation. Global charge conservation would ultimately require sending signals instantaneously, while it turned out that the equations of electromagnetism do not support signals exceeding the speed of light! The full implication of this was realized only with the advent of the special theory of relativity. Conservation laws such as the local invariance of charge play a crucial role in physical theories that are characterized by an interplay among symmetries, conservation laws and dynamics. Such features are found in gauge theories which are of fundamental importance.

4.3 The gauge principle

We already had a brief introduction to the ideas of symmetry in Ch. 3. A symmetry operation is an action which leaves an object unchanged after the operation. Rotating an equilateral triangle by 120 degrees in its plane would be such an operation. The equivalent mathematical operation or transformation, say where positions x, y are changed, can be applied to an equation of physics. That is, if the equation is $F(x,y) = 0$, we can ask whether it is still valid if we put the new, transformed values x', y' into the equation. If the result is the same as before, we say that the physical law is invariant under the given transformation.

In Ch. 3 we noted the translational and rotational invariance of all fundamental laws, and the relativity principle that there is no privileged observer. These already limit the possible type of laws that we can have. Indeed we found that Newton's laws were the simplest set that satisfied those invariances. In the chapter on the principle of least action (Sec. 3.6) we found a perfect harmony between the 'worm's eye view' (i.e., local properties) and the 'bird's eye view' (global properties) enshrouded in the principle of least action. This was associated with the simple nature of the laws of force (potentials) used in the theory. This would then suggest that symmetry constraints may be enough to dictate what the laws of physics should be! Maxwell's theory of electromagnetism is a 'gauge theory'. Its generalization, known as quantum electrodynamics (QED), is the paradigm for the theories of the weak and strong interactions included in the standard model.

Electromagnetism is a theory in which the behaviour of the interactions (or dynamics) is intimately wedded to a symmetry principle. As we already noted, since a symmetry operation leaves the physics unchanged, we call them invariances of the laws. In a global invariance, the same transformation is carried out everywhere, simultaneously. It is like applying the same fixed tax on every household. This does not take account of local features, e.g., that some households are rich, while others are poor. This can be corrected by having not only a tax, but also a locality-dependent subsidy to take account of local (house to house) variations in income.

In the physics context, a globally invariant theory need not be invariant under local transformations. Nevertheless, by introducing compensatory force fields (or equivalently, potentials) that transform in the required way, it might be possible to restore the desired local invariances. When that is possible we have a theory with global and local gauge invariance.

The electromagnetic field consists of $\vec{E}$, and $\vec{B}$. They can be expressed in terms of two potentials ϕ and $\vec{A}$ that obey certain symmetry operations, i.e., local and global transformations (known as gauge transformations) that leave $\vec{E}$ and $\vec{B}$ unchanged. The associated invariance is called gauge invariance. The potential ϕ, a scalar, is called the scalar potential, while $\vec{A}$ is known as the vector potential. If χ is an arbitrary function, changing $\vec{A}$ to $\vec{A}'$ and ϕ to ϕ' such that

$$\vec{A}' = \vec{A} + \nabla\chi; \;\; \phi' = \phi - \partial\chi/\partial t \tag{4.3}$$

leave the electric and magnetic fields unchanged. They are given by

$$\vec{B} = \nabla \times \vec{A}; \;\; \vec{E} = -\nabla\vec{B} - \partial\vec{A}/\partial t \tag{4.4}$$

The gauge invariant aspects of Maxwell's theory and their vast implications were only dimly appreciated in the 19th century. The quantum theory does not deal simultaneously with the momentum $\vec{p}$ and the position $\vec{q}$ of particles. Instead we need a wavefunction $\psi(q)$ or its momentum analogue $\phi(p)$ to describes the physics. Since the physics should be the same for any orientation of the observer, the same physics must be recovered from the wave function $\psi' = \exp\{i\alpha(q)\}\psi$ for the new orientation where ψ has been rotated by the angle $\alpha(q)$. This $U(1)$ symmetry ensures local gauge invariance at each location q. It turns out that this can be true if we assume that there are two potentials, viz., a vector potential $\vec{A}$ and a scalar potential ϕ having the necessary 'gauge freedom' in defining the observed $\vec{E}$ and $\vec{B}$ fields. This process exactly defines how electric charges and currents interact with these fields.

Gauge symmetry results from the invariance of the physics to rotations of the the abstract function space used in the quantum theory. This ensures that the electron charge has the same value when an electromagnetic field couples to a lepton, hadron, nucleus, ion, or an atom.

4.4 Maxwell's equations and the ether (aether)

Maxwell put together all the laws of electricity and magnetism into a compact set of equations which are now known as Maxwell's equations. Although experiments in electromagnetism during the 19th century involved charged pith balls, loadstones, coils carrying current etc., Maxwell considered the limiting case where there were none of these. The equations become very simple in this case, and Maxwell showed that even though there were no magnets or charges, there could be an oscillating electric field and a magnetic field in what is normally called *empty space*. This was called an *electromagnetic field*. Maxwell was able to calculated the speed of propagation of these previously unknown electromagnetic oscillations, and found that they traveled with the known speed of light! It immediately became clear that light is nothing but oscillations of the electric and magnetic components of the electromagnetic field. The case where there are free charges and currents is also easily worked out.

Setting the charge density ρ to zero, Gauss' law becomes $\nabla \cdot \vec{E} = 0$, and setting the current j to zero it is easy to show that:

$$\nabla^2 E - \frac{1}{v^2}\frac{\partial^2 E}{\partial t^2} = 0, \quad v^2 = \mu_0 \varepsilon_0 \tag{4.5}$$

This type of equation is a 'wave equation' for a wave with a velocity v, which the theory gave in terms of the constants μ and ε. Here $\mu_0 = 1.26 \times 10^6$ henrys per meter is the magnetic constant of free space, while $\varepsilon_0 = 8.85 \times 10^{-12}$ farads per meter is the electric constant of free space. Using their known values, the velocity of electromagnetic waves comes out as $\sim 3 \times 10^8$ m/s. This was a phenomenal discovery, identifying v as *the speed of light c*. Its value was well known during Maxwell's time from astronomical observations (see Sec. 4.6).

Maxwell's solution when the charge density and the current are non-zero can be written in symmetrical form in terms of the potentials $\vec{A}$ and ϕ, in the Lorentz gauge:

$$\nabla^2 \phi - \frac{1}{c^2}\frac{\partial^2 \phi}{\partial t^2} = -\rho/\varepsilon; \quad \nabla^2 \vec{A} - \frac{1}{c^2}\frac{\partial^2 \vec{A}}{\partial t^2} = -\mu \vec{j} \tag{4.6}$$

These can be written as a single equation for the four component object $\{\phi, A_x, A_y, A_z\}$, as discussed below, revealing a 4-dimensional structure of physical reality.

Thus Maxwell unified three great branches of physics, *electricity, magnetism*, and *light* in one stroke, into one equation. The hitherto mysterious nature of light is revealed as oscillations of coupled electric and magnetic fields. Maxwell could predict the existence of other types of undetected electromagnetic (EM) radiation at wavelengths outside the visible spectrum — analogous to light but invisible.

Such EM waves had not been found experimentally in 1873. Maxwell predicted them from a theory which started off from the 'iron-filings imagery' of Faraday. The reformulation of Maxwell's equations by Heavyside, Hertz and others led to a clearer picture of the physics, and finally to the experimental discovery of the electromagnetic waves by Heinrich Hertz. He was a Professor of Physics at the Karlsruhe Polytechnic in Germany. During 1885-89 he set up experiments where an induction coil was made to spark at one end of the laboratory. Another coil, set up further away produced a weak spark in response to the first spark. This phenomenon could be explained if EM waves generated by the first coil were picked up by the second coil. Hertz showed that these EM waves obeyed the same laws of reflection and refraction as visible light.

But what was the medium that supported these waves? According to Newton the gravitational force of the earth acts on distant bodies like the moon. This 'action at a distance' was a mystery to Newton. Propagation of sound and other 'elastic waves' was very well understood by 19th century science. In fact, the possibility of an all-pervading but invisible medium, named the *ether*, or *aether* was an old metaphysical idea which could play the role of a medium for propagating electromagnetic fields or even gravity. The lines of stress or tension in the ether could be the lines of force manifested by Faraday's experiments.

The idea of an all-pervading ether allows one to replace the 'action at a distance' phenomenon by 'action via the medium'. When two oppositely charged pith balls attract each other, this happens by elastic stresses which are set up in the ether and propagate from one pith ball to the other. At first there were many attempts to construct models of the ether itself. However, increasingly it was recognized that such descriptions were unnecessary; it was enough to recognize it as a 'field' defined by the theory itself. Many scientists of the time, including young Albert Einstein, were deeply influenced by Maxwell's field concept that could banish the 'action at a distance' picture which baffled Newton.

In Chapter 3 we noted that the principle of the absence of a 'privileged observer' under Galilean transformations leads to physical laws whose simplest forms recover Newtonian physics. In the world view of Galileo and Newton we assume that there is an absolute firmament which can be used as the *reference frame* for measuring the velocity of any given object. The all-pervading ether would be an ideal candidate for such a reference frame. We know from Newton's equations or Kepler's laws that different planets move at different velocities. The earth itself moves at different velocities at different parts of its orbit round the sun. Hence if you are moving with a speed v towards a pulse of light (e.g., a spark as generated by Hertz) moving at the speed c with respect to the ether, the velocity of light relative to you is $c+v$? Similarly, if you are moving away, the relative speed becomes $c-v$? That is, if Newton's ideas of absolute velocities and relative velocities were taken together with Maxwell's theories, then optical observations may provide a means of confirming or falsifying some of these ideas and investigating the ether. Michelson and Morley's experiments paved the way to the theory of relativity. We take up that story in a subsequent section.

4.4.1 *The Aharanov-Bohm effect*

Although the electric field E and the magnetic field B associated with a current carrying coil cannot be seen by the eye, their effects can be made manifest, as known since the days of Michael Faraday. Today most people would grant the 'physical reality' of E and B. However, can we also say that the potentials $\vec{A}$ and ϕ have a 'physical reality'? The Aharanov-Bohm effect was predicted theoretically in 1949 by Ehrenberg and Siday, and then independently by Aharanov and Bohm in 1959. It was confirmed experimentally by Chambers in 1960. This effect predicts that there are phenomena that need the potentials $\vec{A}$ and ϕ for their description. The force fields E and B are an incomplete description of the electromagnetic field interacting with quantum particles like electrons. Such particles have wave properties (see Ch. 6), and the wave acquires a phase change when traveling through a region where the vector potential $\vec{A}$ is non-zero. Hence, if the wave is split into two beams and recombined, an interference pattern is set up depending on the lengths of the two paths traversed by the two beams. The demonstration of such intensity oscillations constitutes verification of the Aharanov-Bohm effect. It also demonstrates that the potentials $\vec{A}$ and ϕ are not mere mathematical concepts, but genuine aspects of our physical reality.

The potentials $\vec{A}$ and ϕ are not completely defined, as seen from Eq. 4.3 where we can add or subtract contributions from a gauge term χ to $\vec{A}$ and ϕ.

On the other hand, we can make a selection of the gauge (i.e., in effect, fix a value for χ), and make the potentials definite for that choice. This is some what similar to fixing a system of coordinates to select our space and time coordinates, or positioning ourselves on a specific mountain lookout to survey a city. A different lookout would show a different, but equally real view. Furthermore, $\vec{A}$ and ϕ occur in the equations of Maxwell and hence they are at least as real as 'electromagnetic waves'. In fact one may say that $\vec{A}(x)$ is the wavefunction of the electromagnetic field, and directly relates to the intensity of the field at x. Hence it is conceptually reasonable to regard $\vec{A}$ and ϕ as objects at least as real as our perception of space and time.

The Arahanov-Bohm effect is an important prediction that was confirmed experimentally, establishing the physicality of the potentials $\vec{A}$ and ϕ. Yakir Aharanov and David Bohm described the effect known by their name in 1959. Unknown to them, Werner Ehrenberg and Raymon Siday had predicted the same phenomenon a decade earlier in a short paragraph contained in a paper on electron optics. This effect says that a charged particle like an electron moving in a region where there is no magnetic field would still acquire a change of phase of its wavefunction ψ, if the vector potential A in the region is non zero. If there are two paths $\vec{L}_1$ and $\vec{L}_2$ available for the electron to arrive at a detector, ψ would acquire different phase shifts on traversing the two paths. Hence if a sufficient number of electron arrivals is recorded by the detector, an interference pattern of intensities would be found. This is the most common form of the Aharanov-Bohm effect. It is completely analogous to the interference pattern expected for light paths in the Michelson-Morley experiment (see Sec. 4.6), or in the two-slit experiment with electrons or photons (see Fig. 6.3).

4.5 The four-dimensional structure of Maxwell's equations

A corner stone of the theory of relativity is the 4-dimensional view of the world. In the world of Galileo and Newton, we can think of ourselves living in space (three dimensions) while this space some how moves along from past to present. The past doesn't exist because it has ceased to exist. The future doesn't exist because it has not yet come to existence.

So the time dimension is some how different to the east-west, north-south and up-down dimensions (x,y,z) of space. This space consists of 'right here' as well as all other parts of space, and all these regions seem to exist in our perception. Unlike the 'past-present', we have no intuitive difficulty in imagining that 'east-west' exists all the time. This is the point of view found in the Newtonian view of nature and even in Maxwell's equations until they were re-interpreted in the light of the experiments of Michelson and Morley (Sec. 4.6). Let us look at Maxwell's equations more closely using a slightly more technical discussion.

It is convenient to bunch the four numbers t, x, y, z into a vector of 4-components, denoted explicitly by x_0, x_1, x_2, x_3, and indicated by the symbol x_μ. Since there are four components, it is called a 4-vector. The time component of this 4-vector is x_0, while x_1 is the value of the x coordinate and so on. The space-like three vector, i.e., x_1, x_2, x_3 is denoted by the bold-face $\mathbf{x}$, or $\vec{x}$.

Electromagnetism involves charges and currents. These can be naturally grouped into another 4-vector. The charge density at the point $\mathbf{x}$ is denoted by $\rho(\mathbf{x})$, while the three components of the current $\mathbf{j}(\mathbf{x})$ are denoted by j_1, j_2, j_3. In fact we can rename $\rho(\mathbf{x})$ as j_0 and and include it in a 4-vector j_0, j_1, j_2, j_3 which can be denoted by j_μ. It is called a 4-current, it being understood that the j_0 is really the charge density. Since the currents flow in a certain direction, they are vectors. The charge density is a scalar. Currents flow in a system because there are potential gradients driving them. Thus the 4-current j_μ implies the existence of a 4-potential A_μ where μ takes the values 0,1,2,3 to specify the four components A_0, A_1, A_2, A_3. The index μ can be thought of as specifying the *four directions in a four dimensional space*, consisting of time ($\mu = 0$), and space ($\mu = 1, 2, 3$). The potential A_0 associated with the charge density (i.e., scalar) component A_0 is the scalar potential that we denoted by ϕ. The three components of $\vec{A}$ are A_1, A_2, A_3. The 4-potential A_μ is sometimes written as $(\phi, \mathbf{A})$.

The equation of continuity that we already discussed briefly relates the time ($\mu = 0$) variations and space ($\mu = 1, 2, 3$) variations of the charges and the currents. Let us denote the derivative of the current j_μ with respect to the μ-th component of x by $\partial_\mu j^\mu$. We need to add up all the changes in all the directions (space directions i.e., μ = 1, 2, 3) and in time (i.e., μ = 0). That is, whenever we see the letter μ repeated as in $\partial_\mu j^\mu$ then we know that we have to add up all the components. The equation of continuity says that the net effect of the sum of all the changes is zero. Thus we have:

$$\partial_\mu j^\mu = 0. \tag{4.7}$$

The continuity equation only looks at the gradient of the component j^μ in the direction μ. If we apply a stress along the x-direction in a rubber sheet, then distortions and stresses are set up in the y-direction as well. Similarly, if there is a potential gradient in some direction μ, this gradient can make currents flow in the direction μ as well as other directions ν. That is, we need to deal with potential gradients like $\partial_\mu A^\nu$ and $\partial_\nu A^\mu$ which depend on two indices μ and ν. Physical quantities which depend on several indices are known as *tensors*. Hence, to get the full picture of the disturbances set up by the potential gradients in the system, we need to look at a more complicated object known as the *stress tensor*, denoted by $F^{\mu\nu}$. One might picture this as a mathematical description of the stresses in the 3-D network of 'lines of force' revealed by Faraday's iron filing experiments. The stress tensor is defined by

$$F^{\mu\nu} = \partial_\mu A^\nu - \partial_\nu A^\mu. \tag{4.8}$$

Maxwell's equations can be restated as asserting that the gradients of the stress tensor generate the electromagnetic currents in the system. That is,

$$\partial_\mu F^{\mu\nu} = j^\nu. \tag{4.9}$$

We recall that repeated indices are summed over in the above equation Maxwell's equation written in this form manifestly satisfies the continuity equation and the gauge principle.

We have a 4-dimensional description of currents and potentials in the previous (small print) paragraphs because we wish to bring out another deep principle known as *Lorentz invariance* that is begging to reveal itself in Maxwell's equations. This is the corner-stone of the special theory of relativity. Maxwell's equations already contain the germ of relativity, and satisfy relativity!

In fact, by looking for the simplest equation which satisfies all these symmetry principles, i.e, Lorentz principle, gauge principles, equation of continuity, we can literally derive Maxwell's equations 'from scratch'. This once again establishes the simplicity of the laws of nature. Just as the equation of continuity is based on the principle of the local conservation of the number of particles in the system, the requirement of global gauge invariance ensures that the electric charge is conserved. The principle of Lorentz invariance is based on the constancy of the velocity of light with respect to any observer what so ever. A key experiment in this context is associated with the names of Michelson and Morley.

4.6 The Michelson-Morley experiment

The history of science is adorned with many key experiments. Of course, with the exponential growth of science in recent times, we have too many key experiments to keep up with. However, the story of Archimedes's experiments on floating bodies, and the account of Galileo's experiments on falling bodies carried out from the tower of Pisa have a special place in the history of science. Such experiments helped to get rid of the Aristotelian concepts (see Ch. 3) of motion in terms of final causes (Sec. 3.6) and divine purpose. Renaissance thinkers ushered in the world view that has become our natural mental baggage. The experiments of Michelson and Morley played a similar role in ushering a new world view. There we have to radically revise our simple everyday intuition about space and time if we wish to deal with energy scales and speeds going well beyond everyday experience.

The objective of the Michelson-Morley experiment was to determine the absolute speed of light c very accurately, taking account of the earth's absolute velocity with respect to the ether. Even the velocity of propagation of (electromagnetic) vibrations in the ether had been calculated by Maxwell from his wave equation. Electric sparks and lightning betrayed the fact that light is also a form of electromagnetic waves. The earliest accurate measurement of the speed of light had been achieved by Olaus Römer in the latter part of the 17th century. He had noticed that the predictions of the eclipses of the Jovian satellite *Io* by Jupiter were quite accurate when the earth was close to Jupiter. However, the predictions needed a correction when the earth was far from Jupiter. He correctly surmised that the difference must arise from the time taken by the light to travel the longer distance to the earth. Using the size of the earth's orbit and Jupiter's orbit available at the time (1675 CE), Römer estimated that the velocity of light is 2×10^5 kilometers per second (km/s). A modern value of c is 299,792.5 km/s.

Michelson and Morley's first experiment, conducted in 1887, used two light beams. The two beams were made by splitting a beam of light at an angle of 45° on a half-silvered mirror, so that one part of the light went right through the mirror, while the reflected part went off at 90° to the transmitted beam. Such a device is called a *beam splitter*. Thus one beam could be directed along the direction of motion of the earth in its orbit, while the other beam would be directed at right angles. The beams were reflected by mirrors after traveling a given distance L and brought back to a common point. If the speed of light in the two directions were different, due to the motion of the earth through the ether, the crests and troughs of one beam would not match those of the other beam, and an interference pattern would arise, due to the tiny difference in time taken by the two beams.

The time taken by light to go a distance L and return at a speed c would be $2L/c$. The velocity of the earth is quite slow compared to that of light. If the velocity of the earth through the ether were v, then the time difference ΔT for doing the two paths (along and at right angles to the motion of the earth) is easily found to be $(2L/c)(v^2/2c^2)$ since v is much smaller than c. If the earth were at rest then this time difference would be zero. For the sort of lengths L of the light path that is practical, using the known orbital speed of the earth, this time difference ΔT is of the order of a fraction of a trillionth of a second. One cannot measure such short times using a stop watch or chronometer. However, if the two (identical) light waves were out of phase by even a tiny time interval, then their waves (troughs and crests) do not fall on top of each other and an interference pattern will be formed. The observation of interference patterns was all that was needed. The experiment could have detected even a one-kilometer earth speed.

Michelson and Morley did the experiment and came up with a null result. No interference pattern was observed. The two beams were perfectly in step!

The experiments revealed that the velocity of light was independent of the direction of propagation to within the accuracy of the experiment. In fact, the measured value of the speed of light, c, was just the same as that predicted by Maxwell's equations, and did not become $\vec{c}+\vec{v}$ as required by the parallelogram law of composition of relative velocities.

The velocity of the earth cannot be zero as there was already enough astronomical evidence from a variety of sources. The time lag in the observation of the eclipses of the moons of Jupiter (cf., the early work of Römer repeated more accurately) fitted in very well with the value of c for the velocity of light, and the dimensions of the earth's orbit which dictated the velocity v of the earth. This means, the velocity of light c for any observer is always the same. That is, *c is a universal constant independent of the state of motion of the observer.*

So it appears that you can never catch up with a ray of light, how ever fast you chase it.

4.7 Lorentz, Fitzgerald, Poincaré and the dawn of relativity

The accuracy of the Michelson-Morley experiment could not be doubted. However, science is a doubt-system and not a belief system. Hence the key experiments continue to be repeated with very high accuracy. Modern experiments by Brillet and Hall using Helium-Neon lasers, and recent work by Hughes and Drever [169] using nuclear-spin precession techniques, have re-confirmed the null result.

Hence, no matter what the observer's velocity is, the velocity of light is going to have the same value c. How do we measure a velocity? We select a direction, and find the distance traveled in a given time. Thus we need a distance measurement and a time measurement. The Scottish physicist Fitzgerald suggested that the velocity of light could appear constant if the measuring unit, e.g., the meter stick would begin to contract as its speed rose to values comparable to the speed of light. This idea may appear to be a rather 'contrived', highly artificial explanation. Meanwhile the Dutch physicist Hendricks Lorentz made the same suggestion, but based on a more detailed concept of material objects. After all, meter sticks are made out of atoms. By then even electrons — the atom-like corpuscle of electric charge — had been discovered by J. J. Thompson. Of course every one did not necessarily believe in atoms or electrons. However, Lorentz took the atomic picture seriously and considered that the meter stick, made up of a collection of charged particles need not remain rigid at all speeds. Joseph Larmor, an Irish physicist, recognized that just as the meter could contract, dilation in the unit of time could also occur and provide a complementary approach. Clearly then, both space and time are conspirators in the null result seen by Michelson and Morley.

Henri Poincaré, the celebrated French mathematician, had also understood the deep implications of the Michelson-Morley experiment and conjured up a mathematical vision of a 4-dimensional world. However, it was left to Minkowski (1909) to explicitly formulate the modern view of spacetime found in the special Theory of Relativity. We go to *Science et Methode* published in 1897, and look at what Poincaré had to say about 'Relativity of Space' [165].

> If I am at a definite point in Paris, at the Place du Panthèon, for example, and say 'I will come back here tomorrow', and if I am asked 'Do you mean that you would be back at the same point in space?' I should be tempted to answer yes. However, I would be wrong, since between now and tomorrow the earth would have moved, carrying with it the Place du Panthèon, which will have traveled more than a million kilometres. And if I were to be more accurate, I should gain nothing, since the millions of kilometres have been covered by our globe in relation to the sun, while the sun in its turn moves in

relation to the Milky Way, and the Milky Way is in motion without our being able to apprehend its velocity. Suppose that in one night all the dimensions of the universe became a thousand times larger. The world would remain similar to itself, if we give the word similitude the meaning it has in Euclid's book III. ... I would not notice anything at all. The most exacting measurements would be incapable of revealing anything of this enormous change since the meter stick that I use would have changed in exactly the same proportion ...

But that is not all ... According to a hypothesis of Lorentz and Fitzgerald, all bodies carried forward in the earth's motion undergo a deformation. This deformation is, in reality, very small, since all dimensions parallel to the earth's motion is reduced by a hundred-millionth, while dimensions perpendicular to this motion remain unaltered. But it matters little that it is small — it is sufficient that it should exist for the conclusions that I propose to draw from it ... It is evident that we cannot demonstrate this deformation. ... my meter stick, ... has in its turn undergone the contraction such that ... I shall not become aware of any change.

What then, I would be asked, is the use of the hypothesis of Lorentz and Fitzgerald if no experiments enable us to verify it? The fact is that my statement has been incomplete. I have only spoken of measurements that can be made with a meter stick. But we can also measure a distance by the time that light takes to traverse it, on condition that we admit that the velocity of light is constant, and independent of its direction. Lorentz could have accounted for the observations by supposing that the velocity of light is greater in the direction parallel to the earth's motion, than in the perpendicular direction. He preferred to accept that the velocity of light is the same in the two directions, but that the material bodies are smaller in the former than in the latter. If the wave fronts of the light had undergone the same deformations as material bodies, we would never have perceived the Lorentz-Fitzgerald contraction.

Poincaré in his technical writings had formulated and presented special Relativity in essentially the same way as Einstein was to do. However, it was more of a brilliant aside from a deep thinker who was on another far-reaching adventure of exploration in the world of mathematics. At this time Poincaré was making big advances connecting non-Euclidean geometry, Fuchsian functions, differential equations and the foundations of topology. Theoretical physics was interesting to him because it churned up some interesting differential equations. It is perhaps not surprising that Poincaré did not make the clean break from the ideas of 19th century Physics that Albert Einstein was to make, in his presentation of the theory of Relativity.

Chapter 5

The Theory of Relativity

The main ideas of special and general relativity are explained, paying attention to conceptual issues, paradoxes and puzzles introduced by this revolution in our belief system. A preliminary look at the nature of *time* is undertaken.

5.1 The historical context

In previous chapters we discussed how the old Aristotelian concepts and church dogmas had been overthrown by the valiant effort of renaissance thinkers. It was not until 1822 that the sacred congregation of the Index (of condemned books) lifted its ban on publications which stated that the earth was not the immobile center of the universe. Finally, in 1992, the Church tendered a convoluted apology to Galileo, stating that the 'Creator stimulated Galileo, anticipating and assisting his intuitions'. Galileo, like Newton, believed in a creator but rejected the Aristotelian theology of the church. Instead, a mechanistic view of the world, with the earth revolving around the sun, and objects obeying various laws of nature began to take shape.

The world view of Galileo and Newton was very successful and held firm almost to the end of the 19th century. *This world view is still the operational world view of our everyday life*. As we saw in our discussion of Clerk Maxwell's work, it was the study of electromagnetism and the universality of the velocity of light which forced the leading thinkers of the late 19th century to seriously question the Newtonian view of the world. We already noted that Poincaré, Lorentz, Fitzgerald and others had already begun to talk in terms of a new view of the world where space and time were no longer what we intuitively understood them to be.

The Copernican revolution had destroyed our reassuring belief that the earth was firmly fixed at the center of the Universe. But there was at least some kind of absolute space through which everything moved forward in time as the planets went around the sun, with the clocks ticking in unison with the uniform flow of time. The definitive experiments of Michelson and Morley, and the nature of light radiation as embodied in Maxwell's equations set the stage for the Lorentzian world. It pulled the rug of space from under our feet and tinkered with time itself, distorting even our notions of *before and after*.

Einstein's 1905 paper which presented the theory of relativity is not only a historic document, but a classic that begins with two very clear, insightful sections that contain little or no mathematics. It deals with what is now known as the special theory of relativity. Special relativity makes us give up our ordinary ideas about space, time, and our normal judgments about when two events occur simultaneously. Space is not to be thought of as an arena where time runs forward, with events occurring in space. Space and time are 'married' to form 'spacetime' where the separate entities, space and time, are mere constructs of an observer's state of motion. The laws of motion in spacetime are the subject of the special theory of relativity. It modifies Newton's theory of motion and the kinetic energy of moving bodies as speeds approach the speed of light which cannot be exceeded by any mover. Examining the kinetic energy within the special theory of relativity led Einstein to his famous equation, $E = mc^2$. Given the fact that the special theory arose from a study of electromagnetism where the Coulomb law of force is an inverse-square law $(1/r^2)$, just as in the case of Newton's law of gravity, it was natural to now ask if gravity also propagated as some kind of waves, and if this too was controlled by the principle of Lorentz invariance which required that the velocity of light was the same for all observers.

About a decade after the 1905 work, Einstein presented his General theory of Relativity which brings gravitation into the Lorentizan framework. Einstein shows that we can explain the motion of planets and other gravitating bodies without postulating the existence of a force of gravity. If no force acts on an object, then it moves at constant speed following the shortest path available to it — namely, a straight line in free space. However, the shortest path on a curved space (e.g., a curved surface) is itself a curve, and each planet is following the shortest path in spacetime which is not flat, but has a curvature caused by the presence of matter! Thus Einstein gets rid of gravity and substitutes a new geometric property — curvature of spacetime, to explain the observed facts. Einstein's theory explains all the observations that had been explained using Newton's theory, and also predicts additional effects which were not predicted by the old theory. For example, if spacetime is curved due to the presence of matter, and the shortest path is a curve (know as the geodesic), then light passing near a massive object like the sun should also feel this curvature and show a deflection from the simple straight-line path.

This, and many other effects predicted by general relativity have been fully confirmed by experiment. On the other hand, it is not at all clear if Einstein's curved spacetime picture applies at very short length and time scales that are relevant to the quantum world of sub-atomic phenomena.

The specific geometric flavour of general relativity (GR) can in fact be exchanged for a very different-looking theory, i.e., a gauge theory, as shown by Utiyama in 1956 [210]. He showed that GR could be derived by generalizing the global (i.e., spacetime independent) coordinate transformations of special relativity to local ones. As was the case in electromagnetic theory, the added restrictions coming from local invariance implied the existence of a new field — i.e., the gravitational field. Even the interaction $1/r^2$ seems to come out of the theory. However, this was not a quantum theory.

While the special theory of relativity (Lorentz invariance) is a fundamental part of quantum mechanics (QM), we lack a form of general relativity valid at the extremely small length scales of QM. We first consider the special theory of relativity which integrates very well with QM.

5.2 Special relativity — Einstein's two postulates

In the introduction to his 1905 paper Einstein presents two postulates. The first, called the Principle of Relativity, states that 'the same laws of electrodynamics and optics will be valid for all frames of reference, for which the equations of mechanics hold good'. The second postulate states that 'light is always propagated in empty space with a definite velocity c which is independent of the state of motion of the emitting body'. The first postulate is already familiar to us. We have already come across the Galilean principle of relativity (Sec. 3.3) which stated that the equations of mechanics hold good for all observers moving at fixed velocities relative to each other (i.e., 'valid frames of reference'). Einstein's relativity requires that all the laws of physics known at the time be the same for all 'valid frames of reference', subject to the additional condition that the velocity of light be the same for all observers, irrespective of their state of motion. Einstein arrives at his theory by noticing that only relative motions among magnets and coils enter into the phenomena of electricity and magnetism. All concepts like absolute velocities, absolute ether, luminiferous ether, ('Lichtäether') etc., seem to be unnecessary.

Normally, two observers can 'synchronize' their watches at a given moment, and when they meet each other again, if their clocks agree, such observers belong to the same 'inertial frame' denoted by A. They are moving together and have the same notion of simultaneity. However, unlike in Galilean relativity, another observer moving at a different fixed velocity (i.e., in a different reference frame B), say at a velocity v with respect to A, would not have the same notion of

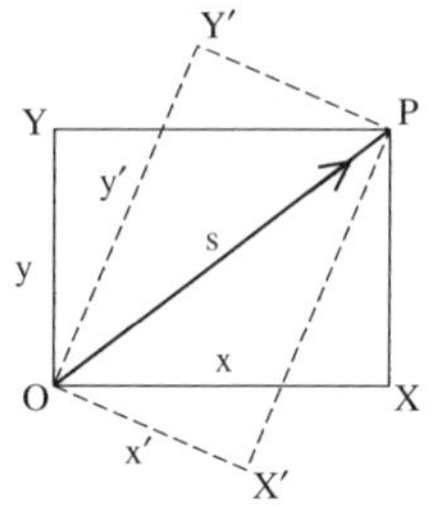

Fig. 5.1 The distance between the two points O and P in a 2-Dimensional world is given by s, calculated from the length x in the X-direction, and the 'width' y in the Y-direction using Eq. 5.1. However, a different selection of x-axis and y-axis, by a different observer, e.g., x' and y' is possible. Nevertheless the length s remains unchanged.

simultaneity as the observers in A. That is, what is judged to be two simultaneous events by an observer in A would not be judged to be simultaneous by an observer in B. In effect, we have to abandon some of our ingrained notions of every-day clock time if there is to be a fixed velocity of light for all observers. In our everyday world, time 'runs like clockwork' and time intervals are not dependent on how fast you are moving. We return to this in a later section (Sec. 5.10) when we discuss the clock paradox.

If Alice throws a ball to Bob at time t_a, and if Bob were to catch it at time t_b, we expect that $t_b > t_a$ in our ordinary experience. The ball play is seen by an observer U, moving at the speeds u. The speed may be close to the velocity of light c. Special relativity requires that lengths contract and clocks slow down or speed up to ensure that the speed of light is seen to be constant by U, Alice and Bob. This allows the possibility that U may observe Bob catching the ball before it was thrown by Alice if Alice, Bob and the ball were quantum particles! This paradox does not occur in ordinary experience or in special relativity by itself since causality (time order) has to be respected and speeds of the observers and the ball remain small compared to c, the speed of light. However, subatomic particles obeying QM also satisfy causality but in a sneaky way. Thus, Alice may be an atom A which moves at high speed and throws an electron, and Bob might be another atom B. Quantum mechanics fully conforms to special relativity, but it is possible for the atom B to seem to catch (absorb) an electron before it was thrown (emitted) by the atom A. This is interpreted in QM as a case where B emits an *antiparticle* and A absorbs it! Thus Feynman graphs have anti-particles going backward in time, without violating causality. Lorentz invariance, coupled with Quantum Mechanics will lead us to a new fundamental symmetry — the duality between particles and antiparticles.

This "entanglement" of space and time is better understood within the 4-dimensional (Minkowski) picture of the world involving t,x,y,z with t as the fourth dimension. A 4-dimensional description already appeared in the context of Maxwell's equation (Sec. 4.5).

5.3 Marrying space and time to get spacetime

In the new picture of spacetime, two observers can agree on the *event separation*, but not on the space separation or time separation, as is possible in the everyday world of human experience.

Let us first consider two-dimensional space, as we have on a plane sheet of paper. We can draw a set of axes x and y, and mark a point P (see Fig. 5.1). The separation (i.e., distance) between the origin O and the point P is denoted by s. The value of s can be calculated using the theorem of Pythagoras. That is, the square of s is given by:

$$s^2 = x^2 + y^2 \text{ in 2-D; } s^2 = x^2 + y^2 + z^2 \text{ in 3-D.} \tag{5.1}$$

This distance does not depend on how we draw the axes, but only on the points O and P (see Fig. 5.1). That is, x, y themselves have no basic significance, while s is the 'invariant' quantity which remains independent of the way we draw the axes. Two 'observers' can set up their axes in two different ways, and yet they would agree on the value of s. Instead of considering a plane (i.e., 2-dimensional space), we can consider three-dimensional space where we have three axes x, y, and z. The distance between O and P is the diagonal of the 3-D box instead of the diagonal of the rectangle given Fig. 5.1.

We note that the three components x^2, y^2, z^2 are joined together with three plus signs, i.e., we have the 'signature' (+,+,+) in our intuitive geometry which is Euclidean. Here again, just as in the 2-D case, we can redraw the axes and make the x-direction align with the east-west direction, or the up-down direction etc. However, the distance s remains independent of how we draw the axes. The work of Lorentz, Fitzgerald, Poincaré and Einstein himself implied that we should consider a 4-dimensional world where the time t is one of the dimensions somewhat similar to x, y, and z. If time t were multiplied by the velocity of light c, the quantity ct is a distance. Hence we might think of just adding a $(ct)^2$ to Eq. 5.1 to make it 4-dimensional. However we know that we cannot make a space direction (e.g., the east-west direction) identical with the 'direction' of time by any known physical means. Hence it is clear that in marrying space with time, we need to recognize that although we can change the x, y, z directions among themselves, the time axis has to enter into the picture somewhat differently. Hermann Minkowski, a mathematician colleague of Hilbert and a teacher of Einstein introduced a subtle distinction between, x, y, z, and time t by regarding the time as an imaginary axis! That is, use ict instead of ct as the 'fourth dimension'! Here $i = \sqrt{-1}$, and hence $(ict)^2$ is $-c^2t^2$. So we have spacetime distances involving four squared quantities, unlike in ordinary space where we use three squared quantities in the Pythagoras theorem.

In constructing the 4-dimensional event-separation ζ between two point in spacetime, we use a generalization of the Pythagorus theorem. The 2-D and 3-D distance s is already in Eq. 5.1. In Minkowski's 4-dimensional world the spacetime interval ζ may be written as:

$$s^2 = x^2 + y^2 + z^2 - c^2t^2. \tag{5.2}$$

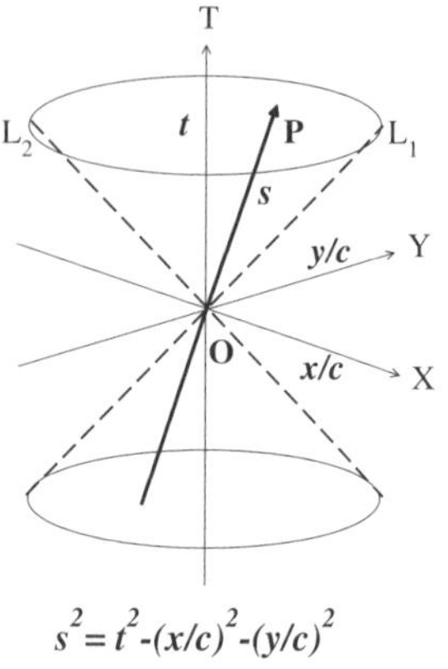

Fig. 5.2 The vertical axis OT is the 'worldline', i.e., the flow of time for a stationary person at O in the 2-D world with X and Y directions only. The distances x and y are divided by c, i.e., x/c, y/c where c is the speed of light. In a 3-D world there is a z/c which cannot be depicted here. OL_1 and OL_2 are the worldlines for two photons traveling at the speed of light. All objects with finite mass move more slowly, and their world lines, e.g., OP, lie inside the 'light cones'. The light cones define the causally connected, 'knowable' part of space-time around an observer at O. S/he can affect only events in the upper cone; and s/he is affected only by those in the lower cone.

The above form is said to ave a $(+++-)$ signature. However, since $x/c, y/c, z/c$ are time-like quantities, we may also write ζ to be timelike:

$$\tau^2 = t^2 - (x/c)^2 - (y/c)^2 - (z/c)^2. \tag{5.3}$$

This has the signature $(+---)$. Both forms are in use, and serve equally well to give the event separation. Minkowski's 4-D world geometry embeds the ideas of Lorentz, Poincaré and Einstein since the velocity of light c is the same for all observers in it.

One can also use a system of units where $c = 1$ to avoid having to write c in the above equations. A two-dimensional space-time is depicted in Fig. 5.2 and has a $(+--)$ signature. In 3-D ζ is a quantity where the time and 3-D-space parts are joined together with the $(+---)$. This form is used in Bjorken and Drell [36], Aitchison [4] and other well-known texts. The interval ζ written as τ is really an 'effective time' between the two events. In our, everyday notion of the external world, we assume that the distance s is space-like and three-dimensional. When two terrestrial surveyors measure a spatial distance, they come out with the same value of s because their relative motion is zero or usually quite negligible when compared to the speed of light. All inertial observers would get the same answer for the interval ζ if they used Eq. 5.3, or Eq. 5.2, while the individual components x, y, x, t would differ for each observer.

The intuitive picture of the world as floating forward in time is not the world-view of relativity. Instead, observers move in spacetime along trajectories called *worldlines*. This is the spacetime path from one event to another. The implication of all this is that *it is the event separation that all observers agree upon, and not the space separation or the time separation of everyday usage*.

This is the essential content of *Lorentz invariance*. The contraction of a fast-moving measuring stick discussed under the Lorentz-Fitzgerald contraction is a non-intuitive result of Lorentz invariance. Various other 'paradoxes' arise from the requirements of Lorentz invariance and they are well known in the popular literature. Let us review these issues at a slightly more technical level.

If we consider two observers A and B in relative motion, then in converting the distances and times measured by the moving observer, to the values measured in the other system, we need a transformation rule which ensures that the velocity of light is measured to be the same in both coordinate systems. This rule is known as the Lorentz transformation. In the usual (Galilean) world view (Sec. 3.3), we consider 'observers' A and B who are at the position $P = (x, y, z)$ at time $t = 0$ and B is moving at a constant speed v in the x-direction, A's position becomes $x' = x - vt, y' = y$ and $z' = z$ as seen by B. They are in the same time zone and their clocks times agree, with $t' = t$. This change from $x \to x', y \to y'$ and $z \to z'$ as seen by B is what happens in our everyday world. It is called a 'Galilean transformation'.

Einstein discusses the Lorentz transformation in section 3 of his paper under the title *Theorie der Koordinaten und Zeit-transfromation*... etc., and shows that

$$t' = \beta(t - vx/c^2),\ \ x' = \beta(x - vt),\ \ y' = y,\ \ z' = z. \tag{5.4}$$

Here the all important *Lorentz factor* β appears. It is called the 'dilation factor'. Just as we need a conversion rate for going from dollars to euros, the conversion factor between the two frames of reference A and B is simply β. It is given by

$$\beta = 1/(1 - v^2/c^2)^{1/2};\quad \beta \to \infty \ \text{ when } v \to c. \tag{5.5}$$

As v becomes more and more close to the velocity of light c, the term v^2/c^2 in the Lorentz factor β becomes increasingly closer to unity. Then the Lorentz factor becomes infinitely large. Hence, a small time interval $\delta t'$ in B's world appears as a very large, dilated time interval $\delta t = \beta\delta t'$ in the world of A. Normally, v is much smaller than the velocity of light. Then Eq. (5.5) may be written as

$$\beta = 1 + (1/2)v^2/c^2 + \text{smaller correction terms.} \tag{5.6}$$

If the velocity of B relative to A, i.e., v were small compared to c, then v^2/c^2 is negligible and β becomes unity. Then the Lorentz transformation (Eq. 5.5) reduces to the Galilean transformation. That is, special relativity reduces to common experience for normal terrestrial speeds.

When the length of a meter stick moving at the velocity v contracts (Lorentz-Fitzgerald contraction), as seen by the observer at rest, the time interval (e.g., one hour) is observed to dilate (become bigger). Similarly, if the mass of an object is denoted by m_0 when weighed in a system in which it is at rest, and if it were then weighed in a system moving at the velocity v relative to the first, its mass m would seem to have increased. This increase is also given by the Lorentz factor, via the formula

$$m = \beta m_0 = m_0/(1 - v^2/c^2)^{1/2}. \tag{5.7}$$

We know that $\beta \to \infty$ as the velocity $v \to c$, and hence the mass of a particle $m \to \infty$ when approaching the speed of light, unless its rest mass m_0 itself were zero.

The photon is an example of a zero-mass particle — it is just the quantum of light itself and hence it is not surprising that it can travel at the speed of light! Thus no signal or 'information carrier' can travel faster than light. In Fig. 5.2 we consider an observer O and his time line OT. Since we cannot 'draw' a 3-D world and a time line, we draw a 2-D world (in perspective) with 'space' axes X, Y and the time axis t, along OT. All observers who are stationary with respect to O will have worldlines parallel to OT. Consider a ray of light traveling away from O at the speed of light, measured in units such that $c = 1$ (in common metric units, $c \simeq 3x10^{10}$ cm/sec). Choosing $c = 1$ in an equation like $r = r_0 + ct$ simply means that the slope of r with respect to the t axis is unity, i.e., makes a 45° angle. Since the x, y distances of the photon from the origin increase with time, we draw the world line OL at 45° to the OT axis. Thus we can define a cone for the fastest travelers (i.e., photons and other zero-mass particles).

All observers move slower than photons (light) and hence their world lines will be contained entirely inside the light cone. OP (see Fig. 5.2) is the world line of such a moving object. Clearly, all the events that happen in the light cone 'below O' are the only 'past events' which can possibly influence O. Similarly, O can only influence the future events that are inside his upper light cone. In effect, these light cones define the event region for information transmission, and the extent of the universe that O (at time zero) can know of, or influence.

A very important conclusion that flows from Einstein's special relativity is the equivalence of energy and matter. Here again, if energy and matter are just two forms of currencies, then the conversion rate between them is governed by an 'exchange rate'. This happens to be c^2, and hence

$$E = mc^2 . \tag{5.8}$$

Although everything needed to prove this was in Einstein's first 1905 paper, this world-famous equation appeared as a second paper in 1905. There are a number of subtleties (see for instance, Ehlers, Rindler and Penrose *et al.* [68]) in proving Eq. (5.8), and Einstein himself gave a more careful discussion in 1935.

We can get a 'feel' for the equation if we write E for the energy equivalent of a mass m, and E_0 for the energy when it is at rest and having a mass m_0 in its rest frame. Let the velocity v be small compared to c. When we begin to move the mass at a speed v, we add kinetic energy to its initial energy. This kinetic energy is $(1/2)m_0v^2$. Hence the total energy is

$$E = E_0 + (1/2)m_0v^2 .$$

On the other hand, if we consider that the mass $m = \beta m_0$, and use the expression for β given in Eq. (5.5) we get

$$m = m_0\{1 + (1/2)v^2/c^2\} .$$

This can be rearranged to

$$mc^2 = m_0c^2 + (1/2)m_0v^2 .$$

On comparing these two equations it is clear that the famous equation, viz., $E_0 = m_0c^2$ and $E = mc^2$ are consistent with the above.

Matter is merely a concentrated form of energy with every unit of matter giving c^2 units of energy. It is this energy which fuels the heat of the sun, or the fury of atomic explosions. In the General Theory of Relativity Einstein shows that this 'concentrated form of energy' (i.e., matter), is essentially contained in the stress associated with the curvature of spacetime. This point of view gives a geometric meaning to both matter and energy. However, Einstein himself, and also scientifically oriented philosophers like Bertrand Russell preferred to give the primacy to energy — that is, matter is regarded as a concentrated form of energy. According to Russell, 'atoms' are simply regions where there is a high localization of energy. In reality, the duality of energy and matter has merged into one entity, and matter manifests itself as energetic excitations of an underlying quantum field.

If matter is indeed a concentrated form of energy, let us look at the interplay between gravitational energy and the energy contained in matter. It turns out that the special theory of relativity already hints at the possible existence of exotic objects like black holes. Let us summon up our handy language of elementary mathematics to help us here.

The gravitational force F between two objects, with masses M and m, is given by

$$F = G\frac{Mm}{r^2}\,. \tag{5.9}$$

The factor G is a proportionality constant which enables us to use our favorite units, e.g., kilograms and meters, pounds and feet or whatever, on the right hand side, and get the force in Newtons or foot-pounds per sec per sec, or whatever units of force we desire on the left hand side. Every force is a gradient of a potential. The gravitational force corresponds to a gravitational potential. The energy of a unit of mass (i.e., set $m = 1$) placed at a distance r from the mass M is known as the gravitational potential of the mass M. Just as the electrostatic force (Coulomb's law) leads to an electrostatic potential, the gravitational potential $V(r)$ has the form

$$V(r) = G\frac{M}{r}\,. \tag{5.10}$$

The gravitational energy E of a mass M placed in this potential V is MV, i.e., GM^2/r, since V is the energy for a unit mass. If the gravitational mass becomes comparable to the rest mass m_0 (also known as the Lorentz mass), then

$$Mc^2 = GM^2/r\,.$$

Hence, when

$$r = GM/c^2\,, \tag{5.11}$$

the energy of gravitational compression and the energy encapsulated in the mass M match up.

When two mighty forces match each other, we must regard it as a warning sign that something drastic should happen. Gravitational energy and the energy contained in matter itself match each other when the radius of the material object becomes of the order of GM/c^2. However, the special theory of relativity does not force us to recognize any problem here or the existence of singularities like *black holes*. Although Einstein himself never accepted such a possibility, this becomes unavoidable in the general theory of relativity.

5.4 General relativity

The ideas associated with the special theory of relativity unified all the laws of physics known to nineteenth century physics within the framework of a four-dimensional spacetime, except for gravitation. Gravitation treated x, y, z and t separately. In modern terms, this unification means that the laws of mechanics, electricity, magnetism and optics flow from the gauge principle (Sec. 4.3) and the principle of Lorentz invariance. That is, the known laws of physics appear to be the same to all observers moving at constant relative velocities (i.e., no acceleration or retardation) with respect to each other. These laws respect the 4-dimensional event structure of the world associated with the universal value of the velocity of light. Hence they are said to be Lorentz invariant.

Thus an immediate problem was recasting Newton's law of Gravitation into a Lorentz invariant form, i.e., a form which recognizes that no signals can be transmitted faster than the speed of light.

What can we say about observers accelerating under gravity? Such observers are known as non-inertial observers. Are they going to find that the laws of physics have changed because they are non-inertial? General relativity applies to accelerating systems. The most obvious form of acceleration is gravitational acceleration of a falling body. The acceleration of a falling body is ascribed to the force of gravity. In 1907 Einstein realized that a freely falling person would not observe the existence of a gravitational field. That is, if you are locked in a cabin without any windows, and if the cabin is falling freely, you would not know that you are falling. However, an external observer on the ground would see you falling and think that there is a gravitational field pulling you down. The difference in the conclusions arises purely from the different 'frames of reference'. Hence Einstein proposed the 'principle of equivalence' which states that *a gravitational field and a uniformly accelerating frame of reference are physically equivalent*.

What is being asserted is that there is a point of view where there is no force of gravity. That is, 'gravity is a fictitious force' like the 'centrifugal force' which is superfluous in a global view of the problem of a rotating bucket full of water.

Objects move along 'straight' paths when no force acts on them. The 'straight' path is the shortest path, and such motion satisfies the principle of least action (Sec. 3.6). The shortest path between two points is called a *geodesic*. The geodesic on a plane is what we normally call a straight line. The geodesic on a curved surface (e.g., the surface of a sphere) would also be a curve in the geometric sense. That is, a stone thrown at an angle to the earth's surface follows a parabolic path, the parabola is the 'shortest path' or geodesic which satisfies the least action principle. The presence of the earth has produced a curvature in spacetime. Because of this curvature, the geodesics are ellipses, parabolas etc., as seen in the orbits of planets and comets. Clearly, a revolutionary new 'geometric explanation' of gravity that does not require 'action at a distance' has appeared.

Einstein's 1916 paper formalized this revolutionary idea which has been aptly paraphrased by John Wheeler [224] as *Matter tells spacetime how to curve, and spacetime tells matter how to move*. Einstein's new theory removed several weaknesses of Newtonian gravity. How does the sun which is millions of kilometers away act on the earth? The problem of 'action at a distance' was completely removed, since the motion of a planet is determined by the local curvature of spacetime. This curvature, produced by any mass (e.g., the sun) becomes weaker as we move away from it. Newton's form of the law of gravitation was not consistent with Lorentz invariance. Space and time had their separate, independent existence and obeyed Galilean relativity. The new picture of gravity presented by Einstein was Lorentz invariant, in that it respected the 4-dimensional spacetime concept which preserved the fixed velocity of light for all observers.

5.4.1 *Black-holes, white dwarfs and neutron stars*

Let us take up John Wheeler's 'Matter tells spacetime how to curve, and spacetime tells matter how to move' as a very concise non-mathematical statement of general relativity (GR), and follow the consequences. In our discussion of the special theory of relativity, we arrived at Eq. (5.11) where we realized that if matter gets compressed to a radius smaller than GM/c^2 then the gravitational compression energy becomes larger than the energy stored as matter. In GR, matter and gravitational curvature becomes the 'same thing', and hence we might guess that a run away situation can arise, giving rise to a blackhole.

In fact, if an object (e.g., a star) of mass M were to shrink to a radius equal to or smaller than

$$r_s = 2GM/c^2 \tag{5.12}$$

then, using Wheeler's terminology, 'matter tells space to curve' even further. This curved spacetime swallows up the matter which can never return to the world outside the radius r_s. This radius, known as *the Schwarzschild radius*, defines the size of a blackhole in general relativity. Notice that this value is twice the value inferred by us using simple arguments from special theory of relativity.

Karl Schwarzschild was a German officer at the Russian front where in 1916, he found a solution to Einstein's equations of general relativity. He took the special case of the spacetime geometry around a spherical mass M placed at the origin of coordinates. Clearly, even the conditions at the battle front cannot deter a true genius! These solutions provide an excellent description of the spacetime around the sun, the earth and similar objects. In fact, modern implementations of the GPS (Geodesic Positioning System) using satellites orbiting the earth provide a very practical application of General Relativity and a running test of its validity.

The Schwarzchild solutions showed that if $r_s = 2GM/c^2$ then the mass would collapse under its own gravity. In the case of the earth, r_s turns out to be about 3 cm. Hence the earth would collapse into a blackhole if it could be compressed to the size of a muffin! The Schwarzschild radius r_s defines a spherical surface around a massive object such that if you get closer than r_s, i.e., $r < r_s$, then there is absolutely no escape for you. It is called a black hole (BH) because even light or electrons cannot escape it (unless you invoke the quantum theory, which always allows a tiny amount of indeterminacy). Mathematically, solutions of Einstein's equations which permit objects like black holes exhibit *singularities in spacetime*. There are other singularities like 'worm holes' which connect two disjointed regions of space. Similarly, 'white holes' are the opposite of 'black holes'. Instead of collapsing inwards, they explode outwards — just like in the big bang. The spherical surface defined by $r = r_s$ is also known as the event horizon, since no information can pass from inside it to the outside world.

The Schwarzchild solution can be used to establish three important results which have been verified by observation (i) The advance of the perihelion of the orbit of mercury. The existence of an error in Newtonian predictions was already recognized by 19th century astronomers who wondered if there was a hidden planet causing the advance. GR accurately accounts for the observed 43 arc-sec per century advance. (ii) A ray of light passing near a massive object is 'bent' by the gravitational field. Sir Arthur Eddington made the first observation of the bending of star light near the occluded solar disk, during an eclipse in May 1919. As usual, the early observations and calculations could always be questioned, and needed further confirmation. More accurate modern satellite-based measurements (using radar ranging of planets etc.) have confirmed agreement with the predictions of GR to 10 parts in a million. (iii) Clocks subject to a gravitational field are slowed down. This is not just the time dilation associated with the Lorentz-Fitzgerald contraction (secs. 4.6, 4.7). We are comparing time measured on two clocks near a mass M, e.g., near a blackhole. The clock closer to M runs more slowly. If the distance of the clock from the center of the massive object is r, the clock reading is found to be proportional to the square root of $(1 - 2MG/c^2r)$. Hence, when $r = 2MG/c^2$ the clock appears to have stopped! This radius is just the Schwarzchild radius r_s. Thus the event horizon of the blackhole splits space into two regions. Time has any sense only for $r > r_s$, i.e., outside the blackhole. In fact, physical theory as it exists applies only outside the blackhole.

If the inside of the black hole is inaccessible, the BH is identical with its surface as far as its physics goes. Stephen Hawking considered the possibility of quantum mechanical creation of particle-antiparticle pairs (see Sec. 6.10.2) in the strong gravitational field near a BH. He found that pair-processes at the BH surface allow the possibility of some radiation to come from black holes. This radiation, known as *Hawking radiation* has a temperature T_B which is inversely proportional to the mass of the black hole. Thus black holes emit radiation and evaporate, becoming hotter and hotter as they diminish in size. Bekenstein had meanwhile established that the surface area of a black hole had all the properties of entropy (Sec. 9.2.4) while the mass of the black hole plays the role of heat energy. Thus a thermodynamic description of black holes has become possible!

Einstein regarded the singularities in his equations as unphysical mathematical issues. However, modern work on neutron stars, pulsars and black holes use Schwarzschild's work as the springboard for the analysis of observational cosmological data. In effect, all the predictions of the general theory of relativity, including the existence of black holes in many galaxies have been confirmed by observations [17]. Hence general relativity, together with its singularities have to be part of any future version of, say, quantum gravity.

The Quantum Theory (QT) does not allow the show to go on without its own say. As Einstein said in 1954, one must not 'seem like an ostrich who forever buries its head in the relativistic sand in order not to face the evil quanta'. The evolution of a star to become a blackhole is modified by the quantum theory.

Chandrasekhar [46] had already, in 1936, considered the effect of quantum statistics on the gravitational collapse of stars. As a star collapses under its own weight, the matter (e.g., hydrogen) gets ionized into electrons and protons. These particles obey Fermi statistics. That is, an electron must have its own unique energy state and spin state (designated by the momentum k and the spin s). This is known as the *Pauli Exclusion Principle*, and ensures that electrons cannot be crushed together. The most one can do is to put the electrons, each into available (k, s) states, with just one particle per state, filling up to a maximum energy known as the Fermi energy E_F, corresponding to the Fermi momentum k_F. Such an electron system is said to be a degenerate system. The filled system of electrons is known as the Fermi sea (Fig. 8.2). Thus the collapse of the star is halted by the electron degeneracy effect. Such contracted stars are known as *white dwarfs*. If the mass of the white dwarf is greater than 1.44 solar masses, the electrons get crushed into the protons which together form neutrons. Thus *neutron stars* are more massive and denser than white dwarfs. If the neutron stars are heavier than 2-3 solar masses, then the neutron degeneracy effect cannot halt the gravitational collapse of the neutron star. Such stars become black holes and provide extreme tests of GR and QT.

While the special theory of relativity has been tested and confirmed at every length scale including the tiniest dimensions of elementary particle physics, we cannot say the same for general relativity. General relativity regards matter as causing a curvature in spacetime. If there is only gravitational attraction, matter will collapse or expand and there could be no stable universe. Einstein recognized this and realized a repulsive energy has to be included in his equations to balance the forces. The scale factor attached to this cosmic repulsive energy is called the cosmological constant (Sec. 5.8) and it is usually denoted by the symbol λ which has to be positive for repulsive energies.

Why should the forces be balanced? When Einstein presented GR in 1916, the common belief was that the universe was an eternal firmament which was nicely in equilibrium. Edwin Hubble's work on the expansion of the Universe as shown by the red shift of spectral lines came only in 1929. If the expansion arose from a big bang (Gamow *et al.*, [14]), then an after-glow of the big bang should exist. Such a cosmic microwave background was proposed by Alpher and Herman in 1949. This was detected only in 1960, independently of the work of Alpher *et al.*, by Penzias and Wilson.

However, even an empty universe with a positive λ has some curvature. Such an empty universe is known as a de Sitter universe. The quantum theory also begins with a matter-free universe — the vacuum state. The excitations of the vacuum appear as matter. That is, if the 'vacuum' is a tuning fork, matter is the sound of the tuning fork. Thus we have two seemingly different pictures of the nature of matter — matter as an energy curving spacetime, and as excitations. So far, the quantum theory has withstood all experimental tests and many scientists would assert that it is the most successful theory that science has ever produced. If the empty universe of Einstein is identical with the vacuum of the quantum theory, then the energy of the vacuum, known as the zero-point energy, has to appear in the equations of general relativity as the energy of the empty universe.

Einstein's equations of general relativity, containing repulsive and attractive (gravitational energy) terms gives a very rich model of the universe. Depending on the balance between the attractive and repulsive terms, we can now have an expanding or collapsing universe, or universes which oscillate between collapse and expansion. When Einstein wrote down his equations of general relativity, the universe was believed to be static, and hence the cosmological constant could be adjusted to balance the gravitational attraction. Experimental astronomy, beginning with the work of Hubble has shown that light from distant stellar objects shows a red shift -a definitive sign that the light sources are moving away from us. A whole host of other data, including those from supernova observations [179] are available nowadays for calibrating the expansion of the universe. If we think of spacetime as the surface of a rubber balloon, and the galaxies as spots marked on the balloon, then the distances between all objects increase as the balloon is being blown up. This picture of the expanding universe enables us to obtain an actual estimate of the value of λ, the cosmological constant of Einstein.

The quantum theory also provides an estimate of the zero-point energy of the vacuum and make an estimate (Sec. 5.8) of the value of λ that may be used in Einstein's equations. Unfortunately, the two estimates, i.e., from general relativity and from the quantum theory, differ by at least a factor of 10^{120}. That is an incredibly large number (1 followed by 120 zeros!). There is clearly something seriously wrong with our understanding of the quantum vacuum and the relativistic description of the empty universe. If matter is viewed as excitations of the vacuum, the 'quantized curvature' generated by these excitations should be the *gravitons* of general relativity. Hence one should be able to 'unify' the two theories and express the concepts of both theories in a common language. This unification of gravity and the quantum theory has not been achieved as yet. Efforts at 'grand unification' (Weinberg [220], Witten [230]) within supersymmetry, string theory etc., or more heretical approaches where all elementary particles are regarded as collective excitations of an even deeper level of reality (Volovik [213]) are all parts of the on going research in physics. Let us now take a slightly more technical view of the general introduction given in this section.

5.5 The metric of spacetime

We should give a slightly more technical discussion to do full justice to the underlying ideas of general relativity. It is often stated that general relativity is a very difficult subject. Actually, the basic ideas are simple although the technical manipulations can be over-whelming.

We already know that the four-dimensional distance s, evaluated from the Minkowski rule is $s^2 = (t)^2 - x^2 - y^2 - z^2$, where we use distance units such that the speed of light $c = 1$. This 4-dimensional spacetime interval, Eq.(5.3) has the interesting feature that it has terms like $x \times x$, but no cross terms like $x \times y$ etc., appear. In effect, x, y, z are said to be orthogonal directions and have *no overlap*. In order to discuss x, y, z and also the time dimension t in the same manner, it is useful to introduce similar symbols for all four dimensions of spacetime. Let us set $x^0 = t$, and $x = x^1, y = x^2, z = x^3$. Here the superscripts do not indicate powers, but count from 0 to 3, and in general, any one of them is x^μ, with $\mu = 0, 1, 2, 3$. Thus $\{x^\mu\}$ is a 4-vector of the sort already encountered in Sec. 4.5. An ordinary 3-D vector would then be $(x = x^1, y = y^2, z = z^3)$. We also have vectors which are like gradients (e.g., of a potential). Such vectors are denoted with lower indices, (e.g., $E_x = E_1$ etc).

The 4-dimensional 'interval separation' ζ introduced in Eq. 5.3 is like a scalar product (i.e., dot product) $\mathbf{a} \cdot \mathbf{b}$ between two vectors, except for the need to include the correct sequence of $(+ - - -)$ signs, i.e., the Minkowski signature. The signature determines the matrices which define the 'metric' to be used. Thus, the metric $g^{\mu\nu}$ of ordinary Euclidean space, used in the standard dot product is a matrix where all the diagonal elements are unity, while the off-diagonal elements are zero. The metric of Minkowski space, which obeys the requirements of Lorentz invariance, is denoted by $g_{\mu\nu}$ and has the element g_{11} equal to 1, while the other three diagonal elements $g_{22} = g_{33} = g_{44} = -1$. All the off-diagonal elements are zero. This is a 4×4 tensor. Its inverse is denoted by $g^{\mu\nu}$, and turns out to be identical with $g_{\mu\nu}$.

It is known as the Lorentzian metric of 'flat' spacetime, or the metric of Minkowski spacetime. Instead of using Cartesian coordinates, one may use polar coordinates (ct, r, θ, ϕ) and this is used in discussing spacetime near a spherically symmetric object like a planet or a black hole. We can use the metric to write the spacetime 'interval' as in the form

$$\begin{aligned}\zeta^2 &= x^0x^0 - x^1x^1 - x^2x^2 - x^3x^3 \\ &= \Sigma_\mu x_\mu x^\mu = \Sigma_{\mu\nu} g_{\mu\nu} x^\mu x^\nu \\ &= g_{\mu\nu} x^\mu x^\nu . \end{aligned} \tag{5.13}$$

In the above equation the first and second lines explicitly indicates summations over all the components. In the last version the summation sign (sigma) is dropped and we assumed the sign convention that all repeated indices are summed. The metric $g_{\mu\nu}$ relates vectors with upper indices with those indicated with lower indices:

$$x_\mu = g_{\mu\nu} x^\nu .$$

This introduction of the metric tensor opens the door for including the effects of curvature in spacetime. If spacetime were curved, then the elements $g_{\mu\nu}$ of the metric would no longer be simply made up of 1, −1 or zeros. If we take the 'worm's eye view' in any locality, spacetime would indeed appear flat, just as the earth would seem to be flat if we explore only a dime-sized area. But if spacetime were curved, this will become important as we extend our local view to encompass bigger and bigger regions of spacetime. If spacetime were absolutely flat then the derivatives of $g_{\mu\nu}$ with respect any variables x_η would be zero. Thus the curvature of spacetime can be related to the nonzero derivatives of the tensor $\mathbf{g}$ with respect to any x^μ.

5.6 Specifying the curvature of spacetime

There are many ways of specifying the curvature of a surface. Locally, the ground around you may seem flat. But looking towards the horizon, you need to modify your first estimate. Hence additional correction terms are needed. Mathematicians had developed convenient concepts and mathematical techniques for studying curvature, long before the advent of relativity.

In fact it is customary to define an object known as a Christoffel symbol (also called an affine connection coefficient) denoted by the symbol $\Gamma^{\sigma}_{\mu\nu}$ which has three indices and takes the form

$$\Gamma^{\sigma}_{\mu\nu} = (1/2)g^{\sigma\rho}(\partial_{\mu}g_{\nu\rho} + \partial_{\nu}g_{\rho\mu} - \partial_{\rho}g_{\mu\nu}). \quad (5.14)$$

These are then used to define an object known as the Riemann curvature tensor $R^{\sigma}_{\mu\nu\rho}$ which depends on 4 indices. Since each index can take four values 0,1,2,3, the Riemann tensor for specifying the curvature of spacetime involves $4\times4\times4\times4$, i.e., 256 components. All these components vanish (i.e., become zero) for a flat spacetime. When we go to curved spacetime, it turns out that we do not really have to work with equations dealing with 256 curvature components, because of certain inherent symmetries found in the Riemann tensor. We can then use a reduced form of the Riemann tensor known as the Ricci tensor and the Ricci scalar R defined by

$$R_{\mu\rho} = R^{\nu}_{\mu\nu\rho}, \quad R = g^{\sigma\rho}R_{\sigma\rho}. \quad (5.15)$$

We recall that in these equations the repeated indices are summed over and hence disappear. Finally, the Einstein curvature tensor $G_{\sigma\rho}$ is defined in terms of the Ricci tensor and the Ricci scalar as follows.

$$G_{\mu\nu} = R_{\mu\nu} - (1/2)Rg_{\mu\nu}. \quad (5.16)$$

It turns out that this choice for specifying the curvature of spacetime has a very special property analogous to the conservation law contained in the equation of continuity, viz., the property specified by the equation 4.7. Recalling that repeated indices are summed:.

$$\partial^{\mu}G_{\mu\nu} = 0, \quad \text{compare with } \partial_{\mu}j^{\mu} = 0. \quad (5.17)$$

This suggests that the Einstein curvature tensor is related to some flow pattern. What ever flows has energy and momentum. Hence it should not be surprising that $G_{\mu\nu}$ is related to the energy-momentum tensor usually denoted by $T_{\mu\nu}$. This is entirely analogous to the stress tensor that we encountered in Maxwell's equations (Eq. 4.8). In fact, the Einstein energy -momentum tensor is also known as the energy-stress tensor, and contains all the details of the energy and momentum (p_x, p_y, p_z) of matter causing the curvature of spacetime, as specified by the Einstein curvature tensor. The energy-momentum tensor $T_{\mu\nu}$ of a uniform distribution of matter with a rest-frame energy density ρ, and pressure p is given by the 4 by 4 array where all the off-diagonal elements are zero.

Thus general relativity relates the curvature of spacetime to the energy-momentum tensor.

5.7 Einstein's equations of general relativity

Einstein's equations simply state that the curvature of spacetime is proportional to the energy-momentum tensor.

$$G_{\mu\nu} = (8\pi G)T_{\mu\nu} . \tag{5.18}$$

Here G is just the usual gravitational constant of Newton, as in Eq. 5.9. Basically $8\pi G$ ensures that Einstein's equation becomes Newton's equation in the non-relativistic limit where velocities are small compared to that of light.

If we write Eq. 5.18 in terms of the Ricci tensor, it becomes

$$R_{\mu\nu} - (1/2)Rg_{\mu\nu} = (8\pi G)T_{\mu\nu} . \tag{5.19}$$

The left-hand side describes the curvature of spacetime, while the right-hand side describes the 'content' of spacetime (i.e., energy-momentum). The energy-momentum tensor also obeys a conservation law (i.e., a continuity equation) of the form

$$\partial^{\mu} T_{\mu\nu} = 0 . \tag{5.20}$$

This is really saying something we already know — that the energy and the momentum are conserved. We also know from Eq. 5.17 that a similar conservation law holds for the Einstein curvature — this conservation law had been known to mathematicians under the name of the *Bianchi identity*.

5.7.1 *The Friedmann-Robertson-Walker equations*

In physics we like to work with 'reductionist models', i.e., usually very simplified systems. If we consider a uniform 3-D space, it is best to use a system of coordinates which reflect that symmetry.

The usual distance element of 3-D space

$$ds^2 = dx^2 + dy^2 + dz^2, \quad \text{or} \quad (ds)^2 = (dx^1)^2 + (dx^2)^2 + (dx^3)^2 , \tag{5.21}$$

can be expressed using spherical coordinates and assuming spherical symmetry. In spherical coordinates we specify an origin, and an arbitrary point P at a radial distance R, polar angle θ, and azimuthal angle ϕ. Here the 'vertical axis' is the z-axis, and θ is the angle measured from the north pole to the point P, while ϕ is like the angle specifying the 'longitude' of a location (the latitude is simply $\pi/2 - \theta$ in radians). The radial distance R may also be written as $R = ar$ where a is a scale factor. Given small changes $d\theta, d\phi$ and dr, the change ds for 'flat' space is

$$ds^2 = a^2 \left[dr^2 + r^2 d\Omega^2\right], \quad d\Omega^2 = d\theta^2 + \sin^2\theta d\phi^2 . \tag{5.22}$$

This can be generalized to include the effect of local curvature on the line element. Thus using $k = 1, 0$ and -1 to indicate positive curvature, flat space or negative curvature, the most general, isotropic metric in 3-D has the length element

$$ds^2 = a^2 \left[dr^2/\{1 - kr^2\} + r^2 d\Omega^2\right] . \tag{5.23}$$

In 4-dimensional spacetime ds is the element of event separation. Since we assumed uniformity everywhere, we can choose the same time coordinate t at each point in space. The scale factor a can itself be a function of time. The time element dt contributes cdt to the metric. However, we can measure our distances in units such that $c = 1$. Thus we have the event element ds of the Lorentz invariant Friedmann-Robertson-Walker (FRW) metric in the form:

$$ds^2 = dt^2 - a(t)^2\left[\frac{dr^2}{1-kr^2} + r^2 d\Omega^2\right]. \tag{5.24}$$

As before, $k = 0$ for flat geometries, while $k = 1$ for a closed universe, and $k = -1$ for an open universe (negative curvature). If we take different objects labeled i at different 4-D locations $(t_i, r_i, \theta_i, \phi_i)$, they are said to be co-moving, as they remain 'for ever' at those co-moving positions, while the distance between them will change with time due to the scale factor $a(t)$. Books on general relativity show that light gets red-shifted in a FRW universe due to the time dependence of the scale factor $a(t)$. The Hubble parameter H is

$$H = a^{-1}(da/dt). \tag{5.25}$$

Einstein's equations can be used to derive equations for $a(t)$. These depend on the density of matter ρ and the pressure p. The word 'matter' in the relativistic sense means 'mass-energy'. The resulting equations, derived in standard texts on relativity, are known as *the Friedman equations*.

$$\left[\frac{da}{dt}\right]^2 = \frac{a^2(t)}{a_0^2} - k, \quad a_0^2 = \frac{3}{8\pi G\rho} \tag{5.26}$$

$$\frac{d^2a}{dt^2} = -\frac{\rho + 3p}{2\rho a_0^2}a(t). \tag{5.27}$$

Thus a_0 is a typical length scale of the universe for the mass-energy density ρ. The mass-energy could be mostly radiation, with very little mass and mostly photons, or it could be a mass dominated with very little radiation ($p \ll \rho$). Using the first of the equations, the density of the Friedman universe can be written in terms of the Hubble constant and the curvature. The critical density ρ_c for a 'flat' universe is obtained by setting $k = 0$ and evaluating the density from Eq. 5.26. With the present estimates of $H \simeq 1/14$ billion years, $\rho_c \sim 9.5 \times 10^{-30}$ g/cm^3.

The FRW-equations provide a heuristic way of incorporating the quantum theory into cosmology. Thus Wheeler, deWitt and others used these equations for introducing a 'wavefunction of the universe' (see Sec. 6.10.4).

5.8 The cosmological constant

After deriving the beautiful equation (5.18) of General Relativity which provides a geometrical meaning to the mystery of matter, Einstein considered its implications. At the time the universe was believed to be a stable system in equilibrium, as the stars of the Milky Way were found to have negligible velocities. Einstein felt that he should add a repulsive term into his equations to balance the gravitational attraction of matter — i.e., to prevent the universe from collapsing inwards.

Einstein modified his Eq. 5.19 by adding a term linear in $g_{\mu\nu}$, the metric of spacetime.

$$R_{\mu\nu} - (1/2)Rg_{\mu\nu} + \lambda g_{\mu\nu} = 8\pi G T_{\mu\nu} \tag{5.28}$$

The proportionality factor λ is Einstein's famous cosmological constant, introduced essentially as an *ad hoc* parameter. Here λ was chosen to balance the gravitational attraction of mass in the universe.

However, the tiniest fluctuation in energy or curvature would push such a model of the universe out of equilibrium. Today we know that the universe is rapidly expanding at an accelerating rate. Thus the cosmological constant has become relevant. Also, the total 'visible' mass does not account for the mass deduced from observations of cosmic microwave background and deuterium abundances. The observed perturbations of dynamics of stars in galaxies, and galaxies in galactic clusters also indicated the existence of hitherto unknown 'dark matter'. Hence a component called 'dark energy' has been proposed. Einstein's cosmological constant is the simplest way of taking account of 'dark energy'.

Modern cosmology, based on some form of 'inflation + big bang' theory, supported by observations of an expanding universe, presents a system which is far from equilibrium. Thus the cosmological constant, put into the equation *by hand* by Einstein can be given a new role. It signifies the energy of the 'empty universe', while the $T_{\mu\nu}$ term represents the usual energy-momentum contribution. The 'state of lowest energy' is often known as the vacuum state in quantum physics (QP). Thus the quantum vacuum is a 'vacuum' only in an average sense, and allows all sorts of 'vacuum fluctuations' or 'excitations' in the 'empty system'. In fact, these excitations are precisely what are known as the 'fundamental particles' of nature. Thus the vacuum is 'full' of particles and their anti-particles, and energy fields associated with them. This is just another example of the capacity of QP to unify both the vacuum and its duality, i.e., a fullness of excitations which becomes the 'fundamental particles' of physics. The energy of spontaneous fluctuations of the vacuum constitutes the 'zero-point energy' of the vacuum.

Hence we may attempt to estimate the cosmological constant by an evaluation of the zero-point energy of the vacuum using the Quantum Theory. Using the standard model of particle physics, we have positive contributions from the zero-point energy $E(k)/2$ of quantum fluctuations from Bosonic fields (e.g., photons, or possibly particles described by the Klein-Gordon equation, Eq. 6.63), and a negative term from the occupied negative Fermionic states in the Dirac sea (Sec. 6.10.2).

Since the largest contributions come from the very high momenta, the energy spectrum of the particles can be considered as $E(k) = ck$ or $E(k) = k$ in units where $c = 1$. The total zero-point energy is estimated by summing over all values of momenta k, with a cut off at the Planck energy $E_{Pl} = (hc^5/G)^{1/2} \simeq 10^{19}$ GeV. This energy scale is associated with the Planck length, $l_{Pl} = (hG/2\pi c^3)^{1/2}$ which is $\sim 10^{-35}$ m. Unfortunately, GR is not expected to hold at such small distances. We review these energy and length scales within the limits set by the Heisenberg uncertainty relations in Sec. 6.6.

The above calculation of the energy of the empty universe leads to a value which is about 10^{123} bigger than the value inferred from astrophysical observations! Clearly, 'the empty universe' of General Relativity and the 'vacuum state' of the Quantum Theory are not the same thing. We clearly don't know how to put these two theories, i.e., GR and QT together. We need a theory of gravity (i.e., quantum gravity) valid for length scales smaller than the Planck length. We return to these topics, when we discuss the quantum theory more fully.

5.9 The Inflationary Universe

Our attempt to understand the energy of the 'empty universe', i.e., the cosmological constant of General Relativity using the Quantum Theory led to a major difficulty. Modern approaches begin from quantum-field concepts that attempt to accommodate gravity. 'Inflationary theory' is such an example (Guth [93]), with some of its predictions verified to astonishing accuracy by observations. In fact, inflationary cosmology has begun to replace big bang cosmology as the mainstream point of view (cf. Linde [134]). The basic idea behind inflation theory involves the existence of a 'false vacuum' (i.e., a metastable state which is not the true vacuum) set by the high energy density of the nascent universe.

The false vacuum is set by the high energy density ρ_{qf} of various scalar quantum fields in the very early universe (a scalar field can be specified by just one number, unlike a vector field (see sec. 4.5). This energy density dominates all other mass-energy densities, and remains more or less unchanged when the universe expands. This is contrary to the behaviour of ordinary matter whose density decreases when the universe expands. It turns out that this ρ_{qf} when included in the equations of GR drives the system into a cosmic expansion which grows rapidly.

This metastable state drives the expansion described as 'inflationary'. It is in fact mind boggling. A small spot of the size of 10^{-26} meters, i.e., one thousand-trillion times smaller than an atom, blows up to the size of a cow in about 10^{-36} s — i.e., in a trillion-trillion-trillionth fraction of a second.

In fact, even the need for a metastable initial vacuum state seems to be avoided in more recent models known as 'chaotic inflation'. The scalar quantum field appearing in these theories has sometimes been called the 'quintessence field'. Even if the potential defined by the scalar quantum field ϕ were like a simple harmonic oscillator, when the expansion of the universe is taken into account, an additional friction-like term appears. This, when included in the Einstein field equations mimics the job done by the 'false vacuum' concept of the earlier theory, in a transparent manner. The duration of the inflation may be as short as 10^{-36} s. At the end of the expansion, the quantum field ϕ simply oscillates near its minimum value, and loses energy by creating particle-antiparticle pairs. These interact and equilibrate to a temperature T. By then, the universe has grown to the enormous $r_U \sim 10^{100,000,000,000}$ km, i.e., inconceivably bigger than the size of the universe visible to us, $r_u \sim 10^{23}$ km. (Our universe is just the region within which we can communicate by light signals, since the theory of relativity forbids communication faster than light. Thus $r_u \sim 1/H$, the inverse of the Hubble constant.). We see that our universe has to be homogeneous and isotropic because all uneven features were exponentially stretched out. From this time onwards, the cosmology developed for the big bang theory applies.

If the initial universe contained many regions with different magnitudes of the scalar quantum field ϕ distributed at random ('chaotically'), then some regions where ϕ is too small would not inflate. Those regions which inflate produce homogeneous, isotropic domains out of the initial 'chaos'.

One would think that this gigantic explosion must surely need a large amount of energy. However, according to Guth, a gravitational field carries negative potential energy [93]). The inflationary process produces a large amount of matter, and hence a corresponding gain in negative gravitational energy which offsets the explosive energy. Thus, as cosmic inflation takes place, the total energy remains essentially constant! After the initial inflationary phase, the universe continues in its more leisurely Hubble type of expansion. The current age of the universe, reckoning from the big bang, is estimated to be about 14 billion years.

This type of inflationary theory leads to verifiable predictions about today's condition of the universe. An important parameter is the ratio between actual mass density ρ in the universe, and a critical density ρ_c, (see Sec. 5.7.1) denoted in the cosmology literature by the letter Ω. If $\rho = \rho_c$, i.e., if $\Omega = 1$, spacetime will be essential 'flat'. That is, spacetime will then look the same in all directions — i.e., isotropic and homogeneous. On the other hand, if $\Omega > 1$, Einstein's equations imply that spacetime will be curved and would be like the earth's surface where parallel lines running north-south meet at the north and south poles. However, if $\Omega < 1$, spacetime will be a saddle-shaped open surface.

The critical density $\rho_c = 3H^2/(8\pi G)$. Here H is the Hubble constant and G is the usual Gravitational constant of Newton. It is worth noting that during the exponential inflationary period, the separation $s(t)$ between any two points in spacetime at time t expanded as $s(t) \simeq \exp(H_p t)$, where the prevailing Hubble constant $H_p = (8G\pi\rho_{qf}/3)^{1/2}$. While General Relativity does not tell us what Ω should be, inflation insists that Ω should be unity!

The inflationary expansion stretched out the initial curvature of spacetime, making it essentially flat. Similarly, the after-glow of the inflation and big bang event, i.e., the cosmic microwave background (CMB) is highly uniform and virtually isotropic. This is because parts of the sky which appear to be widely separated were initially in close contact, and hence well equilibrated with each other. Similarly, we observe no magnetic monopoles because the inflation drove their density to an infinitesimally low level. Recent observational data from the Wilkinson Microwave Anisotropy Probe (WMAP, see Sec. 2.6) of CMB, and other data show that Ω is indeed unity within 1% according to Tegmark [206].

In Ch. 3, and Sec. 3.6 we argued that the laws of physics and the inherent symmetries are consistent with a simple, isotropic arena where these laws play out. We also noted that even the sub-atomic particles described by the Standard Model are governed by various abstract symmetries. Inflation gives a neat explanation for the observed simplicity of physical law — it is not just a matter of aesthetic preference, as it is often made out to be. We already noted in Sec. 2.2 that many philosophers of science take a different attitude. Thus Kuhn believed that nature is ultimately too complex for any paradigm to succeed.

5.9.1 *M-branes and cosmology*

Inflation cosmology (IC) indicate that it can, by itself account for the CMB and other residual signs ascribed to the big bang. In fact, we may not need the concept of a big bang at all. In any case, General Relativity does not apply to length scales of the order of the Planck length, $L_p \sim 10^{-35}$ m. So we need a quantum theory. An assumption of IC is that initially the nascent universe was in some 'false vacuum', associated with a scalar quantum field, or that a chaotic distribution of such fields existed. Some of these regions inflated, while others where ϕ was too small aborted. What is the nature of this 'scalar quantum field' ϕ?

The resolution of such questions rests on formulating a quantum theory of gravity which replaces Einstein's gravity at small length scales, but merges with Einstein's GR at bigger length scales. Thus a form of GR must come out of a grand unified theory (GUT) of a quantum field. One of the candidate theories is known as string theory. In Fig. 1.3 we suggested that particles can be thought of as superpositions of waves. However, fundamental particles need to have the property that these superpositions do not die out as the waves propagate. Hence a more subtle approach is needed. Clearly, one cannot just consider them to be points, since a point of zero dimensions is a mathematically singular object.

Even the Indian philosophers of the Buddha's time, and the early Greeks like Zeno understood the conceptual difficulties associated with conceiving a 'point' like object. Mathematically, if an electron is a point particle, then its Coulomb potential $1/r$ becomes infinite at the location of the electron (i.e., at $r = 0$), and much of the physics involving electrons leads to infinite 'self-energies'. The point object is said to have a singularity at $r = 0$. These problems can be avoided if the theory is based on some (topological) concept which does not admit of point objects. A point is a zero-dimensional object. The simplest candidate which avoids point-singularities is a one-dimensional object, i.e., a string.

String theory initially explored the possibility that fundamental particles are excitations (wavelike vibrations) in string-like (1-dimensional) objects with lengths of about $L_p \sim 10^{-35}$ meters, i.e., the Planck scale of length.

This theory succeeded in showing that the standard model of fundamental particles can be realized from string theory in the relevant low energy sector, far away from the Planck regime. The theory required a 10 dimensional spacetime, with six of the space dimensions curled up ('compacted') close to the Planck scale of length, so that we do not even see them. An interesting new avenue in string theory came when it was realized that the fundamental vibrations which mimic the elementary particles need not be confined to strings. We could use

2-D membranes ('one face of a drum'), and coupled membranes, as the objects supporting the vibrations. This type of theory is known as M-brane theory. Thus a mystic might say that our spacetime can be played out with strings, drums or other instruments to utter the sacred word *Om*! One of these sheets or branes, might influence our brane by pushing the vacuum state of our brane into a metastable state (false vacuum) — i.e., generate the scalar quantum field ϕ needed in inflation models of cosmogenesis. This type of brane theory and inflation cosmology is rich enough to produce many initial vacuum states and many inflation scenarios.

That is, we can envisage all types of universes, with all types of elementary particles and interactions — not just those shown in Fig. 1.4. Some of these universes would be eternally in the inflationary state, while others would stop and cool down to form galaxies, stars, planets and living beings. Many of them could be very large, although the size visible to a technological being may be limited by an inverse of the corresponding Hubble constant, $1/H$, (e.g., 10^{23} km for us). Thus we have many many universes (multiverse) in the cosmos, causally disconnected from each other as even light (or other zero-mass particles) cannot go from one universe to another. Weinberg [218] has pointed out that if a multiverse with all conceivable universes came to be formed in the eternally inflating cosmos, then surely galaxies and life as we know them would appear in those very rare regions of the cosmos where the laws of physics are as we know them. This is the natural-anthropic principle, consistent with Nature's way where a few seeds sprout from millions of pollen scattered by the many flowers of a plant. It differs sharply from the divinely-ordained anthropic principles of Carter [43] and others.

5.10 The 'paradoxes' of relativity

Modern physics deals with length scales and time scales far removed from our ordinary 'quotidian' experiences. Thus some physical phenomena can be utterly counter-intuitive to our ordinary ways of thinking. Just as it was difficult for an ancient mind to conceive of the earth as a globe, our quotidian senses cannot intuitively conceive the sapcetime of relativity. Seemingly 'paradoxical, counter-intuitive' or 'baffling' conclusions can easily arise and their explanation is often non-trivial. These paradoxes do not imply inconsistencies in the theory. They raise instructive questions, engaging the attention of many leading thinkers.

The Michelson-Morley experiment (Sec. 4.6) revealed that no observer could 'catch up' with a pulse of light, whatever be his speed. The 'common sense' person would already find this very paradoxical. The resolution of this 'paradox' spawned the theory of relativity which entangled space with time. Thus relativity requires us to cultivate a turn of mind where spacetime is the true reality. Whenever we become sloppy in our thinking and revert to dealing with space and time as two separate entities, then we may end up in 'paradoxes'. Another class of paradoxes appears when we go to general relativity, with its blackholes and other space-time singularities. An even more demanding set of paradoxes arises when the theory of relativity is coupled with the quantum theory which has its own proper set of highly counter-intuitive phenomena. The quantum world and its seemingly paradoxical behavior will be discussed in chapter 6. In this section typical 'paradoxes' attributed to the relativistic character of nature are examined.

5.10.1 *Paradoxes involving the size of fast-moving objects*

A class of paradoxes arises from the breakdown of the quotidian sense of simultaneity when space and time become 'spacetime'. We consider a paradox known under various names, viz., the 'barn-pole' paradox, 'the ladder paradox' etc. A pole (or ladder) *ab* which is six meters long would not fit into the barn or tent *tu* which is 4 meters long (see Fig. 5.3). However, if the pole were moving with a sufficiently high velocity *v* relative to the tent, then its length would shrink as seen by an observer seated in the tent, as discussed under the Fitzgerald contraction (Sec. 4.7). If this velocity were sufficiently high to shrink it to exactly 4 meters, the pole, when moving through the tent *tu*, would fit in it perfectly with the front end *a* at *t* and the rear end *b* at *u*, as seen by an observer sitting in the tent. In this case he thinks that the pole has the same length as the tent.

However, relative to an observer P sitting on the pole, the tent is moving with a velocity *v* in the opposite direction. He would see that the tent has shrunk to become 8/3 meters, and hence the pole would be even more of a misfit, while the observer O would claim it to be a perfect fit. Here P and O totally disagree and we have a paradox. The resolution of the paradox involves working out the invariant length in 4-D spacetime (Eq. 5.3). This is the Minkowski distance *between events* which works out to be the same value for both observers O and P. What is judged to be 'simultaneous' by the observer O sitting in the tent is not seen as 'simultaneous' by the observer P. Different observers would agree if they consider fitting the 'event length' seen by the observer associated with the pole to the 'event length' seen by the observer sitting in the tent, and there would be no paradox.

A variant of the above paradox involves two trains of length L (when at rest) moving towards each other on the same track. A collision can be avoided by shunting to a siding of length S which is shorter than the length L. However, the trains are moving at relativistic speeds, and can avoid the collision as they shrink to the size S, as seen by an observer sitting near the siding. In his view, one train is on the main track, and the other train is simultaneously completely on the siding, with L

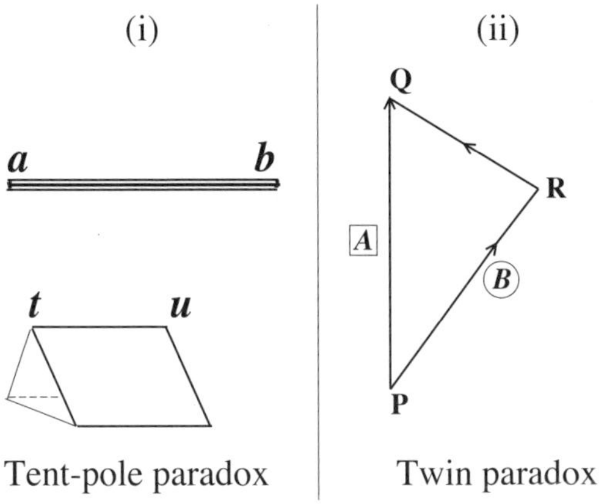

Fig. 5.3 (i) The stationary pole *ab* is too long to go into the tent *tu* and fit into it. A moving pole may be seen as fitting into *tu* by an observer sitting in the tent. (ii) In the twin paradox, one of the twins, A, stays at home. His world line PQ is vertical as in Fig. 5.2. The second twin B goes of in a space capsule. His world line PR is slanted as he is moving with a speed v with respect to A. Then B's world line becomes RQ as he turns towards the earth to meet with A at Q.

having shrunk to the size S. On the other hand, an observer on one of the trains sees the siding as having contracted so that $L > S$, and there is no question of completely placing one of the trains on the siding. However, the on-coming train has a higher relative velocity than the siding, and it is even more length-contracted than the siding. In this case, the tail end of the observer's train reaches the end of the siding just when the cow-catcher (front end) of the shortened end reaches the siding - hence there is no collision. In this paradox we see that the simple idea of 'simultaneity' is not tenable in a general sense. The simple idea of rigidity of an object is also found to be wanting. A rigid object is something which preserves the spatial distances between its parts. The generalized concept of rigidity implies that 'distances' in spacetime are preserved. These 'distances' are found to be just the time as recorded by a clock attached to the moving object. The 'clock paradox' deals precisely with this aspect of spacetime.

5.10.2 *The Andromeda-attack paradox*

A simple paradox involving the violation of our innate sense of 'right-now' and 'time order' should be instructive. All healthy terrestrial beings have the same concept of 'right-now' even though it may be night in New York and morning in Mumbai, as seen by TV newscasters who appear on the same screen from different time zones. The largest distances on earth, and our relative speeds are minuscule in comparison to astronomical length scales and velocity scales, where our intuitive ideas fail. And yet, as Roger Penrose [160] dramatically points out, even with quite slow relative velocities, as for two individuals A and B walking past each other, 'the events on the Andromeda galaxy judged by the two people to be simultaneous with the instant they pass one another could amount to a difference of several days. For one person (A), the space fleet launched with the intent to wipe out life on the planet Earth is already on its way; while for the other (B), the very decision about whether or not to launch that fleet has not yet even been made!'

However, for both A and B, the decision to attack was made before the actual launch of the fleet. No violation of causality is involved. A more 'realistic example' would be the appearance, growth and disappearance of a supernova, as recorded by two telescopes T_1 and T_2 in space with a constant velocity between them. By the time T_2 reports the appearance of the supernova, T_1 may have already reported both the appearance and the disappearance. However, for both telescopes, the appearance of the supernova would preceded maximum brilliance and final disappearance, in the 'right' (causal) time order.

5.10.3 *The clock paradox or the traveling-twin paradox*

This is a celebrated paradox that drew the attention of Einstein, Langevin, Bohm, Wheeler and other leading thinkers who have contributed to 20th century physics. In Einstein's paper on Special Relativity, Einstein discussed what he called a theorem about synchronized clocks. That is: 'if two synchronized clocks A and B are initially at the same location, and if B travels round a closed path back to A, the clock B will run slow relative to A'. Just as the 'Fitzgerald contraction' of length was paramount in the barn-pole paradox, the 'Larmor dilation' of time, first discussed by the Irish physicist Joseph Larmor plays the key role here.

From Eq. 5.4 we know that the time unit t_A on the clock A, and the time unit t_B on the clock B are related by $t_A = t_B\beta$, where $\beta = 1/(1-v^2/c^2)^{1/2}$ is the dilation factor, and v/c is the speed of the traveling clock in units of the velocity of light c.. It is this slowing down of the traveling clock that Einstein elevated to the rank of 'theorem'. Time dilation has actually been tested experimentally using unstable elementary particles like muons. Their half-life is a measure of the duration of their motion. Muons when at rest decay into an electron and two neutrinos with a half-life of 2.2 microseconds. If they are moving at relativistic speeds, as is the case when they are formed in the upper atmosphere (due to the action of cosmic rays), their half-life gets extended due to time dilation. Hence, instead of decaying in a few microseconds, they survive for much longer periods sufficient to guarantee their arrival on earth. Thus the time dilation effect is not some virtual effect; nor an inconsistency in the theory of special relativity; it is a real physical effect which has been predicted and also verified experimentally.

Devices using the global-positioning system (GPS) use distances and times relayed from several satellites to compute the GPS's location using Einstein's equations in the computations, thus providing a constant 'experimental' check of the theory. Since the satellites are moving high above the earth at high speeds, both the special theory of relativity and the general theory of relativity come into play.

If the clock B were moving at 3/5 the velocity of light, then the dilation factor is 5/4. Thus when the clock B shows an elapsed time of 4 hours, the clock A will already show 5 elapsed hours. This 'clock paradox', was cast into a very striking statement by the French Physicist Paul Langevin in 1911 who replaced the synchronized clocks by identical twins A and B. The twin A stays at home, while the twin B flies out at the time instant P in a rocket and returns to earth at the moment Q, only to find that his twin A has aged by the time dilation effect. It is rather hard for us to intuitively accept that one can 'defeat' the aging process by going on a rocket flight. Thus the clock paradox has now taken on a more painful human character! It is clearly not surprising that initially there was much opposition to the theory of relativity from religious and other absolute-time advocacy groups.

The easiest way to 'understand' the 'twin paradox' is to calculate the 'proper time' of the two twins named A and B. Let the 4-D position of A, who stays 'at home' be $(t,x,x,z) = (0,0,0,0)$ initially. After 10 years it becomes $(10,0,0,0)$. The moving twin B has 4-D positions given by $(0,0,0,0)$ initially; he travels in the x-direction at a speed of $v_x = (\sqrt{3})c/2$ per year for five years. Then his position would be $(5,5v_x,0,0)$. Then he turns around (i.e., an acceleration to make v_x to $-v_x$) and travels back to reach the 4-D point $(10,0,0,0)$. The proper time intervals τ for each twin can be calculated as:

$$\tau = \sqrt{(\Delta T)^2 - (v_x\Delta T/c)^2 - (v_z\Delta T/c)^2) - v_z\Delta T/c)^2} \tag{5.29}$$

$$\tau_A = = 10 \text{ years} \tag{5.30}$$

$$\tau_B = \tau_{go} + \tau_{come} = 2\sqrt{5^2 - (3/2)5^2} = 5 \text{ years} \tag{5.31}$$

Thus the proper time of the stay-at-home twin is twice that of the traveling twin. This is a result of the 4-D spacetime of special relativity.

Normally, one would think that B's motion relative to A should be equivalent to A's motion relative to B, and hence the twins would age in the same manner.

The 'clock theorem' of Einstein — moving clocks tick more slowly than a 'stationary' clock — as well as the 'Jumaux voyageurs' paradox of Paul Langevin are cast within the special theory of relativity (SR). The stay-at-home twin remains in the inertial frame of the earth. The traveler leaves it and re-enters it when $v_x \to -v_x$. Thus there is a real asymmetry between them. One might think that what is *observed* relative to A is also seen relative to B, and so both twins would age in the same way. It turns out that this is not so. The easiest way to reveal this is for A and B to attempt to monitor each other's clocks using light beams. Now one has to include the time delay in receiving these light signals in the analysis. (e.g., see Davis [52]). Then it becomes clear that the traveling twin did have a slow clock, and presumably a slowed heart-beat as well! The explanations of the paradox as envisaged in the earliest versions of the twin paradox are purely a matter of SR. Nevertheless, there is some room for further analysis. The traveling twin B has to change course to return, and hence subjects himself to accelerations. Thus A and B are living out quite different lives. While this might be significant from a 'biological' point of view, the physics of acceleration has to be taken seriously. Since such accelerations and the effects of gravitational fields are indistinguishable in general relativity (GR), some authors have invoked GR as well, to give a more complete discussion of the 'twin paradox'. Thus today there is a rich literature which discusses many variants of the twin paradox. These may involve flat geometries, twins falling into black holes and other potentially very interesting situations. For some of these discussion, we refer the reader to Wheeler [224], Shultz [196], Unruh [209] and references therein.

5.11 Our quotidian sense of space and time

Our every day sense of space and time are that of two distinct realities. Space is like a boat where we are sitting in, while time is a relentless stream carrying the boat forward. There is a distinct 'arrow' of time, in the sense that we cannot 'go back' in time, although we can go back and forth in 'space'. We as sentient beings have sampled our world, and in the course of evolution, developed innate concepts which are adequate tools for organizing the information collected from our sensory organs. That is, our innate concepts of space and time are components of the algorithm (model of reality) used by the software of the brain. This software controls our conscious and unconscious actions.

In an earlier age, this sort of discussion was cast in philosophical or theological language. St. Augustine and St. Thomas Aquinas grappled with problems of an abrupt beginning for time, or if there was an eternal flow of time into which God created the world from an eternal and unchanging heaven. Did He stand out of space and time (whatever that means) when He created space-time?

Early Indian philosophers and Buddhists in particular argued that there are only 'naama' (roughly translated as 'sounds') and 'roopa' (images) co-existing to form events. Modern philosophical discussions may begin with those of Immanuel Kant, where space and time are regarded as *a priori* concepts formed in the mind to arrange the sensory inputs from the external world perceived by us.

Humans have evolved in their brains neural nets rendering a mapping (i.e, projection) of the more complete reality external to it, interpreting it using its own inputs. This is similar to some observer taking a cone and selecting to make conic sections (Sec. 2.4.2) of the reality (cone) in front of him. The conic sections may be circles, ellipses, or straight lines, and hence may seem starkly different (Fig. 2.4) from the more complete reality of a 'solid cone'. However, they are valid (2-dimensional) 'mappings' or projections of the 3-dimensional object. In the same sense, our quotidian sense of space (three dimensions), and time as a separate dimension, form a valid practical mapping of the 4-dimensional world of spacetime. Concepts like simultaneity, past and present become elementary and unambiguous in this simplified quotidian world. Einstein has not overthrown the ordinary world of space and time. He merely showed how it should be modified to include other energy and length scales.

5.11.1 *Physics and our innate notion of time*

The theory of relativity says that the world is four-dimensional, without a unique separation of space and time. Instead, at each point in spacetime there are many time-like 'directions' and many space-like directions. Our physical sense of 'space' and 'time' results from an 'information gathering process' that samples this 4-D spacetime based on our 'experience' that are stored in our brain cells. We discuss in Sec. 14.2.1 how spatial information is recorded in 'place cells' of the brain. Such recording involves changes in the synapses of neurons in the hippocampus. These records constitute a 'memory' that defines the *past*, because making records dissipates biological energy, costs entropy and *sets a time direction*. Thus an organism forms a localized notion of 'now' (time), and 'here' (space) at each point along its world line, and holds them in memory.

That is, we have an innate sense of an absolute past and an absolute future, with the future inexorably 'becoming' the past. The biological processes of the body, e.g., eating, digestion, work and sleep are chemical processes that have their own time scales. In effect, living organisms, whether they be simple cyanobacteria, or humans, have evolved to respond optimally to the cyclic process of day and night rhythm. This is achieved by genes which are rhythmically expressed ('turned on or off') to make chemicals in the brain and various tissues of the body. Their net effect is to rhythmically alter the concentration of hormones and other essential chemicals in cells providing a built-in clock (the 'circadian clock', see Sec. 1.4.1) which responds to the day-night cycle of our environment. Cells replicate, grow and die, and this too has an arrow of time. All these processes depend on the legacy of free energy infused into the universe during its early inflationary stage. Biological processes are also a path for the degradation of this free energy.

On the other hand, the mechanical world of physics revealed by Galileo and Newton describes the world in terms of positions of particles (denoted by $\vec{q}$) and their motion (i.e., momenta, denoted by $\vec{p}$). Time is only a 'parameter' t which is inferred from the motion of some object. Indeed even the behaviour of the quantum world (Sec. 6.1) is discussed using the mathematical equivalent of operations carried out on the observed system by 'operators'. Thus we have a position operator $\hat{q}$, and another operator $\hat{p}$ to refer to the momentum, but no need to directly invoke 'time'. Of course, momentum implicitly involves time. Hence 'time' in physics is a 'deduced reality', or a 'parameter' (denoted by t) which occurs in the model. This is also true in Hamiltonian or Lagrangian formulations of classical mechanics. The quantum theory has no 'operator' whose mean value or eigenvalue gives the measured time. Wolfgang Pauli (part I-[156]) pointed out that t must be an ordinary quantity (i.e., a c-number instead of an operator). According to quantum mechanics, if we could treat t as an operator, the commutator of t with the Hamiltonian should give the 'time dependence of t', while its mean value should be the time observable. However, we have no such operator or time observable. In fact, the 'time' taken by a quantum particle to cross a barrier (traversal time) is thus not a measurable quantity. Various attempts were made in the 1990s by Rolf Landauer, Marcus Buttiker and others to define how long an electron or photon would take to traverse a barrier. In fact , it does not have a unique specification. Probably the best one can do is to use Bohmian trajectory calculations of the sort carried out by Leavens and Aers [130]. In effect, quantum processes can be discussed without any reference to time, just as economic transactions can be rewritten without any reference to money!

Our biological sense of time has a commonality because we are all 'made' in the same way. But, just as the size of different people's 'foot' is not quite the same, our sense of time also differs somewhat. Astrology and time keeping in ancient times were 'useful' sciences associated with magic, religion, and prophesy. Even today, many astrologers in India prepare their own 'ephemeris' using ancient calculational methods. Navigators had also found the need for accurate measurements of time, distance and location. A more reproducible method of time measurement became possible when person-centered or 'cultural' (subjective) approaches were replaced by an 'objective' method based on simple cyclic processes which could be better controlled. Here again we see that the degree of success of the 'scientific method' involves the establishment and acceptance of common procedures and tools independent of the political, cultural, biological and other factors that weigh in on the 'subjective approach'.

Galileo, as well as some earlier Arab scientists had recognized that the duration taken for the back and forth motion of a simple harmonic oscillator, e.g., the oscillations of a pendulum could serve as a reproducible unit-measure of time. Today the oscillations of light (light waves) emitted by excited cesium atoms are used to define a unit of time (the second). Such a device is called a Cesium atomic clock. Its astounding accuracy may be unnecessary for dealing with human agendas; but such accuracy is needed in astronomy and in scientific research. Thus the standard-clock unit of time is an objective idealization (or reductionist extraction) from the initial, rough biological or psychological unit of 'subjective' time based on the variation of day and night. Note that this gives a *unit of time*, while the direction of time resides in the *direction* set by irreversible processes.

Although the simple idea of simultaneity is replaced in the theory of relativity by a more sophisticated concept based on event separation, there is still a meaning to a causally connected past, present and future. In Fig. 5.2 we discussed how our 'knowable world' and its causal past and future are contained in the 'light cones' associated with our location in space-time. Although our equations of physics work equally well with time t running forwards, or with $-t$, i.e., time running backwards, we have no way of traveling into the past. Although our differential equations have time symmetry, *the boundary conditions* which determine their solutions are not symmetric. Thus there is an arrow of time, inexorably pointing to the 'future'. One obvious reason for such an arrow is the initial cosmological asymmetry. The universe started from a very low entropy, pre-inflationary state. Inflation and the big bang selected a highly asymmetric evolution triggered by the initial reservoir of free energy. The second law of thermodynamics (Sec. 9.1.1) sets a direction for all processes, where the reservoir of free energy inherited from the early universe is degraded and the entropy of the universe increases. This is the cosmological arrow of time. Irreversibility is also a property of many-particle systems due to the very long Poincaré recurrence times. However, these times greatly exceed cosmic times, and hence periodicity becomes a one-way trip. We meet these ideas again in Chapter 9.

Chapter 6

The Quantum World and 'Reality'

This chapter examines the nature of quantum reality by looking at basic concepts of the quantum theory of non-interacting particles, as understood within the conventional 'Copenhagen', or Bohmian and other interpretations of the quantum theory. Relativistic quantum theories, and essential cosmological applications are also discussed, since they shape our understanding of physical reality.

6.1 Innate reality and deduced reality

We have a sense of an 'external reality' which we know intuitively, and use in our day to day tasks. Anticipation of the immediate future requires us to involuntarily grasp the mechanical laws of motion as well as the behavior of living creatures which share our environment. Thus we have a sense of other beings — an innate 'theory of the mind' [171] which enables us to guess the intent of others. Even today, for the vast majority of human beings, this intuitive reality includes the effect of 'invisible' beings, astral influences, divine and satanic forces, ancestral spirits, karma, and so forth. These 'non-rational' factors, spooky action at a distance, etc., are invoked to 'make sense' of things that one may not comprehend — terrifying events, strange coincidences and other 'intangible' aspects of daily life. Why was Rama spared when Sita was washed away by a Tsunami? Such emotional questions are as real as 'why is the tap leaking'?

This 'intuitive reality' is an evolutionary dividend possessed by the primitive man, and to varying extents by other animals [172]. Hence we may call it an 'innate reality'. This innate reality summarizes local experience, and involves a more or less 'flat' earth which is stationary.

However, this intuitive reality got modified to what we may call the first-level of deduced reality already by the Hellenic period. Land surveying, engineering works, ocean navigation and warfare by early man, ushered in this first-level of 'deduced reality' where the shape of the earth, its relation to the planets etc., became clearer with the invention of new tools like the telescope. This deduced reality developed into a sophisticated framework with the sun as the center of the solar system, complete with Newtonian mechanics. This Newtonian world-view is at first sight strikingly counter-intuitive and psychologically inconsistent with the 'innate reality' that we described before. Yet, today every school child assimilates it and accepts it, with little effort, and some writers are wont to claim that 'classical mechanics does not need any interpretation' because it is so intuitive. One has to just go back to pre-Newtonian ideas about force, space, motion, teleology, fate, divine wrath etc., to understand that this was not so.

The deduced reality based on Newton has banished capricious spooky action at a distance (e.g, by magic) which violates physical law, while allowing for a form of action at a distance via gravitation. Electromagnetism brought another extension of this world-view by introducing the concept of action via fields. This too follows well-defined laws, and did not have the personalized power ascribed to magical processes. Thus we 'understand' how radio waves and other energy fields can propagate and act at a distance; this is not spooky.

The beginning of the twentieth century brought about two new revolutions in our view of deduced reality, i.e., Relativity (Ch. 5) arising from the finiteness of the velocity of light, and Quantum Mechanics (QM) arising from the finiteness of quanta. Relativity has revolutionized our understanding of the large-scale structure of our world, without substantially changing *the local, i.e., short-ranged structure of the deduced reality* which is essentially as understood by Galileo, Newton, and Boltzmann. Even the advent of the GPS ('global positioning system') which uses general relativity is merely a modern means of getting at a local map of a neighborhood previously supplied free by automobile associations.

The study of matter within the Newtonian viewpoint had increasingly suggested that matter could be considered as a collection of tiny interacting particles named 'atoms'. This was the 'corpuscular theory of matter, with a pedigree going back to the Greeks and the Indians. A *new revolution in physics came about when very small length scales typical of atomic and subatomic phenomena became accessible*. Just as the telescope and the spectroscope opened up large length scales, spectroscopes, particle accelerators and detectors like *Gargamelle* (see Fig. 2.5) were the new tools that ushered in the sub-atomic world.

Quantum mechanics deals primarily with the very small scale building blocks of the macroscopic world. These objects, called elementary particles (e.g., electrons, photons, quarks, etc.) are completely remote from our 'innate-intuitive' sense of reality and from the 'deduced reality' of Newtonian physics. In fact, distinguished scientists like Wilhelm Ostwald, Nobel Laureate in chemistry (1909), and philosophers like Ernst Mach doubted even the very existence of this atomic and sub-atomic world until overwhelming evidence piled up by the end of the 19th century. Today we have a detailed, well-tested theory of this sub-atomic world which is far more counter-intuitive than the world of elementary atoms that Ostwald and others objected to. The 'deduced reality' embodied in Newtonian physics has come to be called 'classical physics', while the world of elementary particles is said to be governed by 'quantum physics'.

Many popular writers like to claim that the world of modern physics is 'weird'. We are often told that even Feynman had stated that trying to 'understand' the quantum world is futile, while Einstein struggled with it. In this book we are more optimistic and follow Einstein's own aphorism that 'the most incomprehensible thing about the world is that it is comprehensible'!

Given that quantum mechanics governs the world of sub-atomic particles, it is not surprising that the 'reality' revealed by QM is very different to the innate, 'intuitive' reality of our everyday experience, or the more accurate macroscopic reality used by the civil engineer. However, quantum reality is in some sense the fundamental reality of the world, and we need to assimilate it. A working physicist who uses quantum mechanics regularly gets quite accustomed to it, and soon finds that it is not strange or 'weird'. QM reduces to the working reality of the engineer at everyday length scales, and has so far proved to be an astonishingly accurate theory without any disagreement with experiment.

However, what if QM actually forbids a deduced reality? What if the very process of observation changes the world in an essential way, preventing us from identifying a unique reality external to us? What if a 'conscious mind' has to intervene to give 'existence' to what we perceive, as Eugene Wigner and several other physicists were wont to believe [214, 228, 160, 201]? The quantum theory has actually opened up the possibility of asking just such questions in a formal way. For some, there is no external reality other than a reality rooted in human consciousness. For others, there are multiple layers of realities co-existing in separate many-worlds and new branches breeding with every act of observation [56]. However, a more 'usual' world-view is also supported by many scientists who have examined the quantum theory [26, 211, 141].

In Sec. 2.1 we discussed what some thinkers have identified as fundamental questions. Thus Wheeler [225] in his intriguing way wrote in 1992, 'Among all the mysteries that still confront us in our probing of nature, none present more challenge than these',

- How come the quantum?

- How come one world 'from the registrations of many observer-participants'?
- Could the physical world be an 'it from bit'?

In this chapter we examine the quantum theory while taking up these questions and related issues stated in a less mysterious manner. We compare and contrast the 'classical reality' of the everyday world with the 'reality' that is implied by the quantum theory, viz., 'quantum reality'.

6.2 Quantum reality as a 'prepared' reality

What are the main characteristics of quantum reality that we can compare or contrast with classical reality? It turns out that the prediction of properties of quantum systems is similar to predictions in statistics; i.e., what is reported is a mean value or an 'expectation value'. However, the knowledge of the quantum system becomes available only after a very non-intuitive process of 'information processing'. Let us compare the process of normal statistical information gathering with an analogous quantum problem. Let us consider 'grading' a set of objects. In one case we deal with school pupils graded according to their marks, while in the other we consider electrons graded according to their energies.

For instance, given 100 students sitting for an examination, we may find that, say 18 students got a 40% mark while 5 students got a 15% mark etc. In general, given a sample of N_s students, we take the number $N(x)$ who have scored the mark x and plot a graph of $N(x)$ against x. In general the probability $W(x)$ of scoring the mark x is $N(x)/N_s$ for large N_s. Since the probability that the student would score some mark whatever is unity, the sum of $W(x)$ for every value of x must add up to unity. That is, the integral of $W(x)$ over x in the range 0–100 is unity. We look at a classical probability 'weight function' $W(x)$ of this sort in more detail below (Sec. 6.2.1).

In QM, instead of a probability $W(x)$, we have to work with a probability *amplitude* $\psi(x)$ whose square gives the probability.

Consider grading electrons, i.e., quantum particles, according to their energy E when emitted from a slit to arrive at a detector. The energy E is directly related to the momentum $\vec{p}$ since $E = p^2/2m$ where m is the mass of the electron. Hence we need the probability weight function $W(p)$ for the momentum distribution. A classical particle has a trajectory which gives the momentum $\vec{p}$ at any location $\vec{q}$, given its initial location and momentum at the slit. However, *quantum mechanics does not allow us to specify positions and momenta simultaneously*. In Fig. 1.3 we thought of a particle (i.e., a localized excitation) as a superposition of waves with many momenta. Hence there is no unique momentum associated with the localized particle. Thus no specifiable path (trajectory) is followed by a quantum particle in the usual picture of QM. If we could draw the path followed by a particle, its gradient at time t would give us the velocity $\vec{v}$, and hence the moment $\vec{p} = m\vec{v}$. Thus we would simultaneously know both $\vec{q}$, and $\vec{p}$. This is not allowed.

Instead, QM allows all possible momenta, and attaches a probability to each such momentum via a probability amplitude. This is known as the 'momentum wavefunction' of the particle, $\psi(\vec{p})$ when the particle momentum is used, with no specification of the position $\vec{q}$. Alternatively, we may use the wavefunction $\psi(\vec{q})$ where the position appears, with no specification of the momentum. This is known as the *position representation* of the wavefunction.

The probability amplitude $\psi(\vec{q})$ is calculated by solving the Schrödinger equation of QM for the boundary conditions used in the experiment. Once $\psi(\vec{q})$ is obtained, its square $|\psi|^2$ gives the probability of finding an electron at the location $\vec{q}$. If we calculate $\phi(\vec{p})$, its square can be used to calculate the probability $|\phi(p)|^2$ for finding an electron in the momentum state $\vec{p}$ and hence in the energy state E.

The position representation $\psi(\vec{q})$ and the momentum representation $\phi(\vec{p})$ are related to each other by a Fourier transform. Thus $\phi(\vec{p}) = \int \psi(\vec{q}) \exp(-i\vec{p}\cdot\vec{q}) d\vec{q}/V^{3/2}$, where V is the volume of the system. This can be written compactly as $\langle p|\psi\rangle$ where the complex conjugate of the plane wave state is indicated as a state vector $\langle p|$. A dynamical property $A(\vec{q},\vec{p})$ is represented by an operator $\hat{A}\,[\vec{q},(1/i)\partial/\partial\vec{q}]$ in QM. The expectation value of the dynamical property in QM is $\int \psi^*(\vec{q})\hat{A}\,[\vec{q},(1/i)\partial/\partial\vec{q}]\,\psi(\vec{q})d\vec{q}$. This expectation value is written as $\langle\psi|\hat{A}|\psi\rangle$, with the wavefunctions normalized, i.e., $\langle\psi|\psi\rangle = 1$.

The intrusion of the wavefunction ψ into the statistical description leads to several very important characteristics of quantum systems. Some such characteristics may be found in classical systems of interacting particles, but not in the over-whelming way typical of quantum systems. These characteristics are:

- Usually the expectation values of quantum systems are *non-local*, in the sense that they depend on all parts of the system including the nature of the boundary defining the system via ψ.
- A quantum system which has many possible states of 'existence', usually manifests as a *superposition* of all the states. Similarly, two or more quantum systems after interaction may form superposition states known as 'entangled' states.
- The observed value of a property A depends on the sequence of the operations carried out on the system — including the preparation for the measurement. Thus the expectation value is *contextual*. Hence the very measurement process is intrinsically included in the expectation value. In effect, the final ψ used in obtaining the average value $\langle\psi|\hat{A}|\psi\rangle$ contains the effect of the

final conditions imposed on the system. How a system (e.g., a two-slit experimental setup) is prepared, e.g., with one slit open, or two slits open, gets included in the wavefunction. The setup for measurement, i.e., with an interference setup, or with detectors placed near one of the slits etc., also gets included in the final wave function. However, textbook discussions may use simplified idealized accounts of isolated unobserved systems. The discussion of such ideal systems contains many pitfalls. These have led to a vast literature on interpreting QM. Furthermore, the name 'average value' truly means that if only a single experiment is done, the observed result may be wildly different from the predicted expectation value.

- There is no limiting process where by we can reduce the 'disturbing' effect of the measurement to zero by, say, reducing the intensity or power of the probe. This is because a quantum system responds as a holistic system *via finite quantum jumps* whose magnitude cannot be made as small as we please. However, a 'weak-measurement' theory has been proposed (see Sec. 7.5.2), but this is highly controversial at present.

The strong dependence of quantum systems on the way they are prepared, i.e., the boundary conditions in space and time imposed on the system, makes it appropriate to refer to quantum reality as a *prepared reality*.

An experimentalist proposing to examine some facet of nature has to make a demarcation or selection of a part of the external world. Certain boundary conditions have to be set up. Certain variables have to be kept under control while others are allowed to vary. Experimental devices, i.e., apparatus, have to be set up within the periphery of the demarcation. This is what we call a process of *reduction* or *preparing the system*. The experimentalist does not study the whole universe in all its glory. However, some theorists do (see Sec. 6.10.4).

Even in classical mechanics, the selection is further simplified by various constraints applied to it, to make it some sort of an 'ideal system' amenable to physics. If we are studying the behavior of gases, a sample of the gas is contained in a suitable chamber. Its thermodynamic properties depend on the pressure P, volume V, temperature T, and the chemical constituents of the gas. Ideally, something very close to 'Boyle's law', (i.e., the product PV is approximately constant) can be established if the temperature is held constant. Such observations of the properties of gases can be understood within *the kinetic theory of gases*, where the gas is considered to be made up of atomic particles undergoing random collisions with one another, and with the measuring probe. The particle motions and collisions obey Newton's laws. The temperature T becomes a *theory laden*, highly enriched concept. It becomes a measure of the mean kinetic energy of the gas atoms.

All properties of the system are obtained as probability distributions constructed from the appropriate many-particle wavefunction $\psi(\vec{q}_1, \vec{q}_2, \dots)$ or its momentum representation. The classical probability distribution function W is much more direct, and can depend on the positions and momenta simultaneously.

6.2.1 *The probability weight function*

Let us consider a very classical system, namely a large sample of N_s school children, say a thousand (i.e., $N_s = 1000$) preparing for a qualifying examination. Variants of the method described here are followed in many educational institutions, e.g., the method known as the *stanine* method. We count the number of students $N(m)$ getting a mark m, and construct the percentage number $n(m) = N(m) \times 100/N$ with the mark m. This is plotted as the vertical axis against m along the horizontal axis. This gives a graph of the *distribution function* $n(m)$. An important property of the distribution is the average mark $\bar{m}$, i.e., the mean value, obtained by adding all the marks and dividing by N_s.

It is common to scale all marks (say different teachers from several grade 12 classes in several schools) so that all curves have the median at 50% with the 20% near the median falling within the 45%–55% middle bin, with all marks scaled correspondingly. This enables one to compare different groups of students coming from different sources on a somewhat equitable basis.

The test data can be scaled to show an average mark of $\bar{m}$ equal to 50%. Plotting $n(m)$ for each mark is equivalent to dividing the range 0–100 into 100 bins, and finding how many students score in each bin. To simplify matters, we can partition the range into just 9 bins, with the bins arranged symmetrically around the mean value $\bar{m}$. For example, taking $\bar{m} = 50$, we can partition the range of marks, in the range 0–100 into 9 bins for marks as follows. All marks below 15% are put into bin 1, while all marks above 85% are put into bin 9. The in-between range is split into seven equal-sized bins, i.e., 15–25%, is the second bin, and then 25–35% 35–45% 45–55%, 55–65%, 65–75%, with 75–85% being the eighth bin. Thus the first bin and the last bin (9th bin) are the 'tail ends'. Of course, it is better to use 100-bins rather than this *coarse graining* of the data by using only 9 bins. On the other hand, we may argue that the subjectivity of the marking of examination questions is such that an examiner cannot really evaluate an answer sheet to better than an uncertainty of 10 marks.

A typical set of examination results may look like the data points shown as boxes in Fig. 6.1 where we have used a marking range 0–100. The fractional number of students falling into each bin gives us a probability weight $W(x) = n(x)/100$ for finding a given mark. The assumption we make is that the marks would be distributed randomly in the range 0–100. It turns out that such data

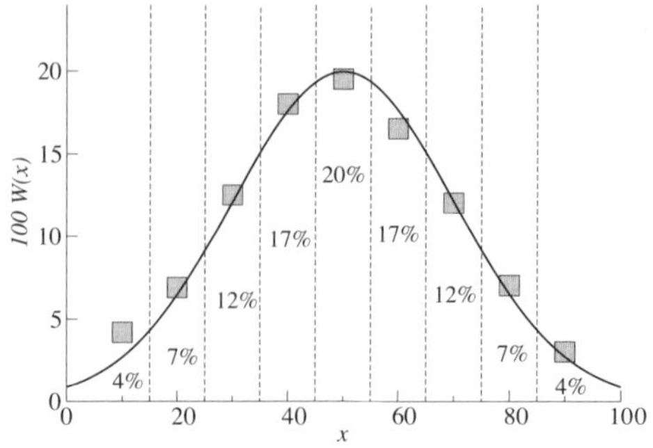

Fig. 6.1 Examination marks (x-axis) in the range 0 to 100 are analyzed into 7 bins of equal size, and two tail-end bins which are bin 1 and bin 9. The number of students $N(x)$ scoring marks in each bin is plotted (vertical axis) as a percentage along the vertical axis. The resulting data points fall approximately on a bell-shaped curve known as the normal distribution, or the Gaussian probability weight function $W(x)$.

approximately fit a bell-shaped curve known as a 'Gaussian distribution', or a 'normal distribution'. The curve shown here has a mean or median value $\bar{x}$ equal to 50, and a standard deviation σ of 20. The 'standard deviation' is the off-set from the average mark (median) such that approximately 2/3 of the sample would fall within it. That is, roughly the middle bin and 1.5 bins left and right of it (in the figure) would be within one standard deviation from the peak. The bell-shaped curve becomes flatter when the standard deviation σ becomes larger, and more steep when σ becomes small. So, if all the students got the median mark, the curve would become a sharp spike at $x = 50$, and σ would be zero. The bell shaped distribution is ubiquitous in nature.

The bell curve is a main-stay of statistical distributions, and is given by the form

$$W(x) = \frac{1}{\sqrt{(2\pi\sigma^2)}} \exp\left[-\frac{(x-\bar{x})^2}{2\sigma^2}\right] \tag{6.1}$$

Here $\bar{x}$ is the mean or median value of the distribution. The standard deviation σ is a measure of the width of the Gaussian. Carl Friedrich Gauss is usually credited with discovering the 'normal distribution' around ~1809. Hence it is also known also as the Gaussian distribution. In fact, usually, any function of the from $\exp(-ax^2)$ is known as a Gaussian. The justification for the now widely used method of least-square fitting of data was given by Gauss, for errors obeying the normal distribution, while Laplace established the 'central limit theorem' which states that generally, the average value of a large number of random draws from a given sample is distributed approximately as in the bell curve. The examination marks discussed previously were data drawn from such a sample.

A striking example of the application of the Gaussian distribution is found in the kinetic theory of gases. James Clerk Maxwell and Ludwig Boltzmann showed that the velocities, momenta and other physical properties can be expressed in terms of the Gaussian distribution. Hence the Gaussian forms used in the kinetic theory of fluids are often referred to as 'Maxwell-Boltzmann distributions'.

The kinetic theory assumes that each gas atom is like a billiard ball of mass m, moving with a definite velocity $\vec{v} = (v_x, v_y, v_z)$ at each point x, y, z on a trajectory, and suffering collisions that lead to a sharing of the energy and setting up a common temperature. This common temperature T given in energy units (i.e., with the Boltzmann constant $k_b = 1$) is proportional to the mean kinetic energy.

The Maxwell-Boltzmann distribution of velocities in an ideal gas is given by the product of the distribution in v_x, v_y and v_z. In a gas at rest the molecules can go in positive and negative directions, and hence the mean value of v in any direction is zero. The probability distribution of any arbitrary v_x (positive or negative) is given by:

$$W(v_x) = \left[\frac{m}{(2\pi T)}\right]^{1/2} \exp\left(-\frac{mv_x^2}{2T}\right) \tag{6.2}$$

$$W(v_x, v_y, v_z) = \left[\frac{m}{(2\pi T)}\right]^{3/2} \exp\left[-\frac{m(v_x^2+v_y^2+v_z^2)}{2T}\right] \tag{6.3}$$

Thus $\sqrt{(T/m)}$ is the standard deviation in any direction x, y or z. The integral of $W(v)$ over all values of v is unity as the total probability weight must be unity. Hence W is said to be a *normalized distribution*.

Kinetic theory of gases is a statistical theory. The measured, or 'observed' quantities are the statistically averaged properties of the gas. Individually, a gas atom may have any kinetic energy k say, $k(v) = (1/2)mv^2$, where v is the velocity of the particle. Thus, if the probability of finding a velocity $\vec{v}$ is $W(\vec{v})$, then the temperature is simply the mean-value $\langle W(\vec{v})v^2/(2m)\rangle$. Here we have used the angular-brackets $\langle\cdots\rangle$ symbol to indicate that the weighted average is taken over all the particles. That is, for any property $A(\vec{v})$,

$$\langle A\rangle = \int A(\vec{v})W(\vec{v})d\vec{v} \tag{6.4}$$

Here the integration is taken over all velocities ($-\infty$ to ∞) in x, y and z directions. There is no dependence on position because the gas is uniform.

In the previous paragraphs (in small print) we indicated how Newton's mechanics may be used to determine the form of the probability distribution W, needed for evaluating the observed mean value of any physical property. If we take a simple event in ordinary life, e.g., a coin toss, we just talk of the probability $P(1)$ of observing a head, and this is in the end the same as our $W(1)$. The probability of observing a head is 1/2, and observing a 3 in a dice throw is 1/6, If we toss a coin and a dice together, the joint probability of getting a head and a three is (1/2)×(1/6). In the same way if A and B are two operations with probabilities P_A and P_B. the joint probability for both A and B occurring is is the product $P_A P_B$. Of course, if the events A and B are correlated, this is not true. It turns out that events have special correlations built into them in the quantum world via the ψ function, and the simple probability product law does not hold.

In dealing with the properties of an ordinary gas or liquid, the walls of the container and other boundary effects act *locally*. If a large hole were made on one wall, the effect is not felt 'instantaneously' in the bulk, as would be the case for a wavefunction fitted to the boundary conditions prevailing at the wall. This local character of observable properties is what we might call 'near-here-ness', or 'localness', in the sense that any implicit correlations are short-ranged compared to the size of the sample being studied. If the sample is large enough, one end of a classical system can be studied without reference to the other end.

One may study a number of properties of the gas, e.g., its viscosity, heat conductivity, magnetic properties, compressibility etc. The different measurements of different properties are independent of each other. One may measure the pressure, the temperature, the viscosity etc., in an arbitrary sequence of experiments and expect to get the same results, to within the expected precision of the instruments. That is, such measurements are *non-contextual*.

6.2.2 *A new ball game: amplitudes and probabilities*

The quantum theory is regarded as a statistical theory of nature. Just as in the classical kinetic theory, we deal with probability distributions of an ensemble of data. However, it is precisely the key properties of localness and non-contextuality that are found to be absent in the quantum description of the world. A quantum system is highly dependent on the boundary conditions imposed in preparing the system for measurement. As already noted, the probability weight function W applicable to quantum systems depends in an essential way on the whole system through the wavefunction of the system. Wavefunctions, being amplitudes of various possible states available to the system, can superpose and entangle, as discussed further in this chapter and the next chapter.

If we trap an electron in a one-dimensional parabolic well, we have constructed a 'reduced system' that we can study. The probability weight function for locating the lowest energy state of the electron at some point x becomes a Guassian because the wavefunction ψ itself is a simple Gaussian $C\exp(-\alpha x^2)$. Here C is a constant chosen to ensure that the total probability is unity, while α is a measure of the steepness of the parabolic confining potential. We discuss parabolic confinement in detail in Sec. 6.4.

When some property A is to be measured, using a given experimental set up Exp_A, the wavefunction takes the form $\psi(\vec{X}, Exp_A)$. If a new experimental setup Exp_B is put in place to measure a different property B, the whole system and the wavefunction change to accommodate the new boundary conditions associated with Exp_B. Thus, the system has to be prepared for each measurement, and each of them would have their specific ψ-functions. A lot of discussion has taken place as to weather the 'consciousness' of the observer has to be included as part of the 'experimental set up'. Furthermore, the observed system is left totally changed by the very act of measurement, preventing further investigation of that system. Such 'epistemological' issues regarding quantum measurement would be taken up in Ch. 7.

We already remarked that the measured values of various properties A or B may depend on the order in which they were measured – i.e., contextuality.

Putting a sock first (A), and then the shoe (B), is not the same as putting the shoe first, and then putting the sock next. If the system were initially in the state ψ_0, let the action of the operator A put it into the state ψ_A. Then, when B acts on it after A, it now acts on ψ_A. On the other hand, if B acted first, i.e., on state ψ_0 and puts it into the state ψ_B, and hence when A acts on it after B, it is confronting a new physical situation. Thus the actions AB and BA are not equivalent.

That is, as already noted, *quantum reality is both non-local and contextual*, except in some special or trivial cases.

Most of the alleged 'weirdness' or 'paradoxes' of quantum reality arise from the holistic character of quantum systems, arising from their non-locality and contextuality. Classical systems also may have highly correlated components, and hence may be holistic and contextual. However, John Bell showed that quantum non-locality and quantum contextuality exceed classically realizable levels of non-locality and contextuality [18] in entangled systems (Ch. 7). This is summarized in two famous theorems that we discuss further in section 7.4. They are known as the Bell-Kochen-Specker theorem, dealing mainly with contextuality, and the Bell-inequalities theorem, dealing with non-locality. As befits holistic systems, the two theorems themselves are interconnected [141].

Even the wavefunction of a single particle can have two aspects superposed in it depending on the setup. The two-component whole made up of A and B is the crux of the famous 'two-slit' experiment. If a classical particle (e.g., a bullet, or a gas molecule considered in the kinetic theory of gases) were sent towards a hard screen with two slits (i.e., two small openings), than it can only go through one of the openings (if the path of the particle lines up with one of the slits). On the other hand, a quantum particle (e.g., an electron) need not be aligned to the slits, and *may seem to* go through both slits at once [75], and also be reflected from the screen.

Why can't we almost naively claim that the electron goes through both slits at once? It is easy to show that such an assumption contravenes the equation of continuity, local gauge symmetry and other matters that one has to respect in a good physical theory.

We encountered the idea of fields in discussing Maxwell's electromagnetic theory, and in discussing the metastable vacuum states in pre-inflationary cosmology (Sec. 1.3). One can think of the quantum field, characterized by the amplitude ψ as an entity that pervades through both slits and 'feels' (i.e., subject to) all the boundary conditions of the system. The electron itself is a localized excitation of the field that can occur at any location x, but subject to the probability $|\psi(x)|^2$. Thus an electron can be emitted at x_e in the vicinity of the emitter, with the probability $|\psi(x_e)|^2$. That is, the initial starting point itself is subject to an uncertainty. Once emitted, the electron has to go through *one or the other of the two slits* or be reflected back by the walls or the screen. The pattern detected on the screen after many electron have arrived on the screen also obeys $|\psi(x_s)|^2$, where x_s is any point on the screen. The formal existence of a trajectory connecting the emitter point x_e and the screen point x_s was shown in detail by Bohm, by simply re-writing Schrödinger's equation in a slightly different manner, but without any modifications to quantum mechanics itself. However, it is not an observable, but a conceptual tool (see Secs. 6.2.5, 6.7).

Would this behaviour be possible for a macroscopic object as large as a soccer ball? In principle, if one can create a quantum state ψ_{ball} of such a large object and maintain the boundary conditions needed to uphold the coherence of the system, it should show up at x with $W(x) = |\psi_{ball}(x)|^2$. Thus the problem becomes a matter of our 'engineering' competence in setting up suitable boundary conditions in preparing the system in ψ_{ball}.

When we are dealing with two or more quantum systems, a new effect known as entanglement where the the two systems join together to develop interference terms, or 'cross-terms' among their amplitudes. The resulting state cannot be written separated (factorized) back into their individual states. They remain entangled even if they are separated by astronomical distances, with no intruding perturbations.

Consider an electron at x_1 in a hydrogen atom A described by $\psi_A(x_1)$, and a second electron in another hydrogen atom B, in the quantum state $\psi_B(x_2)$. The probability of locating the 1st electron at some x is $P_A = |\psi_a(x)|^2$, and we also have a P_B for the second electron in the system B. If we look at both electrons x_1 and x_2, the usual 'classical' answer for the joint probability $P_{A,B}(x_1,x_2))$ is given by the product $P_a(x_1)P_b(x_2)$. It is possible to bring the atoms together and allow them to interact by 'jiggling them' with an external disturbance, for example by shining a light pulse in an optical cavity. Then the amplitudes ψ join to make superpositions, and we get a whole slew of possibilities in the quantum problem, that are not available in the classical problem.

For instance, we can put the two electrons in the two states ψ_A, ψ_B as $\psi_A(x_1)\psi_B(x_2)$, which is what we have to begin with, when the system A is independent of the system B. Since the electrons are indistinguishable, we can also have $\psi_A(x_2)\psi_B(x_1)$. These then are the possible *amplitudes* for the two-electron system. But then, amplitudes can, and do superpose. Hence the state of the two systems A and B brought together could be

$$\psi_{A,B} = C\psi_A(x_1)\psi_B(x_2) + D\psi_A(x_2)\psi_B(x_1) \tag{6.5}$$

$$|\psi_{A,B}|^2 = (|C|^2 + |D|^2)|\psi_A(x_1)|^2|\psi_B(x_2)|^2 + \tag{6.6}$$

$$CD^*\psi_A(x_1)\psi_B^*(x_1)\psi_A^*(x_2)\psi_B(x_2) + h.c. \tag{6.7}$$

Here $|C|^2$ and $|D|^2$ are the fractional amounts from the two separate components, that add up to unity since the total probability has to be one when integrated over all configurational space (the cross term evaluates to zero because the states ψ_A and ψ_B are assumed to be orthogonal). The C and D are complex numbers. The probability $P_{A,B}(x_1,x_2)$ is $|\psi_{A,B}|^2$ and this is not equal to $P_A(x_1)P_B(x_2)$, but contains interference terms involving CD^* and DC^*.

These cross-terms ensure that there is no way we can separate the system A from system B in the combined state described by $\psi_{A,B}$. This process extends the idea of superposition within a single quantum system, to construct a new 'holism' with two originally independent systems. It is essentially a quantum phenomenon named *entanglement* by Schrödinger. Half a century later, such 'pure science' has given rise to quantum information theory as well as possibilities of quantum computers etc. Those topics are briefly mentioned below, but dealt with in more detail in Ch. 7.

Two objects (in effect, quantum states) A and B 'entangle' to form the quantum system C. If we do something to one end of C which is (let us say) mostly A, we may affect the other end (mostly B) which is far away, and so, what seems to be 'action at a distance' becomes possible. This is the essential content of the celebrated EPR (Einstein, Podolsky, Rosen, see [70]) paradox. The paradox is that we can affect something far away with no evident agent for doing it. This was what Einstein termed 'spooky action' at a distance. However, there is nothing spooky here since *we set up a system with boundary conditions extending very far out*, and set up a coherent quantum field pervading through out the system, excluding any external purturbations. It is the quantum field that is the 'agent' that links the far-away parts together. In the language of Bohm (see Secs. 6.2.5, 6.7), we would say that a new quantum potential was created that incorporates a new holistic system containing A and B. The quantum potential Q cannot be written as $Q_A(x_1)+Q_B(x_2)$, as a sum of separate parts purely associated with A, or B. Real experiments using the EPR ideas have been performed using entangled particles and large molecules, and QM is found to be valid [12] in all cases. Experiments with entangled photons have shown entanglement even at kilometer distances, although the low efficiency of the particle detectors has raised concerns. Subsequent, more refined experiments have upheld the non-local predictions of the quantum theory.

6.2.3 *Leggett's views regarding quantum reality*

Classical reality, or 'everyday reality' assumes that what is happening here can be dissociated from what is happening at a point sufficiently far away. That is, localness, or near-here-ness is expected to hold. Such a concept of reality is implicit in the 'local-reality' ideas inherent in EPR. Non-locality arises from the boundary conditions which enter into the wave-like solutions of Schrödinger's differential equation describing the system where there are built in particle correlations that have been painstakingly safeguarded in the experimental set up.

Anthony Leggett, a distinguished theorist who made seminal contributions to our understanding of quantum liquids, suggests three broad points of view about quantum reality:

(a) 'QM is the complete truth about the physical world at all levels, and describes an external reality'.
(b) 'QM is the complete truth giving correct predictions of experiments, but describes no external reality'.
(c) 'QM is not the complete truth about the world; at some level between that

of the atom and that of human consciousness, other *non-quantum* principles intervene'.

Leggett claims that a majority of physicists who would accept the applicability of wave-particle duality to an electron would decline to do so in regard to a macroscopic object like a cat. Noting that very large molecules like the Fullerene C_{60} molecule have been found to show wave-particle duality, Leggett rejects options (a) and (c), and claims that 'at the end of the day we have no alternative but to live with option (b)', i.e., QM is purely a superbly successful calculational tool but describes no external reality! Leggett's actual position may be more nuanced than that. However, at one extreme we might think of Father Ingoli (Sec. 3.3) who claimed that the heliocentric Copernican system is just a successful calculational tool which does not show the external reality innate in the divinely and morally consistent geocentric system.

Much of the discussion on macroscopic Cat states ignores that such systems are at finite temperatures. We calculate the thermal de Broglie wavelength of a 1 kg cat at room temperature in Eq. 7.56, and find that it is less than a millionth of the radius of a proton. Hence entangled states of macroscopic objects only concern those who limit their quantum theory to absolute zero. The experiment on C_{60} referred to by Leggett involved a de Broglie wavelength of $\sim$25 picometers (Ref. [9]), and was a *tour de force* where the system was protected from virtually instantaneous decoherence. In our view, this is hardly a reason to conclude that only option (b) is possible.

6.2.4 *Copenhagen school and hidden variables*

The option of claiming that QM provides a superb calculational algorithm without revealing the detailed dynamics of 'external reality' is the position established by Heisenberg, Bohr, Schrödinger, Dirac, Pauli and other founding fathers (see

Fig. 6.2 Niels Bohr (1885–1962), Erwin Schrödinger (1887–1961) and Werner Heisenberg (1901–1976), together with Paul Dirac (1902–1984, see Fig. 7.4) could be regarded as the founding fathers of quantum mechanics, while Planck, Einstein, Bohr and de Broglie had contributed to the early quantum theory. Spectroscopy provided the experimental basis for these theories.

Fig. 6.2) who laid out the 'orthodox interpretation' of QM, sometimes called the 'Copenhagen interpretation', or the 'conventional interpretation' of QM. Niels Bohr was a dominant figure at the Salvay conferences where fundamental issues were discussed, and most scientists went along with Bohr. Schrödinger subscribed to Bohr's views, although with reservations. Einstein was quite unhappy with a theory that gave only statistical predictions.

The wave-nature and corpuscular nature of quantum objects are 'complementary characteristics', determined by the experimental set up which measures 'expectation values' (i.e., statistical mean values) of properties of systems of particles that have neither a definite position $\vec{q}$, nor a definite momentum $\vec{p}$, but defined only by amplitudes $\psi(\vec{q})$ and $\phi(\vec{p})$. Observations depend on the experimental context, and unlike in classical physics, simultaneous observations of certain properties are impossible in quantum mechanics, because one observation disturbs the other. Further, in the words of Bohr (p. 26 of [32]),

> ... (these *complementary* observations) ... 'cannot be combined into a single picture by means of ordinary concepts, they represent equally essential aspects of any knowledge of the object in question that can be obtained $\cdots$'.

An important element of the Copenhagen interpretation is that we need not 'discuss the underlying reality' in order to function as scientists. The equations of QM are solved with the effect of the measuring apparatus included in the calculation. That is, in the most common 'interpretation', if the measurement operator of some property is $\hat{A}$, the inclusion of the measurement operator converts the wave-function ψ to $\xi = \hat{A}\psi$, and the expectation value is take to be $a = \langle\psi|\xi\rangle$. An actual experiment may not give this value, but some value a_i. However, many accurate repetitions of the experiment reveal the mean value of the set a_i to be a. Thus the predicted 'expectation values' of observables are statistical averages. At the moment, we have no theory that predicts the result of the i-th actual individual event to be a_i. For instance, given the half-life of a radioactive material to be one hour, and given one gram of it, we know that half the number of atoms would decay in an hour; however, if we select a single atom, we cannot say when it would decay. Whether this statistical nature is intrinsic or a result of our ignorance of a more fundamental reality, associated with 'hidden variables', is left as a metaphysical issue which is meaningless and untestable. It is here that Einstein departed from Bohr and his colleagues. He felt that the statistical nature of predictions of the quantum theory indicated that it was *an incomplete theory*.

One could take the position that a complete theory is one which is capable of answering any precisely formulated question that one can ask of nature, and from the theory. In that sense, QM is indeed complete, and no experimental disagreement has been found.

However, one may object with Einstein and others that quantum mechanics is not complete. When electrons are observed as scintillations on a screen, these are part of our nature, and we are unable to predict them. Dürr, Goldstein and Zanghi state that 'time measurements, e.g, escape times or decay times, are particularly embarrassing for the quantum formalism ...' [66].

However, this objection can be leveled equally well against classical physics. Although Poincaré's results on the extreme sensitivity of trajectories in classical dynamics had been known since the end of the 19th century, their pervasive importance was not appreciated in Einstein's time. In Sec. 8.6 and Ch. 9 we discuss how effective indeterminism (in the sense of non-predictability from dynamical laws) arises from the determinism of classical physics. Nevertheless, Einstein did not claim that classical physics was incomplete. In fact, Einstein was looking for a formulation of QM that looked like classical physics. David Bohm presented such a formulation during 1952–53. Given his theory, we ascribed our inability to predict individual scintillations to our ignorance of the precise initial positions of the particles. In Bohm's theory (see Secs. 6.2.5, 6.7), he assumes that we know only their statistical distribution $|\psi(q_1 \cdots q_n, t_0)|^2$ at the moment t_0 when the electrons were emitted by, say, a filament. Thus Bohmian mechanics (which should be regarded as a better restatement of QM for non-relativistic systems) provides at least a partial answer to these objections. Similarly, Bohmian mechanics (BM) provides the best available (i.e., most physically appealing and mathematically consistent) discussion of arrival times, escape times, barrier traversal times etc., as discussed in the review by Leavens and Aers [131].

Our observable world is a coarse-grained, experimentally prepared average of the quantum world. The principle tenet of experimental physics that must not be forgotten is as follows (cf., van Kampen [212]).

Principal Tenet: *QM deals with robust macroscopic properties of quantum systems which are given as statistical averages associated with each specified experimental setup. Such average values are independent of the observer.*

This may sound very different to what is stated in the popular literature [201], as well as in some what technical works [226, 207], regarding the role of the observer or the need for a direct non-linear intervention of a 'conscious mind' in quantum processes. A lot of confusion is avoided if we ask what we mean by quantum systems, and what type of phenomena are to be studied. QM was introduced to deal with the observed spectra of atoms and molecules, photoemission, nature of black-body radiation, specific heats of solids etc., i.e., *macroscopic* observer-independent properties where classical theory had failed. Today we apply it to superconductors, quantum fluids, nanostructures, lasers etc. We expect QM to apply to any system seamlessly, as long as we are able to appropriately 'prepare' the system for observation. Specifying the experimental setup is also an important part of the above tenet, and brings in the *participatory aspect of the experimental preparation in the observed reality*.

The experimental observations are gathered and reported using equipment which work in the regime of classical physics, as noted by Bohr [33]. The human observer interacts with the macroscopic apparatus, and hence the division between him and the experimental set up can be located in the macroscopic domain itself. This is just a boundary condition and nothing mysterious. We hope to develop these ideas in our discussions of various interpretations of the quantum theory and its so-called paradoxes.

A fundamental theory attempts to describe reality using fundamental building blocks, e.g., elementary particles, which are usually many orders of magnitude smaller than our measuring equipment. Thus experimental results are always the average values obtained for a large number of samples of the system. We deal with a beam of N electrons where N is very large, or we deal with many repetitions of very nearly the same experiment to accumulate an average result. This average is a macroscopic property, as mentioned in the principal tenet that we enunciated.

If we attempt to look at individual electrons, the data appear to be erratic, unpredictable, or stochastic. Just as in the kinetic theory of gases, the statements of QM are about statistical averages and not for definite outcomes of individual events. QM predicts which *type* of individual event could occur, and assigns a probability to such events. It is as if there are *hidden variables* that we do not know about, controlling the individual events. The objective of hidden variable theories is to replace the apparent statistical nature of quantum phenomena (implicit in the wavefunction which defines the probability $W = |\psi|^2$) by an explicit averaging over the hidden variables, just like in the kinetic theory of gases. However, if such hidden variables exist, they have to recover the quantum correlations (non-locality and contextuality) introduced by the Schrödinger equation. Bohm's hidden-variable theory (see Secs. 6.2.5, 6.7) is manifestly non-local and those who looked for local realism (e.g., Einstein) did not support it.

One of the famous mathematical problems posed by Hilbert at the turn of the 20th century was the nature of the mathematical foundations of the quantum theory. John von Neumann presented such a formulation [214] in 1932, and sharpened the discussion of many aspects of the quantum theory. He argued that the mathematical structure of QM is such that there cannot be a 'hidden variable' formulation which could give a kinetic-theory like deterministic under-pinning of QM. The reputation of von Neumann was such that virtually everyone accepted that hidden-variable theories (HVT) are not possible. However, some three decades later John Bell showed that von Neumann's 'proof' of the impossibility of HVT was based on a simple inadmissible assumption.

If $\hat{A}$ and $\hat{B}$ are operators corresponding to measurements of the properties A and B, and if the results $v(A)$ and $v(B)$ are obtained irrespective of the sequence of application (i.e., $\hat{A}\hat{B}$ or $\hat{B}\hat{A}$), then such 'non-interfering' operators are known as 'commuting' operators. They are non-contextual since the order, i.e., the context, does not matter. On the other hand, we may have two observables, P and Q which interfere with each other and hence the order of operations is very important. They are 'contextual' in the nomenclature used by some philosophers and physicists. It is enough for us to call them observables associated with 'non-commuting' operators. That is, $\hat{P}\hat{Q} \neq \hat{Q}\hat{P}$ where we use the 'hat' when we wish to emphasize operators. Such 'non-commuting' 'operators' are very common in every-day life. Thus, withdrawing money (W), and depositing money (D), can have very different results for D followed by W, as opposed to W followed by D when operating on a null bank account (a vacuum state!).

If $\hat{A}$ and $\hat{B}$ are commuting operators, with expectation values $v(A), v(B)$ then if $\hat{C} = \hat{A} + \hat{B}$, where the '+' sign means simultaneous action, it is clear that the expectation value of $\hat{C}$, i.e., $v(C)$ should be the sum of the expectation values $v(A)$ and $v(B)$. However, if $\hat{R} = \hat{P} + \hat{Q}$, where $\hat{P}, \hat{Q}$ are non-commuting quantum operators, there is no reason for the expectation value for R, say $v(R)$, to be the sum of the independently observed values $v(P)$ of P and v(Q) of Q. In fact, $\hat{P}$ and $\hat{Q}$ disturb the quantum state and hence their simultaneous observation R cannot be carried out (except as some mean value). According to John Bell, the fatal and rather trivial error in von Neumann's proof that HVT is impossible assumed that the $v(R) = v(P) + v(Q)$!

6.2.5 *Bohm's causal quantum dynamics*

John Bell's analysis cleared the shadow of von Neumann that hung over the work of David Bohm who had produced, in 1952, a hidden-variable theory apparantly equivalent to Schrödinger's quantum mechanics as far as experimental observations are concerned. Bohm (see Sec. 6.7) re-wrote the Schrödinger equation as a classical dynamical equation (as in the kinetic theory of gases), except that the particles were subject to an additional potential known as the *quantum potential* Q, arising from the wavefunction interpreted as an actual field. As would be discussed in Sec. 6.7, Bohm wrote the wave function in the polar form $\psi = R(\vec{x})\exp(iS(\vec{x}))$ where R and S are real-valued fields. If we note that a 1-D plane-wave state has the form $A\exp(iS)$ where $S = kx$, it is clear that the gradient $dS/dx = k$ gives the momentum. Bohm showed that such a momentum k and the position x could be used to define a classical-like dynamics consistent with the Schödinger equation. Thus Bohm's quantum particles move in well defined trajectories, subject to all the physical potentials as well as the quantum potential. It is Q which confers various quantum properties like non-locality (in entangled systems) to the 'classical-like' description. The statistical nature of observed quantities arises, just as in the classical kinetic theory, from the uncertainties in the initial positions of the particles in the n-body system, and in the many repetitions of the experiment.

Bohm's analysis provided a rigorous basis for an intuitive idea of Louie de Broglie, proposed in 1927, that a quantum particle is guided by a *pilot wave* that ensures conformity with quantum behaviour. According to Bohm and Hiley [29], the quantum potential provides 'active information' to guide the electron, but does not input or extract energy from the electron. This is analogous to a motorist getting information from the field associated with his Global Positioning System. Hence de Broglie's usage of the name 'pilot wave' was appropriate.

The manifestation of quantum effects in a classical-like formulation via a potential of quantum origin is reminiscent of the way relativistic effects manifest themselves in non-relativistic formulations (e.g., the spin-orbit interaction). Properties like the electron spin is part of an abstract inner reality. The wavefunction too could be considered to have an abstract physical reality, and not just a mere calculational tool. This topic is discussed further in Sec. 6.5. In fact, Bohm's picture prompts one to (a) re-interpret the meaning of ψ given by Max Born where $|\psi(x)|^2$ is the probability of *finding* the particle at x as the probability of the particle *being* at x, and (b) regard ψ as an actual physical field generating the potential Q that moves the particle to regions with higher values of $|\psi(x)|^2$.

Thus Bohm's reformulation of the Schrödinger equation (see Sec. 6.7) provides a basis for a deduced quantum reality which is on the whole quite similar to our normal concept of deduced reality, although some of this reality is submerged in an abstract function space. Hence our principal Tenet: *Quantum Mechanics deals with robust macroscopic properties of quantum systems which are given as statistical averages associated with each specified experimental setup, independent of the observer*, may if necessary be more easily understood within Bohm's reformulation of the Schrödinger equation.

The standard formulation, as given by Bohr, Schrördinger, Heisenberg, Dirac and others provides a language and a calculus which has now been applied to every aspect of physics including relativistic quantum fields, the physics of the standard model and beyond. This is not the case with Bohm's formulation. The Bohm-de Broglie approach is best suited for laboratory-scale quantum physics. Formal studies of general quantum problems will continue within the standard approach. Meanwhile the measurement of a quantum property can be re-written with clarity in terms of Bohm's theory, since all measurements (Ch. 7) fall into the laboratory world of pointers and gauges. This ensures that irrelevant metaphysics is not woven into the measurement process and the clear position stated in our principle tenet would not be obfuscated. Philosophical considerations of QM can also profit by remembering that quantum reality is very like classical reality, except for an additional quantum potential which can easily suffer decoherence and become irrelevant. It is the quantum potential which brings in the effects of imposed experimental boundary conditions and quantum behaviour.

6.2.6 *The intuitive ontology of the working physicist*

A significantly empiricist view where reality is associated with measurable (i.e., observable) attributes of a 'real' external world is found in the intuitive *ontology of the working physicist* (OWP). It involves a sense of reality where 'things' exist in 'quantum states', and they make transitions between these states. They evolve in time, decay, scatter etc., *independently* of the observer. Quantum-ψ states are like fields which pervade the system under study (SUS), demarcated from the external world by boundary conditions so that the phase relations of the field components do not decohere. This SUS is a holistic object where all parts are connected by the quantum fields which are the basic reality, although subsumed in an abstract inner space. The quantum fields can have sharply localized excitations (called particles) that can be record on measuring equipment. They can also have wave like excitations that are not localized. The particle is *not* some 'spread-out' object which pervades the same extension as the wavefunction.

Roger Penrose considers that a particle is 'spread out', much like its wavefunction. He contends that, e.g., when the wavefunction goes through two slits, the electron is also simultaneously going through the two slits. Penrose [160] says 'the physical reality of the particle's location is, indeed, $\psi(x)$'. See also his discussion on page 250 of Ref. [160]. This is usually not the intuitive view of the working physicist, the formal positions of Bohmian mechanics (BM), or of conventional QM. Furthermore, it is difficult to reconcile this view with the fact that if the probability distribution for the particle positions is p at some initial time, then it will be equal to p at all future times. This is a property of the Schrödinger equation, and it is sometimes called the principle of 'equi-variance'.

A measurement is equivalent to changing some boundary conditions and details defining the system under study, and hence the measurement changes the quantum state of the system. Thus, the detection of an atom in an excited state by making it emit a photon irrevocably brings the atom to its ground state. The post-measurement state ψ' is irrevocably different from the state ψ of the original SUS. It is as if a cameraman, in photographing a caterpillar, converts it into a butterfly in the process. However, the working physicist is intuitively aware that all caterpillars (excited states, ψ_e) will change into butterflies (ground states, ψ_0) under an observation. It is as if the butterfly appeared only when the camera was pointed at the subject, just so that it can appear in the photograph! The OWP has no difficulty with this fleeting butterfly. In this analogy (not to be pushed too far), the butterflies are the end result, i.e., the ground state and the emitted photons.

The experimentalist observes that the photons emitted by the H-atoms when they decayed had the Lyman-α energy E_α, and concludes that the atoms were in the $2p$-excited state. The purpose of his experiment was to get that information. However, the textbook account of measurement (see Ch. 7), e.g., according to von Neumann, seems very different.

Every observer obtains the same statistical average for the energy E_α and lifetime of the quantum state studied. There is no trace of a 'participatory' subjective effect by the observer. On the other hand, the participatory nature of an observation can be accentuated to affect the lifetime of the quantum state of the SUS by a suitable setup, as seen from the quantum Zeno effect (Sec. 7.7.6).

The ontology of the working physicist is easily understood within Bohm's interpretation of the Schrödinger equation, since Bohm (Sec. 6.7) succeeded in constructing non-local hidden variables which recover the average values predicted by the QM of Schrödinger's equation. This enables one to talk of trajectories of electrons determined by a classical equation, but containing an effective potential that includes quantum effects. This picture clearly shows that even though the quantum field is everywhere in the system under study, the particles are not spread-out objects, as contended by Penrose and others. Hence, in this interpretation of QM, there are no counter-intuitive paradoxes based on the same particle being in two places, going simultaneously through two slits, or being in two contradictory states, etc., as claimed in the cat paradox or the delayed-choice paradox (see Sec. 7.7).

In some of the following sections we derive some elementary results used in quantum mechanics, mainly for later reference, and to flesh out the conceptual implications of the theory. Thus readers may choose to skim over them or skip them to go on to subsequent sections.

6.3 Quantum particles and wave fields

We began this book with a discussion of duality, and we find that quantum reality admits of a duality, but only in a complementary sense. Quantum entities behave either as localized excitations (particles), or wave-like excitations of an underlying quantum field, but *not both* simultaneously

6.3.1 *Electrons as quantum particles*

The first quantum particle to be recognized as a sub-atomic particle was the electron. J. J. Thompson working in the Cavendish laboratory studied Cathode Ray Tubes (CRT). These are evacuated glass tubes fitted with metal electrodes. One end of the CRT has a metal stub connected to the negative end (cathode) of a battery. Usually, the other end has a positive metal plate (anode). Thompson showed that negatively charged particles emanated from the (negatively charged) cathode, and were attracted to the positively charged anode. These particles were identified as electrons, a name already used (in 1891, by G. J. Stoney) for the unit of charge found in electrochemical experiments. Some of the electrons hit the end of the CRT, forming a luminous spot. This was in 1897, and it was the earliest 'television tube' ever built!

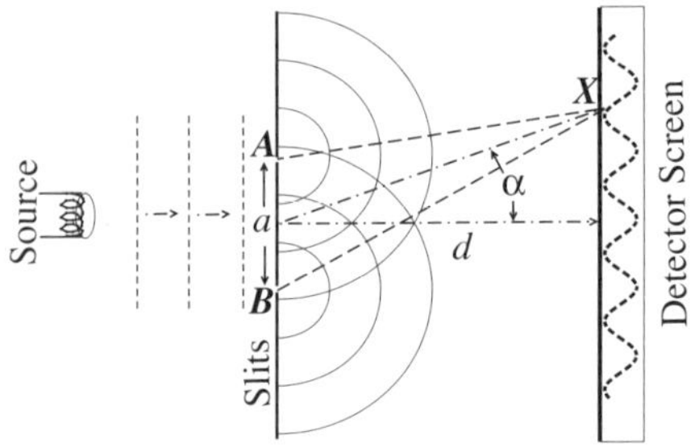

Fig. 6.3 Photons or electrons emitted from the source emerge from the two slits A, B as diffracted waves. The wave amplitudes superpose to produce an interference pattern on the screen which is at a distance d which is usually much larger than the slit separation a. Reflected waves are not shown. (The figure is not to scale).

Thompson, Lenard, Milliken and others studied the electron and claimed that it is almost 2000 times lighter than a hydrogen atom, and hence must be a sub-atomic particle. This was a revolutionary suggestion at a time when some leading scientists (e.g., W. Oswald) and philosophers (e.g., Ernest Mach) even refused to believe in the atomic theory of matter. Even those who accepted the atomic theory claimed that any discussion of the 'inside' of an atom was metaphysics.

6.3.2 *Light waves and quantum particles*

If an atomic theory of matter seemed an unlikely idea, any suggestion that light is made up of particles would be revolutionary. However, such an audacious step was taken by Einstein in 1905. That old Newtonian hypothesis had already been fully countered. Maxwell had given a quantitative wave equation for light. A vast amount of experimental results was available by the end of the 19th century to demonstrate the wave nature of light.

It was understood, even by the Greeks, that light propagated in straight paths and obeyed Fermat's principle of least time (Sec. 3.6). The laws of geometrical optics were based on the concept of light rays. However, starting from Newton's demonstration of the dispersion of rays by a prism, experiments revealed that the ray concept fails if you probe at length scales similar to that of the wavelengths of light (e.g., wavelength of green light is 510 nanometers). The definitive experiment which reveals the wave character of light is due to Thomas Young. Let us review the Young double-slit experiment (Fig. 6.3). Thomas Young described his experiments in 1801 at a meeting of the Royal Society in London. The observation of an intensity pattern on the screen due to the interference of the two wave trains at any given point on the screen can be quantitatively explained if light is assumed to be made up of waves.

A light wave is an electromagnetic wave, as discussed in Sec. 4.2. That is, if the wave is propagating in the x direction, then there is an electric field $E(x,t)$, and a magnetic field $H(x,t)$ at the point x and time t. The electric field at x points in the z-direction, and the magnetic field points in the y-direction, i.e., transverse to the direction of propagation. The electric field $E(x,t)$ varies like a wave, e.g., $E(x,t) = E_0 \exp(ik.x - i\omega t)$ where $k = 2\pi/\lambda$ and $\omega = 2\pi\nu$. The wavelength is λ and the frequency is ν, while ω is called the circular frequency. Here k is called the wave vector. The intensity of the field (brightness of light) at a point x, at time t, is given by $I(x,t) = |E(x,t)|^2$. That is, the square of the electric-field amplitude gives the intensity or brightness of the wave.

The two rays AX and BX in Fig. 6.3 arrive at the point X on the screen, with electric fields $E_A(x,t)$ and $E_B(x,t)$. Hence the total field amplitude is the superposition of the two amplitudes:

$$E(x,t) = E_A(x,t) + E_B(x,t)\,. \tag{6.8}$$

The total light intensity at X is given by:

$$\begin{aligned} I(x,t) &= |E(x,t)|^2 = |E_A(x,t) + E_B(x,t)|^2 && (6.9)\\ &= |E_A(x,t)|^2 + |E_B(x,t)|^2 + \{E_A^*(x,t)E_B(x,t) + E_B^*(x,t)E_A(x,t)\}\,. && (6.10) \end{aligned}$$

The first two terms are just the intensities at X due to the light coming from A and B, while the term in curly brackets is the interference term. This interference term $\xi(x,t)$ is the 'finger-print' of the superposition of the field amplitudes $E_A(x,t), E_B(x,t)$ that produces the field intensity $I(x,t)$.

$$\xi(x,t) = E_A^*(x,t)E_B(x,t) + E_B^*(x,t)E_A(x,t)\,. \tag{6.11}$$

The origin of interference becomes clear from the difference of the lengths of the two paths AX and BX in the figure. If the screen is assumed to be at a distance d which is much larger than the slit separation a, then a simple approximation holds. The path difference $\Delta\ell \approx a\sin(\alpha)$. Two waves which were in phase on emerging from the slits would still be in phase if $\Delta\ell$ were a multiple of the wavelength λ of the light. Thus the condition for seeing bright fringes on the screen is

$$\sin(\theta) = m\lambda/a$$

where m is any integer $\pm 1, \pm 2$, etc. Similarly, the condition for destructive interference is that crests match with troughs. That is,

$$\sin(\theta) = (m + 1/2)\lambda/a$$

where m is any integer $\pm 1, \pm 2$, etc.

These two equations differ by the quantity (1/2) λ/a. Hence if the slit separation a is itself large compared to the wavelength of the light, the interference effect is not noticeable, and the simple geometric (ray-like) propagation of light is recovered. Thus *the wave effects of light manifest only at sufficiently small length scales*. The position of the fringes can be used to determine the wavelength of the light. Maxwell's theory of electro-magnetic radiation provided a quantitatively accurate wave picture of light. And yet, Einstein was lead to propose that light behaves like little bullets, each containing just a quantum of energy, proportional to the frequency ν of the light, in attempting to understand how electrons are ejected from metals like Na when light shines on them. This is called the photo-electric effect. The quantum of light energy, called a photon, is given by:

$$E = h\nu \tag{6.12}$$

where h is a proportionality constant. Millikan experimentally determined h from photo-electric experiments and was amazed to find that it agreed accurately with h, the Planck constant, used in the quantum theory of Black-body radiation. The above relation is called the Einstein-Planck relation.

The Einstein-Planck relation for a quantum of light energy is $E = h\nu$, or equivalently $E = \hbar\omega$. Since the wavelength λ and the frequency ν are related by $\nu = c/\lambda$, we also have $E = h(c/\lambda)$. The energy E of light is proportional to its momentum p, the velocity of light being the proportionality constant (i.e., $E = cp$). Thus we see that the wavelength and momentum are connected by:

$$\lambda = h/p. \tag{6.13}$$

This is the key relation which inspired de Broglie to attribute a wavelength to any quantum particle with a momentum p.

Einstein submitted his research paper to Max Planck, the Editor of the journal *Annalen der Physik* in 1905. Fortunately for Einstein, Planck himself had come to the radical conclusion that light seems to behave like little packets of energy (i.e., quanta of energy) in his study of the radiation emitted by hot bodies. As an object is heated, it appears red hot, then white hot and finally bluish hot. The spectrum of radiation from hot bodies had been a great puzzle and Planck had successfully shown that the quantum hypothesis (i.e., that light energy was emitted in little packets or 'quanta' of energy) was necessary to explain the observations. Planck approved the publication of Einstein's paper without submitting it for independent review. In fact, if it had been sent for review, it would surely have been rejected!

The photo-electric effect implies that although light has a wave character, it can also assume a particle-like nature when it is interacting with matter. When a photon is detected, the detection occurs by the absorption of the photon by a light-sensitive atom. The photon in free space cannot be localized to a length scale smaller than a few undulations of its wavelength λ, (which may be about a thousand atomic units, while an atom or a light-sensitive Zinc Blend molecule is just a few atomic units in size — for atomic units, see Sec. 1.3). *Hence the detection process (the photo-electric or photo-chemical effect) involves a photon that is of atomic dimensions. In effect the observational set up, (i.e., whether we are setting up to hear a click in a photo-detector, or see an interference pattern) determines whether we deal with photons or light waves.*

The particle-like behaviour of light is seen only in the presence of material fields. The Compton effect, where a photon collides with an electron in just the same way as a billiard-ball collision, is also evidence for the particle-like nature of light. This happens when the photon is interacting with matter. Hence the presence of material fields and the boundary condition used to isolate or reduce the system for observation etc., change the character of the electromagnetic field to manifest itself as 'photons' when observed at a detector.

Attempts have been made to find an entirely wave-like description (e.g, at one time by Bohr) of photon effects, without using the photon concept. Lamb [124] even proposed giving up the name 'photon'. Photoelectric experiments can be explained by assuming that the metal surface can absorb energy in quanta, while

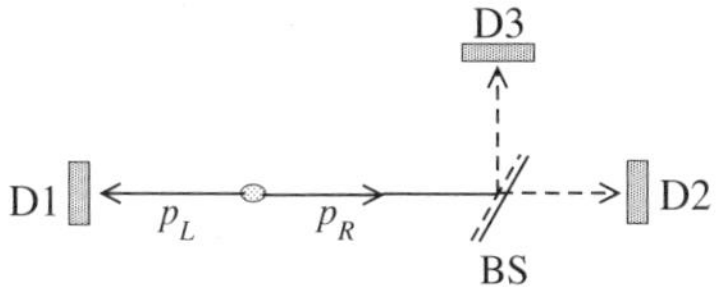

Fig. 6.4 The entangled photon p_L is detected at D1, while the right moving entangled photon p_R is beam-split at a half-silvered mirror and detected at D2 or D3.

the radiation field itself is not quantized. However, the advent of experiments with entangled photons (Sec. 7.3.4), single-photon experiments and quantum optics have made the photon concept indispensable.

6.3.2.1 *Single photons*

Modern replications of Young's two-slit experiment have demonstrated a remarkable result. According to the quantum theory, the intensity $I(x)$ of light at the point x is proportional to the number of photons, while the energy in each photon depends only on the frequency ν. Hence, by reducing the intensity of the light sent to the two-slit screen, we arrive at the case where we only have, on the average, just one photon presented with the possibility of two slits. Even such a weak beam of light shows the interference pattern typical of its wave nature. So, if the photon is a particle, did it interfere with itself? A particle, defined as a sharply localized excitation, cannot take both paths at once. One may well ask, 'which path did the photon take to interfere with itself'? The photon takes only one path, and not both. The interference emerges from many single-photon events. This is discussed below and in the section on Wheeler's delayed-choice experiment (Sec. 7.7.5).

A photon source may emit two entangled photons, which move in opposite directions (Fig. 6.4). One of the photons is detected on the left at detector D1. The right-moving photon is sent through a beam splitter (BS) which transmits or reflects it to two isolated detectors D2 or D3. When the left-moving photon is detected, only one of the detectors D2, or D3, but not both light up. This fact is most easily explained if we agree that a single indivisible photon (rather than a wave) moved to the right [48]. Such an analysis of the experiment appears inconsistent with Penrose's view that a quantum particle has the same extension or distribution as its wavefunction (Ch. 6 of Ref. [160]).

In observing the interference pattern arising from a single photon, we record the arrival of the individual photons on our photographic screen S (see Fig. 6.3) over a period of time. During this time we ensure that our light source is very weak, so that only one photon or perhaps no photons at all are present. The accumulation of a large number of detection events on the screen builds up to the $\psi|^2$ distribution predicted by QM to show the interference pattern. The individual arrivals, taken in themselves, might seem to be stochastic. This can be handled in a variety of ways. A 'safe' approach, typical of the 'conventional interpretation' of the Copenhagen school, is to say that we do not ask 'the welcher weg' (which-way) question, since it is unobservable. All we know is the probability of finding a photon at the point $\vec{x}$ on the screen, at time t as given by the square of the amplitude of the light wave, viz., Eq. 6.9. The detected position of the photon has become an observable, but its previous positions are considered devoid of meaning. This is the 'bare calculational algorithm' considered unsatisfactory by many, including Einstein, since it does not 'provide an understanding of what is going on'. However, even though the arrival of photons seems 'stochastic', each photon 'knows' to avoid the dark fringe-areas of the final interference pattern that emerges only at the end of many observations. Thus, John Bell characterized this property as a 'beable'. This seemingly deterministic, beable-like origin of each event is easier to understand within the Bohm-de Borglie interpretation of quantum mechanics (Sec. 6.7) where the electron motion is determined by the quantum potential Q which carries the needed information about the wavefunction of the system and guides the photons to the interference fringes. The 'stochastic' character of individual detections is ascribed (in Bohmian approaches) to our ignorance of the exact initial time and position of the photon at emission.

6.3.3 *Polaritons as detectable photons*

Light moving in free space can be described as particles (photons), or as classical electromagnetic waves without any contradiction. The particle-like picture only becomes essential in describing the interaction of light and matter, although the semi-classical theory of radiation had been very useful. Many of the 'paradoxes' relating to the photon, (gleefully recounted by writers of popular news articles who wish to emphasize the 'weirdness' of the quantum theory) can be resolved if we recognized that in most cases the quantized particle is not a photon but a highly localized dielectric excitation known as a polariton.

The photon (in the presence of a single slit, or two slits, and a screen) is an excitation in a quantum field which recognizes the boundary conditions of the set up. It is not a free field. If the photon propagator of the free field is $D_0(\vec{k},\omega)$, this becomes a new propagator $D(\vec{k},\omega)$ in the presence of a matter field. In the simple case where the material system is translationally invariant, we can write $D(\vec{k},\omega)$ in terms of the polarization operator $\Pi(\vec{k},\omega)$ of the matter field.

$$D(\vec{k},\omega) = D^0(\vec{k},\omega)/\left[1-\Pi(\vec{k},\omega)D^0(\vec{k},\omega)\right]. \tag{6.14}$$

A real experimental system is not translationally invariant, and $D(\vec{k},\omega)$ etc., have to be expressed in terms of real space variables that include the existence of slits, detectors, screens etc. Then the above equation becomes an integral equation which has to be solved self-consistently with the polarization of the matter fields. The contribution to the polarization operator made by the screen is very important as the photons interact with the slits, screen etc and are detected at the screen. The poles of the photon propagator inclusive of the polarization corrections define the polariton field. *It is the polaritons which are detected, and these are very localized excitations.* Hence we may say that the use of simple momentum representations of the photon field, without explicitly including the details of the experimental system can be very misleading. The use of real-space representations using wavefunctions (instead of propagators), is also somewhat unsatisfactory due to the possibility of a variety of forms for the 'photon wavefunction' ([24], [64]) while each of them has its merit. They do not adequately include the the matter fields which are essential in forming polaritons.

Bohm's picture using a point particle and the wavefunction works well for non-relativistic QM. In more common QM it is sometimes said that although a 'true' particle, e.g., an electron, could be 'localized', a photon cannot be localized and this is presented as an objection to its 'particulate' nature. However, the photon is a 'relativistic' particle. Unlike a 'point' particle which may be localized at the point $\vec{x}$ in a non-relativistic theory, the photon (which is a massless relativistic particle moving with the speed of light) can only be localized to within a few wavelengths. Even a 'nearly point-like' particle with a finite mass cannot be localized more precisely than its Compton radius if the relativistic picture were used. Any attempt at closer localization will lead to the production of particle-antiparticle pairs from the localizing field. Thus, within the full relativistic picture there is no conceptual difference in the nature of the localizability of photons and other particles.

Given a photon moving in the z-direction in a cylindrical waveguide, the x and y directions are confined. The mode confinement energy E_0 associated with the confined motion plays the role of a mass for the photon such that $m_0 = E_0/c^2$, where c is the velocity of light. Hence it is fair to say that most of the conceptual and epistemological difficulties associated with the photon can be overcome by suitably including the boundary conditions associated with the experimental situation, and recognizing that the polariton-like excitation is what is recorded by the detectors.

Unlike free photons, polaritons (i.e., photon-like excitations which arise in the surface layers of the detector) can be localized to regions much smaller than the wavelength λ of the photon. Thus the photographic screen detects the particle-like polaritons which are the true elementary excitations of the quantum field in the vicinity of the screen. Within this picture, it is the polaritons which show particle-like behaviour, while free-field photons behave like waves. They are never localizable to better than a few wavelengths. Thus one may take the point of view that the 'welcher weg' (which way) question does not arise, and the screen simply detects the polariton excitations in its surface layers, induced by the wave-like photon field modified by the slits (or slit) downstream.

6.3.4 *The wave nature of particles*

The experiments of J. J. Thompson and others established the existence of a sub-atomic charged particle named the electron. The discovery of radioactivity, i.e., the spontaneous decay of materials like Uranium and Radium had been discovered. Rutherford's experiments had shown that the atom consists of a very small nucleus surrounded by a number of electrons which move around the nucleus like the planets around a microscopic sun. Within the first two decades of the 20th century, sub-atomic particles like electrons, protons, and α-particles (He nuclei) had been discovered and identified. Light itself had joined the ranks of particles, under the name of photons, while still retaining membership in the 'wave club'.

Rutherford's model of the atom implied that electrons are whizzing around the positively charged nuclei. Thus for example, the Helium nucleus, having two units of positive charge, would have two electrons (negative charges) revolving around it. Opposite charges attract, and accelerated charges radiate out their energy. Hence the electrons should spiral into the nucleus in a flash of light, and He atoms cannot be stable. In fact He atoms are extremely stable, and were first discovered in the sun by spectroscopic observation!

In discussing scientific revolutions (Sec. 2.4.1), we noted the importance of improved tools like telescopes and spectroscopes. These push the accessible energy and length scales to new regimes which demand new concepts. Spectroscopic observation of sunlight had revealed that in addition to the continuous variation in colour (i.e., frequency) first observed by Newton using a prism, sharp spectral lines (i.e., pure pitch lines, or dark fringes corresponding to a pure pitch) are also observed. Fraunhofer who started off as a skilled glass maker [114] has been credited with making the first diffraction grating, and the observation of the dark spectral lines in the year 1814. A diffraction grating is made by drawing a series of parallel lines on a glass surface. The lines are opaque, and the spaces between them act as slits allowing the passage of light. The dark spectral lines are now known as Fraunhofer lines. Such sharp spectral lines are seen in the glow of hot substances. For example, glowing common salt shows two sharp yellow spectral lines, called D lines by Fraunhofer. The line spectrum of hydrogen had been extensively studied by Balmer, a German high-school teacher. He identified the Balmer series of lines. In 1864 Norman Lockyer found that the light from the sun showed spectral lines that did not correspond to any lines emitted by terrestrial substances known at the time. Lockyer made the bold suggestion that there is a hitherto unknown element in the solar corona, emitting these new, sharp spectral lines. The new element, named Helium, was later discovered by Sir William Ramsay a quarter century later, on earth.

Thus we have two intriguing questions: How is it that negatively charged electrons don't spiral into the positive nucleus? How can electrons emit sharp spectral lines given that electrons are expected from Maxwell's equations to lose energy by spiraling into the nucleus, emitting a continuous spectrum of light? Niels Bohr (1912) proposed that electrons don't spiral into the nucleus, but instead move in *stable orbits*. The sharp spectral lines arise when electrons jump from one orbit to another, emitting 'quanta' of light. The Bohr model of the atom (see Fig. 6.5) used such assumptions and Newton's laws to *quantitatively predict* the observed frequencies of the spectral lines of Hydrogen and Helium.

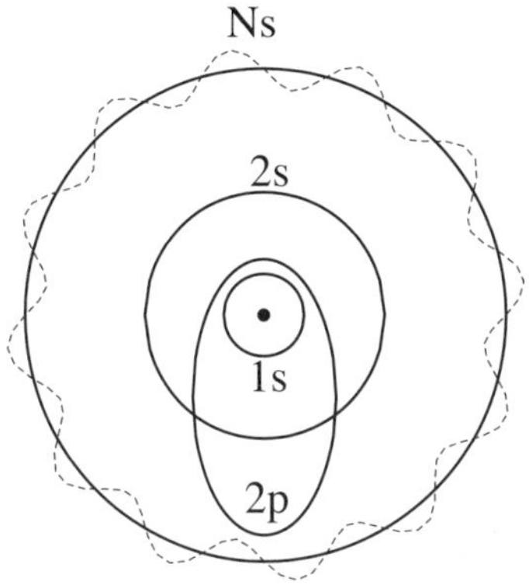

Fig. 6.5 The Bohr model (not to scale). The 1s orbital (or energy level) is the ground state, with a radius of $\sim 10^{-10}$ cm. The 2s, 2p (with three copies, p_x, p_y, p_z) are excited states. An infinity of levels exists, with quantum numbers N = 1,2, etc. The 'quantization condition' defining a level was interpreted by Louie de Broglie as the condition for standing waves of electrons, as in the orbital Ns.

Louie de Broglie in Paris noticed that the conditions imposed by Bohr to construct stable orbits for electrons revolving around the nucleus were just what were needed to form standing waves which connected smoothly (as shown for the wave marked on the orbital marked N_s in Fig. 6.5). Louie de Broglie was also struck by the suggestion of Planck and Einstein that light of frequency ν was made up of photons with energy $h\nu$. If waves could behave as particles, and if these particles could give rise to discrete energy levels as in the Bohr atom, he conjectured that particles may have hitherto unsuspected wave character.

The Planck-Einstein relation for a quantum of light energy (Sec. 6.3.2) posits that the wavelength and momentum are connected by

$$\lambda = h/p. \tag{6.15}$$

Louie de Broglie proposed in 1923 that such a relation should hold not only for light, but also for elementary particles like electrons, where p is now the actual momentum mv of a particle with velocity v.

Louie de Broglie's idea that material objects have a wave property implies that even a base ball or an elephant has some *wave character*! However, while the wavelength of an electron moving at one-hundredth of the velocity of light is about 25 nanometers (25×10^{-9} m), the wavelengths that are attributed to ordinary objects like baseballs become extraordinarily small, and lead to no practical consequences. We show in Eq. 7.56 that a one kilogram cat at room temperature or 1 K has a de Broglie wavelength that is a hundred-millionth of the radius of a proton. On the other hand, this wave character is the basis of various 'paradoxes' in quantum *thought experiments*. However, such effects are utterly negligible at every-day length and density scales. Even more troubling is the explicit construction of *superfluous* decoherence theories to ensure that these subfemtometer-scale wave properties do not generate macroscopic 'cat states'.

If an electron actually has some sort of wave character, it should show interference if the associated waves could go through two slits at the same time (Fig. 6.3). However, since electron wavelengths are very small, the slits have to be positioned very close to each other. G. P. Thompson, the son of the discoverer of the particle-like electron realized that the metal atoms in a thin gold film have the right sort of separation (i.e., 20 nanometers) to match the wavelengths associated with electrons. Gold can be made into very thin films, and the array of atoms is effectively an array of slits. Thompson's experiment at the University of Cambridge using a very thin gold film established that electrons do behave as waves when presented with a set of slits! Meanwhile, Davisson and Germer, working in USA found that electrons scattered from a clean Ni surface showed a diffraction pattern, independently confirming the wave theory of matter.

What does the wave nature of matter imply? It implies that any object (e.g., an electron) is 'sensitive' to its 'environment'. That is, a quantum object feels the boundary conditions imposed on the space available to it, and adjusts its state to accommodate the externally imposed boundary conditions. In this sense, elementary particles are also 'simple adaptive systems'.

Let us consider 'trapping' a quantum particle by confining it in some manner.

6.4 Schrödinger's equation for a trapped particle

The need to satisfy boundary conditions is the crux of quantization. We consider a 'square well', i.e., a deep flat-bottomed one-dimensional well with essentially 'vertical' walls (Fig. 6.6). Only the x direction is considered for simplicity. If a ball of mass m is in the well, with zero kinetic energy, it has no preferred position and will randomly locate somewhere on the flat bottom of the well.

Its potential energy V is set to zero inside the well, and practically 'infinite' outside. The left and right edges of the well are at $x = 0$, and $x = a$. If the ball had a momentum p, its kinetic energy $T = p^2/2m$, and the total energy is $E = T + V$ remains constant, assuming no dissipation (e.g., no friction). The ball is intuitively imagined as bouncing 'too and fro', with its momentum p reversed at the confining walls. The walls have no other role. The momentum can take any arbitrary value. The function $H(x, p)$ specifying the energy is known as the Hamiltonian of the system, with $H = E$.

On the other hand, an electron (or any other quantum particle), having a wave character, must fit into one of the waveforms that are consistent with the position of the walls which confine the electron. Thus in Fig. 6.6 we consider a 'quantum well' for trapping electrons. One can actually make such quantum wells that trap electrons using two Si slabs to sandwich a slab of Ge which may be, e.g., 5 nanometers wide, using a device known as an MBE machine (Fig. 2.6). The electrons get trapped in the Ge layer where the potential is lower. It is possible to experimentally verify that the light energies needed to excite electrons from the ground level to higher states follow the predictions of the quantum theory of matter, known in the early days as 'wave mechanics'. Electronic devices based on quantum wells and other nanostructures are used in infra-red detectors, night-vision goggles, field-effect transistors, and other components used in computers, TV sets etc. That is, quantum effects manifests themselves in every-day reality.

In Fig. 6.6 (a) we depict a few of the low-energy quantum states of a quantum well. Its state is characterized by a waveform, as shown in the figure. This waveform is known as the wavefunction, traditionally denoted by $\psi(x)$, or $\phi(x)$, and gives the 'amplitude' of the wave at the point x. The form of the wave is similar to that of a vibrating string slung between two posts a distance a apart. The wavelength λ can only be such that $\lambda/2 = a, a/2, \ldots, a/n, \ldots$, where n is an integer. That is, the allowed waves $\phi_n(x)$ are such that the integer n changes by discrete units. This is the essential basis of 'quantization' and n is the quantum number labeling each wave. It is a direct result of the need for the wavefunction to fit between the two posts. *Quantization arises from the boundary conditions imposed on the problem.* Each wave (eigenfunction) has a pitch, and a characteristic energy (an *eigenvalue*). Although a vibrating violin wire is usually thought of as a 'classical mechanics' problem, we see that it is a 'quantized system'.

In fact, the seminal papers [187] of Schrödinger, written in 1926, carried the title *Quantization as an eigenvalue problem*. It is said that Schrödinger was inspired to look for a wave equation after studying the thesis of Louie de Broglie.

It is easy to write down the mathematical functions representing the wave forms by inspection, for a simple system like that of Fig. 6.6. The momentum $p(x)$ of a wave at the point x is a physical quantity in classical physics. However, in QM, both the momentum p and the position x cannot be specified at the same time. The momentum p is replaced by the operator $\hat{p} = -i\hbar d/dx$. The wavefunction in the n-th quantum state is described by a position-dependent function $\psi_n(x)$ *or* a momentum dependent wavefunction $\phi_n(p)$ — they are related to each other by a Fourier transform. That is, in the same way as a crystal lattice is related to its 'reciprocal lattice'. The latter is revealed by X-ray diffraction. We recollect that the total energy is also the Hamiltonian H. In QM this becomes a mathematical operator $\hat{H}$ and the wavefunctions are eigenfunctions of $\hat{H}$. Its eigenvalue is the energy.

$$\hat{H}(x)\psi(x)_n = E_n\psi_n(x). \tag{6.16}$$

This is the Schrödinger equation. We need to solve this equation of the eigenvalue type satisfying the required boundary conditions of the well. The sum of the kinetic energy $(p^2/2m)$ and the potential energy V is H. From now on we use atomic units where $\hbar = 1$. Consider the classical form: $H = [p^2/2m + V(x)]$. The p in H has to be replaced by $(-i\hbar d/dx)$.

$$\hat{H} = -(\hbar/2m)(d/dx)^2 + V(x). \tag{6.17}$$

Since $V(x)$ is zero inside the well, we have

$$\psi_n''(x) + (2m/\hbar)E_n\psi_n(x) = 0. \tag{6.18}$$

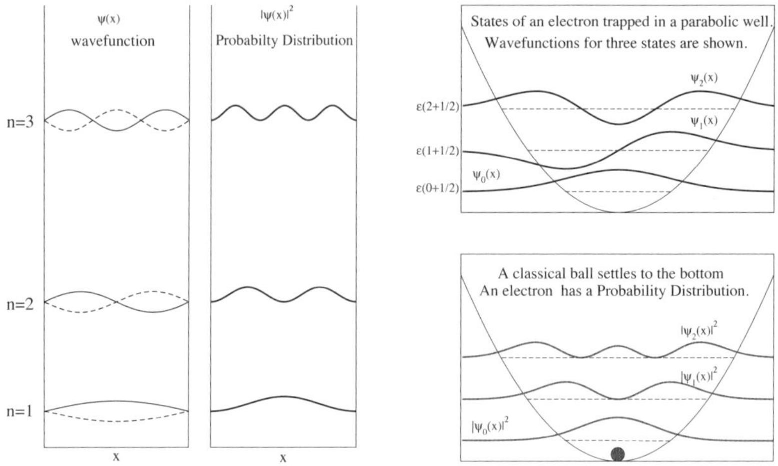

(a) sq. well (b) parabolic well

Fig. 6.6 An electron trapped in a well is a wavelike excitation. Only waves which fit the boundary conditions imposed by the well are possible, and hence there are discrete quantum energy levels, labeled by the index $n = 1, 2, \ldots$ etc.. The higher energy states have shorter wavelengths. The square of the wavefunction, viz., $|\psi_n(x)|^2$ gives the probability of finding an electron at some location x. A decaying wave tunnels into the 'classically forbidden' region of the parabolic well.

The second derivative of $\psi_n(x)$ is denoted with a double prime. The wavefunctions are zero outside the well where the potential energy is infinite. Each wavefunction $\psi_n(x)$ has an energy E_n. Each of these functions is independent of the other functions, and hence may be thought of as 'directions' (vectors) in an abstract function space known as 'Hilbert space'. They are said to be orthogonal because the overlap integral between any two of them is zero.. Paul Dirac began to denote the quantum state ψ_n by the symbol known as a 'ket', viz., $|\psi_n\rangle$, or simply $|n\rangle$, and think of them as vectors in Hilbert space. The complex conjugate ψ_n^* is denoted by the 'bra' symbol $\langle n|$. Thus, if two such vectors n, m are orthogonal we can simply write, $\langle n|m\rangle = 0$, in much the same way as the usual scalar product of two orthogonal vectors, i.e., $\vec{x} \cdot \vec{y} = 0$. The wavefunctions in the position representation are easily shown to be:

$$\psi_n(x) \equiv \langle x|n\rangle = \sqrt{(2/a)}\sin(\pi n x/a), n = 1, 2, 3, \ldots \tag{6.19}$$

with energies $E_n = E_1 n^2$, and $E_1 = h/(2ma^2)$. If the well width a becomes very large, the particle energy is practically continuous, as for a free particle, since the discreteness in the energy spectrum decreases as $1/a^2$. In the free-particle limit, the eigenfunctions become plane waves having the form $\psi_p(x) = A\exp(ip.x/\hbar)$, where A is a constant. If this is acted upon by the momentum operator $\hat{p} = -i\hbar d/dx$, by taking the derivative, we correctly recover the eigenvalue p for that state. This is in fact the basis for the choice of the form of the momentum operator used in the Schrödinger equation.

A classical wave of frequency ω moving in the x-direction with the wavevector k has the form:

$$\phi_k(x,t) = A\exp(ik.x - i\omega t)\,. \tag{6.20}$$

Hence we see that the operator $i\hbar d/dt$ acting on $\phi_k(x)$ yields $\hbar\omega\phi_k(x)$. Since we identify $\hbar\omega$ as E (see Eq. 6.12) the time-dependent form of the Schrödinger equation for the quantum state $\psi(x,t)$ is

$$\hat{H}\psi(x,t) = i\hbar d\psi(x,t)/dt\,. \tag{6.21}$$

Thus the time dimension is intimately connected with the energy. Furthermore, if the Hamiltonian is time independent,

$$\psi(x,t) = \psi(x,t_0)\exp\{-iH(t-t_0)\} \tag{6.22}$$

$$|\psi(x,t)|^2 = |\psi(x,t_0)|^2; \int |\psi(x,t_0)|^2 dx = 1\,. \tag{6.23}$$

Hence the square of the wavefunction remains unchanged as time evolves. Since $|\psi|^2$ is a probability of finding an electron at x, the total probability when integrated over x should be unity. Thus the wave function evolves in time conserving its normalization. This is known as 'unitary evolution'.

In Eq. 6.17 we saw the emergence of the quantum number n from the boundary conditions imposed in the direction x. Each of the space directions x, y, and z, on quantization (i.e., on applying suitable boundary conditions to the differential equation) yields quantum numbers n_x, n_y and n_z. If the system is treated using polar coordinates (r, θ, φ), as is typical in atomic physics, two of these quantum numbers become associated with the angular degrees of freedom. Then we use the quantum numbers n, l, m, where l, m are the 'angular momentum' quantum numbers. n is called the principal quantum number. No quantum number arises from the time dimension t, since no 'boundary conditions' in the time domain can be imposed.

An electron must occupy a state with a unique label of quantum numbers not assigned to any other electron. Particles which occupy quantum states exclusively by themselves are called *Fermions*. Thus only one electron can have the quantum numbers $n = 1, l = 0, m = 0$ assigned to it. These quantum numbers correspond to the lowest energy state of an electron in a hydrogen atom. This state is called the $1s$ state. Experimentally, two electrons are found in the $1s$ state, which means they differ by *some additional, intrinsic attribute*. This attribute is called the 'spin' of the electron, although it is not to be thought of as spinning around its axis. 'Spin' turns out to be an abstract property of elementary particles emerging from Dirac's relativistic quantum mechanics (Sec. 6.10.2). This feature can be incorporated into non-relativistic QM by introducing a new 'spin quantum number' usually denoted by σ. It can take two values for electrons, pictured as 'up spin', and 'down spin', corresponding to half a quantum of angular momentum, i.e., $\pm(1/2)\hbar$.

If we attempt to model the spin of an electron by a small spinning sphere of mass m, and charge e, we need to assign a finite radius to it. This radius does not agree with experimental electron collision data. Furthermore there is no reason why such a 'spin angular momentum' should remain invariant, as is the case of the electron spin. In order to recover the observed angular momentum of the electron, the surface layers of this model of the electron need to move faster than light.

Unlike the energy E of a classical particle, the trapped quantum particle can exist only in specific energy states which are described by the wavefunctions $\phi_n(x)$. This does not mean that the electron in the state $\phi_n(x)$ is a 'charge cloud' having the same distribution as the wave function. Penrose [160] appears to advocate such a view, but most physicists would disagree. The medium supporting the wave is also not specified. Here we recall that Einstein found it useless to attempt to specify the medium that supports the light waves of Maxwell's theory (the hypothetical aether, see Sec. 4.4). However, it is customary to say that electrons and other fundamental particles are the elementary excitations of an 'underlying quantum field'. When this quantum field is free of any excitations (on the average) it is called the vacuum state. Just as our knowledge of a distant drum is based entirely on the sound of the drum, our knowledge of the vacuum state is based entirely on its excitation spectrum.

Max Born proposed that the square of the electron wavefunction, $|\psi_n(x)|^2$, is proportional to the *probability of finding an electron at the point* x, if it is occupying the quantum state (eigenstate) corresponding to $\psi_n(x)$. Since the electron is definitely trapped within the region 0 to a of the well with infinitely high walls, then the sum of all probabilities within $[0,a]$ must be unity. Thus the integral of $|\psi_n(x)|^2$ over the interval $[0,a]$ should be unity. Then the eigenfunctions given in Eq. 6.19 are said to be normalized, i.e., $\langle n|n\rangle = 1$.

Max Born's suggestion is entirely parallel to the fact that if $E(x)$ is the field amplitude of a light wave, then the light intensity is proportional to $|E(x)|^2$, as we discussed (Eq. 6.9) in the context of the two-slit experiment. In 'free-field' situations, the vector-potential $A(x)$ rather than E which also satisfies Maxwell's equations, plays the role of a 'wavefunction of the photon'. However, the 4-component relativistic nature of Maxwell's equations makes simple comparisons with the Schrödinger equation misleading. Furthermore, photons in the presence of matter become polaritons, i.e., a superposition of electronic and photon-like excitations. These are what are detected experimentally.

The probability interpretation of the wavefunction has deep epistemological consequences. One point of view, e.g., of Heisenberg and Pauli, is to abandon the possibility, even in principle, of specifying the trajectory of an electron as this requires one to know its exact position and momentum at all times. Instead only the probability of finding an electron at a given location can be known, if a suitable measurement of its position were undertaken, while nothing definite can be said about the momentum. The ancient Greek philosopher Zeno would have been extremely happy about this, as he had asserted that if a particle is known to be at a definite location, then it cannot partake in any 'motion'. Einstein and others who came with a lot of Newtonian baggage were deeply troubled that this implies the end of causality and predictability. It was in regard to this that Einstein said *God does not play dice*. For Einstein, God was the orderly law-abiding behaviour of the universe, and not the capricious God of the scriptures. He wanted a causal world. This is the classical causality as understood in the world of levers, pulleys and pendulum clocks. As discussed in Sec. 6.7, Bohm has shown how classical causality can be regained and a trajectory specified, wherein the quantum uncertainty is relegated to uncertainties in initial conditions.

Let us try to look at the 'position' and 'momentum' of a quantum system in a somewhat artful manner. Suppose we attempt to 'fix', i.e., measure or delimit the position of the particle by shrinking the width a of the quantum well. Then it is easy to show that we make the momentum of the particle correspondingly uncertain. This is a manifestation of the 'uncertainty principle' of quantum mechanics. It may be understood as a consequence of the wave character of matter.

Since the electron can be anywhere within a (see Fig. 6.6-a), the uncertainty in its position x, denoted by Δx is clearly a. The momentum of the particle is given by de Broglie's relation $p = h/\lambda$.

The largest value λ can take is $2a$, (see Fig. 6.6) and the momentum of the particle may be $p = \pm h/2a$. Hence the momentum uncertainty is $\Delta p = h/a$. Thus we see that the product of the uncertainties in the position and the momentum of the quantum particle in this extremal case is:

$$\Delta x \Delta p = h. \tag{6.24}$$

This implies that if we try to localize the particle by making the well narrower and narrower, its momentum becomes increasingly uncertain. That is, we cannot know the position and the momentum of a quantum particle at the same time. This is also intuitively clear from Fig. 1.3 where a localized quantum particle is depicted as being made up of a superposition of many plane waves $\exp(ik.x)$ of wavelength $\lambda = 2\pi/k$. Let Δk be the extent of variation of k. If ψ is restricted to a region Δx, the plane waves must add constructively only in that region, and destructively elsewhere. The number of wavelengths λ contained in $\delta x = k\Delta x/(2\pi)$. This number must be at least unity for distructive interference at the edges of the raange Δx. Thus, writing $\Delta p = \hbar \Delta k$ we have $\Delta x \cdot \Delta p \geq h$.

Instead of looking at an electron trapped in an infinitely deep well with sharp walls, let us consider an electron confined to a harmonic well (i.e., a parabolic potential, see Fig. 6.6). The *quantum dots* of nanotechnology are actual realizations of such systems. A classical particle (e.g., a billiard ball) dropped into the harmonic well would oscillate along the wall of the parabola, like a pendulum bob. If the particle had zero energy, it would come to rest at the bottom of the well.

6.4.1 *Harmonic wells, zero-point energy, and tunneling*

A quantum particle in a harmonic well (parabolic well) can only assume quantum states allowed by the waveforms that are appropriate to the shape of that potential. That is, the quantum particle 'senses' the boundary conditions externally imposed on it. The eigenstates (energy states) of the electron (quantum particle) are determined by the reductionist process which isolates a section of reality as defined by the manner of confinement of the electron. In the usual quantum picture (n.b., the Bohm picture is different), the electron occupying the ground state is not at rest, but has kinetic energy and potential energy, adding up to a total of $\hbar\omega/2$, where ω is the frequency of the classical oscillation in the harmonic well. This is called the zero-point energy of the quantum oscillator. Thus a quantum particle does not localize at a point, but retains a zero-point motion associated with the extension of the ground state wavefunction. This zero-point energy cannot be extracted from the particle even by cooling to absolute zero, since the particle cannot fall below its lowest energy state.

One can experimentally measure the existence of a zero-point energy by measuring heat capacity and internal energy data of quantum crystals (e.g., hydrogen crystals) where the effect is most evident. The vibrational energy levels of simple diatomic molecules also clearly reveal the existence of zero-point motion. We can actually fabricate, using modern crystal-growth techniques, nanostructures called 'quantum dots' which are really parabolic energy pits which confine electrons. Their spectra as well as their current-voltage characteristics can also be used to establish the existence of a zero-point energy, as well as the energy-level structure shown in Fig. 6.6.

Instantaneous, dynamic dipoles exist in atoms since the negatively charged electrons revolve around the positive nuclei at an 'orbital distance'. If two metal plates were brought together, the instantaneous dipoles interact and produce an attractive force known as the van der Waals force. This can also be reinterpreted in terms of the effect of the metal plates in confining the zero-point motion of the excitations of the vacuum to the region between the two plates. This interpretation of the van der Waals force is known as the Casimir effect. The interaction of the instantaneous dipoles is electromagnetic and the effect may be discussed entirely in one picture, or in the other (just as some problems requiring a volume integral can be equivalently stated in terms of a surface integral and sources).

The energy of empty spacetime (Sec. 5.8) is of great cosmological interest. For instance, the sum of all the zero-point energies of elementary particles (which are viewed as excitations of the vacuum) might correspond to the energy of empty spacetime. Since this is also a measure of Einstein's cosmological constant, it seemed possible to estimate the cosmological constant from quantum arguments. However, the quantum estimate of the total zero-point energy is far in excess of the value implied by the cosmological constant obtained from astrophysical data. This mismatch in energies is perhaps a symptom of the current mismatch between general relativity and the quantum theory.

A sound wave cannot penetrate a rigid wall. However, if the wall were somewhat compliant, the wave penetrates the wall and rapidly attenuates itself. The sharply rising walls of the potential in Fig. 6.6 are rigid and the quantum waves cannot penetrate into them. However, the walls of the parabolic well (Fig. 6.6) rise gently. The quantum waves can penetrate into regions totally forbidden to a classical particle, but not forbidden to waves. This phenomenon is known as *quantum tunneling*. It is no more surprising than the penetration of sound through a wall. This implies that if we put in some electrons into a 'container' (i.e., a potential well with thin confining barriers, as found in some nanostructures, see Fig.7.3), the electrons would tunnel through the walls (barriers) and escape. In fact, in heavy nuclei like Uranium, the nucleons are not strongly confined by the nuclear potential, and some constituent particles have a finite probability of tunneling out, leading to radioactive decay of such heavy nuclei.

6.5 Is the wavefunction a real physical object?

So far we have presented the wavefunction ψ as the amplitude of a statistical probability $W = |\psi^2|$. This is the 'safest' or 'minimalist' position regarding the wavefunction. Should we ascribe some reality to ψ? This is a question that the Copenhagen school decided to avoid by simply labeling it as metaphysical, and Niels Bohr (see Fig. 6.2) was very wise in choosing this approach at the time. In effect, the answer does not matter for getting predictions out of the quantum theory. However, both Schrödinger and de Broglie were prepared to concede some degree of reality to ψ, at least in the Solvay-conference discussions.

If the electron spin is a part of reality, the wavefunction too partakes in an intrinsic reality described by vectors in an abstract space known as Hilbert space. The only way we know about ψ, or the electron spin, is via their role in the expectation values of physical properties that exist in our own physical space.

So, is ψ a part of an intrinsic abstract reality, or just a calculational tool?

When complex numbers were introduced, numbers like $\sqrt{(-1)}$ were regarded as 'imaginary', while ordinary numbers were 'real'. Many people regard the magnetic field $\vec{B}$ and the electric field $\vec{E}$ near a wire carrying a current to be real physical fields, while the electromagnetic potentials $\vec{A}$ and ϕ (discussed in Sec. 4.3) are often considered to be mere mathematical constructs. They are uniquely defined only when a gauge (i.e., a certain frame of reference or view point) is selected. Certain experimental phenomena like the Aharanov-Bohm effect (see Sec. 4.4.1) imply that we cannot deny the physical reality of these potentials.

The Aharanov-Bohm effect says that if an electron is presented with two different paths to a detector in a region with non-zero $\vec{A}$, the phase difference acquired by the electron includes an effect from the vector potential even if the physical field $\vec{B}$ is zero. The ψ amplitudes of the two paths interfere as in the two-slit experiment. If we collect enough electrons in a detector, we should obtain an interference pattern showing Aharanov-Bohm oscillations in the intensity.

In our view, if we are ready to accord a physicality to the vector potential $\vec{A}$ of the electromagnetic field, there seems to be no reason to deny the same degree of physicality to the the wavefunction ψ. In fact we may note that $\vec{A}$ may be regarded as a wavefunction of the photon, with the equations of Maxwell playing a role somewhat similar to the role of the Schrödinger equation for ψ, although the latter is non-relativistic. The ontology of the working physicist is completely happy with treating ψ as a real entity. Bohm's interpretation of quantum mechanics assigns a reality to ψ as an actual filed, and also to the quantum potentials that are directly related to ψ (see Eq. 6.37). The ψ is also the Born probability amplitude as discussed in Sec. 6.7. Hence Bohmian mechanics retains the statistical description and provides a physical meaning to the wavefunction in a consistent manner (see Sec. 6.7.1 regarding reservations of some scientists regarding the use of the quantum potential).

The classical Hamiltonian $H(x,p)$ resides entirely in our ordinary space and time (vector notation suppressed for brevity). On replacing p by the operator $-i\hbar d/dx$ we leave the ordinary space-time and enter a new, abstract spacetime where the vector representing ψ resides. We know that the plane-wave set $|\vec{k}\rangle$ is a basis that can be used to expand *any* $|\psi\rangle$. However, this k-space is nothing but the 'reciprocal space' used by crystallographers. The reciprocal space of crystals is also an abstract, intrinsic space that can be accessed only via certain experiments. In our view, we may regard ψ as a real entity that exists in a space similar to k-space of reciprocal lattices of crystals. Special experiments using ultrafast lasers can be used for 'molecular-orbital tomography', i.e., for 'displaying' the active orbitals of molecules [155] just as X-rays are used for 'displaying' reciprocal lattices.

In some formulations of QM ψ is traded away for other functions like propagators, sets of Feynman paths etc. Density-functional theory (DFT) by-passes the wavefunction and a *functional of the one-particle density* is shown to be necessary and sufficient (see Sec. 8.3) for treating the physics of many-particle systems. Although most Bohmian theorists may not say this, it is possible in Bohm's approach (Sec. 6.7) to do *without* ψ if the quantum potential Q is specified. Then Q can be recast into a differential equation for $R(x,t) = |\psi(x,t)|$. Knowing $R(x,t)$, the continuity equation may be treated as a differential equation for $S(x,t)$ such that $\psi = R\exp(iS)$. Here we draw the reader's attention to the fact that we can have many seemingly different perspectives of reality (see Sec. 2.4.2), and all may be regarded as alternative descriptions of various aspects of the same reality.

Physicists who decline to give a reality to the wavefunction ψ usually subscribe to the collapse of ψ when a quantum system is observed. The collapse of ψ is very easy to understand if it is an aspect of our thought process (i.e., a calculational concept). This point of view fits in with the subjective interpretations of Wigner and others where a conscious mind is needed to give reality to any aspect of the observed world. The Copenhagen school accepted the von Neumann collapse model without any specific point of view about ontological questions.

6.6 The uncertainty principle and particle trajectories

Quantum particles are often like waves, and waves of quantum excitations are often like particles. This 'neither fish, nor fowl' character arises from the two 'complementary' realities viz., real-space and momentum space well known to mathematicians and crystallographers. It is not specific to QM. The real-space description of a crystal lattice, or the momentum-space description of a crystal lattice, are equally real and complementary. They can both be specified simultaneously. However, this is not so for objects at atomic length scales.

We take it for granted that we can specify the position $\vec{q} = (q_x, q_y, q_z)$ of a runner, and that he is at rest (zero momentum, i.e., $\vec{p} = 0$) at the moment t just before the firing of the starter gun. In classical physics we assume that we may determine the position and the momentum of a particle without disturbing the particle in any significant way. In the case of the runner, we see that he is on the starting block. We can see the runner because the arena is illuminated. Let us say that the light had a particularly unusual tinge to it and that one of the runners, extra-sensitive, became agitated. This agitation leads to some uncertainty in his position (i.e., not stationary at the starter block, $\Delta\vec{q} \neq 0$) and his momentum (which is no longer strictly zero, $\Delta\vec{p} \neq 0$). Nevertheless, using a movie camera we may keep track of his movements accurately. We might also lower the intensity of the light and virtually eliminate the athlete's agitation.

The light illuminating the athlete is made up of photons, the quantum particles of light. These are like tiny bullets that hit the athlete, and some get reflected. Let us get a magic potion from *Alice in wonderland* and reduce the size of the athlete to that of an electron. The 'tiny bullets' of light are no longer tiny. The photon is now like a missile and he gets blasted out of the vision of the camera. Lowering the intensity of the light is futile because that only reduces the number of bullets coming in, without reducing their energy. The the wavelength λ of the light has to become longer (colour becomes redder) to reduce the photon energy. Not disturbing the electron implies the use of light of a sufficiently long wavelength. Unfortunately, just as a kilometre-long wave, however strong, is useless to a surfer, the long wavelength light fails to even notice the electron-sized athlete. In effect, if the wavelength is λ, then such light cannot be used to pin-point an object to better than an uncertainty of the order of λ. Thus we are unable, even in principle, to experimentally determine both the position $\vec{q}$ and the momentum $\vec{p}$ of the electron using any sort of probe. Hence we cannot know the trajectory of a particle, as is possible for every-day objects that follow Newtonian mechanics.

One might at this point continue to insist that the electron (or any other elementary particle) has a definite position $\vec{q}$ and a definite momentum $\vec{p}$, even though we cannot determine them. That is, they are 'hidden variables'. Such hidden variables, if introduced into the theory, must be consistent with the quantum theory, i.e., more specifically, the the Schrödinger equation (SE). Bohm showed how such 'hidden' variables p, q could be introduced, where-in the uncertainty is transferred to our lack of a precise knowledge of the initial state. This uncertainty is precisely such that all physical predictions are a mean value taken over the probability weight $|\psi(\vec{q},t)|^2$, or its momentum analogue $|\phi(\vec{p},t)|^2$.

If the uncertainty in $\vec{p}$ and $\vec{q}$ are $\Delta\vec{p}$, and $\Delta\vec{q}$, then the Heisenberg uncertainty principle states that their product has to be at least as large as $\hbar$

$$\Delta\vec{p}\Delta\vec{q} \geq \hbar. \tag{6.25}$$

We studied this explicitly for electrons trapped in a well (eq. 6.24). It is true in general that if we make $\Delta\vec{p}$ small, then $\Delta\vec{q}$ has to be proportionately large to keep the magnitude of the product essentially the same. The value of $\hbar$ depends on the units (centimeters or other) used for the position $\vec{q}$, and the momentum $\vec{p}$. In elementary particle physics it is convenient to use a system of units where $\hbar$ and c are unity. Atomic units are used in atomic/molecular physics, where the unit of length is not the centimeter, but the 'atomic unit of length' (a.u.), which is the radius of a hydrogen atom, i.e., about 5.29×10^{-9} centimeters (see Sec. 1.3). It follows from Eq. 6.25 that the position of a particle (i.e., the value of $\vec{q}$) can be determined to arbitrary accuracy by allowing the complementary property, i.e., the momentum $\vec{p}$ to be correspondingly less accurate.

This type of arbitrary accuracy is not possible in the relativistic theory where particles are associated with antiparticles. The de Broglie wavelength $\hbar/p$ is a measure of the localization of a quantum particle. Any attempt to localize an electron will increase Δp when the energy ($E = p^2/2m$) could increase such that an electron-positron pair of mass M could be formed, where E is changed to mass ($E = Mc^2$). This property is called the *Klein Paradox*. Zero-mass particles like the photon (which is its own anti-particle) cannot be localized to better than a few wavelengths associated with the allowed modes of the optical cavity trapping it. That is, photons are 'particles' with no definite position, but occupying a definite momentum state or cavity mode. In effect, photons are waves until you detect them as a 'particle' by its death signal – the bright scintillation on a photo-detector screen.

If the photon were a classical particle with a definite trajectory and energy, then its energy and arrival time at a detector would be known precisely. The wave character of a particle is embodied in the energy-time uncertainty relations.

$$\Delta E \Delta t \geq \hbar\,. \tag{6.26}$$

The energy-time uncertainty principle can be understood by noting that an energy eigenstate of a photon or an elementary particle is a monochromatic wave, i.e., an infinite plane wave, which does not have a well-defined time of arrival at a given point. Conversely a very narrow wave packet possess a well-defined arrival time (assuming a simple-minded approach to arrival times). However the wave packet contains many different frequencies, and is therefore a superposition of many waves with many different energy states (cf., Fig. 1.3).

A more rigorous derivation can be constructed (without appeal to arrival times) using the equation of motion (time evolution) of an arbitrary operator $\hat{A}$, e.g., in terms of its commutator with the Hamiltonian $\hat{H}$. A characteristic time $\Delta\tau$ for the width of $\langle\vec{A}\rangle$ to become of the order of $\langle\vec{A}\rangle$ exists. This time width can be applied to the energy itself (setting $\vec{A} = \vec{H}$), and we obtain Eq. 6.26.

6.6.1 *The Planck scales of length and time*

The quantum uncertainties Δp and Δx, and also energy and time measurements ΔE and Δt can be used to understand the limiting size scale known as the Planck scale, using the limits set by the existence of blackholes in defining mass and density relationships. In effect, the smallest possible length, the Planck length, corresponds to a momentum uncertainty spanning on to the energy of a blackhole.

Consider the energy-time uncertainty relation $\Delta E \Delta t \geq \hbar$. For a body moving with a velocity v near the speed of light, Δt would be $\Delta L/c$. Then, if the uncertainty in its energy can be pushed to be as large as the order of its mass-energy mc^2, we see that a minimum length scale ΔL of the order of

$$\Delta L = \hbar c/mc^2 \tag{6.27}$$

begins to emerge. If ΔL is further reduced, we reach the blackhole radius $\Delta L \sim R = Gm/c^2$. Substituting for m and setting the Planck length ℓ_p as the minimum possible ΔL, we have

$$\ell_p = (\hbar G/c^3)^{1/2}\,. \tag{6.28}$$

That is, if we attempt to measure a distance smaller than a Planck length $\ell_p = 1.6 \times 10^{-35}$ meters, the uncertainty in our time intervals, rest masses, etc., would push us into the regime of a blackhole. The Planck length ℓ_P is associated with the Planck time $t_P = \ell_P/c = 5.4 \times 10^{-44}$ sec., which is the smallest time interval that can be meaningfully discussed. The Planck time corresponds to a Planck energy $E_P = \hbar/t_p$ which is of the order of 10^{19} GeV. Thus, initially, within the time interval $t = 0$ to $t = t_p$, such an energy would be spread within a tiny ball of radius equal to the Planck length, setting the stage for the explosive inflation (see Sec. 5.9) discussed in cosmology.

Hence even cosmogenesis is determined by the quantum theory because it started at quantum length scales.

6.6.2 *Heisenberg Operators and state vectors*

The uncertainty principle of Heisenberg states that we cannot precisely measure both the position and the momentum of a quantum particle simultaneously. This property can be neatly embedded in the idea of operations which disturb each other, as already alluded to in earlier discussions (e.g., Secs. 1.1.1 and 3.1).

Such operations can be represented by mathematical operators whose action can depend on the order of execution of the operations. Let us think of the operations of measuring the position q and the momentum p. If we confine ourselves to one dimensional motion in the x direction, we can use $p = p_x$, and $q = q_x$ in the following discussion. Just as we talked of symmetry operations (see section 3.2) and symmetry operators, we can talk of the act (operation) of measuring the position q and denote this operation by an letter carrying a hat, as in $\hat{q}$ (or tilde, viz., $\tilde{c}$ for everyday operations). Similarly, we can denote the operator signifying the measurement of momentum by $\hat{p}$. Before we talk of these operators, let us consider some everyday operations and their 'operators'. We can refer to the operation of someone putting on a cap by $\tilde{c}$, and putting on a glove by $\tilde{g}$. Clearly, the order of the operations, i.e., putting on the cap first and then the glove, i.e., $\tilde{c}\tilde{g}$, is immaterial since the result is the same if we first put on the glove and then the cap, $\tilde{g}\tilde{c}$. This is sometimes referred to as 'noncontextuality', esp. by philosophers. We say that $\tilde{c}$ and $\tilde{g}$ are commuting operators, and that $\tilde{c}\tilde{g} = \tilde{g}\tilde{c}$. In effect, the two operations of putting on a glove and putting on a cap do not interfere with each other. On the other hand, the operation of putting on a sock, denoted by the symbol $\tilde{s}$ and the operation of putting on a boot $\tilde{b}$ on the same foot as the sock interfere with each other. That is, putting on the boot after putting on the sock is not the same as putting on the sock after putting on the boot.

$$\tilde{s}\tilde{b} \neq \tilde{b}\tilde{s}. \tag{6.29}$$

That is, $\tilde{s}$ and $\tilde{b}$ are non-commuting operators. The operators corresponding to the momentum $\hat{p}$ and the position $\hat{q}$ are such non-commuting i.e., *contextual* operators in QM.

The mathematical operation corresponding to measuring the momentum in the x-direction of a quantum system in the state $\phi(x)$ involves taking the gradient $-i\hbar d/dx$ of the wavefunction $\phi(x)$. The measured average value is the average value of the operator evaluated with the probability distribution $|\phi(x)|^2$. If we denote the average value (mean value) by $\langle p_x \rangle$, then

$$\langle p_x \rangle = \int \phi^*(x)\{-i\hbar\frac{\partial}{\partial x}\phi(x)\}dx. \tag{6.30}$$

As noted (Sec. 6.4), the wavefunctions are treated as vectors $|\phi\rangle$ in an abstract vector space whose directions are set by orthogonal functions which are unit vectors. In the Schrödinger picture the stationary wavefunctions ϕ carry a time dependence of the form $\phi = \phi(x)\exp(-iEt/\hbar)$. In the Heisenberg picture, the state vectors define a time-independent geometry while the operators become time dependent. The net effect of the two pictures is to yield exactly the same mean-values for the observable results.

Quantum mechanics says that only properties associated with commuting operators can be measured at the same time — an eminently reasonable point of view. In fact as one becomes more and more accustomed to quantum mechanics, it becomes a part of 'common sense'!

6.7 Bohm's interpretation of the Schrödinger equation

The rationale for Bohm's approach can be best expressed, surprisingly enough by quoting Bohr [33], who pointed out that 'the account of the experimental arrangement , and the result of observation must be expressed in unambiguous language with suitable application of the terminology of classical physics'. Bohm's formulation usually provides a picture of quantum phenomena close to our ordinary intuition which is based on the classical world.

Bohm introduced point-like particles having position and momenta, with trajectories that evolve under the action of the external potential and the quantum potential Q. Intuitively, one would imagine that a free particle (e.g, external potential is zero) would not turn around 'by itself', but a Bohmian particle can do so due to the action of Q.

Einstein and many others (e.g., Madelung) were deeply troubled by the philosophical implications of the uncertainty principles. If we could say nothing about the trajectory of each individual electron, and if we can only give answers which are only statistically true, then there must be deeper dynamical laws which controlled the motion of individual electrons. In classical mechanics, as used in the kinetic theory of gases, the trajectory of a particle moving under imposed forces (i.e., any potential) was explicitly given by Newton's laws.

Given a particle of mass m in a potential $V(x)$, these laws asserted that the rate of change of momentum dp/dt is equal to the force $F(x) = -dV(x)/dx$ acting on the particle. The momentum $p = mv$, where v is the velocity of the particle at any instant t

$$dp/dt = mdv/dt = F(x) = -dV(x)/dx. \tag{6.31}$$

Thus, given the initial position x_0, and the initial momentum p_0, the position after a very small time interval δt is given by

$$x = x_0 + \delta x; \quad p = p_0 + \delta p$$

where the small changes $\delta x, \delta p$ are calculated using the initial velocity and acceleration given by Eq. 6.31. For example; if the force is a constant F and the initial velocity $p_0 = 0$, it is easy to show that $x(t) = x_0 + (F/2m)t^2$ and $p(t) = p_0 + Ft$. Thus the whole trajectory $x(t), p(t)$ as a function of time can be calculated.

The whole trajectory is determined, backwards or forwards. Einstein felt that QM, aspiring to be *the* fundamental theory of nature is not complete unless such a calculation was possible — *God does not play dice*. Thus, at first Einstein sought to disprove QM. However, following his debates with Bohr at the Solvay conferences [200] in 1927 and 1930 he began to claim that QM was 'an incomplete theory'. The EPR paradox (Sec. 7.7.2) was motivated by such thinking.

David Bohm working at Princeton was deeply influenced by Einstein's concerns and made a careful study of QM from the Copenhagen point of view and published his excellent 1951 book on Quantum Mechanics [25]. However, during the period 1952–1954 Bohm wrote several papers [26] in the *Physical Review* where he showed that the Schrödinger equation can be cast into the form of a Newton's equation, but with an additional effective potential called the *quantum potential*, denoted by $Q(x,t)$, acting on the particle. He asserted that 'an electron is a particle following a continuous and causally defined trajectory with a well defined position, accompanied by a physically real wavefield $\phi(x,t)$'.

Bohm's approach requires that: (i) $\phi(x,t)$ satisfies the Schrödinger equation. (ii) The velocity of the particle v is given by $v = \nabla S(x,t)/m$ where the wavefield $\phi(x,t)$ is of the form $R(x,t)\exp\{iS(x,t)/\hbar\}$, where R, S are real valued. (iii) The associated initial (i.e., $x = x_0, t = t_0$) probability distribution of particles ρ is $|\phi(x_0,t_0|^2$.

The last statement means that the initial conditions available for the particle trajectories agree with the probability distribution ρ due to the boundary conditions and other details that went into the preparation of the system under study. These three assumptions were shown to be consistent, and were designed to give results identical to Schrödinger's QM. Let as take a more detailed look at Bohm's approach. If $V(x)$ is an external potential, the Schrödinger equation (SE) is,

$$\hat{H}\phi(x,t) = [-\frac{\hbar^2}{2m}\nabla^2 + V(x)]\,\phi(x,t) = i\hbar\frac{\partial\phi(x,t)}{\partial t}. \tag{6.32}$$

For a 1-D problem ∇^2 is $\partial^2/\partial x^2$. The wavefunction $\phi(x,t)$ must be determined by solving the above equation subject to various boundary conditions. For example, if V is the 1-D form $V(x)$ which changes its value due to the existence of material interfaces at the locations x_1, x_2 etc., the wavefunction must be continuous, and it first derivative $\partial\phi/\partial x$ must match at each interface. If the external potential $V(x)$ tends to infinity as $x \to \infty$, then $\phi(x,t)$ must tend to zero for large x. Thus the wavefunctions 'knows', i.e., contains all relevant information about the system. Analogously, when an electron passes through two slits (as opposed to one slit) in a 3-dimensional problem, the corresponding boundary conditions are contained in any wavefunction which is a solution of the Schrödinger equation.

Bohm [26] writes the wavefunction in the form:

$$\phi(x,t) = R(x,t)\exp\{iS(x,t)/\hbar\}, \tag{6.33}$$

$$R(x,t) = \sqrt{\rho(x,t)};\ \rho = |\phi(x,t)|^2, \tag{6.34}$$

so that R (positive), and S are real functions. Substituting this form of the wavefunction into the SE, one obtains Bohm's form of the equations as:

$$\frac{\partial\rho}{\partial t} + \nabla\cdot\left[\rho\frac{\nabla S}{m}\right] = 0, \tag{6.35}$$

$$\frac{\partial S}{\partial t} + V + \frac{(\nabla S)^2}{2m} + Q(x,t) = 0, \tag{6.36}$$

$$Q(x,t) = \frac{-\hbar^2}{2m}\frac{\nabla^2 R}{R}, \tag{6.37}$$

Eq. 6.35 is simply the equation of continuity. The term $(\nabla S)^2/2m$ in Eq. 6.36 is interpreted as the kinetic energy of the point-like particles with momentum $p = \nabla S$. In effect, Eq. 6.36 has the form:

$$\partial S/\partial t + H(p,x) + Q = 0. \tag{6.38}$$

This is exactly like a classical Hamilton-Jacobi equation with the extra potential Q, called the quantum potential Q, or the Bohm potential. The 'particle' at the location x has a momentum $\vec{p} = \nabla S$.

The theory describes a point-particle with a definite momentum and position, moving on a trajectory determined by Newton's equations based on the total potential including the quantum potential.

$$m\frac{d^2\vec{r}}{dt^2} = -\nabla\left[V(\vec{r}) + Q(\vec{r})\right], \text{ or } d\vec{r}/dt = \nabla S(\vec{r},t)/m. \tag{6.39}$$

Thus (a) the quantum potential Q, (b) the Bohm point-particle, and (c) the well-defined classical trajectory are features of Bohm's original(1952–1953) reformulation of the SE. A 'complete description' of the system is taken to include the wavefunction *and* the positions of all the particles in the system at all times. The latter may be regarded as a set of hidden variables, and finally, only a statistical average over the density distribution $\rho = R^2$ is observable. In more recent discussions of *Bohmian mechanics*, the quantum potential has been de-emphasized, while the particle current and dynamical aspects are emphasized [107]. The advantages and disadvantages of the quantum potential are discussed in Sec. 6.7.1. Here we remain close to Bohm's early work and his last book [28] with Hiley, and use the language of the quantum potential.

Since Bohm's approach specifies where the particle would 'be', *via* explicit trajectory calculations, while conventional QM gives only the mean value of observations, it is sometimes known as a theory of *beables*.

If the 'classical limit' where the magnitudes associated with the Planck constant $\hbar$ is deemed to be negligible, the term $Q(x)$, i.e., the quantum potential in Eq. 6.39 also becomes negligible, and only the standard Newtonian's equation remains. Sometimes $Q(x)$ is expanded in powers of $\hbar$ and only the terms up to $\hbar^2$ are retained. This is a simple but useful 'quasi-classical' approximation to quantum problems known as the WKB approximation, named after G. Wentzel, H. A. Kramers and L. Brillouin who introduced it in 1926. Of course, WKB was introduced directly, as an approximation to the Schrödinger equation, and not via the concept of the Bohm potential. A WKB-type approximation is used in various cosmological applications, e.g., by Hawking and Hartel, in discussions of a 'wavefunction of the universe' (Sec. 6.10.4).

In a number publications Cushing, e.g., Ref. [50] speculates that if Bohm's version of the quantum theory had been properly formulated in 1927, it might have become the 'standard interpretation', while the Copenhagen interpretation would have had great difficulty in gaining ground. It should however be noted that the Copenhagen interpretation is a very convenient linear theory whose formal simplicity and mathematical austerity facilitated the formulations of modern relativistic quantum-field theories (e.g., quantum-electrodynamics, gauge fields, chromodynamics) without being tied down to a specific inertial frame defined by a non-local potential. It is mostly in the intuitive understanding of quantum phenomena, measurements etc., that the Bohm approach is most illuminating.

Fig. 6.7 David Bohm was a young assistant professor at Princeton when he presented a formulation (1951–52) of QM which was 'complete' in at least one of the aspects demanded by Einstein. He provided individual trajectories for quantum-particle dynamics by re-writing the Schrödinger equation in a physically transparent but mathematically equivalent form. Although this did not satisfy Einstein, 'Bohmain mechanics' has today become one of the most important interpretations of the quantum theory.

6.7.1 *Bohm's quantum potential*

The Bohm potential has raised much discussion since its formulation, and after the revival of interest in causal interpretations brought about by the work of John Bell [18]. One of the objections adduced against the quantum potential is that it has the 'unusual effect' of making the particle density ρ influence the dynamics of individual particles. However, we do not see anything unusual in that, since it is typical of many-particle physics that the individual dynamics is influenced by the over-all particle distribution. For instance, a low-order (mean-field) many-body theory like the Hartree theory of many electrons requires that an electron moves in the many-body field of the other electrons given by ρ.

If we treat the quantum potentials as an effective many-body field, then it is objected that this requires raising the role of ρ from an 'epistemic' to a 'physical' status. However, ρ does play such a physical role not only in Hartree theory, but also in density functional theory where a one-body density functional is said to be necessary and sufficient to do quantum mechanics (see Sec. 8.3).

Dúrr, Goldstein and Zhangi [66] summarize their attitude to the quantum potential with the statement, 'We believe that the most serious flaw in the quantum-potential formulation of Bohmian mechanics is that it gives a completely wrong impression of the lengths to which we must go in order to convert orthodox quantum theory into something more rational. The quantum potential suggests … we must incorporate into the theory a quantum potential of a *grossly* nonlocal character'. We emphasized the word 'grossly' in the above to high-light where the attitudinal squemishness resides. Most physicists who use the Hartree-Fock equations as a basic first step in their calculations have given up their squeamishness of non-locality, just as young medical students who see blood become quite sanguine to it. Regarding 'the lengths' needed to reach the classical Hamilton-Jacobi form, we need a few lines, from Eq. 6.33 to Eq. 6.38, where most of the manipulations are mere re-definitions.

Hence, in our view, these objections are rooted in attitudinal and philosophical prejudices.

Another objection raised against the concept of the quantum potential is that it has the character of a kinetic energy, and that the particle has 'no back reaction' on the field associated with the quantum potential. By the virial theorem, we see that potentials and kinetic energies have an equivalent nature, and the objection that Q has the character of a kinetic energy is of no importance. Since the quantum potential Q has been derived from the wavefunction ϕ of a many-particle system (say), then its 'reaction' is already contained in its contribution to ϕ, exactly as in Hartee-Fock theory.

The dependence of the quantum potential on ψ, and the claim that it 'acts instantly' and 'without a source', have made it into an intriguing object. Louie de Broglie's discussions in terms of 'pilot waves' and 'a particle' moving under the 'guidance of the pilot wave' are consistent with more recent discussions [29] of 'active information' being provided to the particle by the Schrödinger field. The Bohm potential enshrouds information about the quantum system and the boundary conditions that the quantum particles must satisfy, via its dependence on the wavefunction. It is set up at the same speed as the time taken by the experimentalist to prepare his experimental system. That the quantum potential 'acts instantly' is an 'after the fact' perception.

The non-locality of the quantum potential poses no problem in the context of the Schrödinger equation which is non-relativistic. One is familiar with electrostatic potentials of classical charges which set up 'instantly'. For instance, the Coulomb potential between two particles of charge q and located at $\vec{r}_1$ and $\vec{r}_2$, how ever far apart, is given by $q^2/|\vec{r}_2 - \vec{r}_1|$ and this is assumed to be instantly set up. In practice this has to be at the speed taken up by the experimentalist to position those charges and 'prepare' the experiment.

Moving charges cannot be used to send signals faster than that of the electromagnetic field (i.e., the velocity of light). When the Schrödinger equation for two particles is considered, as in the EPR experiment (Sec. 7.7.2), non-local potentials that instantly connect and correlate them are found to appear. Even the Pauli anti-symmetrization of two fermions is such a non-local effect. However, it can be shown that such non-local interactions cannot be used to exercise a remote influence on the far away particle. That is, 'remote steering' using these instantaneous quantum correlations is not possible, as required by the special theory of relativity. Instead of using the Schrödinger equation, one may use the Dirac equation and a Bohmian formulation of it can be given. These generalizations are discussed in Bohm and Hiley [29].

It has been argued that the velocity field $\vec{v}(x,t) = \vec{J}(x,t)/\rho(x,t)$ accomplishes what Q and $\vec{\nabla}S$ does, in a pedagogically cleaner manner without a return to Newton's equations. If one uses the proper expression for $\vec{J}$ (e.g., that given by Holland), the same equation of motion which is first-order in time can be used even for the Dirac equation, and for clarifying hard problems in the theory of transmission, reflection — and dwell times in QM [131]. However, in 1952–53, Bohm *was* trying to show to Einstein and other critics that QM was as good as returning to Newton's equations! Furthermore, the formulation in terms of Newton's equations is extremely useful as it is the natural framework for Verlet-type molecular dynamics simulations where the system is defined by the initial conditions and step-wise updated potentials.

The emphasis on $\vec{J}(x,t)$ is also found in Erwin Madelung's work (1927). In his hydrodynamic interpretation of Schrödinger's equation Madelung does not have particle trajectories. He recasts the Schrödinger equation into a flow equation containing a quantum pressure tensor closely related to the Bohm potential. Other statistical or classical reformulations of QM involve recasting the Schrödinger equation into a diffusion equation, e.g., that of Edward Nelson (1966). It is not clear if such diffusion theories succeed in violating Bell's theorem.

6.7.2 *Bohm's picture of a particle trapped in a well*

An intriguing result of Bohm's theory is that the velocity of an electron in a square-well potential (see Fig. 6.6) in any stable quantum state would be zero. All the eigenfunctions written in the polar form are such that the momentum (the gradient of S) would be zero. Thus in BM the particle is *simply sitting still* in some location in the well. This aspects of Bohm's theory did not please Einstein.

Einstein [67], in the 1953 *Festschrift* in honour of Max Born criticized Bohm stating that 'the vanishing of the velocity contradicts the well-founded requirement, that in the case of a macro-system the motion should agree approximately with the motion following from classical mechanics'. Pauli made similar objections to Bohm's theory, and also expressed his aversion to unobservable entities. In his reply to Einstein, published in the same *Festschrift* Bohm argued that his theory 'is self-consistent and does not contradict any known experimental facts'.

There is no way of directly 'detecting' this motionless state of the electron proposed by Bohm, or demonstrating the intuitive idea that the particle is moving 'too and fro', and bouncing between the walls of the well.

An experiment can be imagined where the infinite walls on both sides are suddenly removed at the time t_s. Then the particle will be ejected from the well with a time-dependent momentum distribution which asymptotically becomes consistent with the momentum distribution of the stationary eigenstate $\psi_n(x)$, occupied by the particle at the instant t_s, viz., Eq. 6.40, in agreement with the predictions of standard QM.

The s-states of atoms, i.e., $\psi_{ns}(r,t)$, are real-valued and spherically symmetrical. They are also found in BM to have the electron (Bohm point particle) 'sitting still in an arbitrary position' as the net force on the particle is zero. There is a statistical ensemble of possible static locations for the particle, given by $|\psi_{1s}(r,t)|^2$.

Although Einstein did not quite say so, he probably expected to see a 'too and fro' motion of the particle in any quantum state of the well, as in 'zero-point motion'. We show that this intuitive picture is not natural or evident even in conventional QM, and hence not peculiar to BM. Also, it appears inconsistent with the existence of nodes in ψ. However, in the following we present a consistent picture where conventional QM, intuition, and BM all agree, if particle currents are *re-written* to fit in with the intuitive picture in a mathematically consistent manner.

The intuitive picture is consistent with the force on the walls due to the 'too and fro' impact of the particle, and the value obtained from the Hellmann-Feynman theorem. However, if the wavefunction $\psi_n(x)$ is decomposed into plane waves, the most probable momentum p is indeed found to be $p = 0$ (as in the Bohm picture) and not the 'too and fro' $\pm k$, for $n = 1$. With $\hbar = 1$, we have

$$\phi_n(p) = \int_0^a \psi_n(x) e^{-ip.x} dx/(2\pi), \tag{6.40}$$

$$\text{then } |\phi(p)|^2 = (2\pi a)\frac{n^2 - n^2(-1)^n \cos(pa)}{(p^2a^2 - n^2\pi^2)^2}. \tag{6.41}$$

The modulating factor in the numerator also goes to zero at $k = \pm k_n$ and the intuitive picture is not held, even in conventional quantum mechanics. The real-space probability distribution $\rho(x) = |\psi_n(x)|^2 = (2/a)\sin^2(k_n x)$ becomes zero at the nodes of the sine wave (see Fig. 6.6) and also at the walls; i.e., the probability of impact seems to be zero.

The particle cannot pass through the nodes of ψ_n where $\rho = 0$. Clearly the momentum cannot be $\pm k_n$ at the nodes. Also, $\rho = 0$ at the walls and the intuitive idea of momentum reversal fails. The quantum mechanically evaluated current is also zero. The average position of the particle obtained from the expectation value of x is $a/2$ for any odd n; i.e., the particle sits at the mid point, both in QM and in BM. Thus the common 'intuitive picture' seems to be internally inconsistent.

6.7.2.1 *State of rest as a superposition of 'too and fro' motion.*

We attempt to give a possible resolution of the intuitive idea of 'too and fro motion' in a well applicable to both BM and conventional QM.

In the following we sometimes suppress the index n for brevity. The particle current in the usual Schrödinger formalism is given by given by:

$$j(x,t) = [\psi^*(x,t)\nabla\psi(x,t) - \{\nabla\psi^*(x,t)\}\psi(x,t)]/(2mi). \tag{6.42}$$

This is indeed zero for particles trapped in the well, with wavefunctions $\psi_n(x)$ given by Eq. 6.19, i.e.,

$$\psi_n(x) = \sqrt{(2/a)}\sin(k_n x); \quad n = 1, 2, \ldots. \tag{6.43}$$

We treat ψ_n as a state formed by a superposition of two eigenstates $\xi_\pm(x)$, with equal and opposite momenta, as suggested by the form $\sin(k_n x)$ contained in the wavefunction, and from Eq. 6.40 where $\pm k_n$ occurs as a hidden pole. Also, if a magnetic field is applied (along the z-direction), normal to the well, these two states become 'edge states' pushed to the edges of the sample in the y-direction due to the Hall effect, and hence this decomposition has a good physical basis.

$$\psi_n(x) = (2/a)^{1/2}\{e^{ik_nx} - e^{-ik_nx}\}/(2i) \tag{6.44}$$
$$= \xi_+(x) - \xi_-(x);\ \text{more precisely}\ \xi_\pm = \theta(x)\theta(a-x)e^{\pm ikx}. \tag{6.45}$$

Then we have

$$\rho_n(x) = |\psi_n|^2 = \rho_+(x) - \rho_X(x) + \rho_-(x), \tag{6.46}$$
$$\rho_\pm(x) = |\xi_\pm(x)|^2;\ \rho_X(x) = (\xi_+(x)^*\xi_-(x) + h.c.) \tag{6.47}$$

The density contribution due to the interference term is $\rho_X(x)$. The current may also be written as:

$$j_\pm(x) = Im\left[\xi_\pm^*(x)\frac{d}{dx}(\xi_\pm(x))\right] \tag{6.48}$$
$$j(x) = j_+(x) + j_X(x) + j_-(x). \tag{6.49}$$

Here $j_X(x)$ is the contribution to the current from the interference terms in ρ_X. We have:

$$j_+(x) = \frac{k_n}{2a} = -j_-(x), \tag{6.50}$$
$$j_\leftrightarrow \equiv j_+(x) + j_-(x) = 0. \tag{6.51}$$

The total current calculated directly from $\psi_n(x)$ is zero, and $j_\leftrightarrow$ is also zero. Hence the cross-term $j_X(x) = 0$, as may be confirmed by a direct calculation. These results are obtained even if we retained the step-functions $\theta(x)\theta(a-x)$ in the wavefunction. Thus we only have the current components $j_\pm(x)$ that are equal, opposite, and add to zero as required within the intuitive picture.

The probability current $j(x,t)$ in BM is identical with that of the standard analysis. The partial currents $j_\pm(x)$ follow very easily in BM:

$$\xi_\pm(x) = \sqrt{(1/2a)}\exp^{i(-\pi/2\pm kx)} = R_\pm e^{iS_\pm(x)}, \tag{6.52}$$
$$p_\pm = \nabla S_\pm = \pm k_n;\ \ j_\pm = |R_\pm(x)|^2 p_\pm = k_n/(2a). \tag{6.53}$$

In addition, BM requires that *the continuity equation is also satisfied by the partial currents* $j_\pm(x)$. The continuity equation can be written using partial currents $j_\pm(x)$, j_X and partial densities $\rho_\pm$, ρ_X as:

$$\partial\left[\rho_+(x,t) - \rho_X(x,t) + \rho_-(x,t)\right]/\partial t \tag{6.54}$$
$$+\partial\left[j_+(x,t) + j_X(x,t) + j_-(x,t)\right]/\partial x = 0.$$

As already noted, the current j_X from the interference term is zero. If the interference-density term ρ_X has no time dependence, then this can be written as two separate continuity equations. In the case of an electron in a quantum well this is indeed the case. Since $j_\pm(x,t)$ has no spatial dependence, and since $\rho_\pm$ has no time dependence either, the two continuity equations are trivially satisfied.

In the standard Bohm picture, the particle sits with zero momentum and probability $|\psi(x)|^2$ at the location x, and its kinetic energy has been fully transferred to the quantum potential. In the modified picture where we have 'too and from' motion, the particle carries the kinetic energy while the corresponding quantum potential is zero. That is, the Bohm quantum potential can also be written as two components using the arbitrariness of the phase of ψ.

$$Q = -(\hbar^2/2m)\{\partial^2\text{Re}(\xi_+ - \xi_-)/\partial x^2\}/R_n = Q_+ + Q_-, \tag{6.55}$$
$$\xi_\pm = \sqrt{(1/2a)}\exp\{i(-\pi/2 \pm k_n x)\}\ \text{hence}\ Q_\pm = 0. \tag{6.56}$$

We conclude that zero-current quantum-well states of Bohmian particles, or the zero-current states in conventional QM can be transformed to satisfy the intuitive picture of being in a superposition of the 'too and fro' motion of a particle with zero net current. In the Bohm picture the 'too and fro' motion is associated with definite trajectories. The currents $j_{\pm}$ do not confront nodes since $\xi_{\pm}$ do not have nodes; the nodes are in the contribution to ψ^2 arising from the interference term that has no current. The 'too and fro' motion impacts on the walls to produce a force agreeing with the Hellmann-Feynman result. Hence Einstein's objections in regard to the Bohmian mechanics of a particle in a well can be answered, and the intuitive picture is established for both conventional QM and BM.

6.8 Feynman-path formulation of quantum mechanics

In 1948 Feynman published in the *Reviews of Modern Physics* his insightful formulation of quantum mechanics. It is valued today both conceptually and as a framework for numerical simulations of quantum systems on computers. Feynman's ideas were in fact already fore-shadowed in his 1942 doctoral thesis [73]. The classical concept of a particle taking a trajectory from point A to point B is generalized to include all possible paths, and this quantum analogue is shown to obey the same principles of least action as in the classical domain!

In section 3.6 we discussed how the principle of least action governs the motion of a particle and concluded that the laws of mechanics could be viewed as a result of this principle. The paths taken by a ray of light in different media could also be explained by a principle of least time (Fermat's principle). The trajectory of a particle going from a point A to another point B was always the 'path of least action'. In flat spaces, this path turns out to be a straight line. In curved spaces, or media where the physical characteristics (e.g., the refractive index and density) change with position, the path of least action would be suitably curved. The simplicity of the laws of nature was such that local properties acting on the particle ('worm's eye view') were exactly those that matched the global ('bird's eye view') characteristics of the final trajectory (see Sec. 3.9). It was not a case of the particle 'knowing' the path with the least action, in the sense of a destination-directed (teleological) principle. If the laws of nature were more complex (e.g., acceleration not directly proportional to the force, as in Newton's laws), then such a dovetailing of local and global behaviour would not occur.

When we reduce the size scales to reach quantum processes, the 'trajectory' of a quantum particle is not even a possibility in standard quantum mechanics. If a particle starts at A, all we know is the probability that we would detect an equivalent particle at B. This probability is $|\psi(B)|^2$. Feynman gets around this by not allowing a unique path, but using all possible paths (including even those one might intuitively consider highly unlikely) and ensuring that these paths add up to get the particle from A to B with exactly the QM probability $|\psi(B)|^2$.

The classical Lagrangian $L(p,x)$ for a particle of momentum p and position x is simply the difference between the kinetic energy (K) and the potential energy V. Given a path $x_p(t)$, starting at x_0 and currently at x, the action S is the sum of $L\{x_p(t)\}$ over all the times steps δt taken along the path.

$$S(x) = \int_{x_0}^{x} L\{x_p(t)\}dt\,. \tag{6.57}$$

In linking these ideas with a wavefunction, it is helpful to look at the analogy between the wave description of light, and the ray description of geometrical optics. Let u be the electric field associated with the electromagnetic field. It can be written in the form $u = a\exp(i\theta)$, where a and θ are real numbers. The a is the amplitude and θ is the phase of the wave. The limiting case of geometrical optics applies if θ varies by large amounts over short distances.

Given the analogy between ray optics and wave optics, the classical limit corresponds to a quantum dynamics with wavefunctions of the form $\psi = a\exp(i\theta)$. In the limit of classical motion, the phase θ is the action S calculated along the trajectory, using the Lagrangian L. In the quantum case also θ is proportional to the action S, with $\hbar$ the proportionality constant. Hence $\psi = ae^{iS/\hbar}$, a form used in Bohmian mechanics as well. This is a key idea that Feynman incorporated into his path integral formulation of QM, beautifully explained in Ch. 19, vol. 2 of Feynman's *Lectures on Physics* [74]. The path-integral formulation of QM considers *all* possible classical trajectories for a particle to go from A to B, and associate a probability amplitude $e^{iS/\hbar}$ where the action is evaluated from the Lagrangian taken along each trajectory, as in Eq. 6.57. The classical trajectories have different phases and hence interference effects arise. The sum (i.e., integral) over the contributions from all possible trajectories has to be carried out in order to get the full amplitude of the wave at the final destination.

In effect, Feynman has demonstrated that the principle of least action still determines the dynamics, but the single-trajectory of classical dynamics is replaced by a dynamics of many paths that pervades all space since the paths go every where. Interference effects can arise among different paths, since it is the amplitudes which are additive. Hence non-locality and contextuality are built into this scheme in a manner quite different to that of Bohmian mechanics. The latter assigns definite trajectories to particles that follow Newton's equation (see Eq. 6.39) while the initial particle distribution at (x_0,t_0) obeys the probability density $|\psi(x_0,t_0)|^2$. Feynman paths are continuous (but not differentiable almost everywhere), and initial positions and momenta are specified. The method does not define unique trajectories; instead, all possible trajectories are considered.

6.9 Quantum mechanics as a theory about information

The development of computers, communication theory etc., has led to information theoretic formulations of QM. Experimental results are units of information displayed as 'bits' and represented in a computer memory. Unlike classical systems, quantum systems require that if we make predictions of certain outcomes with complete certainty, then complementary outcomes are left as indefinite. Thus the information content that can be extracted from a quantum system is restricted.

Claude Shannon [186] proved in 1948 that the only function suitable as a measure of the uncertainty or 'error' in a message (a 'message' is a string of binary bits) has the form

$$S = -k\sum_i P_i \log(P_i)\,. \tag{6.58}$$

P_i is the probability for a given state of the bit to occur, and k is a constant. This is Boltzmann's expression for the entropy of a system which can occur with probabilities P_i in a statistical ensemble, and k is the Boltzmann constant. If we had complete information about the bit i, then $P_i = 1$, and hence $S = 0$. On the other hand, a loss of information or uncertainty in the state of the bit produces a fractional value for P_i. Then S is positive, since the logarithm of a fraction is itself negative. The logarithm occurring here is the natural logarithm, i.e., to the base e.

John Wheeler [225] had already asked if the physical world could be an 'it from bit'. Thus, it has been claimed [49, 77, 96, 234] that QM can be given an information theoretic basis by using such ideas as the framework of the theory. The principle of superposition of states is interpreted as a 'no cloning' restriction, since any attempt to carry out a cloning is thwarted by the 'interference terms' familiar from standard formulations (see Eq. 6.11). Similarly, two quantum systems when in contact with each other 'interact' by their Bohm quantum potentials and get entangled (see Sec. 7.1). The correlations set up by such entanglement do not decrease with distance since such quantum systems are nonlocal and contextual. Hence, one may ask if a remote system may be 'controlled' via quantum interactions which are not available to classical systems. This was termed 'remote steering' by Schrödinger, or 'spooky action at a distance' by Einstein.

Entanglement prevents what is known as strict *bit commitment* in communication theory. Consider Adam who encodes a bet about the outcome of a coin flip that is yet to be carried out by Eve who is far away. The encoded bet says that the coin will come out heads. This encodes as a choice between 0 (tails), and 1 (head) of a bit. Thus Adam had made a 'bit commitment', choosing 1. He informs Eve that he has made a commitment, but does not reveal the commitment since he does not want her to manipulate the coin flip. The commitment is recorded and mailed to Eve, but she cannot look at it as she does not have the password to the message. Eve flips the coin, and broadcasts the result. Then Adam reveals the password so that Eve can read Adam's bit commitment to be heads. This 'fool-proof' bit commitment scheme works because Adam cannot remotely manipulate the bit sent to Eve. QM allows the possibility, at least in principle, of remotely modifying the bit that was sent away, by keeping it entangled with a bit that Adam keeps with himself. Thus, entanglement, a key property of quantum systems, may be stated as a property of nature which prohibits perfect bit commitment.

The information theoretic approaches needs the underpinning of generalized Hilbert spaces (e.g., C^* algebras) etc., to function as physical theories. Although they provides a new way of looking at various constraints on the theory, it is unlikely to displace the existing pictures of quantum reality, either within the statistical interpretation of Bohr *et al.*, or within the observer-independent 'causal' interpretation of Bohm *et al.* The concept of the wavefunction plays a central, unifying role in these well known approaches, and connects conveniently with the concepts of quantized fields of many-particle systems.

More recently, some authors (Pusey *et al.* [176]) have claimed that if a quantum state represents only knowledge, information, or Bayesian belief, but not *ontic* (not linked to a basic reality), then it would contradict the quantum theory!

6.10 Relativistic quantum mechanics

Relativistic QM is relevant to the non-relativistic world as well because additional effective potentials associated with the electron spin, spin-orbit interaction etc., are contributed by components of the wavefunction which 'hide' in the underlying structure of the particle vacuum. Thus we see that effective potentials are used not only in Bohm's theory, but also in standard non-relativistic QM.

Relativistic formulations have added a deep conceptual and factual substratum which at once clarifies, and intrigues our reach towards quantum reality. Today, the modern form of the quantum theory involves relativistic quantum fields for matter and radiation. The wavefunction of the Schrödinger equation is the non-relativistic analogue of the quantized matter fields of quantum field theory. According to Steven Weinberg [221] *in its mature form, the idea of quantum field theory is that quantum fields are the basic ingredients of the universe, and the particles are just bundles of energy and momentum of the fields.*

In the following sections we discuss the relativistic theories that came prior to the development of proper quantum field theories. However, the relevance of the underlying quantum fields is already visible in the existence of various corrections to radiative processes in atomic physics.

In Fig. 6.8 we show a rudimentary 'Feynman diagram' for a process where an electron in the 2p state of an excited H-atom emits a photon and transits to the 1s state which corresponds to the ground state of the H atom, and the emitted photon corresponds to the Lyman-α radiation. Prior to the transition, matter is in an excited state while the electromagnetic field is not even manifest. After the transition, matter has gone to a lower energy state and the energy difference has gone into creating a quantum of electro-magnetic radiation. The field has infinitely many degrees of freedom, being analogous to a vibrating solid which has a large number of normal modes. In Fig. 6.8 we also show the Feynman diagram for the scattering of two electrons by the Coulomb interaction. However, here we think of the Coulomb interaction as being mediated by the emission of a 'longitudinal' photon by the first electron, while the second electron absorbs the photon and moves to a higher momentum state. When we consider relativistic systems, matter itself begins to take the character of excitations of quantized fields. That is, the concept of 'single particles' disappears and we need to deal with quantized fields which are essentially systems with an infinite number of modes. The Lamb shift, an experimental consequence of such underlying quantum fields, is discussed in Sec. 6.10.3. Nevertheless, here we sketch how the early attempts to construct relativistic quantum theories for a particle (e.g., an electron) led naturally to the multi-particle view (or quantized-fields view) of nature.

An everyday phenomenon like magnetism has its origins in relativistic QM. Relativistic effects add 'correction terms' to the non-relativistic quantum equation of Schrödinger. Some theorists (e.g, Roger Penrose [160], Paul Stapp [201]) have claimed that quantum phenomena and even electron spin-effects associated with the brain hold the key to the nature of consciousness. Although we do not share those views, we need a knowledge of such physics.

It was Maxwell's equations, and the motion of light relative to the 'aether' that forced physicists like Einstein to look for a Lorentz invariant form of Newtonian mechanics. The resulting 'special theory of relativity' (Sec. 5.2) required us to treat the four dimensions (t,x,y,z) in a unified way. Probes using light (e.g, spectroscopes) are the basis of much of our information about quantum systems like electrons or atoms. Hence it is natural to look for a Lorentz invariant form of QM that respects the finiteness of the speed of light.

6.10.1 *Klein-Gordon equation*

In our discussion of the non-relativistic Schrödinger equation we used the substitutions

$$p_x \to -i\hbar\partial/\partial x; \;\; E \to i\hbar d/dt \tag{6.59}$$

in the energy expression. For relativistic particles, the relation between E and p for a free particle is

$$(E/c)^2 = p^2 + (mc)^2\,. \tag{6.60}$$

Let us define the four-vectors

$$p_\mu = \{p_1, p_2, p_3, iE/c\}; \;\; x_\mu = \{x, y, z, ict\} \tag{6.61}$$

$$\text{or} \;\; p_\mu \to \hat{p}_\mu = -i\hbar\partial/\partial x_\mu\,. \tag{6.62}$$

The Klein-Gordon equation is obtained by substituting for $E, \vec{p}$ in Eq. 6.60:

$$\{\Sigma_\mu \tilde{p}_\mu^2 + mc^2\}\phi(x_\mu) = 0\,. \tag{6.63}$$

It can be shown that the properties of this wavefunction under coordinate transformations are consistent with those of particles with spin zero. If we consider free motion with a definite momentum $\vec{p}$, the particles can have positive or negative energy with

$$E = \pm c\{p^2 + (mc)^2\}^{1/2}$$

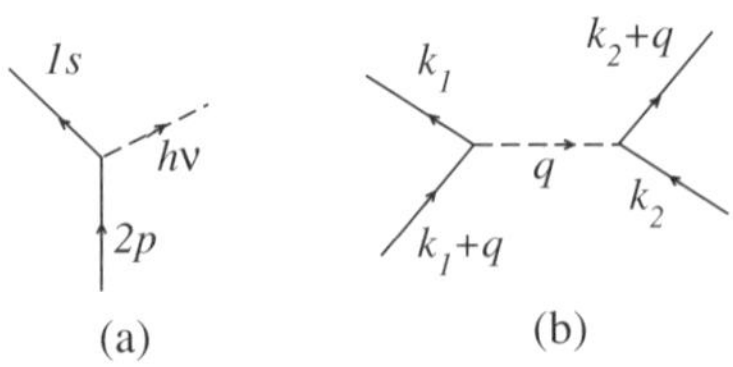

Fig. 6.8 In (a) an electron in the 2p state transits to the 1s state and emits a photon of energy $h\nu$. In (b), an electron in the momentum state $k_1 + q$ repels another electron in the state k_2, transmitting to it a momentum q, via the Coulomb interaction which is mediated by a 'longitudinal photon'.

The negative energy solutions are interpreted as particles of positive charge, while the positive energy solutions are spin-zero negatively charged particles. This equation can also correspond to particles of zero charge. Thus the relativistic description has revealed additional degrees of freedom. In the non-relativistic theory there is only one state of free motion with a well-defined value of the momentum. In the relativistic theory of zero-spin particles, there are three possible values of the particle charge for free motion with a well defined momentum.

The probability density ρ obtained from the Klein-Gordon equation is not positive definite.

However, this difficulty is overcome if the charge density $e\rho$ and the charge current are considered, since the total charge is conserved. Pions, and possibly kaons are described by the Klein-Gordon equation. These are particles which interact very strongly with other particles, indicating that the idea of free motion of 'individual' relativistic particles is an extreme idealization.

6.10.2 *Dirac equation*

In 1928 Paul Dirac succeeded in constructing a relativistic equation suitable for describing electrons and other spin-1/2 particles. Dirac looked for a Lorentz invariant formalism where the equation of continuity contained a positive definite probability density. Given the 4-dimensional nature of space-time, Dirac had to introduce a four-component wavefunction. This allowed him to reveal that electrons come hand in hand with positrons, and that each has two spin states, thus completing the four states of the four-component wavefunction.

Let us denote this by $\tilde{\phi}$, with components ϕ_1, ϕ_2, ϕ_3, and ϕ_4. Hence the Dirac equation itself is a matrix equation, containing the Dirac matrices $\tilde{\alpha}$ and $\tilde{\beta}$ that combine the various terms in the equation. Here $\tilde{\alpha}$ is a vector operator whose components are 4×4 matrices. The Dirac equation can be written in various equivalent forms:

$$\{c\tilde{\alpha}\cdot\tilde{\mathbf{p}}+\tilde{\beta}mc^2\}\tilde{\phi} = i\hbar\frac{\partial\tilde{\phi}}{\partial t} \tag{6.64}$$

$$\text{or}\quad \{(-1/c)\partial/\partial t+\Sigma_j\tilde{\alpha}_j d/dx_j+(imc/\hbar)\tilde{\beta}\}\tilde{\phi} = 0. \tag{6.65}$$

The first equation is in the form presented by Dirac in 1928. Here we easily recognize the time-dependent term, a term dependent on the momentum operator $\tilde{\mathbf{p}}$, as well as a term that is an mc^2 energy associated with m the rest mass of the electron. The Dirac matrices combine these terms to form the Dirac equation. These matrices can be expressed in terms of simple 2x2 matrices known as Pauli matrices, and are related to the Clifford C*-algebras. However, all physical consequences of the Dirac equation are independent of the actual representation used for the matrices. The equation satisfies Lorentz invariance and the continuity equation.

The Dirac equation has negative and positive energy solutions. Such pairs were interpreted as particle-antiparticle pairs created by an excitation of the underlying 'vacuum' (see also Sec. 6.10.3). The reverse process, where a particle annihilates itself by combining with an anti-particle, leads to the emission of a photon (a γ-ray). The negative energy solutions were interpreted as states of particles of positive charge, i.e., positrons which are the anti-particles of electrons. The positive energy states were interpreted as particles of negative charge, i.e., electrons.

Thus electrons and positrons are excitations of the underlying vacuum. This 'vacuum' is sometimes called the 'Dirac sea'. Particle-antiparticle annihilation needs a third body to establish momentum conservation. Predictions of the Dirac equation have been verified subsequently. For instance, the positron was discovered by Carl Anderson in 1933 five years later.

The type of particles described by Dirac's 4-component wavefunction $\tilde{\phi}$ was revealed by its behaviour under rotations etc.. The particles are fermions with charge $\pm e$ and spin $(1/2)\hbar$, resolving a deep puzzle about the nature of electron spin. Thus the Dirac equation describes electrons, muons, protons, neutrons, and neutrinos as well as their anti-particles. The $(1/2)\hbar$ spin, incomprehensible within the non-relativistic theory emerged naturally as an intrinsic property of the eigenfunctions of the Dirac equation. The 'spin' does not correspond to an actual spinning of the electron in real space, but a property of an abstract reality (see Sec. 6.5).

The Dirac equation, when applied to electrons in heavy atoms, e.g., gold, has given detailed and quantitative explanations of the fine structure of their spectra. *Even the colour of solid gold, and the liquid nature of mercury at room temperature, etc., can be explained using calculations based on the Dirac equation.* In fact, the Dirac equation and the physics flowing from it are believed to be sufficient for a complete discussion of all chemical and biological phenomena. That is the basis of the claim, attributed to Dirac, that the 'rest (of nature) is chemistry'.

Electrons in 2-D films of graphite (graphene) have unusual properties. The electrons fill two electrons in each level (because of spin), up to the Fermi Energy. The relationship of the energies of such electrons with their momenta for motion within the 2-D sheet is of the form $E = v_F p$ where v_F is the velocity at the Fermi energy. This energy-momentum relation is similar to that of photons, where we have $E = cp$ with c the velocity of light of momentum p. The electrons in graphene sheets thus seem to effectively have a mass of zero (like photons), but are Fermions. The Fermi velocity plays a role similar to that of c for photons. These massless Dirac particles obey an equation known as the *Weyl equation* that had been previously used in neutrino physics. The physics of graphene [82] and its many-body problem [60] have thus become a laboratory for studying relativistic particles where the effective energy scales are much smaller and within the techniques used in condensed-matter physics.

6.10.3 *Vacuum fluctuations and the Lamb shift*

The energy positions of the $2s$ and $2p$ levels of the hydrogen atom coincide if calculated using the Dirac theory, Schrödinger or the simple Bohr theory. On the other hand, Lamb and Retherford showed experimentally that there was a small energy difference of 1057 MHz ($\sim 4.37 \times 10^{-6}$ eV) between the $2s_{1/2}$ and $2p_{1/2}$ levels, with the 2s having a slightly higher energy. The energy shift between the $2s_{1/2}$ and $2p_{3/2}$ levels is significantly bigger. This type of energy difference is attributed to aspects of the vacuum not contained in the Dirac equation. The vacuum is a quantum many-body system described by theories of quantized fields. Thus 'radiative corrections' to the Dirac theory are needed.

As briefly alluded to earlier, the vacuum is not empty space. Even in the absence of any particles or radiation, virtual particles are constantly created and annihilated. Thus, while the vacuum is charge neutral on the average, virtual electron-positron pairs, and other types of pair excitations, give rise to ephemeral electric fields $\Delta\varepsilon(r)$. These incessant charge fluctuations are similar to the zero-point oscillations of the lowest level of the parabolic well that we studied in Sec. 6.4. The interaction of atomic electrons with these fluctuating electric fields is the cause of spontaneous emission of radiation from any excited quantum system. The same effect also triggers the well-known process of vacuum Rabi oscillations in cavity quantum electrodynamics [97]. In addition, quantum 'vacuum fluctuations' are found to determine the spontaneous emission lifetime of superconducting quantum bits (qubits or qbits). Thus vacuum fluctuation effects are ubiquitous, and are a fundamental source of decoherence of quantum systems. The $\psi(r)$ functions of 2s and 2p levels of the hydrogen atom differ significantly in their behaviour close to the nucleus, for small r. Thus ψ_{2s} rises to a constant value at $r = 0$, while ψ_{2p} goes to zero. Hence the coupling of these charge distributions with the vacuum fluctuations, e.g., $|\psi(r)|^2\Delta\varepsilon(r)$ differ. The resulting energy shift of the 2s and 2p levels can be calculated in second-order perturbation theory, and these differ by the experimentally observed amount. The shifts of a spectral line can be related to the line widths using the Kramers-Krönig relations. So far there has been excellent agreement between QM and experiments.

An energy shift ΔE is directly related to a lifetime Δt of a quantum state (or transition) by the uncertainty relation $\delta E \Delta t \geq h$. Hence the Lamb shift is related to the excited-state life time τ set by spontaneous emission. The line width due to spontaneous emission is the 'natural width' of a spectral line. The lifetime of the $2p$ state of Hydrogen is 0.16×10^{-8} s, while the $2s$ has the incredibly long life of 1/7 seconds, since the $2s \rightarrow 1s$ dipole transition is symmetry forbidden.

6.10.4 *Wavefunction of the universe*

In discussing the inflationary origins of the universe (Sec. 5.9) we recounted how it is believed that all matter-energy in the universe that we can observe seems to have existed in extreme close proximity, packed into length scales measured by the Planck length $\ell_p = (hG/2\pi c^3)^{1/2}$ which is $\sim 10^{-35}$ m. Already when length scales go below $\sim 10^{-9}$ m we are in the nanoscale range where quantum effects become important. Hence the very early universe cannot conceivably be governed by classical mechanics. Clearly, we need a theory which is at once the progenitor of the quantum field theory and gravity. The theory, central to quantum cosmology, should spell out how the various particles, fields, as well as space-time come out as consequences of the theory. While string theory and brane theories (Sec. 5.9.1) have been proposed, no unique GUT has emerged. Meanwhile, already in the middle of the 20th century, various less systematic but simpler approaches [13] had been presented by cosmologists.

In 1967 deWitt published what he called the 'Einstein-Schrödinger equation', by simply starting from one of the FRW equations that we already discussed in Sec. 5.7.1:

$$\left[\frac{da}{dt}\right]^2 = \frac{a^2(t)}{a_0^2} - k, \quad a_0^2 = \frac{3}{8\pi G\rho}. \tag{6.66}$$

Here $k = 0, 1$ or -1 for a flat, closed, or open universe. Using units where the speed of light $c = 1$, a_0 is a length scale of a universe with a mass-energy density ρ. Quantization is achieved by replacing the momentum da/dt by $-i\hbar\partial/\partial a$ as required by the Schrödinger prescription. That is, we begin with the classical Hamiltonian H of the FRW model. It is found to be zero on comparison with the Friedman equation, as seen from the first two equations given below. The third equation is in the schördinger form ($\hbar$ set to unity) and the eigenfunction is denoted by ψ and depends on a.

$$k + (da/dt)^2 - a^2(t)/a_0^2 = 0, \quad a_0^2 = 3/(8\pi G\rho) \tag{6.67}$$

$$-(a/b_0)\left[k + (da/dt)^2 - (a/a_0)^2\right] = H \quad b_0 = 8/(3\pi) \tag{6.68}$$

$$\left[-\frac{d^2}{da^2} + a^2\{k - (a/a_0)^2\}\right]\psi = 0. \tag{6.69}$$

Since the energy associated with ψ is zero, there is no time evolution or any aspect of ψ that can be associated with the time parameter! It should also be noted that ψ is effectively a functional that describes the geometry of the system via the parameter a. This simple scale factor a is like a one-dimensional variable describing the motion of a particle of mass $m = 1/2$ given in Planck-scale units of mass-energy. The particle moves in the potential:

$$V(a) = a^2(k - a^2/a_0^2). \tag{6.70}$$

It is usual to consider a closed universe, i.e., $k = 1$. If $a < a_0$ then $V(a)$ is positive and finite. A classical particle is confronted with a barrier and there is no physical solution for the classical Friedman equation for $a(t)$ where the time t is real. Mathematically, a solution with an imaginary time can be constructed.

$$a(t) = a_0 \cos(it/a_0). \tag{6.71}$$

On the other hand, since the potential is not infinite, quantum particles can tunnel into a barrier region and hence there is a finite probability of finding a particle in that region. Thus the wave function $\psi(a)$ is non-zero both inside the barrier ($a < a_0$), and the classically accessible region outside the barrier ($a > a_0$). The wavefunctions inside and outside can be calculated and matched at the common meeting point at $a = a_0$. The boundary conditions at the matching point usually ensures the continuity of the wavefunction and its derivative. If an accurate solution (analytic or numerical) is not used, the two approximate functions have to be matched using some heuristic boundary conditions. Instead of doing an accurate calculation, Hartle and Hawking in 1983 constructed an approximate solution, i.e., the WKB solution which keeps terms in the quantum corrections up to second order (Sec. 6.37) in $\hbar$ and mixes equal amounts of the inner and outer solutions, to present what they called the 'no boundary model'. The two parts of the Hartle-Hawking WKB approximation to $\psi(a)$ are

$$\psi(0 < a < a_0) = |aF(a)|^{-1/2}\exp\{-(\pi/2)a_0^2F(a)^3\} \tag{6.72}$$

$$\psi(a > a_0) = |aF(a)|^{-1/2}\cos\{(\pi/2)a_0^2F(a0^3\}; \; F(a) = (a^2/a_0^2 - 1)^{1/2}. \tag{6.73}$$

The cosine function is of the form $(\exp(ix)+\exp(-ix))/2$ and contains equal amounts of an incoming and outgoing wave. This type of tunneling is possible for the barrier form of a closed universe ($k=1$), where as inflation predicts $k=0$, i.e., a flat universe. However, a very large universe is locally flat, and the heuristic nature of quantum cosmology is such that this is only a minor problem.

This WKB-type of solution gives the scale factor a, and hence, parametrically, the time t appears as a/a_0. Given the extreme simplicity of the assumptions made, and since the model is of any value only in the very early life of the universe, it appears that time emerges simply as a parameter associated with the evolution of the radius of the universe. For times prior to $a<a_0$, 'time' loses any physical meaning, as the parameter becomes $\tau=it$. If the Universe were to move into a contracting phase, even though time in the sense of 'age' would increase, whereas a time based on a/a_0 would reverse itself. Hence this model is clearly only of some use in the early history of the system.

We noted that quantum mechanics makes no predictions of individual cases, even for the Universe. Only a mean value, taken over many measurements, can have any statistical significance. Hence we are confronted with a difficulty when we talk of the 'wavefunction of the universe'. We can conceive of a multiverse, where there are many inflationary processes going on, and the statistical average arises if a suitably large part of the multiverse is considered.

Hartle, and Gell-Mann have developed a scheme known as the 'consistent histories' (see [83] [29]) approach, where the notion of many repetitions of a measurement is replaced by a summation over the many possible histories. The probabilities of the histories are used as weights in the summation. This is like a generalization of the Feynman path-integral approach, and presented as a version of Everett's many-worlds interpretation of QM. The summations over the histories can take account of interactions which lead to decoherence and coarse-graining. The alternative histories approach has considerable advantages over standard formulations, when it comes to applications in cosmology.

Alternatively, some authors have invoked a Bohmian picture, when the corresponding epistemological difficulties become less acute. The mean values and uncertainties are easily associated with the launching of the density fluctuation in the barrier potential $V(a)$ that we discussed in the paragraph in small print. However, current formulations of Bohmian mechanics do not have the full generality of the Poincaré-Lorentz transformations for them to be seriously applicable to quantum cosmology. Nevertheless, in spite of such lacunae, there is already a number of papers ([41]) that apply Bohmian concepts to quantum cosmology.

The concept of a wavefunction of the universe would be justified if it leads to new observable physics or a clarification of basic cosmological concepts.

Chapter 7

Entanglement, Measurement and Quantum Paradoxes

This is a challenging chapter. Some readers may jump from the introduction to 'quantum muddles' (Sec. 7.6), and return to previous sections referred from there on. We review superposition and entanglement as being basic to quantum systems. The nature of qbits, chemical bonds, entangled photons and quantum computers is discussed. Pair-distribution functions, Bell inequalities etc., expose the nature of quantum correlations. The theories of measurement and decoherence are treated. The possibility of using samples of a system for measurement, even for entangled many-body systems emerges as a function of their correlation lengths. Finite-temperature effects are shown to provide a natural source of decoherence for macroscopic superpositions and cat states. We examine 'quantum muddles' found in the philosophical literature, and review key quantum paradoxes using conventional and Bohmian quantum mechanics.

7.1 Introduction

Popular accounts of quantum phenomena have a tendency to emphasize what is claimed to be the 'weird' nature of the quantum world. Einstein himself dabbed some quantum phenomena as 'spooky action at a distance' and set the stage for such an attitude. 'Entanglement' (Verschränkung) is the specific word, introduced by Schrödinger [188] to characterize this aspect of quantum phenomena. Entanglement is actually a very common everyday phenomenon. In fact, as discussed below, 'entangled' electrons are the basis of chemical bonding and the very stuff of life. Entangled states are sometimes known as 'cat states' after a 'tongue-in-cheek' discussion of such quantum states by Schrödinger (Sec. 7.7.3). They are much less weird than a lot of things we readily take for granted!

What you do in your own house is, generally speaking, quite independent of what someone else in another house on your street may be doing. That is, what affects you is 'local' to you, and effects which are sufficiently far away do not affect you. We call this character 'near-hereness', localness or *local realism*. The latter will be defined in more detail in Sec. 7.3.7. One of the essential features of quantum systems (except for some simple systems) is that this 'local property' of near-hereness is replaced by a property of 'every-whereness' (nonlocality) which involves all components of the system. These 'correlations' are not a result of long-range Coulomb interactions or gravitational fields, but arise from the wave nature of matter. In the early language of Bohm, they are a result of 'quantum potentials'. Waves spread throughout the system and 'feel' the boundaries of the space containing them as well as other particles. Thus, quantum particles (e.g., electrons) remain 'correlated' even if they are far away if suitable boundary conditions are maintained, while classical particles (e.g., marbles) can be regarded as independent objects when they are only a few marble-diameters away.

Erwin Schrödinger, introducing the concept of entanglement writes [188]: *when two systems, of which we know the states by their respective representatives* (i.e., wavefunctions), *enter into temporary interaction, ... (and) separate again, they can no longer be described in the same way as before, viz., by endowing each of them with a representative of its own. I would not call that one, but rather the characteristic trait of quantum mechanics, the one that enforces its entire departure from classical lines of thought.*

Consider two electrons that are isolated for observation by setting up suitable boundary conditions. They interact by the Coulomb repulsion, and also by their exclusion of each other because they are 'Fermions'. In effect, two Fermions cannot be in the same quantum state. This is called the Pauli exclusion principle. Hence, if two electrons have the same space wavefunction, they must differ in their spins; if one has an up-spin the other would have a down spin.

Normally the two electrons in the He atom are in the lowest orbital (i.e., have the space wave function $\phi_{1s}(r)$, with quantum numbers $n = 1, l = 0, m = 0$, see Fig. 6.5); the two electrons must differ in the spin state to ensure that their overall quantum states, designated by the four quantum numbers n, l, m, σ are not identical. Consequently the two electrons in the lowest state of the He atom have opposite spins, with $\sigma = +1/2$, and $-1/2$, corresponding to the names 'up-spin' and 'down-spin'.

Consider two electrons, one on earth and the other far away, on the moon, forming a quantum state. The Pauli exclusion effect acts instantly, just like the non-local quantum potential of Bohm. Although theorists use the word 'instantly', the experimentalist takes his time to set up the system, when the interactions also get set up. Once the two electrons are separated by a lunar distance, one may ask if the wavefunction still has to satisfy the Pauli Exclusion Principle. The answer is 'yes'. We are saying that we have set up boundary conditions to establish a finite wavefunction connecting the electron on the earth, and the electron on the

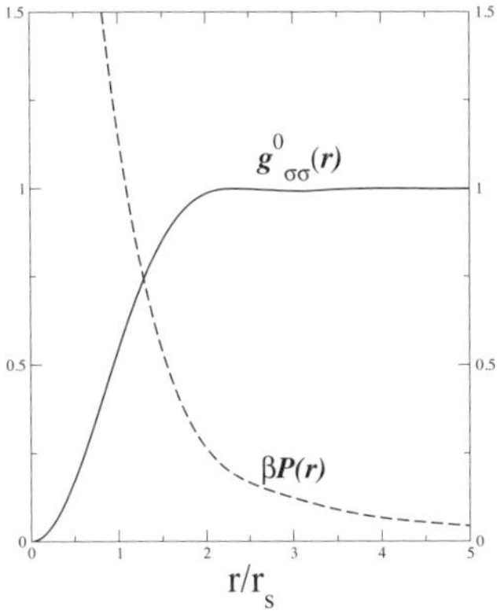

Fig. 7.1 The pair-distribution function $g^0_{\sigma\sigma}(r)$ gives the probability of finding a σ-spin electron at r if there is already a σ-spin electron at the origin (Coulomb repulsion neglected). We plot r/r_s on the x axis, and the result is valid for any density, r_s being the sphere-radius containing an average of one electron. Thus the 'entanglement' due to Pauli exclusion holds even if r_s is a lunar distance. The Pauli exclusion is exactly mimicked by a scale-independent repulsive potential $\beta P(r)$.

moon. In an actual experiment, it is impossible to exclude the rest of the world and have just two electrons separated by such long distances. Hence, in practice, such 'spooky action' at a distance, enforcing the exclusion principle does not occur. Terrestrial electrons are not entangled with lunar electrons. However, *quantum mechanics allows us the theoretical possibility of engineering such macroscopic quantum states which defy our normal intuition* about near-hereness.

If the electrons have the same spin, their spatial wavefunctions must differ to satisfy the Pauli exclusion principle. Hence they cannot occupy the same space orbital. Consider a gas of up-spin electrons, so dilute that there is on average just one electron in a sphere of radius r_s, where $r_s = 385{,}000$ km, which is the distance between the earth and moon. Let us place our origin of coordinates on one such electron, and consider a very large volume (e.g., the size of the solar system), empty except for these non-interacting electrons and a neutralizing uniform charge which is completely static and pervading the whole system. Since there is one electron per volume $4\pi r_s^3/3$, the electron density n is given by $n = 3/(4\pi r_s^3)$. Nevertheless, the Pauli principle acts to exclude the approach of a neighbouring electrons into its own space. This is best seen from the PDF, i.e., the radial *pair-distribution function* $g_{\sigma,\sigma}(r)$ which gives the probability of finding an electron of spin σ at the radial distance r, given that there is already an electron of spin σ at the origin. Neglecting Coulomb interactions, as indicated by a superfix zero, we plot $g^0_{\sigma,\sigma}(r)$ in Fig 7.1. A classical potential which can mimic the Pauli exclusion effect, known as the Pauli potential $\beta P(r)$ is also shown (see Sec. 8.4.1). The PDF of a uniform fluid is also simply the two-body density $n(r)$, i.e., the distribution as seen when sitting on an electron which is taken as the origin, scaled by the average density n. That is

$$g(r) = n(r)/n, \quad n = 4\pi r_s^3/3\,. \tag{7.1}$$

For a quantum system the electron-density distribution $n(\vec{r})$ has to be calculated using quantum mechanics. In dealing with interacting systems, the simulation technique known as the quantum Monte Carlo (QMC) method becomes necessary for accurate results. If the Coulomb interaction is neglected, analytical results become available. If the spins of the electron pair were different there is no Pauli exclusion. Hence the PDF of non-interacting anti-parallel electrons in the absence of the Coulomb interaction is unity for all r and all densities (i.e., for all r_s). However, if the spins are the same, the electron at the origin 'excludes' the other electrons and forms a 'Fermi hole' around itself.

Analytical and numerical results for the pair-distribution function using a many-body wavefunction (e.g., a Slater determinant) where all the electrons are entangled are available (see Sec. 8.4.1). A typical result is shown in Fig. 7.1. What is remarkable about this result is that it is 'universal', for all r_s, and holds equally well at all length scales. Entanglement *via* the Pauli exclusion effect maintains the same relative importance at all length scales as it depends on r/r_s. That is, there can be no *local realism* even if the electrons are separated by astronomically large r_s separations. The 'local region' has an extent R which is the'healing distance' such that $g(r)$ becomes essentially unity. The range of R, and the Pauli potential $P(r)$ are proportional to the particle-separation scale given by r_s.

The heart of the Einstein-Podolsky- Rosen (EPR) paradox (Sec. 7.7) is the 'persistence' of the correlations even when the two objects that became 'entangled' are subsequently separated very far apart. However, this 'persistence' depends on our engineering ability to shield against the disturbing effects of the external world and maintain the integrity of our boundary conditions that define the entangled pair. These effects, which destroy entanglement or superpositions, are loosely termed 'decoherence'.

7.2 Superposition and interference

Let us consider the simpler phenomenon of superposition. We recapitulate the discussion in the last chapter regarding a beam of quantum particles (e.g., electrons or photons) confronted with two slits (i.e., openings) A and B. There we used amplitudes of light waves. Here we use wavefunctions of electrons. The quantum particles are detected on a screen at some point x. The particles that come through do not just pile up around the openings A and B. They are found in locations that would normally be blocked to 'classical' particles (e.g., bullets). In fact, the observed distribution of particles displays the intensity pattern of two interfering waves coming out of the two openings (see Eq. 6.11).

If a quantum particle has the possibility of being in two possible states, the wavefunction ϕ becomes a superposition of the two possible wavefunctions. Waves when superposed add in intensity when the crests overlap, and become nullified when a crest falls on a trough. This interference appears as an intensity pattern if it is captured on a screen. The crossing of the waves to form the interference corresponds to mathematical terms in the equations that are known as 'cross terms'.

The probability, being the square of ϕ, develops cross terms between the two states, and these are the interference terms that lead to effects typical of quantum systems. Let us consider the two-slit problem as in fig. 6.3, except that we consider a beam of electrons arriving at the screen surface where detectors are positioned. For simplicity, the detector/screen is assumed to be non-reflecting. Let the slit B in the two-slit experiment, Fig. 6.3, be closed. Let electrons going through the single open slit A arrive at a detector placed at x far away from the slit. This defines a single-slit probability P_a for slit A. Similarly we can measure a probability P_b with A closed and B open. The probability of finding an electron at x is given by the square of the wavefunction, evaluated at the detector, and similarly for P_b

$$P_a = |\phi_a(x)|^2; \quad P_b = |\phi_b(x)|^2.$$

Now let us change the boundary conditions of the system. Both slits A and B are opened, recovering the configuration of Fig. 6.3. Thus an electron passing through the slit A and/or B are now part of the same quantum system. Electron states are waves which can now 'spread' over the spaces around the open slits A as well as B. A possible quantum state of an electron in the system is obtained as a superposition of the two waves. That is, for any detector location x

$$\phi_{AB}(x) = \{\phi_a(x) + \phi_b(x)\}/\sqrt{2}. \tag{7.2}$$

The reader would perhaps note that the above equation is only an approximation The Schrödinger equation has to be solved anew taking account of the new boundary conditions. The opening up of both slits allows $\phi_a(x)$ to modify slightly to yield a perturbed form $\psi_a(x)$. Similarly, $\phi_b(x)$ has its modified counterpart $\psi_b(x)$. In principle we may need an expansion in a larger *basis set* of functions (see Sec. 7.3.2) to account for all the details of the modified situation. However, the modified functions are usually very close to the original ones.

The probability of observing an electron at the detector A is now modified because the electron is in the superposition state $\phi_{AB}(x)$. The new probability $P_{ab}(x)$ is the square of the new wavefunction.

$$\begin{aligned} P_{ab}(x) &= (1/2)|\phi_a(x) + \phi_b(x)|^2 & (7.3)\\ &= \{|\phi_a(x)|^2 + |\phi_b(x)|^2 + |\phi_a^*(x)\phi_b(x) + h.c|\}/2 & (7.4)\\ &= \{P_a + P_b + P_{int}\}/2. & (7.5) \end{aligned}$$

In addition to the individual probabilities P_a and P_b, associated with the slits A and B, we now have the interference term P_{int}. It gives rise to regions of high probability (anti-nodes), and low probability (nodes) of finding an electron (see also Sec. 6.3.4) on the screen. If the detector happened to be exactly on an anti-node, one may never detect an electron there. $P_{ab}(x)$ includes information from everywhere; i.e., all of phase space is sampled by ϕ_{AB}. This information need not be 'transmitted' from one region to another as *it is already contained in the wavefunction*. The Bohm quantum potential also gets modified when the boundary conditions of the system (i.e., opening both slits etc.), are being modified by the experimentalist, over a finite length of time.

Although the wavefunction is zero at the nodes of the interference pattern, it is useless to position a fine wire mesh exactly on the nodes, and hope that the wire mesh 'does not get noticed' by the wavefunction, or equivalently by the Bohm quantum potential $Q(x)$.

$Q(x) \sim \{\partial^2 R(x)/\partial x^2\}/R(x)$, where R is $\sqrt{|\phi(x)|^2}$, and hence could be very large when $R \to 0$.

This is the basic error in experiments which attempt to beat non-locality by exploiting regions where the fields are zero, although the potentials are non-zero. Thus Afshar [1] attempts to 'beat complementarity' and examines the possibility of determining the path taken by photons that form an interference pattern by placing a wire-mesh at the zero-field regions to gather added information. Here we may also note the Aharanov-Bohm effect, where an electron moves on a path limited to a spatial region which is far away from a solenoid, and where the magnetic field is exactly zero. Although the fields are zero in such regions, the vector potential of the electro-magnetic field is not zero, and the Aharanov-Bohm effect prevails. Needless to say, Ashfar's experiments reveal nothing new.

David Bohm, in his book, *Wholeness and the Implicate order* [27] attempted to use non-locality ideas to construct a conception of nature and external reality (known as 'implicate order') as some sort of hologram that reaches beyond local realism. In our view, such philosophical extensions are not necessary to 'explain' the existence of novel correlations in quantum systems.

7.2.1 *Qubits, qbits or quantum binary states*

A classical object like a simple switch can exist in two states, viz., the 0, or 'off' state, and 1, the 'on' state. Such zero-one states are the 'bits' (or *c*-bits) of classical binary representations used in computer memories. Quantum particles with two states $|0\rangle$ and $|1\rangle$ can, in addition exist in superposition states; these are known as 'qbits', sometimes also written as 'qubits' (David Mermin has recommended the form 'qbit'). They are the units of 'quantum computation'.

How does one convert two quantum states into a superposition? This is best understood by looking at a two-level system, e.g., a magnet that can align *with* a magnetic field, or *against* the field, applied along the z-direction. A classical magnet undergoes a precessional motion about the z-direction at a circular frequency ω_0, known as the 'Larmour frequency' (n.b., circular frequency ω is $2\pi\nu$ where ν is the frequency with $E = h\nu$, and ω is the circular frequency, with $E = \hbar\omega$).

Let the two states (up, down) be denoted by the symbols $|u\rangle$, and $|d\rangle$. These can be the two spin states $S_z = \pm 1/2$ of an electron. The energy separation of these states is $\hbar\omega_0$. The spin is subject to an oscillatory magnetic field in, say, the x direction. When the energy quanta of the transverse (x) field begin to match the energy gap between 'u' and 'd' levels, the system displays reversible oscillations between them by absorbing quanta from the oscillating field, and emitting them back by the processes of stimulated and spontaneous emission. The system is said to be undergoing *Rabi oscillations*. This involves the system starting in, say, 'd', then forming a superposition state (horizontal arrow in Fig. 7.2), and moving to the 'u' state (up arrow), and returning by the same path to 'd'.

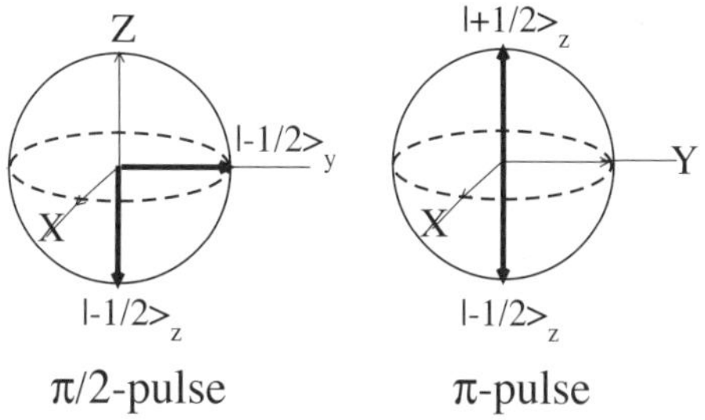

Fig. 7.2 (Left) The Bloch-sphere representation of two-level superpositions: (left) a $\pi/2$-pulse convert the down-spin state $|-1/2\rangle_z$ to a superposition state, Eq. 7.8 (a spin state in the y-direction). (Right) a π-pulse converts a down-spin state to an up-spin state.

The Hamiltonian of any two-level system modeled as a spin-up/down system, together with a transverse oscillating field can be written as:

$$H = H_{ij} = \frac{\hbar}{2}\begin{vmatrix} \omega_0 & 2\omega_x\cos(\omega t) \\ 2\omega_x\cos(\omega t) & \omega_0 \end{vmatrix}. \tag{7.6}$$

Here H_{11} and H_{22} are the matrix elements of the two levels by themselves, and contain the Lamor frequency ω_0 due to the static field along z. The off-diagonals have ω_x, i.e., the Larmor frequency corresponding to the alternating-magnetic field with frequency ω applied along x. The cosine in the transverse term $2\omega_x\cos(\omega t)$ can be split into $\pm\omega$ components. Only the component rotating in the same direction as the Larmor precession is important near resonance, and hence the Hamiltonian can be written in the *rotating-wave approximation*, and in a rotating frame at frequency ω, as

$$H = \frac{\hbar}{2}\begin{vmatrix} \omega_0-\omega & \omega_x \\ \omega_x & \omega_0-\omega \end{vmatrix}. \tag{7.7}$$

Suppose we are at resonance, so that $\omega = \omega_0$, and we start with a spin pointing downwards (negative direction of z), in the eigenstate $|d\rangle$ which is the $|-1/2\rangle$ state of the spin operator S_z. Let the alternating field be applied for a time τ such that $\omega_x\tau = \pi/2$. This is known as a $\pi/2$-pulse, and the spin will rotate in the $Y-Z$ plane by an angle of $\pi/2$ around the x-axis and its final state will be a spin-state pointing in the positive direction of the y-axis, i.e., the eigenstate $|1/2\rangle$ of the spin operator S_y. This is in fact a superposition state, since:

$$|-1/2\rangle_z \xrightarrow{\pi/2\text{ pulse}} |+1/2\rangle_y = \{|-1/2\rangle_z - i|+1/2\rangle_z\}/\sqrt{2}. \tag{7.8}$$

Hence the geometrical meaning of this superposition of the two eigenstates of S_z is that it is an eigenstate of S_y. This is shown in Fig. 7.2.

If we applied the alternating field for a longer time such that $\omega_x t = \pi$ we have a 'π-pulse'. Its effect would be to drive the $|-1/2\rangle_z$ state to the $|+1/2\rangle_z$ eigenstate where the spin is pointing up, i.e., in the positive z-direction. Hence $\pi/2$- and π pulses of electromagnetic radiation are very useful for constructing superpositions of two-level systems, or transforming a given eigenstate of a two-level system into the other eigenstate. In general, an arbitrary electromagnetic pulse can position the end of the spin vector in some arbitrary position on the 'Bloch sphere' shown in Fig. 7.2. If the corresponding polar angles are θ and ϕ, the superposition state has the form:

$$|\psi\rangle = \sin(\theta/2)e^{i\phi}|-1/2\rangle_z + \cos(\theta/2)|+1/2\rangle_z \tag{7.9}$$

$$\text{or } |\psi\rangle = \sin(\theta/2)e^{i\phi/2}|-1/2\rangle_z + \cos(\theta/2)e^{-i\phi/2}|+1/2\rangle_z. \tag{7.10}$$

These two equations emphasize the fact that only the relative phases of the two components are physically important. In the following we can use the compact form

$$|\psi\rangle = C_0|0\rangle + C_1|1\rangle \tag{7.11}$$

where C_0 and C_1 can now be expressed in terms of polar angles of a point on the Bloch sphere.

Thus Eq. 7.11 depicts the action of an optical pulse which mixes the states $|0\rangle$ and $|1\rangle$. If the angle $\theta = \phi$ is zero, the system stays in the ground level $|0\rangle$. If we interact the system with a π-pulse, the system is transferred entirely into the excited state $|1\rangle$. Similarly, if the initial state were $|1\rangle$, the π rotation gives the $|0\rangle$ state. Hence a π rotation acts to convert $|0\rangle$ to $|1\rangle$, and *vice versa*. i.e., it acts like a *negation*, known as a 'NOT' operation of a logic gate. Thus a rotation from $\theta = 0$ to $\theta = \pi/2$, i.e., creating a superposition can be thought of as the 'square root' of the 'NOT' operation.

If moving from one state to another is thought of as the logical operation of negation, then a superposition is neither a negation nor an affirmation, but the square root of a negation. A classical spin can provide a continuum of states between affirmation and negation. The use of qbits instead of arithmetical bits provides us with generalizations of our elementary ideas of logic operations which are tied to frequentist concepts (see Sec. 2.1.2) of probabilities. In fact, just as we have used the properties of rotations in qbit operations, more general symmetry operations can be used to generate novel logic operations.

An operation which is allowed in a classical system is the process of making a 'clone' of an unknown system. This is just not possible for arbitrary quantum systems because QM allows only 'linear', unitary processes which are equivalent to rotating the state vector in its abstract space. In contrast, copying is a non-linear (quadratic) process where $a \to aa$. This impossibility of copying a quantum state onto another is known as the *no-cloning theorem*.

Given a system A in the *unknown* quantum state $|\phi\rangle$, and another system B in the state $|\xi\rangle$, we wish to bring A and B together and clone (i.e., exactly copy) A onto B, by carrying out the operation $|\phi\rangle|\xi\rangle \to |\phi\rangle|\phi\rangle$. There is no operator which can do this. In fact, if $|\phi\rangle$ were $c|a\rangle + d|b\rangle$, instead of getting a copy of $|\phi\rangle$, one might get $c|aa\rangle + c|bb\rangle$, and this is like $|\phi\rangle|\phi\rangle$ only if $\langle a|b\rangle = 0$. Hence we cannot clone arbitrary, unknown states. Instead, the two states would form a superposition containing interference terms [233].

Unlike classical bits, qbits use states superposed together; operations on them can be carried out 'together'. Hence qbit operations are 'intrinsically parallel computations' for all the superposed states. Thus, certain computations which are classically insoluble because they grow too rapidly in complexity can be solved using qbit operations. This is one important reason for developing quantum computers. Such computers have to secure the qbits from undergoing *decoherence* by phase randomization. That is, an entangled state of two particles, or a superposition of two states, prepared carefully using optical pulses, or using two slits, may settle down in one specific componenet of the superposition if the system were disturbed by external factors. External processes which modify the boundary conditions, or introduce external fields in a stochastic manner, destroy the superpositions. 'Decoherence' will be discussed in Sec. 7.5.3.

The index 0 or 1 could obviously stand for the two spin states of electrons, or some other two-valued property like the polarization of photons. Instead of speaking of 'up-spin' and 'down-spin', we talk of the 'horizontal' and 'vertical' polarizations of the light, where we mean that the electric field of the light wave is acting in the vertical (v) or horizontal (h) direction at right angles to the motion of the wave. Instead of Stern-Gerlach magnets (to separate spins), we use a polarizing filter which allows only v or h light to pass through to detect polarized light.

If we denote the 'vertical' and 'horizontal' polarizations by two unit vectors $|e_1\rangle$ and $|e_2\rangle$ in a two-dimensional space, with E_1 and E_2 the components of the electric field $E = \sqrt{(E_1^2 + E_2^2)}$, then the electromagnetic wave can be written as a vector $|w\rangle$ such that

$$|w\rangle = c_1 e^{i\phi_1}|e_1\rangle + c_2 e^{i\phi_2}|e_2\rangle, \quad c_i = E_i/E. \tag{7.12}$$

The phase angles ϕ_1, ϕ_2 are unequal for elliptical polarizations, out of phase by $\pi/2$ for circular polarization, and equal for linear polarization. The direction of propagation is at right angles to the plane containing the two polarization vectors $|e_1\rangle$ and $|e_2\rangle$. A polarizer (i.e., a polarizing filter) will only pass light whose electric field is parallel to its 'pass axis', and the resultant light is linearly polarized. Thus a glucose solution can rotate the plane of polarization of light passing through it to the right, while fructose rotates the plane of polarization to the left, by amounts proportional to the concentration of sugars, and the path length. The capacity of these molecules to rotate the polarization of light is connected with their lack of centers of symmetry, and hence having a 'handedness' (chirality).

7.2.1.1 *A qbit subject to successive optical pulses*

A beam of light split into two paths by two slits can be made to produce an interference pattern by making their path lengths different, as in the two-slit experiment. A variant of this is to use two optical pulses that are allowed to be absorbed by a two-level atom with a time delay. The absorption profile can be made to show interference fringes. They are known as Ramsay fringes.

Consider a two-level system, e.g., an atom with the ground state and just one excited state being relevant to the energy scales of the problem. It is subject to a coherent optical pulse, and its transition dipole moment plays the role of the transverse field in the spin-problem that we discussed earlier. The optical pulse is absorbed and we assume that the absorption profile is an approximate Gaussian. One may subject such test atoms to two identical optical pulses, by making the atoms pass through a second optical cavity. Then it is found that the absorption profile has a deeply fringed fine structure. This is an effect analogous to the two-slit problem; here the atom can start from the same initial state (e.g., the ground state) and reach the final state by two paths. The two paths correspond to excitation by the two pulses where the time delay between the pulses produces a phase difference leading to interference. We may denote the two-levels $|0\rangle, |1\rangle$ by $|-1/2\rangle, |1/2\rangle$ in the following discussion when referring to Fig. 7.2, as they are mathematically equivalent. An optical $\pi/2$ pulse of duration τ acts on the system at time t_1. Then, after a delay larger than the pulse time τ, another identical coherent pulse is sent at time t_2. The atom, in the $|0\rangle$ state, i.e., $|-1/2\rangle$ can absorb the first pulse and move to a superposition

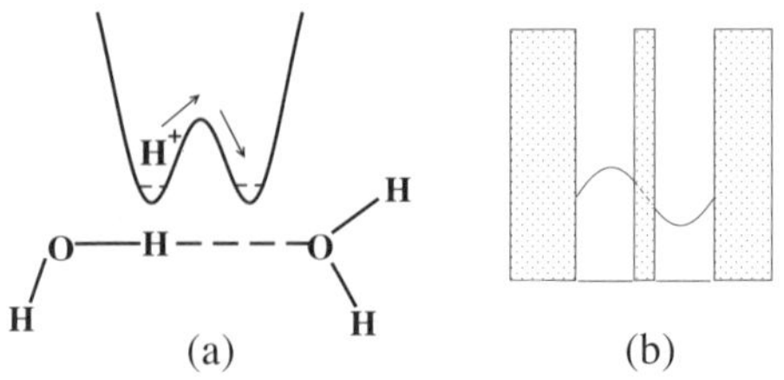

Fig. 7.3 (a) H bonds in water and living matter contain classically coupled double wells for the H-atom positions. H^+ jumps over the barrier classically and cooperatively, and contributes to the electrical and thermal conductivity of water. (b) In a quantum double well with a narrow middle barrier, electrons tunnel through the barrier. The ground-state ψ is indicated.

state, say the state marked $|1/2\rangle_y$ in Fig. 7.2, where it evolves freely (that is, the Bloch vector rotates in the X-Y plane) until the second pulse arrives. If the superposition state is in $|1/2\rangle_y$, the pulse would be absorbed and the system would move into the excited state $|+1/2\rangle_z$ (i.e., the $\pi/2$ pulses add together). However, if the superposition state had rotated to $|-1/2\rangle_y$, it would return to the ground state, and lead to an absorption minimum.

Instead of the absorption profile of a single pulse, we find narrow interference fringes within the Gaussian profile, as first discovered by N. F. Ramsay in 1950.

7.2.2 *Double quantum wells and hydrogen bonds*

A hydrogen atom positioned between two oxygen atoms of two adjacent water molecules forms hydrogen bonds bridging the oxygen atoms. The H atom is actually a proton in a double well potential defined by the two oxygens, as in Fig. 7.3(a). The proton may take an equilibrium position near the first oxygen atom, or the second. However, the H ions are too heavy to behave as quantum particles and tunnel through a potential barrier and form a superposition state, or form entangles states with other nearby protons. That is, roughly speaking, quantum properties manifest themselves only if the proton's de Broglie wavelength $\lambda \approx \hbar/p$ is comparable to the effective tunneling width of the potential barrier. The momentum p is the thermal momentum in water at the room temperature T; and at very low-T it is approximately the proton-gas Fermi momentum. Hence

$$\lambda = h/p \simeq \hbar/\sqrt{(3mk_BT)} \tag{7.13}$$

for a particle of mass m at the temperature T, with k_B the Boltzmann constant. This is quite small for protons at body temperatures where H-bonded protons behave as 'classical particles'. Furthermore, quantum superpositions need well-protected boundaries to prevent decoherence from external influences. Consequently, although hydrogen bonds are abundant in the brain, they cannot be invoked to play the role of qbits in a 'biological quantum computer associated with consciousness', although such speculations exist in the popular press. There is at present no evidence to assert that quantized, discrete energy levels are involved in neuronal processes that take place at the brain temperature. On the other hand, there is a large amount of evidence to assert that even primitive prokaryotic cells carry out 'computations' using molecular configurations that play the role of classical bits, and molecular processes functioning as logic gates. Such computation, e.g., using the multiple (allosteric) configurations of proteins as the 0's and 1's, has been termed *wetware* by Dennis Bray [38]. Protons in double wells of the 'H-bond' can have such allosteric applications, but they cannot be qbits except near $T = 0$.

The situation is different for electrons. An electron in a double-well nanostructures, as shown in Fig. 7.3(b) can be in a superposition involving the two wells. Double quantum wells can be fabricated using techniques of nano-technology, as in the five-layered double-well structure given below.

||Thick AlAs layer||GaAs layer||very thin AlAs layer|| GaAs layer||Thick AlAs layer||

The GaAs layer may be, say, 30 atoms thick, while the epitaxially grown AlAs layer in the middle may be as thin as a monolayer of AlAs. An electron may occupy a quantum level in the left well (state 0), the right well (state 1), or as a superposition state in both wells. This possibility is excluded if the middle barrier layer of AlAs is sufficiently thick, preventing the tunneling of the electron from one well to the other (except in the case of resonant tunneling).

7.3 Entangled states

In discussing superpositions we considered a single quantum system with, e.g., two energy states. Thus superpositions of the up-spin and down-spin states of an electron were considered and geometrically illustrated on the Bloch sphere (Fig. 7.2). In superpositions of two sates, physical properties did not depend just on probabilities P_1 and P_2 of the two individual quantum states, but also on interference terms that depended on the wavefunctions of both states.

Entanglement is a generalization of those ideas to two or more quantum systems that have interacted. The physical properties of the resulting final state cannot be described by those of the individual systems. New holistic effects are found to exist, because the wavefunction of the whole system (i.e., the *entangled system*) is not a simple product of the wavefunctions of the initial systems.

Consider two quantum systems S_1 and S_2, where the quantum states are labeled $\phi_1(i)$ and $\chi_2(j)$. Here i, or j list the eigenstates of each system. An entangled state $\psi_{12}(k)$ of the two systems cannot be written as a product of a state of the system 1 and a state of the system 2. It is a linear superposition of at least two such product states. An example involving just two states would be,

$$\psi_{12}(k) = \{C_1\phi_1(i)\chi_2(j) \pm C_2\phi_1(j)\chi_2(i)\}, \quad |C_1|^2 + |C_2|^2 = 1. \tag{7.14}$$

Such an entangled state remains an entangled state (i.e., cannot be decomposed into a product) if the basis sets $\phi_1(i)$ and $\chi_2(j)$ were replaced by other bases obtained from them by unitary transformations. In practice, if the two systems are interacting strongly, then the state $\psi_{12}(k)$ involving only two basis functions from each system need not be an eigenstate of the combined system since the interaction term in the Hamiltonian of the combined system may mix in other states besides i, j, thus requiring many more terms to properly approximate the eigenstate of the interacting whole. However, the basic idea of entanglement is unchanged, and this implies that the above non-factorizable form holds irrespective of the 'spatial distance' between the two particles. Einstein, Podolsky and Rosen in 1935 presented this correlated behaviour of two entangled quantum particles as a paradox (EPR paradox, see Sec. 7.7), because they remain entangled even when pulled apart to infinity, where as they expected *local realism* to be valid, with no 'spooky' distance action'.

The quantum mechanics used for two electrons, or two photons, entangled in a pair tends to be similar except that photon states do not obey Fermi statistics and the antisymmetrization is not needed. Instead, symmetrization is needed. Photons and phonons (i.e., quantized lattice vibrations) are described by field operators which obey Boson commutation rules while electrons, being Fermions, obey anti-commutation rules. The final effects can be dramatically different. In Fig. 7.1 we showed how two parallel-spin electrons seem to 'repel' each other and anti-correlate. Bosons bunch together and can form a *Bose-Einstein condensate* where all the particles are truly entangled into a quantum droplet. Fermions can also condense into a quantum state if the Coulomb repulsion can be screened out by the presence of ion-density fluctuations which attract electrons and induce a net positive correlation, producing superconductivity. Here many electron-pairs, e.g., in opposite momentum states correlate to form a quantum superposition. Kamerlingh Onnes discovered *Superconductivity* in 1911.

Each photon taken independently is characterized by its polarization. Entangled-photon pairs can be made by generating them in a single quantum process. Thus when two photons are simultaneously emitted from an excited atom, or when 'bi-excitons', i.e., two electron-hole pairs in bound-states of nanostructures (e.g., quantum dots) decay, an entangled pair of photons may be emitted.

In the following we consider the important, ubiquitous example of entangled electron pairs, namely, *chemical bonds*.

7.3.1 *The chemical bond, an ubiquitous entangled pair*

A chemical bond, e.g., between two hydrogen atoms, is an example of a system where the electron from the first H-atom is entangled with the electron in the second H-atom, with their spins oppositely paired. An electronic quantum state contains a space part (the electron 'orbital'), e.g., the 1s orbital of hydrogen, and the spin part, denoted as up (u) or down (d) states, defined with respect to some arbitrary z-axis. Let us first discuss the spin of the combined (i.e., two-electron) system that forms the chemical bond.

Two hydrogen atoms when allowed to interact, form a hydrogen molecule, H_2. A *chemical bond* is formed between the two H atoms. The H_2 system is stable and has a total energy lower than the sum of the energies of the two isolated atoms. Each atom contributes one electron to the bond. A simple quantum description of chemical bonding was given by Dirac, and by Heitler and London in 1927. Although each electron has a spin, the usual chemical bond does not have a net spin. Thus the 'up' spin of one electron is exactly balanced by the 'down' spin of the other spin. Such a pair of electrons with zero net spin is known as a *singlet pair*. If both spins were directed in the same direction ('parallel spins'), it is called a *triplet* state as there are three possible ways of realizing such spin states. These are examples of entangled electron pairs.

The total spin S of two spin-parallel electrons is 2 times $(1/2)\hbar$. Thus, $S = 1$ in units of $\hbar$,. Hence, this total spin can orient with respect to a given reference axis (say, direction z) as $-1, 0$ or $+1$, giving us three possible spin states (triplet) for the combined system of two electrons with parallel spins. If the spins are anti-parallel, the total spin is zero and there is only one possible state (singlet). Thus an electron pair with zero net spin can be in the state $u(1)d(2)$, or it may be in $d(1)u(2)$. When ever two or more alternatives become available for a quantum system, the actual state becomes a superposition of all such states. The equally weighted superposition should have the form

$$\psi_1 = \{u(1)d(2) - d(1)u(2)\}/\sqrt{2}. \tag{7.15}$$

This is a spin wavefunction which changes sign when the electrons are interchanged by changing 1 with 2, as required by the Pauli principle. The Pauli principle merely ensures that each Fermion has a unique set of quantum numbers. Here, the electron's space quantum numbers (e.g., n, l, m) can be the same as long as they differ in their spin quantum number σ which is 'up', or 'down'. As already remarked, the net spin of this state is zero. We cannot say anything definite about the spin of any one of the component electrons, except that it could be 'up', or 'down' on separation. Superposition states of two electrons (be they singlets or triplets), are sometimes known as 'Bell states' after John Bell.

The electron ψ has an extension in space and an intrinsic spin-space as well. The space part of the wavefunction of an electron in a normal hydrogen atom is the 1s wavefunction, $\phi_{1s}(r)$. This may be pictured as a spherical distribution with its maximum at the nucleus, and exponentially decreasing radially outwards. The nucleus is very heavy, and sets up a region of space where the electron resides. The boundary conditions used in solving the Schrödinger equation produces quantization. The 1s wavefunction has spherical (s) symmetry and the principle quantum number $n = 1$. In the simple Bohr atom, such a 1s electron follows a circular orbit around the nucleus. Although ϕ peaks at $r = 0$, the volume element $4\pi r^2 dr$ has to be taken into account in finding the most likely location of the electron. When this is done, the radial probability maximum coincides with the Bohr orbit.

In fact, wavefunctions of electrons in atoms are often called *orbitals*. In Fig. 7.4 Dirac has drawn the 1s orbitals of an H_2 molecule on the black board. These are the orbitals corresponding to $\phi_A(r_1)$ and $\phi_B(r_2)$ of two hydrogen atoms A and B, held at a distance R, i.e., the two nuclei are separated by R, while the two electrons are in the orbitals. The simplest case is when the two H atoms, denoted by A and B are very far apart (i.e., R very large), and the two electrons with nominal positions r_1, r_2 are not entangled. Then we have two independent systems. The spin part can be written separately from the space part, and we concentrate on the space part of the total wavefunction. It is a simple product of 1s wavefunctions ϕ_A and ϕ_B. That is, the un-entangled wavefunction is:

$$\psi(r_1, r_2) = \phi_A(r_1)\phi_B(r_2). \tag{7.16}$$

Fig. 7.4 Paul Dirac has drawn the bond-energy curve for two H-atoms as a function of the nuclear separation. The anti-bonding triplet state (marked sym) is also seen in the diagram. These are entangled states of the two electrons of the two H atoms (courtesy AIP Archives).

The two H atoms begin to interact as they move towards each other. We saw that the energy of an electron in a wide quantum well was smaller than in a narrow quantum well (see Sec. 6.4). The electrons that were confined near one or the other of the nuclei now move over a longer region including the middle region between the nuclei, and lower their energy. The electron density becomes maximal in the region between the atoms. The total energy is lowered until the distance R becomes equal to the equilibrium bond length of the Hydrogen molecule, viz., 1.5 Angstroms, i.e., 1.5×10^{-8} cm. In the entangled state, the electrons with nominal positions r_1, r_2 are in the sphere of influence of both nuclei and form a superposition state. Thus the space part of the entangled-state wavefunction is:

$$\psi(r_1,r_2) = \{\phi_A(r_1)\phi_B(r_2) + \phi_A(r_2)\phi_B(r_1)\}/\sqrt{2}. \tag{7.17}$$

If the electrons are exchanged, ψ does not change sign. However, the spin part of the wavefunction is a singlet (see Eq. 7.15) and the total wavefunction inclusive of spin obeys the Pauli principle.

A superposition state for two electrons in hydrogen using parallel spins (triplet) is not a stable system as the Pauli exclusion prevents the electrons from occupying the space common to both nuclei. Electron density is confined near each nucleus. There is no H-H separation which is a minimum of energy. In the Fig. 7.4 we see the lower curve, with a minimum. This is the bonding curve of the singlet state, while the upper curve is the non-bonding triplet state of H_2. Ordinary hydrogen gas is made of H_2 molecules and not H atoms. The entangled state of singlet pairs is the basis of *the covalent bond* in chemistry. In this state, there is a high probability of finding electrons in the 'bonding region' between the two atoms, and it takes energy to move the two atoms apart, i.e., to stretch the bond distance R. Nevertheless, we can take the two atoms as far apart as we like, to reach the dissociation limit. However, the wavefunction remains a superposition. The wavefunction of such a superposition cannot be written as a simple product as in Eq. 7.16. As Schrödinger stated [188], the electrons can 'no longer be described in the same way as before, viz., by endowing each of them with a representative (i.e., in current language, a wavefunction) of its own'.

Does this imply 'spooky action at a distance'? In practice, if we separate the two atoms forming the bond, the energy stabilization begins to decrease. Almost any perturbation or thermal fluctuation from the surroundings may be enough to 'break the bond', and destroy the entangled state. Notice that the calculation for the chemical bond (entangled state) is a 'zero-temperature' calculation. If temperature is included, the relative vibration of the protons come into play. At sufficiently high T, these vibrations are strong enough to dissociate the atoms and the chemical bond (i.e., the entangled state) breaks up.

The entanglement holds as long as we keep the system well protected. Thus we see that the 'spooky action' at a distance holds only as long as the experimentalist 'engineers' it. *There is nothing spooky about it*. The essential feature of such states is that the phase relations among the superposed components are held intact. Many such engineered coherent states have been achieved in atoms and molecules (interacting among themselves or with lasers), superfluids, superconductors etc. In nature, at least on earth, chemical bonds (and more subtle coherent states) are usually found confined to scale lengths of molecules and nanostructures. The stability of the entangled state in chemical bonds is basic to life itself. A long-chain conjugated molecule, a polymer or a carbon nanotube is essentially made up of many singlet superpositions. That they remain perfectly stable shows that the proposed decoherence by 'spontaneous localization' (see Sec. 8.5) etc., have no effect on these chemically coherent large-scale systems.

7.3.2 *LCAO approaches to chemical bonding*

Chemical bonding can be discussed starting from 'orbitals' of atoms and their interactions (as we did so far), where chemical bonding is forming entanglements. Or else, we can begin with wavefunctions that already straddle all the nuclei in question and are superpositions of atomic orbitals known as *molecular orbitals*. They are then used to form entanglements. This method is discussed below.

In discussing the bonding between two hydrogen atoms, we took one energy state $|1s\rangle$ positioned on each nucleus, at A and B. Products of such states were made for non-interacting atoms (Eq. 7.16), and then they were antisymmetrized (Eq. 7.17), leading to entanglement and bonding. An alternative approach (that becomes equivalent if complete sets of expansion functions are considered) is also often used. It works with *molecular orbitals*, made up of superpositions of atomic orbitals.

Consider a diatomic molecule like H_2. The distance of each $1s$ electron from its nucleus is denoted by r_a and r_B. This defines two functions $\phi_A(r_1)$, and $\phi_B(r_2)$, being the $1s$-state of the atoms A and B. An approximate one-electron molecular-orbital wavefunction is a linear combination of these two *basis*

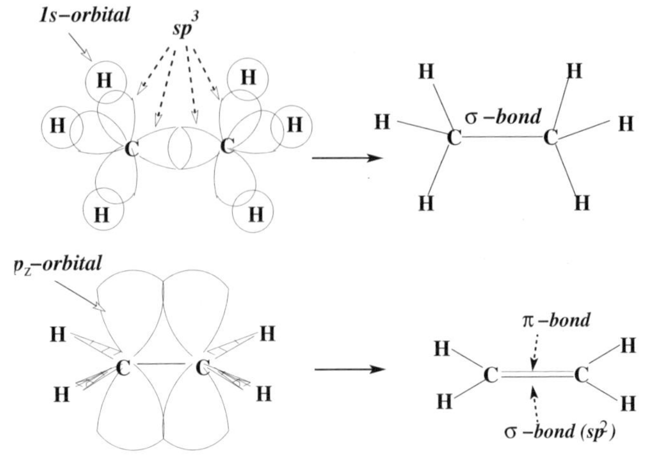

Fig. 7.5 The top part shows the ethane molecule, C_2H_6 with six $1s$-hydrogenic functions, and each C-atom with four sp^3 orbitals. The overlap of these form covalent bonds of the σ-type, with cylinder-symmetry along the bond. The chemical-bond picture is shown on the right. The four bonds at each C atom are directed to the four corners of a tetrahedron. The bottom half shows the ethylene molecule, C_2H_4. Here the C–C bonding involves a σ-bond and a π-bond. The σ-bond is formed by the overlap of two sp^2-orbitals coming together from the two carbon atoms, and the σ-bonding framework gives a planer molecule (x-y plane). The two p_z orbitals overlap and produce the π-bonding.

functions. If $\vec{R}_A$ is the position of the nucleus A, etc., then

$$\psi_1(r) = c_1\phi_A(\vec{r}-\vec{R}_A) + c_2\phi_B(\vec{r}-\vec{R}_B)\,. \tag{7.18}$$

Since $|\psi_1|^2$ is a probability for finding an electron some where in it, its integral is unity. Hence $|c_1|^2 = |c_2|^2 = 1/2$ and $c_1 = c_2 = \pm(1/2)$.

In effect, we have selected two functions, known as 'basis functions' to represent the basic state of the electrons before the bonding (entanglement) occurs. In reality, if the atoms are assumed to be far apart, the electron (or electrons) in the atom A may exist in any of the states $|1s\rangle, |2s\rangle, |2p\rangle, \cdots, |nl\rangle$. Here n, l denote the principle quantum number and the azimuthal quantum number. If we include the spherical angular momentum states denoted by the quantum number m, each atom has the set of atomic states represented by mathematical functions $\phi_k = |nlm\rangle$. Here k is a composite index which represents nlm. We can denote such a state as $|k\rangle$. In general, the principle quantum number n is a discrete number for bound electronic states, with negative energies which vary as $\sim -1/n^2$. The positive energy states are known as continuum states, and these are denoted by the continuous index p instead of n. Then the energies vary as $\sim p^2/2m$ where m is the mass of the electron. Such states are important in plasma physics and in ionization.

Given a set of atoms (say two atoms, with the index $i = 1,2$) infinitely far away, each electron in each atom is in some state $|i,k\rangle$. where k specifies the electronic state. As the atoms are far away, and have not interacted, there is no entanglement. The wavefunctions $|i,k\rangle$ are independent of each other, and there is no overlap. That is, they are effectively *orthogonal*. Such a set of functions, covering all the possible electronic states, is known as a complete *basis set* of functions. However, as the atoms are brought closer to each other, the atomic states overlap and interact; they no longer provide a valid approximation to the electronic states of such atoms. A simple and fruitful approximation is to make linear combinations of overlapping atomic orbitals to make wavefunctions for the ground states (as well as excited states) of the interacting systems. This is the *l*inear-*c*ombination of *a*tomic *o*rbitals (LCAO) method. It was introduced by Heitler, London and Hückel in the late 1920s, and by Felix Bloch under the name *tight-binding method* (TB).

For example, only the *valence electrons* of the carbon atom are used. Thus each carbon atom, having 4-valence electrons per atom, provides a $2s$ orbital and three $2p$ orbitals, while four hydrogen atoms in CH_4 provide four $1s$ orbitals making a total of eight functions. The simplest useful LCAO approximation to an electron state in the methane molecule CH_4 is constructed as a linear combination of neighbouring atomic orbitals, i.e, a superposition ψ with $\psi|^2$ normalized to unity. The first four lower-energy combinations are occupied by two electrons each, and provide an approximation to the molecular-wavefunctions (*molecular orbitals*) of the four chemical bonds of the molecule. The unoccupied upper energy levels are crude approximations to excited states.

The highest occupied molecular orbital (HOMO) and the lowest unoccupied molecular orbital (LUMO) are separated by an energy gap. If this gap is zero, an electric field can move electrons into the LUMO where they travel along the molecule, as in conducting polymers. Such (single-particle) conducting states are made up of superpositions of atomic wavefunctions. If 'spontaneous localization' theories are valid, such states also should undergo decoherence.

Bonds, made mainly of s states of atoms (i.e., angular momentum $l = 0$ states), are known as σ-bonds. If the bonds are mainly overlapping p-states (e.g., p_x, p_y and p_z states), as required in ethylene or benzene, they are known as π-bonds. In reality, the picture is somewhat more complex since orbital symmetries should be taken into consideration explicitly if approximate methods are used. One main objective is to form 'chemically meaningful' combinations of orbitals that mimic chemical bonding. This approach leads to the 'valance-bond' approximation to chemical bonding (the molecular orbital and the valance-bond approaches converge to the same results if a sufficient number of basis functions are used).

Linus Pauling showed in 1927 that the tetravalent nature of carbon is best understood by taking its $2s$ state as being perpetually in a superposition with the $2p_x, 2p_y, 2p_z$ states. Hence the σ bond is made up mostly of s states, but carrying some parts of the three p-states (see Fig. 7.5) to give a total (tetrahedral) symmetry consistent with the s and p states. Such superposition states are known as sp^3 orbitals. This leads to tetravalent carbon, with bonds directed to the corners of a tetrahedron exactly as found in methane, CH_4. Linus Pauling called such superpositions *hybridized orbitals*, and used them with great ingenuity to explain chemical phenomena in organic and inorganic materials, as well as in biomaterials. Pauling's book, *The Nature of the Chemical Bond*, first published in the 1930s, was required reading for students of chemistry and condensed-matter physics well into the 1960s. Pauling was awarded the Nobel Prize for chemistry in 1954.

Just as atoms have stable shells of electrons, with two electrons in the first shell etc., molecular-bond formation occurs with bonds having two singlet electrons (covalent bonds), or four bonds making an octet around each carbon. These simple 'shell filling' rules (e.g., Octet rule etc.) are behind the laws of valence used extensively by chemists as simple rules of chemical bonding.

An *unsaturated hydrocarbon* like ethylene C_2H_4 is short of two hydrogens (and two electrons) to form the saturated ethane molecule where each H-atom has two electrons and each C-atom has eight electrons in the outer shell (which is shared with another neighbouring atom). Thus, instead of making four sp^3 hybridized σ-orbitals containing two electrons at each C-atom, ethylene makes three sp^2-hybridized orbitals at each C, accounting for 6 electrons, and freeing up one p-orbital on each carbon atom, containing one electron. By sharing these, each carbon atom can have an octet of electrons. Let us call this the p_z orbital. In ethylene (see Fig. 7.5) the C atoms now share another pair of electrons forming a π-bond. Hückel-orbital theory often treats electrons in π bonds, leaving the system of electrons in the σ-bonds aside, as the σ electrons are more strongly bonded and form a relatively independent subsystem. The π-electrons of hydrocarbons are more active, easily polarizable, and behave almost like free electrons moving in networks of π-bonds connected with each other. Such connected networks are known as conjugated-hydrocarbon systems. Graphene is an example of such a conjugated network of electrons confined to an atomically thin 2-dimensional membrane. It is made up of a hexagonal network of σ-bonds and π-bonds. A stack of such graphene sheets bond to form graphite.

The theory using atomic functions as basis sets is an approximation. It is better to use as large a set of mathematically convenient orthogonal functions as possible, centered on atoms or otherwise, as the basis set. This is used to construct an expansion of the many-electron wavefunction in Slater determinants (see Sec. 8.2.1). S. F. Boys [34] pioneered the method of Gaussian wavefunctions, as all the many-electron integrals that arise in the many-electron correlation problem (see Sec. 8.2.2.1) could also be handled analytically. The same advantage is available with plane-wave basis sets, although many more plane-wave functions are needed for good convergence in variational algorithms seeking to minimize the total energy. With the availability of fast, large-memory computers, plane wave methods have begun to rival methods based on Gaussian expansions. These approaches to quantum chemistry are known as *ab initio* methods. They give up optimization using 'chemical intuition', in order to exploit computational and conceptual advantages of a formal method. The importance of LCAO (i.e.,

TB) has not diminished, even though *ab initio* quantum chemical methods have become increasingly important for accurate predictions. LCAO/tight-binding methods are useful when chemical insight has to be interpreted and reconciled with quantum theories of bonding.

The LCAO and valance-bond methods clearly show that *chemical reactions are nothing but re-arrangements of 'cat states' or electronic bonds among atoms*, allowing them to change their states of entanglement from one chemical state to another so as to minimize the net free energy of the system.

7.3.3 *Entangling atoms and field modes*

We discussed how the two levels of a spin-1/2 particle could be put into a *superposition* state by interaction with an oscillatory electro-magnetic field (Sec. 7.2.1). Here we consider how to *entangle* a two-state atom with an electromagnetic excitation mode (a photon) using resonant optical cavities.

In our previous discussion the $|-1/2\rangle$ and $|+1/2\rangle$ states were made into a superposition using a $\pi/2$-pulse. The electromagnetic field was treated as a classical field. However, we can use an electromagnetic cavity or optical cavity which allows only a discrete number of confined electromagnetic modes to exist in it. Here the quantum nature of the electromagnetic field has to be taken into account. A resonant optical cavity is conceptually similar to a wind pipe that supports only a specific pitch and a few vibrational overtones. A typical cavity is a hollow conductor blocked at both ends and may be designed to allow only a specific standing wave (principle mode) to exist in it. In such a resonant cavity, the field may contain only one photon ($N = 1$), or no photons ($N = 0$). A classical treatment of the field is applicable only if N, the number of photons is large.

7.3.3.1 *Entangling an atom with a field mode*

If a two-level atom initially in the excited state $|e\rangle$ enters a cavity with no photons, we designate the system by $|e, N = 0\rangle$. However, this state is coupled to the ground state $|g\rangle$ as the excited state can spontaneously emit a photon and create the state $|g, N = 1\rangle$. This regime is referred to in textbooks as *strong-coupling cavity quantum electrodynamics*. Here, if we apply a $\pi/2$-pulse at the appropriate Rabi frequency (known as the vacuum Rabi frequency), the system initially in $|e, N = 0\rangle$ forms a superposition with the $|g, N = 1\rangle$ system.

The formation of the field-atom entangled state can be given as:

$$|e, N = 0\rangle \xrightarrow{\pi/2\text{-pulse}} \psi_{Ap} = \{|e, N = 0\rangle - i|g, N = 1\rangle\}/\sqrt{2}. \tag{7.19}$$

Such mode-atom entangled states are routinely prepared in quantum optics laboratories as intermediates for entangling other quantum systems, cf., Haroche *et al.* [97]. For instance, an atom entangled to a field mode as in Eq. 7.19 can interact with a second atom entering the cavity, even after the first atom has left the cavity, as it remains entangled with the field mode. This interaction can be used to entangle two atoms.

7.3.3.2 *Entangling two atoms*

Ignacio Cirac and Peter Zoller in 1995 entangled two trapped ions using their coupling with phonon modes in an ion trap. The phonons are bosons and the physics is similar to entangling atoms using cavity photons. Hence, although trapped-ion entanglement came earlier, we describe the scheme for atom entanglement, due to E. W. Hagley, Serge Haroche and collaborators (1997).

Entangling two atoms involves entangling the first atom to a field mode, and then allowing the second atom to interact with another optical pulse, with the first atom outside the pulse. However, the latter is still entangled with the field.

Let us prepare an atom-photon entangled state ψ_{Ap} as in Eq. 7.19, and consider a second atom in the ground state $|g_2\rangle$, and not yet entangled with photons or with ψ_{Ap}. This state is

$$|\psi_{Ap}\rangle|g_2\rangle = \{|e, g_2, N=0\rangle - i|g, g_2, N=1\rangle\}/\sqrt{2}. \qquad (7.20)$$

The second atom, in the ground state g_2 is subject to a π-pulse once it enters into the cavity region, while the first atom has left the cavity. The π-pulse couples $g, N=1$ states and $e, N=0$ states of any atom. The first term in Eq. 7.20 has $g_2, N=0$, i.e., a no-photon component and this is unaffected by the pulse. The second part has $g_2, N=1$ which can undergo a rotation under the π-pulse to the state $e_2, N=0$. Hence the effect of the π-pulse when the second atom is in the resonant cavity is:

$$|\psi_{Ap}\rangle|g_2\rangle \xrightarrow{\pi\text{-pulse}} \{|e, g_2, N=0\rangle - |g, e_2, N=0\rangle\}/\sqrt{2}. \qquad (7.21)$$

This result can also be derived more formally using Eq. 7.9.

Thus the excited state of the first atom plus the second atom in the ground state is one entity. This is entangled with the second entity which is the excited state of the second atom plus the first atom in the ground state.

7.3.4 *Entangled photons*

Neutral pions have a half life of 8.41017 s, decaying into two photons. Neutral π-mesons (neutral pions π^0) were first detected in cosmic rays, and led to the discovery of entangled photons in 1949. In 1948, π^0 were artificially produced at the University of California cyclotron by bombarding carbon with α-particles.

$$\pi^0 \to \gamma + \gamma. \qquad (7.22)$$

The two photons fly off with equal and opposite momenta. Their polarizations are entangled as they were born as a pair.

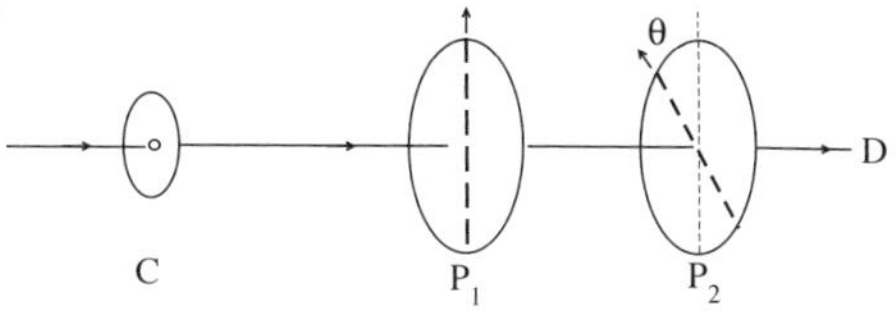

Fig. 7.6 A beam of photons collimated by C passes through two polarizers. The pass axis of P_1 is set vertical, while the pass axis of P_2 is rotated by an angle θ. The photon intensity detected by D is reduced by a factor of $\cos^2\theta$.

In Sec. 7.2.1, and Eq. 7.12 we introduced the unit polarization vectors $|e_1\rangle, |e_2\rangle$ which are at right angles to each other, and to the direction of propagation. Each photon 'has' two such polarization vectors. Experiments show that the entangled photon pair, with momenta $k, -k$, has the form

$$|\psi_u\rangle = \{|k,e_1\rangle|-k,e_2\rangle - |k,e_2\rangle|-k,e_1\rangle\}/\sqrt{2}. \tag{7.23}$$

The label u (ungerade) indicates a function of odd parity. Entangled photons can be produced by other means as well. When positronium (a bound electron-positron pair) decays, photon pairs are formed from the energy $2mc^2$, where m is the mass of an electron or positron. Pions and positronium have odd parity and the photon state in Eq. 7.23, with opposite polarizations has odd parity.

On the other hand, if entangled photon pairs were made using an atomic cascade, one may produce entangled photon states with even parity. An atomic cascade is a series of transitions in an exited atoms. Thus, if a calcium atom were excited to the 6^1S_0 state, it would emit a photon by decaying to the state 4^2P_1 state, and then emit another photon by decaying to the 4^1S_0 state. The two photons (with wavelengths $\lambda = 551.3$ nm, and 4227 nm) are in the visible regime and experiments using linear polarizers and photo-detectors can be conveniently used. Since the initial and final atomic states have zero total angular momentum and the same parity, the entangled pair is expected to have an entangled polarization state $|\psi_g\rangle$ of even parity, as indicated by the suffix g (gerade). Thus for example,

$$|\psi_g\rangle = \{|k_1,e_1\rangle|k_2,e_1\rangle + |k_1,e_2\rangle|k_1,e_2\rangle\}/\sqrt{2}. \tag{7.24}$$

Nanotechnology can be used to make 'quantum dots' which trap electrons. A laser can be used to excite two electrons to an upper energy level E_2, creating a ground-state level E_0 with *two* electron missing. The missing electrons are like a double-positive charge while the excited electrons are like two electrons in a He atom. This system is known as a *bi-exciton*. The excited electron pair can fall back into the ground state with the emission of two photons which are entangled.

In 1949 it was established that the two photons generated from π^0 were entangled photons as in Eq. (7.23), disproving the suggestion by Einstein, that the state would evolve into the simple product states $\{|k,e_1\rangle|-k,e_2\rangle\}$, and $\{|k,e_2\rangle|-k,e_1\rangle\}$ with equal probability, as the photons move apart and become uncorrelated, in keeping with Einstein's ideas of local realism.

Let us consider the action of two polarizers on polarized photons. In Fig. 7.6 we consider a beam of collimated monochromatic photons passing through two polarizers with essentially identical properties, and with their 'pass axes' at an angle θ. Let P_1 and P_2 denote unit vectors along pass axes. The first polarizer P_1 passes that fraction of the beam which is linearly polarized and with the electric field oriented along the vertical, and with amplitude A. The second polarizer modifies the amplitude by the overlap between the two pass axes, i.e., $P_1 \cdot P_2 = \cos\theta$. The overlap is effectively the degree of superposition of one vector on another. The intensity is the square of the amplitude. Hence the intensity of the photon beam now includes the factor $\cos^2\theta$. On the other hand, if we are dealing with a photon source at the quantum limit when only one or two photons are present in the beam, then the intensity itself cannot be scaled by arbitrarily small amounts. The polarizer P_2 has to transmit a minimum of one photon, or no photons. Hence we interpret $\cos^2\theta$ as *the probability* that a photon may go through the polarizers and get detected in D, as we cannot talk of a 'fraction of a photon'.

One may also consider a two-modes system, where the mode A is occupied by N photons, while the mode B is occupied by 0 (zero) photons, giving the state $|N,0\rangle$. Alternatively we could have zero photons in A, and N in B, or $|0,N\rangle$. A superposition of these two sates, $\psi = |N,0\rangle + \langle O,N|$ is known as a NOON state. These states are of interest in photo-lithographic etching since using such entangles photons give greater resolution than with simple photon sources. NOON states may be made with any type of bosons, but exceeding $N \sim 2$ is very hard.

7.3.5 *Pure states and mixed states*

Pure states are single quantum states, while 'mixed states' are mixtures of pure states *without* cross correlations. (see also Sec. 7.5.3 where decoherence is discussed).

If the quantum system can be characterized by a definite wavefunction $\phi_i(r,t)$ then we say that it is in a pure state. However, we may have a 'mixed' state which is a statistical distribution over many possible states where i is the index. A superposition of such pure states may be denoted by

$$\psi(r,\sigma,t) = c_1\phi_1(r,t) + c_2\phi_2(r,t) + \cdots \tag{7.25}$$

where c_1, c_2, etc., are coefficients. The probability distribution is obtained from the square of $\psi(r,t)$. We drop the (r,t) for brevity and write

$$|\psi|^2 = \Sigma_i|c_i|^2|\phi_i|^2 + \Sigma_{ij}c_i^*c_j\phi_i^*\phi_j + h.c. \tag{7.26}$$

$$= \Sigma_i P_i + P_x\,. \tag{7.27}$$

We can also use the coefficients c_i, c_j to define the density matrix ρ. If we take the two level case,

$$\rho = \rho_{ij} = \begin{vmatrix} c_1^*c_1 & c_1^*c_2 \\ c_2^*c_1 & c_2^*c_2 \end{vmatrix}\,. \tag{7.28}$$

The density matrix will be discussed one again in section 7.5.3. It is used to define the 'statistical entropy' also called the quantum entropy, denoted by S and introduced by von Neumann in analogy with the thermodynamic entropy of Boltzmann. However, it is not a mere analogy, but turns out to be conceptually equivalent, as discussed in Ch. 8. It is a measure of the lack of information about the system, in comparison to a pure state with $S = 0$.

$$S = -k_B\mathrm{Tr}(\rho\log\rho) = -k_B\Sigma_i P_i\log P_i\,. \tag{7.29}$$

Here k_B is the Boltzmann constant. If we are dealing with a pure state k, then $P_k = 1$ and all other terms are zero. Hence the statistical entropy of a pure state is zero. If we are dealing with an entangled system U made up of systems 1 and 2, U is a pure state and its entropy $S_U = 0$. However, $S_1 = S_2 = -k_B(P_1\log P_1 + P_2\log P_2)$. Since $P_i < 1$, we see that $S_1 + S_2 - S_U$ is always positive.

If we are dealing with a 'random mixture', the cross terms involving $c_i^*c_j$ and their Hermitian conjugates (h.c), i.e., the off-diagonal terms of ρ whose sum was denoted by P_x average out to zero when summed over many states. If such cross terms are to be preserved, the system has to be carefully engineered. That is, if we are unable to hold onto a careful selection of some states with well-defined c_1, c_2, etc., other uncontrolled states get mixed in and the delicate phase relations get destroyed. Decoherence sets in and entanglement is lost. Then P_x is statistically zero and we only have a sum of probabilities $P_1 + P_2 + P_3 \cdots$ typical of a simple classical mixture.

The probabilities $P_i = |c_i|^2$ define the likelihood of finding the system in the quantum state with the index i. If the distribution is attached to a large thermal bath at the temperature T the probabilities P_i will have the form of Boltzmann factors $exp(-E_i/k_B T)$, where E_i is the energy of the quantum state i. This is a thermalized mixture of states. There are no 'spooky' effects in any sense of the term as the phases generating such correlations have got averaged to zero.

The 'consistent histories' approach to QM, due to Gell-Mann and others, involves a recognition of the existence of decoherence when systems evolve with time, due to coarse graining effects arising from the random interactions suffered by each evolving subsystem. A simple example of coarse graining is the replacement of the continuous range of numbers from 0 to 100 by ten bins, each in the ranges, 0–10, 10–20, 20–30, etc. Each number may represent a dynamical state. In the coarse-grained description, ten such 'fine-grained' states are bundled into one bin of the coarse-grained description. Thus a continuous function $f(x)$ gets replaced by a histogram under the coarse graining process. Unfortunately, how this coarse graining of quantum states arises is not clear and leads to various epistemological and ontological questions that have not been resolved. Is this merely a matter of our 'ignorance' of the detailed fine-grained description, or is it something more?. Similar questions about coarse graining arises in regard to irreversibility (Sec. 9.2.4) and the arrow of time even in classical systems.

7.3.6 *Quantum computers*

Quantum computers (QC) use qbits (i.e., qubits, see 7.2.1) instead of classical binary states ('bits') to build up the basic logic operations required for computing. Quantum computers reached the popular press in the 1990s with promises of supercomputers for the future. We already discussed qbits and noted their parallel-processing capabilities (Sec. 7.2.1). Instead of manipulating numbers one after another, they use arrays of numbers, i.e., matrices. The inherently matrix-based algebra associated with qbit manipulations were already noted. These simplified the problems in cracking secret codes (cryptography) and solving previously intractable problems in number theory. In fact, some prototype QC are known as 'systems for quantum cryptography', or 'quantum key-distribution systems' [22].

The 'quantum key-distribution systems' use qbits made up of entangled photons, with two polarizations per photon. They are in one of four polarizations, e.g., horizontal (0), vertical(1), or diagonal(0), and anti-diagonal(1), that randomly encode 0s and 1s. When a qbit reaches a receiver, a single-qbit operation known as the Hadamard transformation is attempted. This puts suitable 0 and 1 states into entangled states. If this succeeds, what is left behind is a randomly coded unbreakable cipher.

A qbit state $|\psi\rangle$ can be represented geometrically as a point on a Bloch sphere (see Fig. 7.2) by specifying the two polar angles, θ, ϕ together with an unimportant global phase factor.

$$|\psi\rangle = \sin(\theta/2)e^{i\phi}|0\rangle + \cos(\theta/2)|1\rangle. \tag{7.30}$$

Any transformation of a qbit is a norm-conserving transformation, i.e., a rotation of the state vector on the Bloch sphere. The single-qbit Hadamard transformation H_1 transforms the $|0\rangle$ state to $\{|0\rangle +$

$|1\rangle\}\sqrt{2}$, and the state $|1\rangle$ to $\{|0\rangle - |1\rangle\}\sqrt{2}$. Hence it corresponds to the transformation

$$H_1 = \frac{1}{\sqrt{2}}\begin{vmatrix} 1 & 1 \\ 1 & -1 \end{vmatrix} \quad \text{acting on } |0\rangle, |1\rangle\,. \tag{7.31}$$

The n-th order Hadamard transformation is the n-th direct product of H_1 matrices. Many quantum algorithms use the n-th order Hadamard transform as an initial step, since it maps n qbits initialized with $|0\rangle$ to a superposition of all $2n$ orthogonal states in the $|0\rangle, |1\rangle$ basis with equal weight.

The progeny of these early key-distribution 'quantum computers' is expected to make use of 'quantum teleportation' [23]. Here, the information itself is not transmitted, but appears at the receiver *via* the properties of entangled states. A quantum teleporter requires at least two qbit operations to create entangled states. Although photon teleportation has been accomplished for sometime, qbits based on material systems (e.g., squids, quantum dots) are needed to store the entangled pairs long enough prior to teleportation. Thus, the problem of maintaining the entangled states by reducing *decoherence* is a major factor in the development of quantum computers. Material-based qbit systems suffer many-body interactions and this reduces the range of entanglement and increases decoherence.

Developments in quantum computers go together with the development of quantum repeaters needed to make quantum communications work over long networks. These devices are needed for 'entanglement purification', a process which 'corrects' for decoherence. Number theory (e.g., searching for prime numbers) and cryptography are very specialized applications and they cannot perhaps be used to justify research into quantum computers. Just as quantum-key distribution devices reveal the existence of unexpected 'eavesdroppers' by noting an increase in the error rate, quantum computers would probably be used as sensitive if fragile sensors monitoring the security of information processing and transmission. However, research into quantum computers is a part of the study of quantum information, and requires little justification.

7.3.7 *Einstein-Podolsky-Rosen objections to entanglement*

Schrödinger's arguments emphasizing the existence of non-local correlations, viz., entanglement, as a fundamental property of quantum mechanics was not to the liking of Albert Einstein. Einstein felt that if two particles were spatially moved apart, the far away particle should have less and less effect on the 'local' particle. He felt that properties of the local particle were locally determined, and existed independently of other particles which may be far away. This is the concept of *local realism* that Einstein advocated. If there was some non-local action found in nature, then it must be mediated by some 'hidden forces' that have to be described by 'hidden variables'. Since quantum mechanics did not have such a description, but allowed entanglement which seemed to imply 'spooky action at a distance', Einstein felt that QM was incomplete. Einstein's favoured method of investigating nature was *via* 'gedanken experiments'. Einstein, Podolsky and Rosen (EPR) proposed in 1935 such a thought experiment involving two entangled quantum particles. They presented it as a paradox, since their seemingly harmless assumptions led to paradoxical conclusions. We discuss this again in Sec. 7.7.2. Here we discuss David Bohm's more transparent version [25] of the EPR experiment where entangled spin states are used.

Consider two spins in a singlet state, where one spin is in Princeton, with *Eugene*, while the other spin is with Eugene's sister *Manci* in Cambridge, England. On measuring the first spin in Princeton, it is found to be $S_z = +1/2$, and hence Manci finds the second spin to be $S_z = -1/2$ in Cambridge. According to EPR, the measurement of the spin by Eugene who is across the Atlantic cannot influence the spin observed by Manci — this is the basic tenet of local realism. The spin of $-1/2$ observed my Manci is an inherent property of that spin. It is *an element of the physical reality* associated with the second spin.

However, Eugene decides to measure the spin projection in the x-direction, and finds $S_x = +1/2$, and Manci also does an x-direction measurement and finds her $S_x = -1/2$. So, is this also a part of the the intrinsic reality of these spins? The quantum theory says that if the spin in the z-direction is known precisely, then the spin in the x-direction cannot be known as such measurements involve non-commuting operations — that is, such measurements disturb each other. Hence, every time Eugene acts on his end of the entangled pair from Princeton for measurement in a new direction, the other end of the singlet in Cambridge has to change accordingly. Thus local realism is breached and there is 'spooky action' at a distance. There is a paradox unless there are some hidden variables that actuate this 'action at a distance'. Eugene and Manci could work with many singlet pairs prepared in exactly the same manner, and under the same conditions as the first singlet pair, and check if local realism holds on the average, or whether it is actually breached.

An apocryphal story that circulated in Cambridge during the time when the author was a graduate student there went as follows. A rather brash new graduate student in physics nick-named Bull had just arrived in Cambridge and noticed an advertisement for the sale of a chest of drawers. Bull phoned the advertised number, and a feeble voice answered. When Bull asked for details about the chest, the voice answered, 'you should ask Wigner's sister'. Quite undaunted by this answer, Bull had brashly inquired, 'could you not at least give the height, length and width of the chest?'. Apparently, after a long silence, the voice had cryptically retorted 'Do you think they commute'?

The measurements of a_x, a_y, a_z for the dimensions of a chest of drawers, or for a quantum system, can be carried out as these measurements commute. The surprise is in bringing the issues of the quantum domain into the macroscopic domain, exactly as in the EPR and cat paradoxes (Sec. 7.7.3). In general, the three components of a quantum mechanical operator need not commute. Hence if one component is accurately measured, the mean values of the other two components cannot be simultaneously determined for such non-commuting operators. If the expectation value of S_z is measured at time t, then the expectation values of the spin in the x and y directions at time t become uncertain.

It should be remarked that simple calculations of the electron-electron pair distribution function $g(r)$ already existed in the literature, especially by the 1950s. This had all the information necessary to show that local realism has no place in quantum mechanics. The EPR discussion talks of particles having to be 'sufficiently far', for local realism to hold. But EPR do not provide a natural length scale for what they mean by 'sufficiently far', or for the typical radius of 'local realism'. The discussion in terms of $g(r)$ provides the Wigner-Seitz radius r_s as a length scale, and the healing distance of the distribution $g(r)$ as the 'local region'. We showed in Fig. 7.1 that the correlations between two electrons generated by the Pauli exclusion principle remained scale invariant. The correlations could be modeled by a potential ('the Pauli exclusion potential') that was equally effective at all particle separations scaled by the Wigner-Setiz radius r_s. There is no escape from such a non-local interaction. Nevertheless, the researchers in the area of foundations of physics proceeded by another route, viz., the approach due to John Bell that directly addressed the main issues in their own language.

There were no experimental techniques in the 1930s to set up the EPR experiment in the laboratory to check if local realism was upheld or not. Niels Bohr contended that local realism as well as 'hidden variables' were metaphysical matters irrelevant to quantum calculations. Philosophically, Bohr held the empiricist position that an observable had no pre-existing value until it is observed (i.e., measured). However, it was intriguing to ask if the quantum predictions arising from the entangled form of the wavefunction (see Eq. 7.15) could be obtained using a local description applicable to a singlet pair *via* the action of a hidden, random parameter λ. The results when averaged over λ should be equivalent to those from the usual quantum theory. That is, is there a 'local plan' indexed by the hidden variable λ that can capture the results of QM?

The problem remained a matter for philosophical debate until John Bell's work in 1964, discussed below. He assumed the existence of hidden variables, as well as local realism, and derived inequalities that would be satisfied by certain correlations among measurements performed by distant observers studying an entangled pair of particles. If EPR were correct, these inequalities would not be violated. If local realism was breached, as seemed to be the case with QM, Bell's inequalities would be violated and Einstein *et al.* would be wrong.

7.4 The Bell inequalities and quantum correlations

David Mermin [141], a distinguished physicist and savant, writing about 'Hidden variables and the two theorems of John Bell', states that one of them is 'widely known not only among physicists, but also among philosophers, journalists, mystics, novelists, and poets'. The theorem that Mermin refers to is associated mostly with the concept of non-local correlations, and is also known as the 'Bell inequality', or Bell's theorem. The other theorem is associated mostly with contextuality, and is better known as the Kochen-Specker theorem. We had some discussions of these subjects in earlier sections (Secs. 1.1.1, 6.2.4) of the book. Bell's theorem shows that even hidden variables are not enough to describe physical reality in a causal manner, if the hidden variables are purely local.

Everyday experience (at macroscopic length scales) involves correlations between events. These are mostly local correlations. However, we can consider more complicated classical correlations which are non-local. For example, the level of pollution in a city may be correlated with industrial activity in some far off city, due to high-altitude winds which are not known (i.e., hidden variables) to most observers. Events in one part of the globe can have consequences in another, far off part of the globe. A solar flare can have large effects on earth. In effect, such correlations imply that we are dealing with 'holistic' systems where the parts are interrelated, as discussed in Sec. 2.3.2. However, the common sense concept of reality is that if the interactions can be separated far enough, then the correlations should disappear. Thus, if the earth could be moved outside the solar system, there would be no effect on the earth from solar flares. This seemingly harmless assumption of localness is basic to the EPR thought experiment (Secs. 7.3.7, 7.7). Since quantum phenomena show additional correlations beyond local correlations, Einstein felt that there must be hidden variables which bring in these correlations. Hence *Einstein claimed that QM should be deemed incomplete*.

How can we quantify the difference between the correlations among the components of a quantum system, and those of a classical holistic system? Bell's theorem provides an inequality that is obeyed by classical systems, and violated by quantum systems. Bell showed in 1964 that local correlations of the sort claimed by Einstein, Podolsky and Rosen in the EPR paradox (thought experiment) satisfied a certain inequality, while quantum mechanics violated them. Bell, following Bohm, discussed correlations between two entangled spins (see Eq. 7.15) which moved in opposite directions. He considered their detection using a Stern-Gerlach apparatus at each end, looking for spin orientations along the vertical, and along directions at $\pm 2\pi/3$ to the vertical. David Mermin [142], discussing *Quantum mysteries for anyone*, gives an excellent discussion of this. The Bell inequality was generalized by Clauser, Horne, Shimony and Holt (CHSH) [47] in 1969. CHSH provide a more convenient form that we discuss below.

The Bell inequality, or the CHSH inequality is satisfied by classical correlations, and violated by quantum correlations. The origin of this difference can be traced to the nature of quantum (entangled) systems where the state of a system is determined by the superposition of probability amplitudes.

Let us consider two well separated experimentalists, e.g., Bob in the USA and Alice in Orsay, France. A correlated system is prepared, and Alice measures two observables, denoted by A and A', and similarly Bob measures two observables B and B'. As a concrete example, we may think of two correlated photons, since experimental tests [12, 6] on Bell inequalities have mostly used entangled photons. Measurements of the linear polarizations (Sec. 7.2.1, Eq. 7.12) of the light along the polarization directions $|e_1\rangle$ and $|e_2\rangle$ can be made by Alice and also by Bob. We can always choose units of measurement such that the measured values a, b, a', b' are pointer readings which range from -1 to $+1$. Let $P(A,B,a_i,b_i)$ be the probability that Alice and Bob get values a_i, b_i, for the i-th measurement of A, B. In the simplest local picture, $P(A,B,a_i,b_i)$ is of the form $P(A,a_i)P(B,b_i)$ where the probabilities multiply to give the joint probability. We define the correlator between observables A and B as the weighted sum over all pairs of measurements i, j as follows:

$$C(A,B) = \Sigma_{ij} a_i b_j P(A,B,a_i,b_j)\,. \tag{7.32}$$

Clauser *et al.*, considered the pair-wise sum of correlations, named the CHSH-bound B_{chsh}, viz.,

$$B_{chsh} = C(A,B) + C(A',B) - C(A',B') + C(A,B')\,. \tag{7.33}$$

Since a, b, a', b' are restricted to be between -1 to $+1$, one may guess that the mini-maximum value of any $|C|$ is of the order of unity. Hence $|B_{chsh}|$ should be of the same order. We in fact have,

$$-2 \leq B_{chsh} \leq 2\,. \tag{7.34}$$

This result [47] is the CHSH form of the Bell inequality. The single minus sign in Eq. 7.33 can be moved to any term and still the inequalities hold. The CHSH bound can be deduced within a description of local realism, and we give only a brief outline of the method below.

Local realism is understood to mean that all aspects of the system are known, or listed in a set of 'local plans' enumerated by an index λ. The probability that the correlated system is described by the plan with the index λ is denoted by $\rho(\lambda)$. Hence, summing over all possible values of the index,

$$\int d\lambda \rho(\lambda) = 1\,. \tag{7.35}$$

Let $P(A,a,\lambda)$ and $P(B,b,\lambda)$ be the probabilities that measurements of the observables A and B yield the values a, and b for the local plan λ. Similarly, let $P(A,B,a,b,\lambda)$ be the probability that measurements of A, B yield a, b. Since locality is assumed in the plan λ,

$$P(A,B,a,b,\lambda) = P(A,a,\lambda)P(B,b,\lambda)\,. \tag{7.36}$$

This is the probability for the realization λ of the local plan. Summing over the probability distribution of the local plan, Eq. 7.35, we can define the overall probability for the measurements on A and B by

$$P(A,B,a,b) = \int d\lambda \rho(\lambda) P(A,B,a,b,\lambda)\,, \tag{7.37}$$

Eq. 7.32 defines $C(A,B)$ as before. The proof of the CHSH inequality proceeds by keeping λ fixed and looking at

$$\begin{aligned} T_1 &= \Sigma_{ij} a_i P(A, a_i, \lambda)\{b_j P(B, b_j, \lambda) + b'_j P(B', b'_j, \lambda)\}, \\ T_2 &= \Sigma_{ij} a'_i P(A', a'_i, \lambda)\{b_j P(B, b_j, \lambda) - b'_j P(B', b'_j, \lambda)\}. \end{aligned}$$

The absolute values $|\Sigma_i a_i P(A, ai_i, \lambda)|, |\Sigma_j b_j P(B, b_j, \lambda)|$, etc., are demonstrated to be bounded by unity. Hence $|T_1|$ and $|T_2|$ are also bounded by 2. The sum $T_1 + T_2$ is also bounded by 2

$$\begin{aligned} -2 \le T_1 + T_2 \le 2, \\ -2 \le \int d\lambda \rho(\lambda)\{T_1 + T_2\} \le 2. \end{aligned}$$

The last equation is simply the CHSH inequality given in Eq. 7.34.

In dealing with a quantum system we cannot assume the simple product form of the local probabilities, given in Eq. 7.36. Instead, we need the quantum probability $P_q(\tilde{A}, \tilde{B}, a_i, b_i)$, where $\tilde{A}, \tilde{B}$ are operators. Then the quantum correlates are:

$$C_q(\tilde{A}, \tilde{B}) = \Sigma_{ij} a_i b_j P_q(\tilde{A}, \tilde{B}, a_i, b_j). \tag{7.38}$$

Instead of doing the general case, let us consider a case that ties in with experimental tests of the Bell inequalities. The content of the original EPR thought experiment was simplified by Bohm where he considered two electrons entangled *via* their spin states (e.g., as in Eq. 7.15). Experimentally, it is easier to work with polarizing filters and photons, rather than Stern-Gerlach magnets and electrons. Essentially all of the experimental tests on Bell inequalities have used entangled photons.

7.4.1 *Violation of CHSH bounds by entangled photons*

The CHSH-bounds for correlations which obey a localized reality were given in Eq. 7.34, i.e.,

$$-2 \le C(A, B) + C(B, A') - C(A', B') + C(B', A) \le 2. \tag{7.39}$$

The correlators $C(A, B)$ were evaluated on the assumption that a joint Probability $P(A, B, a, b)$ for the observables yielding values a, b is of the product form $P(A, a) \times P(B, b)$. Such a 'local' assumption is not true in QM. We need to calculate the quantum correlator which takes account of the entangled amplitudes characteristic of quantum systems. Entangled systems contain, not only the local plans, but also the interference terms associated with the superpositions of the amplitudes. When the correlator is calculated from the square of the entangled amplitudes, the CHSH bounds are found to be violated.

Let us examine the calculation of the quantum correlator $C_q(\tilde{A}, \tilde{B})$ *via* $P_q(\tilde{A}, \tilde{B}, a, b)$, for the quantum observables A, B now marked with a 'tilde'. As a simple example of a quantum system, we consider two entangled photons, generated using an atomic cascade. Let Alice and Bob, our experimentalists, measure the polarizations in the plane of the unit vectors $|e_1\rangle$ and $|e_2\rangle$. They have the pass axes of their polarizers P_A and P_B at an angle θ. We can indicate the passing or not-passing of a photon by +1 or -1. If bob registers a +1, the photon has passed P_B, and since the two photons in the entangled pair are correlated, the photon coming to Alice's polarizer will also be polarized parallel to Bob's, the probability of it passing through P_A is $\cos^2\theta$. Also, by rotational symmetry about the axis of the photon beam (see Fig. 7.6) we also have:

$$\begin{aligned} P_q(\tilde{A}, \tilde{B}, 1, 1) &= P_q(\tilde{A}, \tilde{B}, -1, -1) = \{(1/\sqrt{2})\cos\theta\}^2, \\ P_q(\tilde{A}, \tilde{B}, 1, -1) &= P_q(\tilde{A}, \tilde{B}, -1, 1) = \{(1/\sqrt{2})\sin\theta\}^2, \\ C_q(\tilde{A}, \tilde{B}) &= \cos^2\theta - \sin^2\theta = \cos(2\theta). \end{aligned}$$

In the CHSH inequality we consider two sets of observations, A, B, and A', B'. Alice does her measurements with the polarizer at $\theta_A = 0, \theta_{A'} = \alpha$ where the angles are measured with respect to the vertical axis. Bob also carries out two measurements, with $\theta_B = \alpha/2, \theta'_B = (3/2)\alpha$. For numerical convenience, we take $\alpha = \pi/4$. This was also the value used in the experimental test of the Bell-CHSH inequalities by Aspect *et al.* [12]. Then, using the result $C_a(\tilde{A},\tilde{B}) = \cos(2\theta)$ it is easily seen that

$$C_a(\tilde{A},\tilde{B}) + C_a(\tilde{A}',\tilde{B}) + C_a(\tilde{A},\tilde{B}') - C_a(\tilde{A}',\tilde{B}') = 2\sqrt{2}.$$

Thus the CHSH bounds are violated. No local plan can capture the correlations contained in the entangled photons whose state is described by, say, Eq. 7.24. The value $2\sqrt{2}$ is sometimes known as the 'Tsirelson's bound', in that QM may violate CHSH upto this value.

The fact that QM violates the Bell-CHSH bounds does not mean that every particle in the Universe is correlated inextricably with every other particle in it. It applies if the particles are entangle. The state of a quantum system may not be describable by a wavefunction. Then we use a description in terms of a mixture (Eq. 7.25), and consider decoherence phenomena that go hand in hand.

7.4.1.1 *Necessary and sufficient conditions for quantum correlations*

The CHSH conditons lay down the bounds for *classical local* correlations. One may look for similar bounds for EPR-type quantum correlations.

That is, given two observers, each with two possible local experiments A_i, B_i, $i = 1, 2$, and outcomes a_i, b_i in the range $[-1, 1]$, and correlators $C(A_i, B_j)$ defined as before, can we establish bounds for all the local and non-local correlations that are allowed by QM.? In fact, such inequalities have been established by Tsirelson in 1993. He gives the necessary and sufficient conditions (see Ref. [40]) in terms of eight inequalities. They are written compactly as:

$$\begin{aligned}&|\arcsin C(A_1,B_1) + \arcsin C(A_1,B_2) + \arcsin C(A_2,B_1) + \arcsin C(A_2,B_2)\\&-2\arcsin C(A_i,B_j)| \leq \pi, \quad i,j = 1-2.\end{aligned}$$

These correlations between two entangles particles persist, even for very large space or time separations between the two components, as long as the experimental boundary conditions (which isolate the two particles from the decoherence effects of the environment) are strictly maintained.

7.5 The 'collapse of the wavefunction' and measurement

Physics is a discipline where predictions of a theory are confronted with experiments. Unlike Thomas Kuhn who would wait for a 'critical mass of deviations', physicists would consider it a momentous event to find even one unexplainable quantum phenomenon, or even the smallest deviation from predictions. Thus *measurement* is central to physics. In fact, quantum mechanics is one of the few disciplines where the nature of measurement is scrutinized in utter detail.

A measurement involves carrying out certain operations on a carefully prepared system. These practical operations have their abstract mathematical counterparts in the theory, known as Hermitian operators. Quite abstract relativistic properties like the 'spin' of an electron are measured by replacing such properties by simpler properties that result from them. Thus electron spin is measured as a macroscopic *distance*, by sending the electrons through a suitable magnetic field, in a 'Stern-Gerlach' apparatus. The 'up-spin' electrons move upwards in the magnetic field, while the down-spin electrons move down. The deviation (i.e., the distance) from the horizontal identifies the spin. A magnetic field can be used to measure the ratio of charge to mass (e/m), as done by J. J. Thompson in his discovery of the electron (Sec. 6.3.1). The earth's magnetic field is in fact used as a natural 'detector' in the study of comic rays showering the earth.

Nevertheless, an almost mystic character has been given to the quantum measurement process by many writers who have used it as an opening for metaphysical discussions about quantum reality. When we measure the state of the spin of a particle in the vertical direction (z-axis) and deem it to have the value +1/2, the quantum theory says that *we have rendered* the value of the spin component in, say the x-direction, totally uncertain. In effect the quantum state corresponding to the z-component has been selected while the wavefunction that existed before our intervention 'collapsed' irrevocably. It may be claimed that when we choose to measure the momentum $\vec{p}$ of an electron, we 'put it' into a plane-wave state, i.e., momentum eigenstate $\sim \exp(i\vec{p}\cdot\vec{r})$, while destroying the wavefunction that existed before. It is argued that *we* have 'created' the reality that we observe!

This view is extended into a subjectivist position giving a direct role to the 'consciousness' of the observer. A most distinguished exponent of this view was Eugene Wigner [228]. Since the system under observation (SUS) as well as the detector both have to obey QM, an entanglement of the wavefunction of the SUS and the detector is claimed to exist. Extending the application of QM further, we need to include the observer also in this composite state denoted by Ψ.

Let us examine such composite states involving the SUS+detectot+observer. If the eigenstates of the SUS, the detector D, and the observer O are denoted by ψ^i_{sus}, ψ^j_D and ψ^k_O, where i etc. are state indices, we may have superpositions of the form $a\psi^i_{sus}+b\psi^j_D+c\psi^k_O$ which are then used in suitable product states to form the entangled composite state Ψ, as done in the molecular-orbital theory of chemical bonding (where one-electron states are used). Alternatively, we may use sums of product states like $\Psi=\sum_{i,j,k}C_iC_jC_k\psi^i_{sus}\psi^j_D\psi^k_O$. Here entanglement is done *via* product states and mixing coefficients C_i, C_j, C_k, as in the 'valance-bond' theory. Both methods lead to equivalent results if complete sets of functions are used. Rigorously, the SUS, D and O eigenstates should all be expanded in a complete orthogonal basis etc., to correctly deal with overlap effects. The detector, and the observer have large densities of states (that is, j,k are almost continuous indices), with essentially 'random phases' associated with the eigenstates.

The probability distribution $|\Psi|^2$ contains numerous 'off-diagonal' terms like

$$|C_i^*C_j^*C_k^*C_{i'}C_{j'}C_{k'}|+h.c.\quad i\neq i',\, j\neq j',\, k\neq k'. \tag{7.40}$$

Clearly, when $i\neq i'$, $j\neq j'$, and $k\neq k'$ the random phase factors in them become rapidly oscillating terms that average to zero. These random phase factors are introduced by the myriads of small uncontrollable perturbations, thermal and other motions of the observers and the detector that arise in performing the measurements. Hence the probability distribution reduces to a sum of terms involving products of diagonal terms of the form $P_S^i \times P_D^j \times P_O^k$ where P_S^i is the probability of finding the system under observation in the state i, and similarly for P_D^j, P_O^k. The latter are essentially in their thermal-equilibrium state, and this returns us to the classical picture of measurement *without* entanglement.

Almost mystically, we are told that 'consciousness', a component of O, is not subject to QM, and that consciousness breaks this chain of entanglements (as well as the unitary evolution Ψ) by selecting a particular property that is to be measured. This mental act, claimed to be outside QM arrests the unitary evolution of Ψ and create the observed reality.

The proof of the proposition that consciousness does not obey QM is best left as an exercise to some savvy subjectivist. The assumption that the SUS must necessarily form a composite Ψ with the observer and the detector cannot be justified except at $T = 0$ where the de Broglie thermal wavelengths λ_T of macroscopic bodies are large enough to permit such interaction with other bodies. Otherwise, the λ_T values of macroscopic objects even at micro-Kelvin temperatures turn out to be smaller than the radius of a proton (see Eq. 7.56). If λ_T were large enough to allow superpositions, the descriptions of the coupling of the observer to the detector, and the detector to the SUS have to include a realistic theory of hybrid (quantum + classical) systems at finite-T.

Wigner and others do not discuss how the composite Ψ is formed. If a small number of atoms or electrons in the SUS interacts with D and O, then the λ_T of the macroscopic body is irrelevant. However, objects placed together in a room do not form superpositions or entanglements on their own. We noted that the ability to form superpositions is mandatory to forming entanglements. Also, we noted that naturally occurring, common entanglements between atoms are nothing but spontaneous chemical reactions. However, Wigner and others are not referring to chemical reactions between the SUS, D and the observer. It is not clear if they even mean the need to antisymmetrize the wavefunction Ψ with respect to electron coordinates. However, from Fig. 7.1, at normal densities (of D and O), the lengthscale r_s is of the order of atomic dimensions, and even wavefunction antisymemtrization between, e.g., D and O is not operative.

Let us examine how we actually make measurements, either on a whole system, or on representative samples drawn from the given system.

7.5.0.2 *The detector as a set of metastable atoms*

In all the accepted experimental observations of physics, the system under study (which may be a *test-sample* of a larger N-particle system), the detector and the experimentalist appear to interact classically. The detector gives pointer readings read by the experimentalist. Here biochemical processes of vision and cognition are involved. The interface between the quantum mechanical evolution of the SUS and the classical world occurs in the detector *where a macroscopic collection of quantum systems (e.g., a gas of excited atoms) all respond to that quantum property of the SUS that we wish to determine*. The test-sample gets transformed in the process and may be useless for further measurements.

How do we detect an electron when it is going past a given slit? We can position a set of excited metastable atoms near the slit.

A hydrogen-like atom with an electron in the $2s$ state decays to the $1s$ ground state only very slowly since the transition is symmetry forbidden (see Sec. 6.10.3). In the Feynman diagram of Fig. 6.8 we showed the decay of a $2p$ electron to the $1s$ ground state which is fully allowed by symmetry). However, if a 'passer-by' electron were to transit near such an atom, it gets distorted by the Coulomb interaction. The $2s$ symmetry is broken by the admixture of $2p$ states and the transition to $1s$ can occur.

The presence of a 'passer-by' electron is enough to perturb many metastable atoms and the electron is immediately detected by the emission of photons from many of the excited H-atoms forming the detector. In practice, experimentalists use other systems than metastable H-atoms. However, they are a convenient model for theoretical discussions [211]. Wilson cloud chambers also detect 'passer-by' elementary particles using the condensation of supersaturated water vapour induced by the passing particle. In fact, the use of a super bubble chamber known as *Gargamelle* (Fig. 2.5) in high-energy physics was discussed in Sec. 2.4.5.

The processes outside the combined system (SUS+Detector) are not important because the signal from the property to be measured has been enormously amplified compared to other possible signals. A good measuring apparatus is like a well tuned radio which latches on to a specific signal rendering the background signals unimportant. In particular, any number of observers can make the observation. The human observers who take note of the relevant classical property of the combined system (SUS+D) may not even be present at the moment of measurement but run the experiment from other cities *via* the internet. Hence 'the existence of an entanglement of the SUS+D+observer' is a hypothesis with little or no basis.

Consider a two-slit experiment where the transmitted parts of an electron wave proceed through the slits A, B on a vertical wall and impinges on a vertical screen at $z = d$. The wavefunction components transmitted by A, and B may be written as $\phi_A(\vec{r})$ and $\phi_B(\vec{r})$. We do not explicitly show their time dependence. These two functions are not perfectly orthogonal but they are sufficiently independent of each other for our purpose, since the total wavefunction $\psi(\vec{r}) \simeq C_A\phi(\vec{r})$ for $|\vec{r}-\vec{r}_A| \ll |\vec{r}-\vec{r}_B|$, and $\simeq C_B\phi_B(\vec{r})$ for $|\vec{r}-\vec{r}_B| \ll |\vec{r}-\vec{r}_A|$. For arbitrary $\vec{r}$ we use the superposition state:

$$\psi(\vec{r}) = C_A\phi_A(\vec{r}) + C_B\phi_B(\vec{r}). \tag{7.41}$$

If the incident wave is symmetric about the horizontal mid line between the two identical slits, then $C_A = C_B = 1/\sqrt{2}$. We place a detector near the slit A, and consider the SUS+detector, inclusive of the wavefunction of the detector, denoted by $\xi(r_D)$ where r_D is a variable centered on the detector (which may be a set of metastable atoms). As long as there is no interaction, the wavefunction of the combined system is a simple separable product of the amplitudes of the SUS and the detector, giving $\psi(r)\xi(r_D)$.

However, when the detector at A interacts with the electron and puts it definitively into $\phi'_A(r)$, at least one metastable atom gets affected and reveals the electron at A by emitting a photon. Other metastable atoms also rapidly emit photons. From then on the detector becomes inactive and there are no further interactions. On completion of this process *the electron wavefunction is said to have collapsed* into the selection $\phi'_A(r)$. Thus

$$C_A\phi_A(\vec{r})+C_B\phi_B(\vec{r}) \rightarrow \phi'_A(\vec{r}). \tag{7.42}$$

This is an irreversible process, the electron wave which had the possibility to go through two slits has 'emerged' from just the single slit A. We say that the electron has been prepared in the state $\phi'_A(\vec{r})$.

What happened to the component $\phi_B(\vec{r})$ of the original wavefunction? Those who believe in 'many-worlds interpretations', or 'many-minds interpretations', of quantum states may claim that the possuble branches of the wavefunction are in parallel universes and/or observer-minds.

Most physicists would argue that the boundary conditions and interactions in the system changed when the electron entered the detector, and the new setup filtered out various components and selected that part of the electron wavefunction which is compatible with the new boundary conditions of the system. Mathematically, since the Schrödinger equation is a differential equation, we need its solution using the new boundary conditions. If the wavefunction is considered a part of physical reality, this leads to no difficulty since the experimentalist is modifying the SUS, and hence the wavefunction gets accordingly modified.

In Bohmian mechanics there is no actual 'collapse' of the wavefunction. The wavefunction as well as particle trajectories exist as an external reality. A transmitted-particle trajectory goes through only one or other of the slits. The trajectory can be calculated from Newton's equations where the particle is subject to Bohm's quantum potential $Q(\vec{r},t)$ and the external potential, with the initial position $(\vec{r},t_0)$ subject to the probability distribution $P_0 = |\psi(r,t_0)|^2$. The specific observation of the electron by the detector removes the uncertainty in the initial conditions (i.e., being known only as a probability), and defines the path of the electron. If the experiment were carried out a large number of times (without a detector positioned at one of the slits), with the initial positions of the particles subject to the probability distribution P_0, different particles would arrive on the screen at $\vec{r}_s = (x_s, y_s, z_s = d)$ at times t_s. Once a large number N_s of particles have reached the screen, then

$$\sum_{s=1}^{s=N_s} |\psi(\vec{r}_s,t_s)|^2/N_s \rightarrow \text{interference pattern} \tag{7.43}$$

The interference pattern appears in the limit of a large N_s, with both slits open.

It is usual to formally discuss the measurement process according to the formulation by von Neumann in 1931, or one of its later forms ([25]). Any possible measurement by the detector corresponds to the action of a Hermitian operator $\hat{D}$. It has a complete set of eigenfunctions $\xi_n(r_D)$ with eigenvalues λ_n. Initially, the electron system evolves in time according to the wavefunction $\psi(\vec{r})$ independently of the detector. Then the detector interacts very briefly but strongly with the system. The detector is often modeled as a system with two canonical variables $\hat{q}_D$ (e.g., position) and $\hat{D}$ (e.g., momentum). They are operators such that their commutator is $i\hbar$, as required by the uncertainty principle. Thus the position of an atom, z, in a magnetic field B of a Stern-Gerlach apparatus (to determine its spin) may serve as $\hat{q}_D$, the pointer-position variable. The quantum property to be measured is the spin σ_z. To be more general we denoted it by the operator $\hat{A}_s$ of the 'system'. The interaction is impulsive, i.e., lasts for a short time Δt centered around the moment t_0, as indicated by $\approx \delta(t-t_0)$. This time is claimed to be so short that we do not worry about the usual time evolution of the system by itself. Thus the new term in the Hamiltonian controlling the time evolution of the combined system inclusive of the measuring instrument, during the short time Δt is

$$\hat{H}_I = -g\delta(t-t_0)\hat{q}_D\hat{A}_s \,. \tag{7.44}$$

Here g is a measure of the coupling between the system and the detector. In the example of the Stern-Gerlach apparatus, g would include the magnetic field B and universal constants like the Bohr magneton $e\hbar/(2mc)$. The minus sign is the standard usage in the literature. The coupling establishes a correlation between the operator of the detector coordinate $\hat{q}_D$ and the spin or other property of the atom given by the operator $\hat{A}_s$. The magnetic field exerts a force on the atom upwards (or downwards) if its spin is 'up (or down)', and thus the momentum operator $\hat{p}_D$ (conjugate with the position $\hat{q}_D$) can be used as a pointer reading proportional to the effect of $\hat{A}_s$. In actual Stern-Gerlach experiments, we allow the momentum to carry the particle during a time τ to a distance and measure the distance d.

The strong interaction does not modify the specific property $\hat{A}_s$ of the system as $\hat{H}_I$ commutes with the operator $\hat{A}_s$. On the other hand, all non-commuting properties will be wildly disturbed by the measurement process. The wavefunction of the combined system (CS) can be expanded in the complete set of eigenfunctions of the operator D of the detector.

$$\psi_{cs} = \Sigma_n \langle \psi(r) | \xi_n(r_D) \rangle \xi_n(r_D) = \Sigma_n D_n \xi_n(r_D) \,. \tag{7.45}$$

The detector state $\xi_n(r_D)$ is identified by its eigenvalue λ_n (i.e., a pointer reading, e.g., a wavelength, or 'up'=$\hbar/2$, 'down'= $-\hbar/2$, property). It will be found with the probability $P_n = |D_n|^2$. If indeed the pointer reading λ_n were found from the experiment, then we say that the wavefunction of the quantum system has *collapsed* definitely into the state $\xi_n(r)$, and we conclude that its fractional participation in the original SUS was P_n while D_n is the coefficient of the amplitude.

As already noted, Eugene Wigner [228] and a few others have drawn the conclusion that the definitive 'coming into being', (or ontology) of the electron state involved the intervention of the observer, and depends directly on a 'consciousness'. According to this view, the quantum states of the detector and the system under observation remain entangled with the consciousness of the experimentalists, assistants, "Wigner's friends", etc., making observations. So, a Stern-Gerlach experiment involves a beam of electrons, a laboratory magnet and a photographic plate in a suitable enclosure, *as well as the minds of experimentalists and onlookers* in a state of entanglement. If such entanglements occurs, the random movements of the experimentalists and others (who are also observing the rest of the external world and hence entangled with it, and randomly moving about) would introduce *incoherent* phase relations into the components of the wave function (or, off-diagonal terms like in Eq. 7.40) containing the coordinates of the experimentalists; such cross terms would average out to zero. Hence even if we take Wigner's point of view at face value, the net quantum effect (i.e., entanglements, superpositions etc.) of all the 'conscious minds' would average out.

If the detector is coupled with the SUS impulsively in a short time Δt, varying energy-uncertainty ($\Delta E \sim \hbar/\Delta t$) phase factors of the order of $i\hbar(\Delta E)t$ would appear in all wavefunction components. These cancel out in the diagonal terms of $|\Psi|^2$, but act destructively in the off-diagonal terms. This erases the quantum superposition and entanglement effects assumed to hold by the subjectivists.

We return to further difficulties in Wigner's point of view in our discussion of the 'mindfull universe' (Sec. 7.6.2), the cat paradox, the thermal de Broglie wavelength of macroscopic bodies and related issues.

7.5.1 *Measurement as a sampling process*

It is instructive to compare the above discussion of measurement in quantum systems with measurements in classical systems as well as in everyday life. Consider a chemist who has to report on the phosphate content of water in a city reservoir. The chemist takes several samples of water, analyzes for phosphates and reports his conclusions. In the process, the samples are irrevocably destroyed. A chemical detection process was used to 'collapse' the phosphate ions into precipitates that were filtered into a Gooch crucible, dried, weighed and the data reported.

The measurement problem in quantum systems is usually posed differently. Most academic discussions are about a system of one or two particles or spins. They are entangled and we have to deal with the whole system. No sampling is possible. The measurement destroys the original system in that it collapses into a specific eigenstate (set by the chosen detector).

However, we can get away from the academic two-spin problems of textbooks and consider, say, an experiment on a gas of atoms whose spin orientation is to be determined. We are dealing with an N-body system, where N is of the order of the Avogadro number ($\sim 10^{23}$), and any one atom is found, on the average in a sphere of radius r_s with the density ρ of atoms such that $\rho = 4\pi r_s^3/3$ with $r_s \sim$ atomic units. A sample of this gas, containing n_s atoms at the density ρ, where n_s is large, is sent through the Stern-Gerlach apparatus. This is done for many samples and an average value for the spin-polarization of the atoms in the gas is obtained. In the process, the spin wavefunctions of *all the samples* have 'collapsed'. But the main body of the atomic gas is still intact. This is like the phosphate-precipitation of the test samples of the analytical chemist. Although the test-samples have 'collapsed', the main body of water still exists in the reservoir.

Here the objectivity of the measurement hinges on the possibility (or impossibility) of taking samples of the quantum system. In Fig. 7.1 we displayed how two electrons in a spin-polarized entangled system of N electrons at a density specified by r_s remain correlated as a function of their separation. Although the N spins makes up *an entangled system*, removing a sample n_s of electrons while keeping ρ fixed does not change the physical properties of the main system. This is because the typical correlation lengths of the system are much smaller than the size of the total system or the size of the samples. Thus we see no difficulty in objectively studying the physics of the spin-polarized system of particles by looking at suitable samples of the system. Similarly, other properties of N-particle systems with $N \to \infty$ can be studied by taking small samples n_s, keeping relevant state variables (e.g., density) unchanged. This is of course not possible if the system under study has the same length scales as the correlation lengths inherent to the system. This is so if the study is restricted to two correlated spins, or two entangled photons, and so on, as in most textbook discussions of measurement theory.

7.5.2 *Weak measurement and post selection*

The measurement theory stemming from von Neumann (1932) was given a careful review by Bohm [25] in Part VI of his book on the Quantum Theory published in 1951. This approach is sometimes known as a theory of 'strong measurement' or 'crisp measurement'. In an 'ideal' von Neumann measurement, the state of the system after the measurement becomes an eigenstate of the measured observable, irrespective of the original state of the system. All non-commuting properties are uncontrollably disturbed. Aharanov, Albert and Vaidman [2] examined an approach to measurement where the interaction between the system (S) and the detector D is very weak. They use a system prepared in the past (pre-selection), and also a post-selection — i.e., two state vectors — to discuss measurement.

They consider an interaction Hamiltonian similar to Eq. 7.44 discussed previously, viz.,

$$\hat{H}_I = -g\delta(t-t_0)\vec{\hat{q}}_D\hat{A}_S\,. \tag{7.46}$$

For weak measurements, the coupling g between the system and the detector (apparatus) is vanishingly weak. Here $\hat{q}_D$ is a position coordinate and its conjugate variable is well defined and would become proportional to $\hat{A}_S$ when measured. Hence it is effectively proportional to a meter reading. This would in effect be a measurement similar to a Stern-Gerlach measurement where the strength of the magnetic field becomes vanishingly small. Thus no observables would be significantly disturbed, and even the operator $\hat{A}_S$ which is to be measured would fail to register a useful value. In effect, the meter reading would at best be a small perturbation no bigger than the experimental noise. However, Aharanov, Albert and Vaidman (AAV) propose a prescription for getting something out of a large number of such measurements. They use the words 'pres-selection' and 'post-selection' where what is meant is that the system is put into an eigenstate of the operator corresponding to a selected property either before, or after the experiment. AAV claim that a large number of such weak measurements on the system, prepared in the initial state ϕ_{in}, pre-selected with respect to the property $\hat{A}_S$, followed by 'post selection' of a final state ψ_f at a later time gives a 'weak value' of the property $\hat{A}_S$ such that:

$$A_w = \langle\psi_f|\hat{A}_s|\phi_{in}\rangle/\langle\psi_f|\phi_{in}\rangle\,. \tag{7.47}$$

It is claimed by AAV that in the limit where the shift (i.e, change) in the position q under the interaction tends to zero, the real part of A_w can be identified as the weak value of the property A. In this sense, if the post-selection is itself ϕ_{in}, then the usual strong expectation value of A is simply an average of weak values rather than an average of eigenvalues. AAV argue that a measurement of a component of a spin 1/2 particle may perfectly well yield a value of 100 within the concept of weak measurements with post-selection. Aharonov [3] would claim that *the weak value of an observable need not be an eigenvalue; indeed, it need not be any classically allowed value. Weak values offer ⋯ a world in which particles travel faster than light, carry unbounded spin, and have negative kinetic energy.*

The subject is at present very controversial. Some authors have argued that the theory of weak measurement is partly a result of incorrect mathematics, and ambiguous even when carefully formulated because explicit examples in finite-dimensional spaces can be formulated where arbitrary results can be obtained [154]. Nori *et al.* [10] have argued that 'weak measurements' which yield 100 for spin-component measurements involve simple misinterpretations. Other authors have claimed that the concept of weak measurement enables a clearer understanding of certain experiments and conceptual puzzles like 'Hardy's paradox' [135]. The latter discusses the 'virtually non-interacting' interaction between a particle and an antiparticle with no annihilation of the pair. Other theorists are critical of these claims, as well as the formulation of the 'paradox'.

7.5.3 *Decoherence of correlations in quantum systems*

Decoherence is the process which destroys the special correlations that exist in a quantum system, and makes it become like any 'non-quantum' system. The wavefunction of a quantum system endows it with coherent phase relations among its different parts, like in an orchestra where different instruments stay in time and remain in tune with each other. The other extreme is where each instrument 'does its own thing' uncorrelated with each other. Thus, a superpositions of two quantum states subject to decoherence looses their superposition.

Two entangled quantum systems become disentangled under decoherence and forms a simple mixture of the two. Thus an entanglement of up-spin atoms and down-spin atoms becomes a simple 50% mixture of up-spin atoms, and down-spin atoms. When chemical bonds, e.g., the bond in H_2 (which is an entanglement of two electrons from two H atoms) becomes disentangled, we have a gas of individual hydrogen atoms. A strong chemical bond would undergo decoherence (i.e., break up) when placed in an environment containing ultraviolet radiation sufficiently energetic to break the bond. However, we normally do not use the word decoherence for such energy demanding processes. It is used mainly for the decay of fragile superpositions or entanglements of quantum states that can be made with little change in energy e.g., between two similar photons which are different (in say, polarization), or two electrons of differing spin. Such systems are usually not energy stabilized, and their liaison is mainly a matter of phase correlations. Hence, if they could be made, they can be conveniently manipulated with low-energy probes in small devices. It is the destruction of such phase correlations when subject to extraneous interactions that is of interest here. Decay of such coherent quantum systems (i.e., decoherence) has also become of great interest as a practical issue in developing quantum computers where it is important to strongly suppress the decoherence of qbits.

Decoherence arises from the interaction of a quantum system with its surroundings. Interactions destroy the special phase relations among interacting quantum states. The surroundings are often modeled by a 'heat bath'.

The system itself is defined *via* its boundary conditions which isolates it from the outside world. But this is never perfect, and we consider a bigger system somewhat as we did in discussing measurement. The enlarged 'total system' has three components. They are (i) the quantum system, specified quantum mechanically by the Hamiltonian $\hat{H}_s$, (ii) the external world (usually termed the 'bath', $\hat{H}_b$), and (iii) the interaction between the system and the bath, *viz.*, $\hat{H}_{sb}$.

$$\hat{H}_T = \hat{H}_s + \hat{H}_{sb} + H_b\,. \tag{7.48}$$

The external world, represented by H_b is a classical system or 'heat bath' with a quasi-continuum of energy levels. H_b is a classical Hamiltonian.

The interaction between the classical energy states of the bath, and the quantum states of the system needs to be treated using QM, just as with the system Hamiltonian $\hat{H}_s$. Hence there are two deep problems that have been largely ignored by writers dealing with decoherence. These are: (a) formulation of a quantum theory of finite-temperature systems, (b) formulation of a consistent theory of *hybrid systems* containing quantum as well as classical subsystems, since the heat bath and even the boundaries of an experimental set up are 'classical components'.

A review of thermo-field quantum theories has been given by Henning [103], while Andre Heslot [104] has reviewed the quantum-hybrid connection, although without a discussion of thermal fields.

The thermal de Broglie wavelength of a particle of mass m is given by $\lambda = h/\sqrt{(3mT)}$ where the temperature T is in energy units, assuming a classical thermal distribution. In other systems, e.g., a degenerate Fermi gas, the corresponding Fermi energy is a measure of T. Hence all systems are nominally quantum systems at $T = 0$. Thus *classicality* can be associated with finite-T, with $\lambda \ll r_s$, where r_s is the average Wigner-Seitz radius or other suitable correlation length. Thus any theory of a hybrid system should be formulated as a quantum theory at finite-T. Matsubara's imaginary-time technique, the Martin-Schwinger-

Kedysh contour technique, and Umezawa's thermofield dynamics, and also finite-T density functional theory (DFT) are methods available for treating finite-T hybrid systems. However, current decoherence theories ignore finite-T effects.

Approaches like Bohm's theory, or DFT using effective potentials and density functionals (instead of wavefunctions) can treat finite-T problems seamlessly. At finite-T, the wavefunction $\psi(T)$ is a sum of eigenstates with mixing coefficients such that the correct thermal distribution is obtained. The Bohm potential $Q(T)$ is related to the modulus $|\psi(T)|$ as usual. In DFT we need finite-T exchange-correlation potentials.

Following Bohm, we can write each of the quantum Hamiltonians as a sum of a classical Hamiltonian H^c and a quantum potential Q. Hence we have

$$H_T = H_s^c + Q_s + H_{sb}^c + Q_{sb} + H_b\,. \tag{7.49}$$

It is the quantum potential Q_s of the isolated system which confers on it any quantum properties (e.g., contextuality and non-locality etc). However, external interactions bring in a new, random quantum potential Q_{sb} which may be large enough to erase the quantum effects if $\langle Q_s + Q_{bs}\rangle$ averages out to zero within some characteristic time and within some relevant coherence length. The interaction of the system with the 'bath', usually occurs *via* absorption and emission of phonons (e.g., thermal vibrations or elastic waves), spin fluctuations (e.g., by coupling to the nuclear magnetic moments of the ambient atoms in the surroundings), or electromagnetic processes (photons). Thus the decoherence of entangled electronic states of quantum dots due to the coupling to nuclear spins has been a serious difficulty in using such states as qbits of proposed quantum computers.

In dealing with an entangled pair of photons, or electrons, as in an EPR experiment, when the separation of the pair increases, the quantum potential from the coupling of the system to bath, Q_{sb} becomes progressively larger. The quantum potential of the system itself, viz., Q_s (that mimic quantum effects) remains scale invariant (e.g., the Pauli potential, see Fig. 7.1) while the boundary effects become bigger as the particle separation is increased. The Bohmian quantum potential for entangled electrons in a quantum well is the kinetic energy which is invariant for different particle separations.

For example, if we take a pair of entangled electrons at x_1,x_2 in a 1-D well of length a, the Bohm quantum potential is the kinetic energy (KE), i.e., $Q(x_1,x_2) = k^2/2$, where k is the relative momentum of the entangled pair (see Eq. 8.10). If the separation s of the pair is increased, the KE is conserved and $Q(s)$ remains constant. Hence these quantum correlations persist irrespective of the pair-separation. The possibility of random intrusions is likely to be proportional to some power of s whose maximum value is a. For instance, (even if there is no entanglement assumed in the y,z directions), the fabrication of the quantum well has to be free of impurities in the growth direction x as well as in the $y-z$ plane. Hence decoherence occurs unless the experimentalist carrying out an EPR-type experiment takes special precautions in fabrication and in keeping the entangled pair isolated from the outside 'bath' (i.e., suitably controlling Q_{sb}). These same considerations are true with respect to more exotic entangled systems discussed in many quantum paradoxes.

If the system Hamiltonian H_s is simple enough (e.g, that of a qbit), then we have a two level system, with levels $|1\rangle, |2\rangle$. Such a system is described by the wavefunction ϕ and the density $|\phi|^2$. Alternatively, 1 and 2 may be two parts of an entangled system, and then $x = (x_1, x_2)$ is a composite.

$$\psi_I(x) = c_1\phi_1(x) + c_2\phi_2(x), \ \langle 1|1\rangle = \langle 2|2\rangle = 1 \tag{7.50}$$

$$|\psi_I(x)|^2 = |c_1|^2|\phi_1(x)|^2 + |c_2|^2|\phi_2(x)|^2 + P_{12} \tag{7.51}$$

$$P_{12}(x) = c_1^* c_2 \phi_1^*(x)\phi_2(x) + h.c. \tag{7.52}$$

The interference terms, or terms associated with the existence of a superposition (or an entanglement) are in the cross term $P_{12}(x)$. If this is integrated over x, the orthogonality $\langle 1|2\rangle = 0$ comes into play. P_{ii} are the diagonal terms, while $P_{i\neq j}$ are the 'off-diagonal' terms that produce interference. They carry the quantum phase-coherence in the system. It is their decay which is most paramount to decoherence.

P_{ij} should not be confused with the density matrix ρ_{ij}. (see Sec. 7.3.5 regarding pure and mixed states). If all possible pure states of the system are labeled by j, then the *density matrix* ρ of the pure state ψ_I is simply a matrix δ_{jI} with just one element, and this is sometimes written as $\rho = |\psi\rangle\langle\psi|$. In dealing with a density matrix, we are dealing with an 'ensemble' of states. If every member of the ensemble is in the state ψ_I, i.e., $\psi_I, I = 0$ say, then we are dealing with a pure state. A superposition of an *up* spin and a *down* spin is a pure state with $|c_1|^2 = |c_2|^2 = 1/2$. This is represented on the Bloch sphere (see Fig. 7.2) as an arrow in the y-direction. However, we can have a 50% mixture of up-spins and down-spins, and they are represented as an equal mixture of random Bloch spheres with the arrow pointing up, or down. The statistical density matrix is most useful in dealing with such mixed states. The pure state (superposition) can be distinguished from the mixed state by, say, their different behaviours under a $\pi/2$ optical pulse.

The superposition state $(|0\rangle + |1\rangle)/\sqrt{2}$ is written as $\psi(\pi/2, 0)$ because it is produced by the action of a $\pi/2$ pulse on the 'down-spin' state $|0\rangle$. If another optical pulse performs a $-\pi/2$ rotation about the y-axis (Fig. 7.2), i.e., $R_y(-\pi/2)\psi(\pi/2, 0)$ we get back $|0\rangle$ with unit probability on readout. If we apply the same R_y to the mixture, we get 0.5 probability for finding *either* the 0, or the 1 state on read out, thus distinguishing the pure state from the mixture.

In a system of mixed states, other states $I > 0$ may be fractionally occupied. If a pure-state ensemble were connected to a thermal bath, the system thermalizes to give such fractional occupations, and we have a 'mixed-state' ensemble. Superposition states like ψ_I would loose their 'off-diagonal' interference terms P_{12} by decoherence and reduce to mixtures. The decay of the diagonal terms to a thermal distribution, i.e., $|c_i|^2 \to \exp(-\beta E_i)$, where β is an inverse temperature, is associated with a thermalization time T_1. The decay of the off-diagonal terms to zero is associated with a time T_2 which is known as the dephasing time, or decoherence time. In dealing with a more general system than a two-level system, the different diagonal terms ρ_{ii} would define a series of relaxation times. The shortest of these would be the analogue of T_1, Similarly, there would be many dephasing times associated with the many off-diagonal terms ρ_{ij}, with $i \neq j$. These would be the analogues of T_2.

One method of following the decay of a mixed quantum state, given as a density matrix $\rho(t)$ would be to look at its *quantum entropy*. The quantum entropy $S(t)$ is defined as

$$S = -\Sigma_i \rho_{ii} \log \rho_{ii} \,. \tag{7.53}$$

Here we have set the Boltzmann constant k_B to unity in energy units. If the system is a pure quantum mechanical (phase coherent) state I, $\rho_{ii} = \delta_{iI}$, and the the quantum entropy is zero.

Any deviation from a pure quantum state due to decoherence leads to an increase in the quantum entropy. Since increase of entropy is associated with loss of information (Sec. 6.9), decoherence may be viewed as a leaking of information from the quantum state to the environment (or bath) containing the quantum system. Another measure used to keep track of decoherence is the *quantum fidelity*. It is the overlap between the actual density matrix and the pure-state reference density matrix, averaged over all the system states.

7.5.4 *Decoherence by localization or quantum gravity*

A number of authors have looked for 'intrinsic decoherence mechanisms' applicable to all quantum systems in general. Such mechanisms ensure the collapse of superpositions of macroscopic extent. If such intrinsic decoherence mechanisms exist, then the EPR paradox could vanish depending on the length scales used in the decoherence mechanisms, and Einstein's 'intuitive ideas of physical reality' would hold. The introduction of non-linear dynamical decoherence terms to the Schrödinger equation (to attenuate macroscopic entanglement) has in fact been the path taken by Ghirardi, Rimini and Weber (GRW), in their 'dynamical-reduction model', or 'spontaneous localization theory'. GRW supplement the Schrödinger equation with random spontaneous localization events to destroy phase coherence. The average effect of the 'spontaneous localization' is negligible for a system with a few particles, but rapidly becomes dominant for a macroscopic number of particles, rapidly distroying any superpositions.

We point out in Eq. 7.56 that finite-temperature effects are sufficient to do this for macroscopic systems, since their wave nature is controlled by their thermal de Broglie wavelengths. These are of the order of micro-femtometers for 1kg objects even at 1 Kelvin. Hence such decoherence theories are needed strictly only at $T = 0$. The GRW model will be discussed in Sec. 8.5.

Penrose has proposed an intrinsic decoherence mechanism using general-relativity [160, 161]. Even a completely isolated entangled system collapses due to *objective reduction* (OR) related to features of fundamental spacetime geometry. We briefly discuss Penrose's gravitational 'objective reduction' model.

7.5.4.1 *Penrose's 'objective reduction' mechanism*

The objective reduction mechanism uses various *assumptions about quantum gravity* that cannot be fully established without a proper theory of quantum gravity. Here we follow the arguments presented by Penrose [161].

Penrose exploits the general relativity notion that mass is associated with spacetime curvature. He also considers that quantum superposition involves an actual redistribution of mass according to the probability distribution $\psi|^2$. This is considered to cause simultaneous spacetime curvatures in opposite directions,

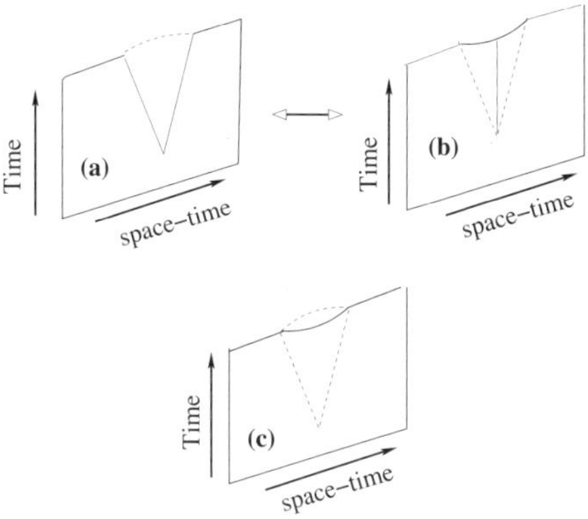

Fig. 7.7 One dimensional quantum states evolving in time are shown in (a) and (b). These two states, on interaction become entangled and form the state (c). Penrose's objective reduction (OR) model claims that quantum states with a space-time extension possess spacetime curvatures associated with their mass distribution (schematically shown as conical distortions in the figure). Thus the entangled state is associated with a 'spacetime bubble' which spontaneously collapses after a time t, inversely proportional to the gravitational self energy ΔE. Thus entangled states which become big undergo state reduction by 'self-collapse'.

causing 'bubbles', or separations in spacetime, as depicted in Fig. 7.7. The bubbles are deemed unstable. When a critical 'objective degree of separation' is reached, instantaneous reduction to non-superposed states is presumed to occur. This process is called 'objective reduction' (OR). Objective reductions are therefore events which reconfigure the distortions in spacetime geometry arising from quantum processes.

The critical amount of spacetime separation causing Penrose's OR is related to the gravitational self-energy $\Delta E = \Delta Mc^2$ of the masses of particles in the superposition states, and the coherence time t, by the uncertainty principle:

$$t \geq \hbar/(\Delta Mc^2)\,. \tag{7.54}$$

The extension of an isolated superposed system is thus inversely related to the length of time for spontaneous collapse. The mass correction ΔM for a superposition of a macroscopic object (e.g., a Schrödinger's cat) is so large that the lifetime of the cat state becomes minuscule.

A large angular momentum Rydberg state of an atom, or their entangled states, would have a significant space extension as well as a mass distribution if we accept Penrose's ideas that $|\psi(\vec{x},t)|^2$ describes the mass distribution. Hence such states should have, besides their natural width, an additional width arising from the OR mechanism. The gravitational self-energy shift ΔE is presumably the real part of the self-energy (SE) operator of the propagator of the state (c) in Fig. 7.7. The SE arises from scattering of gravitons. It is possible to envisage quantum-graviton processes where quantum gravity could perhaps *stabilize* entangled states, forming bound states, rather than add to their decoherence. These cannot be examined rigorously without a theory of quantum gravity.

The value of $|\psi(\vec{x},t)|^2$ gives the probability of finding the particle at $\vec{x},t$. There is no mass redistribution in the usual formulations of QM. In fact, experimental data for electron scattering from atoms show no evidence for scattering from a $|\psi|^2$ mass-distribution of the electrons.

That electron excitations are not mass distributions is confirmed in the possibility of transforming the standard picture into Bohm's picture where the position of a particle at time t can be always specified.

Furthermore, we again note that finite-temperature (T) effects, not included in Penrose's model ensure that the de Broglie wavelength of macroscopic (e.g., 1 kg) objects are of the order of a millionth of a femtometer (see Eq 7.56) even at one Kelvin. If that is so, there should be no macroscopic cat states at $T > 0$.

7.6 Quantum muddles

Niels Bohr, realizing the great difficulties posed by QM, took an extremely cautious approach to the questions raised by the rise of the quantum theory. Discussing Niels Bohr's philosophic stand, David Mermin [143] says that 'As a philosopher Bohr was one great visionary figure of all time, or merely the only person courageous enough to confront head on, whether or not successfully, the most imponderable mystery we have yet unearthed', but adds that 'one wants to shake the author vigorously and demand that he explain himself further …'. Karl Popper, as an eminent philosopher of science, strongly opposed Bohr's interpretation of QM, and tried hard to 'vigorously shake' Bohr. He regarded Bohr's interpretation of QM as 'outrageous'. Popper totally disagreed with Heisenberg, and talked of clarifying the 'quantum muddle'.

7.6.1 *Sir Karl Popper and the 'Quantum Muddle'*

Karl Popper's discussion of the quantum theory simmered for over thirty years in *postscripts*, unpublished notes and *appendices*, and finally saw the light in 1982 under the title 'Quantum Theory and the Schism in Physics' [168]. According to the editor's forward, most of what is in the *Schism* was written in the years 1951–1956 and published in the form it had in 1962. However, by 1956 when Popper's *Logic of Scientific Discovery* [166] was being translated to engish, David Bohm's excellent, careful discussion [25] of the standard interpretation of QM, and Paul Dirac's influential book [65] were well known. Indeed, they are referred to by Popper. By 1982 when his *Schism* appeared, the standard model of particle physics was well known. Mature discussions of quark theories and chromodynamics had appeared by 1982, but not before 1962. However, Popper had been very busy in the 1960–1980 period and had little to do with QM. Popper was still thinking that the quantum theory is 'but a theory of atoms. a theory of their structure as possessing a positively charged nucleus and a shell structure of negative electrons that explained, in principle, very concrete properties of the chemical elements'. By the time Popper wrote these lines, (1951–1982), the quantum theory had been successfully applied to all kinds of length scales which included macromolecules, solids, superconductors, and in the other extreme all the way down to quarks. Nuclear bombs had been exploded. Popper of course know all this in a general sense, and even claimed that the nuclear bomb was a moral argument for realism and objectivism in philosophy. That is, Popper had his idea of nature (a form of local realism) already decided upon when he came to the quantum theory. One cannot help feeling that Popper simply relied on his vintage knowledge of QM and did not give QM all the attention it merited. Popper's approach to clarifying the 'quantum muddle' (term used on p. 5 of *Schism* [168]), suggests that for him, the answer was a matter of restating QM within an improved probability theory.

Einstein believed that the apparent non-locality of QM was a result of unknown hidden variables, and QM was hence an incomplete theory. Popper probably believed that QM is not only incomplete, but also probably wrong. He did not understand why Heisenberg and others felt that QM was a complete theory (i.e., not needing hidden variables). Popper probably took this to be a claim that QM was the final, 'end of the road' theory, and presents that expression to render the German word *Endgültig* into English, while attributing its original use to Heisenberg. Writing in the 1950–1962 period, Popper attempts to identify new particles with hidden variables. The existence of additional particles (e.g., the proton-neutron doublet) had already been simply explained as quantum states of nucleons, and then finally in terms of quark models, well embedded in a beautiful SU(3) analysis of Gell-Mann and others. Thus we see that, unfortunately, Popper's keen mind did not focus on the contemporary QM of his day. Indeed, we pointed out in Secs. 2.4.5, 2.4.6 that philosophers who were studying revolutions and paradigm shifts in physics completely missed the mini-revolutions that took place in the 1970s to 1980s in particle physics and condensed-matter physics. This is not unusual because each subject remains engrossed in its own questions.

Karl Popper expected experimental tests of EPR to finally refute QM. He contends that Bohm's use of two spin states to form the entangled pair for EPR is fundamentally different in some way from the original EPR, 'because we have no realistic theory of spin' (p. 22, [168]). Most physicists would be surprised by that claim. He says, 'When I first heard that ... (they) were intending to test Bell's theorem, I expected their results would refute quantum theory. But my expectation appears to have been mistaken, ..., Nevertheless I have not abandoned my realistic interpretation of physics, and so far I have not even abandoned locality' (p. 25).

Popper states his views about QM in thirteen theses. His main message is that quantum systems have *propensities* which are realized under experiments. These propensities are just probability distributions. In fact, the version of probability theory argued out in 'The logic of scientific discovery' is changed in the *Schism* to a 'probability theory of propensities', as a consequence of Popper's reflections on the quantum theory. Precise values of energy, time, position, momentum are allowed, as in classical reality. When an observation is made, the probability takes a definite value, and that is all there is to the 'collapse' of the wavefunction. The *Schism* states 13 theses about QM that are summarized in (a)–(g) below:

(a) Theses 1–3, 8, assert that the problems treated by QM are essentially statistical problems, and much of 'the muddle' is due to the use of a *subjectivist interpretation* of probability. In subjectivist models, probability is assumed to arise from a lack of certain knowledge, rather than from objective features of reality (e.g., as in a classical or a frequency interpretation for tossing coins, see Sec. 2.1.2).

(b) The 'great quantum muddle' arises from mistaking a probability distribution $p(x)$ of some sample space x as a physical property of the *elements* of that sample. Popper says that physicists 'do not discriminate between utterly different categories..., and rely on the very unsafe assumption that 'my probability of living in the South of England is like my age, which is one of my properties'.

(c) It is stated that energy and time, or position and momentum can be *determined* to arbitrary accuracy. Thus given a bright spot on a screen, and a slit, the path traversed by the photon is simply obtained by connecting the spot to the slit. However, as Heisenberg insists, this is *retrodiction*, and not prediction where the uncertainty principles apply. Even retrodiction in QM is not as simple as envisaged by Popper, as shown in 1964 by, for example, Aharonov, Bergmann and Lebowitz (and further discussed more recently by Albert *et al.* [5]). The uncertainty relations are regarded as limits on scattering processes. They reveal the 'statistical inhomogeneity' of the experimental data.

(d) After discussing probability theory, *propensities* are introduced as the basis of a probability calculus of 'repeatable experiments'. Popper considers $p(a,b)$, the probability of a occurring given b, as the proportion of those cases satisfying b and having equally possible outcomes of a. He generalizes this to weighted outcomes of a, when the possibilities are not equal. These 'weighted possibilities' are 'measures of the propensity, or tendency of a possibility to realize itself on repetition'.

(e) The analogy of a pin board, where small balls roll down to generate, ideally, a Gaussian distribution $p_G(x)$ is used extensively. Probability is regarded as *a real physical property of any concrete unique physical situation*, and of any specific experimental arrangement capable of repetition. In this discussion the probability distribution $p(x)$ is equated with propensity. 'a propensity is thus a somewhat abstract kind of physical property', which can be modified ('it can be kicked, and it can kick back'). The Gaussian curve 'will represent the probability distribution or propensity ($p_G(x)$) for each single experiment with each single ball reaching a resting place'. But if the pin board is slightly slanted ('kicked') then $p(x)$ will change. On removing one pin on the board, the $p(x)$ for every single ball would alter, whether or not the ball actually interacts with the removed pin. Popper argues that this is 'similar to the two-slit' experiment even though we have no superposition of amplitudes. Radioactivity is mentioned as a propensity of each nucleus to disintegrate.

(f) One may feel that Popper wishes to construct a quantum ontology based on probability theory, where the centrality of superpositions of amplitudes is dethroned as if it were a minor technical detail. The pin-ball models may suggest such a thing to a reader. However, W. W. Barteley, the Editor of the *Schism*, explains Popper's position regarding it in a long footnote. Barteley says that 'Popper ... does not explain interference with the pin board example. He is pointing out ... that probability is a quality of a system not ascribable to an element thereof ... Popper... contends that this is all that is involved in the reduction of a wave packet (... 'collapse of the state vector')'. In effect, $p(x)$ is a property of the whole system, i.e., of the holism.

Popper takes the case $p(x,b)$ for reaching various final positions after hitting a definite pin b. This is different from the original normal distribution $p(x)$ and it is presented as an example of a position measurement, as all such balls have been 'checked' at the position of the pin b. The transition from the original $p(x)$ to $p(x,b)$ is said to be 'not merely analogous, but identical with the famous reduction of the wave packet ... (it) ... is not an effect characteristic of quantum theory: it is an effect of probability theory in general' (this is developed in Thesis 9 and elsewhere in the *Schism*).

(g) Finally, having renamed the distribution function as a 'propensity', propensities are identified as physical and real; a propensity amplitude is introduced and even 'propensity waves', and 'propensity fields' which are supposed to be 'real' (ontic) are introduced. The concepts of duality and complementarity are rejected, while instead admitting particles and their associated fields.

The idea of potentialities or propensities was central to Aristotelian physics. Thus an acorn has the potentiality to become an Oak. The conversion of probable knowledge to certain knowledge under observation is also a feature of classical physics, as Popper himself discusses in Sec. 76 of [166]. The key roles of the superposition of amplitudes, contextuality and non-locality in entangled systems etc., are not high-lighted in the *Schism*. Popper may be justified in being critical of many statements of Heisenberg, Pauli and others. However, any one who writes about QM, how ever carefully, would face many difficulties. Popper's claim that 'the great quantum muddle' can be resolved by some clear ideas about probability concepts and propensities may seem far fetched. In fact, the nature of probability itself is a topic of on-going controversy. The need to review the role of probability in understanding the quantum theory has been taken up in more recent times by some writers (e.g., Caves *et al.* [44, 77]) who emphasize a Bayesian, i.e., subjectivist, approach.

7.6.2 *Mindful universes and entangled brains*

Contrary to common intuition, the quantum theory seems to deny even a position and a momentum to a moving particle. Nothing can be said about individual events. Only statistical answers can be given. It seems that the outcome of experiments may be manipulated by the observer at will. If Archibald Wheeler were correct, then we could influence even the past by a suitable 'delayed choice' of our own. There could even be unobserved dead-alive Schrödinger's cats lurking around. Even Einstein was worried about the possibility of 'spooky action' in QM. Has QM finally removed all possibility of an objective reality?

Karl Popper tried to give an essentially realist, observer-independent philosophy of the world revealed by QM, using classical ideas of probabilities and propensities. More successful formulations of objective quantum reality exist. Thus Bohm's causal QM, the many-minds interpretation of Everett, the consistent-histories formulations of Gell-Mann and Hartle, are all attempts to give objective formulations of quantum reality. Several of these invoke decoherence to disentangle the system under study (SUS) from the observer. Most theories assume that what is observed is not just a limited bit of reality, but the whole of reality.

In contrast, Pauli, Heisenberg, and Niels Bohr in particular took a positivist stand, declining to alleviate the epistemic hunger typical of traditional inquiry (Ch. 2). Bohr held that what is 'knowable' is confined to what can be observed in the strict sense of QM; any inferred reality is mere personal belief. Meanwhile, the working scientist follows his (not necessarily precisely worked out) intuitive ontology of the working physicist (IOWP), where quantum systems exist 'for real' in quantum states, get entangled, undergo transitions or propagate as depicted in Feynman diagrams etc. The working physicist has an intuitive description that provides him/her with a useful image of quantum processes.

The human observer is central to Eugene Wigner's view of quantum reality. John von Neumann in his formulation of the measurement process, found it mathematically convenient to construct a superposition of the quantum state of the system under study (SUS) with states of the measuring apparatus, eventually collapsing to the observed result. This is amplified to a grotesque form in the cat paradox of Schrödinger, Sec. 7.7.3. Thus the detector is entangled with the system under study. *However, where is the demarcation between the detector and the human observer*? How is the collapse engineered? Since QM applies to everything, it is claimed that the human observer too must be interacting with the detector, even if there were a time delay. Hence the human observer is said to be in an entangled state with the detector which is entangled with the SUS.

We calculate the de Broglie wavelength of a 1 kg cat in the section on Schrödinger's cat paradox. This is found to be of the order of a millionth of a femtometer. The 'radius' of a proton is also about a femtometer, and corresponds to the type of densities found on a white dwarf. This, if valid, casts serious doubt on the possibility of an entangled state involving the observer and the SUS.

This chain of entanglements is 'broken' by the subjective school of physicists by invoking human consciousness which is said to be able to exercise its volition. Mental processes are *conveniently assumed to be outside the control of QM*. Thus, the demarcation line between the external world (subject to QM) of objects and detectors, and the act of observation, is placed precisely at the doorstep of a conscious mind. In this approach to quantum reality, it is the act of observation which 'selects' a specific state of the entangled system and makes it collapse to the observed state. Henry Stapp [202], a contemporary advocate of this view, talks eloquently of the 'mindful universe'. *The clarion call of Stapp and others seems to be based on a presumed escape from clock-work determinism* that they believe narrowly confines human destiny and classical reality if the universe is not brought under the control of the mind. This also involves an enthronement of an anthropocentric role in the unfolding of reality. Stapp [202] writes:

'Thus (the founding fathers of QM) formulated their new theory ... around the knowledge-acquiring actions of human beings, and the knowledge we acquire by performing these actions, rather than a conjectured causally sufficient mechanical world... it is the revised understanding of the nature of human beings, and of the role of human consciousness in the unfolding of reality, that is, I believe, the most exciting thing about the new physics ...'

'... according to the new conception, the *physically described world* is built not out of bits of matter, as matter was understood in the 19th century, but out of objective *tendencies* — potentialities — for certain discrete, whole *actual events* to occur. Each such event has both a psychologically described aspect, which is essentially an increment in knowledge, and also a physically described aspect, which is an action that *abruptly changes* the mathematically described set of potentialities to one that is concordant with the increase in knowledge.'

'... the boundary between our empirically described selves and the physically described system we are studying is somewhat arbitrary. Th empirically described measuring devices can become very large. This ambiguity was examined by von Neumann (1932) who showed that we can consistently describe the entire physical world, including the brains of the experimenters, as the physically described world, with the actions instigated by an experimenter's stream of consciousness acting directly upon the experimenter's brain.'

Thus we see that the *stream of consciousness* is the final arbiter over the brain as well as the external world. Stapp and others also contend that neuronal circuits are like nano-structures where discrete quantization (e.g, discrete energy level formation, see Sec. 6.4) becomes important. However, it is easily shown that even in quantum point-contacts smaller than the components of neuronal circuits, such quantization disappears long before we reach human body temperatures.

The objections to the type of discussion found in Stapp and other subjectivists are the same as those given in our discussion of Schrödinger's cat paradox (Sec. 7.7.3). Furthermore, classical physics especially when applied to complex adaptive systems, or indeed to most systems outside the usual simple examples of text book mechanics, is found to be *not deterministic* in the sense implied by Laplace long ago, and by Stapp and others today.

Having invoked a quantum model of the brain, with quantum brain states, the strict unitary evolution of QM gives no escape from determinism that Stapp so deeply dreads. He is obliged to call up the von Neumann collapse of the wave-function (called 'process 1') to provide room for freedom of choice which now resides entirely in the unquantized 'stream of consciousness'. Stapp is now faced with the difficulty of making the stream of consciousness interact with the physical world, i.e., just the classical mind-body problem in a new guise. It is here that Stapp has to resort to the quantum Zeno effect, discussed in Sec. 7.7.6.

7.7 Quantum paradoxes

In Sec. 5.10 we discussed paradoxes associated with the theory of relativity where our everyday notions of space and time get replaced by the unfamiliar notion of spacetime. The quantum theory can be equally counter-intuitive and has become a rich mine of paradoxes. A paradox is an 'intellectual riddle' where we begin with seemingly unquestionable premises, and arrive at conclusions which are highly questionable and counter intuitive. In most cases, the resolution of the paradox lies in an examination of one's 'unquestionable premises'.

The paradoxes of QM enabled scientists and philosophers to arrive at a deeper understanding of the nature of QM. However, in our view, just as in the paradoxes of relativity, what we have are *not* unsolved riddles or difficulties compromising physical theory. Thus the attempts by some logicians (for a discussion, see Ch. 7 of Redhead [178]) to construct a new logic to clarify quantum paradoxes, just as new geometry was to special relativity, are probably irrelevant.

Richard Feynman [75], in his BBC broadcasts on the 'Character of Physical law' stated that he can 'safely say that nobody understands quantum mechanics'. It was a time when even Feynman believed that von Neumann had 'definitively shown the impossibility of hidden variable theories'. The 'Copenhagen school', led by Niels Bohr, Heisenberg, Dirac, Pauli and others (except Einstein), held that we need not attempt to infer an underlying reality beyond what is immediately observed. Bell's inequalities had not yet engaged even the leaders of the scientific community. Developments in quantum simulations, quantum information, nature of decoherence etc., had not taken place. Meanwhile, Einstein, Schrödinger, Wigner, Wheeler and others had concocted various seemingly paradoxical thought experiments where quantum reality was presented as totally weird and inconsistent with our intuitive expectations. Niels Bohr and others explained and demystified these paradoxes by pointing out the importance of explicitly including the measuring apparatus in the quantum description. On the other hand, Wigner and others probably added more mystification by suggesting that the world did not exist except as a part of the experimentalist's consciousness!

The Solvay conferences that began in the late 1920s provided a forum for Bohr, Einstein, Heisenberg, Dirac, de Broglie and others to critically examine and discuss intriguing aspects of the newly minted quantum mechanics. Einstein, who led the opposition, at first doubted the validity of some aspects of QM. Subsequently he modified his view to claim that QM was an incomplete theory.

We examine a number of these paradoxes in this section. Some of them involve subtle aspects of the uncertainty principle, complementarity and contextuality. They arise from the persistence of superpositions of quantum states and entangled states, even when large separations in space or time are imposed on the entangled components. The paradoxes arise because intuitively we expect local, non-contextual observations when QM predicts the opposite. The paradoxes ignore the role of decoherence as the intervals in space and time become longer and longer. They also treat very special systems, e.g., just two particles, or two quantum states in isolation. Readers who have worked through this book would not be mystified by these seemingly paradoxical *gedanken* experiments.

7.7.1 *Einstein's clock-photon paradox*

This paradox is also called the 'clock in the box' paradox, or the 'weighable light-flash box (*waegbaren Lichtblitz-Kasten*) paradox. It was presented by Einstein at the 1930 Solvay conference in Brussels to Niels Bohr [33], and others, claiming that this gedanken experiment violated the energy-time uncertainty principle.

$$\Delta E \Delta t \geq \hbar \,. \tag{7.55}$$

This uncertainty principle can be easily understood by noting that an energy eigenstate of a photon is a monochromatic wave, i.e., an infinite plane wave, which does not have a well-defined time of arrival at a given point. Conversely a very narrow wave packet is usually assumed to have a well-defined arrival time; but contains many different frequencies. Therefore it is a superposition of different energy states giving rise to the energy spread in ΔE (see Sec. 6.26 for a more careful discussion).

Einstein probably pictured the photon as a particle with a precise energy at any precisely specified point in time, as he was disputing Eq. 7.55. After all, both energy and time were important concepts in relativity, and there was no hint of any uncertainty in E and t values, as implied by Eq. 7.55. Einstein considered a box containing photons and a clock which would at a precise moment open a shutter releasing some photons (light). Einstein considered the energy change ΔE that occurs in the box at the moment of opening of the shutter to be the mass change Δm. This is related to the energy change by $\Delta E = (\Delta m)c^2$.

The box containing the clock and the photons is hung on an accurate spring balance, with a pointer whose position q indicating the mass m (Fig. 7.8). An initial measurement of m is taken. The photon is emitted at a predetermined time defined by the position of the clock arm $a(t)$. The new position q of the pointer gives the new mass. Given the change in position of the pointer we know the mass change, and hence ΔE. Also, the clock arm gives the exact time when the photon was released, paradoxically violating the energy-time uncertainty principle!

Niels Bohr had no immediate answer. However, the very next day Bohr was ready with his resolution of the paradox. According to the time dilation formula of Einstein's general theory of relativity, two clocks at different heights in the gravitational field of the earth would run at two different rates. When the photon leaves the box, the weight of the clock-box is reduced and the clock and the pointer move to a different height q. Hence the clock runs at a different rate, as if it uses a different unit of time. Hence the clock has to be attributed a slightly different time. In this experiment the gravitational mass is taken to be the same as the inertial mass. Bohr was able to demonstrate in detail that the energy-time uncertainty relations were obeyed by Einstein's gedanken experiment.

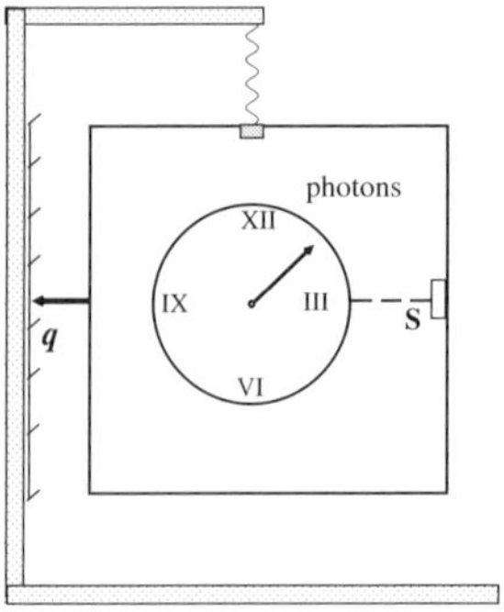

Fig. 7.8 When the time becomes exactly 3 O'Clock, the clock arm triggers open the shutter **S** and releases the photon trapped in the box. The pointer ***q*** reads off the mass of the clock box which is suspended from a spring in the earth's gravitational field.

However, it turns out that this relativistic time dilation can be arrived at without general relativity, as shown by Unruh *et al.* [208]. The Hamiltonian H_{cl} of the clock-box and photons, with total mass m determines both its energy and the rate of change of the clock arm (i.e., the unit of time). The rate of change of the clock arm $a(t)$, where a is a Heisenberg operator, is given by its commutator with the system Hamiltonian H_s. This includes the gravitational field. Thus:

$$\begin{aligned} H_{cl} &= mc^2, \text{ i.e., } M = H_{cl}/c^2 \\ H_s &= H_{cl} + H_g, \quad H_g = (mg)q = (H_{cl}/c^2)gq \\ da(t)/dt &= (i/\hbar)\,[a, H_s]_- = (1 + gq/c^2)H_c\,. \end{aligned}$$

In the above, g is the local acceleration due to gravity. Thus, when the photon leaves the box and reduces the mass, the pointer position moves up, reducing q and the term gq/c^2. The clock at the start of the experiment, and after the loss of the photon, runs at different rates. This produces a time dilation leading to a δT consistent with Eq. 7.55.

Einstein gave up his quest to show that QM is an incorrect theory after the 1930 Solvay conference. Instead he felt that QM may be an 'incomplete theory'. The Heisenberg-type uncertainties in dynamical quantities seemed to him a lacuna *in the theory*, rather than a feature of physical reality. The possibility of the photon emitted from the clock-box remaining entangled with the box seems to have been envisaged by Einstein after the Solvay conference [63]. This testifies to Einstein's deep understanding of QM; it led to his formulation of the EPR paradox.

7.7.2 *The EPR paradox*

We have already referred to the Einstein-Podolsky-Rosen (EPR) [70] gedanken experiment in previous sections of this book (e.g., see Sec. 7.3.7). This was a hypothetical experiment, proposed in 1935 by EPR to show that *QM cannot be a complete theory*. Here the word 'complete' did not imply a theory of 'everything', or even the 'end of the road', as suggested by Popper; rather that it was complete within its own ontology. Thus EPR argued that a complete theory must have a number of features. In addition to the explicitly stated features, Bohm (Ref. [25], Sec. 22.15) pointed out that other implicit assumptions exist. We have included them as well in the following. These assumptions are nowadays stated in terms of the existence of a 'local plan' (see Eq. 7.35 and the associated text):

1. Every *element of physical reality* must have a counterpart in a complete physical theory. It is also implicitly assumed in their paper that this counterpart must always be precisely definable in the mathematical formulation of the theory.
2. If we can predict, without disturbing the system and with certainty (unit probability) the value of a physical quantity, then there exists an element of *physical reality* corresponding to this physical quantity.
3. The world can be analyzed correctly in terms of distinct and separately existing 'elements of reality'.

We have discussed the EPR paradox (Sec. 7.3.7) and its interpretation in our discussions of entangled pairs of electrons, spins and photons. The Bell inequalities (Sec. 7.4) provided a quantitative measure of the extent of the paradox — i.e., the violation of classical correlations by QM. Here we look at the original EPR version for two entangled particles, their positions and momenta.

In the original version of the EPR paradox, two particles are allowed to interact and then separated to a great distance apart, so that they cannot causally interact with each other (if local realism were valid). Let us say that the two particles were prepared in a total momentum state $P = p_1 + p_2 = 0$, and the first particle is observed by Alice, while the second particle is observed by Bob. If Alice measures the momentum of the first particle and finds that to be p_a, then since the total momentum is zero, Bob would find that $p_2 = -p_a$. In effect, Alice can infer what Bob would measure. Normally, within the local concept of reality, we would say that the particle reaching Bob already had the momentum $-p_a$ as an essential element of its reality, even before Bob measured it. It is also not possible to claim that anything Alice does at her end influenced measurements at Bob's end since we assume causal separation.

In QM, the wavefunction gives the most complete description possible of any quantum mechanical system. The wavefunction of the two-particle system is made up of all possible values of p_1 and p_2 which add up to zero. That is, according to Niels Bohr, individual particle momenta have no meaning prior to measurement. However, the moment Alice measured the momentum of the particle at her end to be p what Bob would measure gets fixed to be $-p$. It is this feature that Einstein referred to as 'spooky action at a distance'. EPR expects local realism to be upheld, and paradoxically finds it to be violated.

According to 'local realism', p is an intrinsic property of the particle a, and $-p$ is an intrinsic property of the particle b. Alice and Bob may work with another pair of particles, prepared in exactly the same way, and under the same conditions, and choose to measure *the position* of the particle at her end, and find the value q_a. This is an intrinsic property of the local realism of the particle a. Then, Bob could measure the momentum and find the value $-p$. So, one might (erroneously) assign a momentum p to the particle with Alice, having a position q_a. So we have spooky action at a distance, and also a specification of the position and momentum of Alice's particle, violating the uncertainty relations!

EPR do not precisely define what is meant by two particles being 'far away', since no typical length scale is indicated. In looking at the probability of finding an up-spin electron in a gas of electrons of mean density ρ, containing (on average) one particle in a sphere of radius r_s, we found that the pair-distribution function $g(r)$ was scale invariant. That is, the electron pairs were correlated in the same way whether r_s was small, or spanned an astronomical distance. Thus the spin correlations associated with superpositions that obey the Pauli principle were non-local. The distance R_c taken by $g(r)$ to recover the value of unity is roughly a measure of the range where 'effects prevail locally'. However, this range R was a function of r_s, and hence, when the particle separation ($\sim 2r_s$) is increased, the range of influence increases correspondingly, contrary to the assumptions of EPR. Thus no 'local hidden variables' can describe such correlations.

The Establishment of the Bell inequalities proved conclusively that no local hidden variables can be consistent with QM. In Bohm's interpretation of the quantum theory (Sec. 6.7), local realism is abandoned and a non-local *quantum potential* $Q(x_1,x_2)$ which depends on the wavefunction is introduced. This procedure is a mathematically equivalent restatement of standard quantum mechanics. Within this picture, the particles do have predetermined values of properties before and after the experiment. Thus the 'hidden variables' are non-local. There is no collapse of the wavefunction. The wavefunction, or equivalently the quantum potential, contains the quantum information regarding 'the whole system'. The latter is what is contained within a typical correlation length R_c of the system. In dealing with just two entangled particles, R_c covers the full length L of the system.

Given two entangled particles, their wavefunction can be written in terms of the center of mass (cm) momentum K, cm-position R, the relative momentum k, reduced mass m_r and relative position $\vec{r} = \vec{r}_1 - \vec{r}_2$. Written in the polar form, it is found that the Bohmian quantum potential is simply the relative kinetic energy $k^2/2m_r$ of the pair of particles. This is conserved for any separation of the two particles, and hence there is clearly no local realism since the quantum potential $Q(\vec{r}_1,\vec{r}_2)$ is constant within the Bohm picture.

However, when dealing with large separations L, the difficulty of avoiding intruding perturbers increases at least as the cube of the radius of the system. This sets the stage for rapid decoherence from such perturbations.

7.7.3 *Schrödinger's cat paradox*

What has become known as the cat paradox was introduced by Schrödinger in 1935, and published under the title 'The present situation in Quantum Mechanics'. This has been reprinted in Ref. [226]. The paradox is about QM permitting a cat to be 'alive' *and* 'dead' at the same time, and becoming one or the other only when looked at. Schrödinger wanted to bring out the counter-intuitive nature of (a) superpositions of quantum states, (b) the collapse of the wavefunction to one of the components, on 'just observing the system'.

Schrödinger wrote his papers on entanglement [188] during the period when the EPR paradox appeared. The cat paradox was a warning that naive macroscopic applications of QM may give ridiculous results. However, the possibility of macroscopic 'cat states' like those of Schrödinger has even led to *ad hoc* decoherence theories designed to exclude them.

Schrödinger mentions that the decay of a radioactive atom can have macroscopic consequences, as in forming a cloud-chamber track (cf., Sec. 7.5). He converts the quantum uncertainty of radioactive decay to a paradox about a cat's life or death, to expose the special features of the quantum theory:

> 'One can even set up quite ridiculous cases. A cat is penned up in a steel chamber, along with the following device (which must be secured against direct interference by the cat): in a Geiger counter there is a tiny bit of radioactive substance, so small, that perhaps in the course of the hour one of the atoms decays, but also, with equal probability, perhaps none; if it happens, the counter tube discharges and through a relay releases a hammer which shatters a small flask of hydrocyanic acid. If one has left this entire system to itself for an hour, one would say that the cat still lives if meanwhile no atom has decayed. The psi-function of the entire system would express this by having in it the living and dead cat (pardon the expression) mixed or smeared out in equal parts. It is typical of these cases that an indeterminacy originally restricted to the atomic domain becomes transformed into macroscopic indeterminacy, which can then be resolved by direct observation'

Although this horrific example of a gas chamber for a cat, set against what was going to happen in Germany before long could send painful shivers to pacific readers, we remind them that Schrödinger himself was adversely affected by Nazism and had to flee to England, from where the articles on entanglement were written. The tradition of violence in present-day culture is so strong that writers have little inhibition to using such horrific models. However, Schrödinger does ask for pardon for the language in his text.

The counter-intuitive possibility is that if we can simply apply QM to the states designated $|d\rangle$ = 'dead state' and $|a\rangle$ = 'alive state', then we have a superposition where the cat is neither dead, nor alive. The quantum indeterminacy of the radioactive atom (i.e., $|a\rangle$ = 'atomic state' becomes $|d\rangle$ ='decayed state') is translated into a macroscopic indeterminacy. The paradox asks, what is there in QM to prevent us from making such constructions?

Our *ignorance* of the state of the cat does not give us the *information* to assume that it is in a superposition, or in entanglement with a radio active atom etc. We saw in earlier discussions (Sec. 7.1) that it is non-trivial to create and maintain superposition states or entangled states. Furthermore, a live cat or a dead cat has a quasi-continuous density of quantum states ϕ_i where i is like a continuous variable. Special effects typical of quantum phenomena arise if the various quantum states and their superpositions retain their phase coherence — like groups of dancers who correctly keep step and remain 'entangled' *via* the beat (c.f, de Broglie's pilot wave guiding the particles). However, when such large numbers of quantum states interact randomly, we get a mixed state where all the quantum effects have got washed out, like dancers who have lost their partners, lost the beat and begun to move erratically in a mob, uncorrelated with each other. Then, we have individual systems instead of entangled or superposed systems. When there is no phase coherence, we have classical reality.

Phase coherence is possible because of the wave property of matter. This is common to all matter, be they electrons or cats. It is the wave character of matter that allows the possibility of superpositions. So Schrödinger's cat paradox is more about the question of what would happen if macroscopic objects displayed wave-particle duality to a perceptible extent. If the cat is one kilogram heavy, and at room temperature (300K), its thermal de Broglie wavelength λ_c is given by

$$\lambda_c \simeq h/\sqrt{(3mT)} \simeq 9.45 \times 10^{-23}\text{meters} \tag{7.56}$$

Here we use the following conversions and work *via* atomic units where $\hbar = h/2\pi = 1$, in atomic units. The Hartree or 1 a.u.= 27.2116 eV, 1 eV = 11605 K, 1 kg = 6.0222×10^{26} amu, 1 amu = 1823 m_e with $m_e = 1$ in atomic units. The atomic unit of length (Bohr radius) is $a_0 = 5.2920 \times 10^{-11}$ m.

The 'radius' of a proton is about 0.88 femtometers (1 fm = 10^{-15} meters). The quantum wavelength of the cat is about one hundred-millionth of the size of a proton! Even when cooled to one Kelvin, the de Broglie wavelength remains totally negligible. Even a free electron close to a cat's body would not get entangled with it. Matter has to be squashed to densities where nuclear reactions begin before superpositions are possible at such small values of λ_c. Clearly, Schrödinger's cat states of macroscopic bodies do not exist in nature except at $T = 0$. The cat, the radioactive material, the observer etc., could be in a superposition if they were at $T = 0$, or on white dwarfs and such compressed stellar objects!

One may argue that the center of mass M_{cm} is not the proper mass to use, even though Arndt *et. al.* (Ref. I-[9]) show that M_{cm} is indeed valid for C_{60}, with $\lambda \sim$ in the picometer range. One may claim that QM applies to some molecules of the cat that are specific to it being dead or alive. In that case we are no longer applying QM to a macroscopic system and then there is no difficulty in having superpositions or entangled states. One may also question the use of a momentum calculated from $3k_BT/2$, with $k_B = 1$ in energy units. This is the classical kinetic energy of an ideal gas. We know from Dulong and Petit's law (or from the more modern Einstein-Debye theory of solids) that classically, a system held together by chemical bonds, e.g., a cat, has three degrees of vibration per atom, and three degrees of translation, each having an energy $k_BT/2$. For solids in the low-temperature limit, we need the Fermi energy of the system. Such detailed treatments merely introduce numerical factors of little consequence. The basic conclusion that λ_c is very very small remains firm.

The measurement paradox, where the cat 'collapses' into the $|dead\rangle$ state, or the $|live\rangle$ state on observation, is presented here as typical of quantum measurements. This is far from being the case, as already discussed in Sec. 7.5.0.2, quantum measurements can generally use samples of the main system which is chosen to be very large compared to the typical physical correlation lengths of the system. There we discussed how the state of spin-polarization (up, or down) could be determined by merely sampling bits of the gas. Here we can use the air in the steel chamber to sample for cyanide gas atoms (that may be present or absent). These are purely classical situations controlled by the quantum uncertainty of the radioactive atoms.

Schrödinger implicitly equates this quantum uncertainty to the existence of macroscopic superpositions. This is by no means necessary even for quantum systems. To know that a system is in a superposition is to have definite, detailed knowledge of the system (e.g., the superposition is the square-root of the system's NOT logic operation). Hence assuming that there is a superposition, or a mixture, is to claim more knowledge of the system than is warranted here. Once we grant that λ_T is of the order of micro-femtometers, then superpositions are ruled out, and mixtures become more likely candidates.

A gas analyzer is continually sampling the gas in the steel chamber. The cyanide gas atoms when emitted are numerous enough for small samples to go into the cat's nose, and also into the gas analyzer. The analyzer posts a continuous readout onto a computer, and any 'observer' anywhere in the world can know if any cyanide has been released (i.e., the cat is dead), or not. The quantum uncertainty remains with the radioactive atom and not posited into a hypothetical 'wavefunction of the cat'. In fact the gas analyzer can completely replace the cat. The measurement is entirely classical, and requires no conscious mind to 'bring about' the death of the cat in a superposition.

The dead/alive situation is similar to the choice of a particle going through one or the other of two slits, or being in a superposition that leads to interference. In Bohmian mechanics (Sec. 6.7) the superposed wavefunction ψ exists to define a probability distribution and an additional 'quantum potential' for a classical dynamics. There are no situations in a two-slit experiment where the point-like electron of BM goes through both slits. That is, the cat is always either dead, or alive. The electron always follows one outcome, based on the Bohmian trajectory. The initial conditions for this trajectory are set by the wavefunction which assigns one set of (to be transmitted) trajectories to one slit, while the remaining (to be transmitted) trajectories are set to pass through the other slit. Thus the superposition concept *applies only to the wavefunction* which defines the statistical probabilities of finding the cat to be dead or alive. The cat is never in a dead/alive superposition in Bohmian mechanics.

Other attempts to resolve the paradox, e.g., many-worlds interpretations etc., (while retaining von Neumann's collapse hypothesis and without decoherence, i.e., retaining the unitary evolution of the wave function) are well known. They have also been exploited by tabloid writers who content that QM has opened the door for all types of magical and super-natural worlds. Wigner [228] claims that the experimentalist's mind comes into a superposition state with the system under observation, and the collapse of the wavefunction is produced by an act of consciousness. This latter mental process is not explained and cannot possibly be a unitary transformation. Thus these solutions merely push the problem under the rug of obscurity.

As already discussed in Sec. 7.5.4.1, Roger Penrose has suggested that superpositions involving large objects like cats do not persist because they introduce gravitational instabilities by warping spacetime. This leads to the collapse of such superpositions, known as 'objective reduction'. The spontaneous localization models (e.g., those of Ghirardi, Rimini and Weber) [85] phenomenologically introduce random traps that localize the wave function and suppress superpositions. The localization has little effect until the system becomes macroscopic.

Such attempts to link the observational process in quantum physics with consciousness, quantum gravity or *ad hoc* corrections to the Schrödinger equation have raised the ire of some commentators. This may well be what van Kampan [212] has characterized as the *scandal of physics*. He laments that

> 'eighty years after the development of QM, the literature is still swamped by voluminous discussions about what is called its 'interpretation'. Actually quantum mechanics provides a complete and adequate description of the observed physical phenomena on the atomic scale. What else can one wish? (It is true that the connection with gravity is still a problem, but that is outside this discussion.) The difficulty is that the authors are unable to adjust their way of thinking — and speaking — to the fact that phenomena on the microscopic scale look different from what we are accustomed to in ordinary life. That two electrons far apart may be entangled seems strange to someone who still thinks of electrons as individual particles rather than as manifestations of a wave function.'

In his conclusion he says

> 'in order that a macroscopic apparatus can be influenced by the presence of a microscopic event it has to be prepared in a metastable initial state — think of the Wilson camera and the Geiger counter. The microscopic event triggers a macroscopically visible transition into the stable state. Of course

this is irreversible and is accompanied by a thermodynamic increase of entropy. This is the physics as determined by quantum mechanics. The scandal is that there are still many articles, discussions, and textbooks, which advertise various interpretations and philosophical profundities. In the seventeenth century, Cartesians refused to accept Newtonian attraction because they could not accept a force that was not transmitted by a medium. Even now many physicists have not yet learned that they should adjust their ideas to the observed reality rather than the other way round.'

7.7.4 *How does one know if the cat is dead?*

How does the experimentalist, peering into the steel chamber, know that the cat is dead? It may be simply that the cat is partially poisoned or impaired but not dead. The hydrocyanic acid was not as effective as hoped for. It is now necessary to determine if the cat is dead or not. Most physicists would grant this to be a problem in physiology and not physics.

Nevertheless, some philosophers of science have talked of 'nearly dead' states of the cat, where they consider a superposition $c_1|live\rangle + c_2|dead\rangle$, with $|c_1|^2$ very small. How small should $|c_1|^2$ be before the cat is pronounced dead? This question led to extensive discussions under the name: "the tail of the Schrödinger's cat".

How does one distinguish living and non-living matter? This was precisely one of the question that was posed by the Dalai Lama in the late 1980s to a group of physicists when he visited the Institute of Theoretical Physics (the present Kavali Institute of Physics) at the University of California, Santa Barbara. The group of physicists included the founding director Walter Kohn, Nobelist Robert Schrieffer and other distinguished scientists. The question asked is, in our context, how does one distinguish between $|dead\rangle$ and $|alive\rangle$? Does the wavefunction collapse even if the experimentalist makes a mistake, as it should do if the collapse is determined by the conscious volition of the observer, *a la* Wigner?

With electron spins we have Stern-Gerlach experiments to determine the *up* or *down* spin state. With living matter, the issue becomes very subtle. The lack of a heart beat or a pulse is inadequate, since medical techniques can often revive a patient under such conditions. Living cells that have been cooled to very low temperatures, inhibiting biochemical processes, can be revived to life using suitable techniques. Functional NMR has been used on patients who have been in a vegetative condition for years, to detect patterns of brain activity, enabling doctors to make rudimentary communications with such patients. A substantial fraction of the cells and biological processes of the cat have to perish before it can be pronounced dead. If the medical (veterinary) help available to revive the unconscious cat is technically highly advanced, even a seriously impaired cat can be made to survive. That is, the fraction of impaired cells that ensures the death of the cat depends on the technical capacity of the medical intervention to save the cat.

This is essentially the answer that Niels Bohr is said to have given in this context. *The decision as to whether an organism is alive or dead depends on the precision and sophistication of the measuring instruments and remedial techniques available to the technical team carrying out the probe*. An advanced civilization may consider a cat to be only impaired and revivable, while a more primitive technological civilization would consider it dead.

7.7.4.1 *Creating cat states of large objects*

The essential features of non-classical correlations of QM come about *via* the superpositions of states. Such states, when confined to simple spins, photons, etc., were discussed under names such as 'Bell states', qbits, covalent bonds etc. However, following the Schrödinger's-cat example, the name 'cat state' is also used to describe superposition states, especially of large systems. In any case, even the fullerene molecule C_{60}, when constructed in a superposition state [9], is a very far cry from a superposition state based on truly macroscopic objects which contains millions of atoms. If a fullerene molecule can be put into a superposition state, macromolecular objects like viruses having several characteristics of living organism, may also be considered for the construction of superposition states, at sufficiently low temperatures where decoherence is minimized and the thermal de Broglie wavelength increased. However, even if we work at $T = 1$ K, once the de Broglie wavelength λ_T becomes less than, say, 0.5 Bohr radii (1 a.u.), the possibility of engineering superpositions disappears. At 300 K, this sets an upper limit of a mass of ~ 77 times bigger than C_{60}, for observing superpositions.

Thus Schrödinger's cat states of even the smallest protein, leave aside 'living organisms' or viruses seem an impossibility. Non-living large objects cooled to micro-Kelvin temperatures are not limited by the above calculation of λ_T at $T = 1$ K. Considerable progress has been achieved with other macroscopic quantum systems like superconducting loops. An experimental device familiar to low-temperature physicists and engineers is known as the *superconducting quantum interference device* (SQUID). A superconducting state is a coherent superposition of pairs of electrons (known as Cooper pairs) existing in a single quantum state, known as a Bardeen-Schrieffer-Cooper (BCS) state. The next excited state is separated from the BCS state by a gap. This BCS energy gap prevents the electrons in a BCS state from scattering under the influence of a weak applied field. Thus dissipation-less electron motion (superconductivity) appears in superconducting metals like Lead or Niobium, when cooled to sufficiently low temperatures. Such superconducting currents can be induced in a superconducting metallic ring or loop by applying an external magnetic field. SQUIDS are superconducting loops of macroscopic size, containing a barrier through which Cooper pairs can tunnel (Josephson junction). Such a system is found to exist in many different quantum states corresponding to integral units of the magnetic flux passing through the loop. This raises the possibility of superpositions of such states. That is, the objective is to make cat states of these macroscopic quantum states (identified by their quantized flux and current). A number of groups have successfully constructed such cat states of SQUIDS where the superposition is between clockwise and anti-clockwise current-flow states, providing a direct analogy with qbits. Such cat-state SQUIDS can hence be used to construct quantum logic gates and quantum computers.

7.7.5 *Wheeler's delayed-choice paradox*

Wheeler's [223] delayed choice *gedanken* experiment is based on wave-particle duality and complementarity. Wheeler asks the questions:

> Does the past exist in some objective sense, or is it created by our act of observation?
> Does QM put us in the paradoxical situation of allowing us to change the past?

Whether a photon or an electron would appear as a particle or a wave depends on the experimental setup that we choose. In the two-slit experiment, we may set up to observe from *which slit* the particle is emerging. If the detector is sufficiently far away from the slits, we may quickly change the experimental set up to observe interference, *after* the electron is deemed to have passed the slit, but *before* it arrived at the *which-slit* detector. The electron now faces a new detection system, and behaves 'as if it arrived *via* both slits', and creates an interference pattern.

Hence Wheeler [223, 115] claims that the past does not exist in an objective sense. He presents the 'delayed choice' gedanken experiment to claim that 'no phenomenon is a phenomenon until it is an observed phenomenon', and that

> the universe does not 'exist out there' independently of all acts of observation. It is in some strange sense a participatory universe. (If) the present choice of the mode of observation... should influence what we say about the past ..., the past is undefined and undefinable without the observation.

The delayed-choice experiment is realized using the experimental setup shown in Fig. 7.9. A single photon pulse is split by the beam splitter BS1 into two beams taking paths 1 and 2. Two mirrors (M) recombine the beams at the lower-right end of the interferometer. At first the beam splitter BS2 is not in place, and the two beams are detected by D_1 and D_2. Then, if D_1 lights up we know that the photon has taken the path 1, and similarly if D_2 lights up, the photon has taken path 2. In a single-photon experiment, both detectors cannot light up at the same time. Hence this arrangement provides 'which-way' (welcher weg) path information.

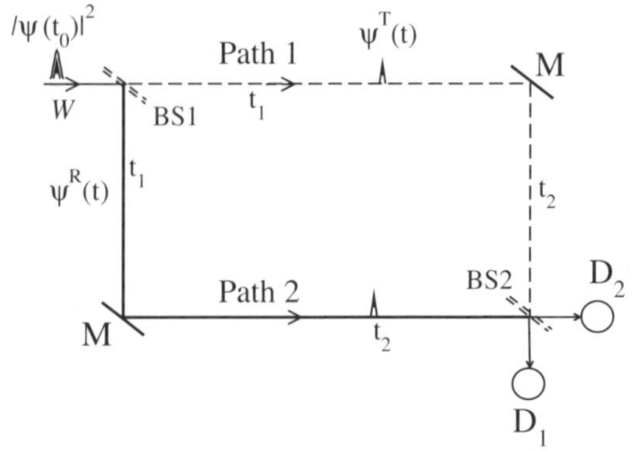

Fig. 7.9 Wheeler's delayed-choice experiment. A single photon pulse W with the distribution $|\psi(t_0)|^2$ at time t_0 is split into two paths by the beam splitter BS1, and directed by mirrors M to the two detectors D_1 and D_2. If the second beam splitter BS2 is not inserted, one or the other detector would light up and reveal the path of the photon. If BS2 is put in place, path information is lost and interference occurs. In delayed choice, BS2 may be put in place or removed at t_2, after the pulse has passed BS1, but before arriving at BS2.

Now let us introduce the second beam splitter BS2 at the exit end, at the point of intersection of the two paths, as shown in the Figure. When the path lengths P1 and P2 are correctly adjusted, one may obtain perfect destructive interference such that no signal is observed in one of the counters (D_1), while a full-strength signal is observed in D_2, due to constructive interference.

The routes to both detectors involve one mirror reflection that adds a phase factor $\exp(i\theta_m)$ to *both* beams initially with amplitude $A(t_0)$ at time t_0. Assuming that the beam splitters split the beams 50/50, both routes to D_2 involve one BS reflection. Hence the two amplitudes carry phase factors $\exp(i\pi/2)$ and sum to $2\exp(i\pi/2)A(t)$ at D_2. On the other hand, one of the two paths to D_1 does not involve a BS-reflection, while the other is reflected by both beam splitters, leading to a phase factor $1+\exp(i\pi/2)\exp(i\pi/2) = 1+(i)^2 = 0$. Hence no signal is observed in D_1. The wavefunction at any time instant may be expressed using such phase information and the transmitted and reflected components of the incident ψ. Thus, at t_1, the total wavefunction $\psi(t_1,\vec{r}) = \{\psi^T(t_1,\vec{r}) + i\psi^R(t_1,\vec{r})\}/\sqrt{2}$.

Just as in the two-slit experiment, the observation of interference is usually (e.g., in Wheeler's discussion) taken to be evidence that the photon has *come by both routes*. However, this view is not obligatory (see Sec. 6.3.2). So far, the experiment is simply a conventional experiment on complementarity. In the delayed-choice version of the experiment, one decides whether to put in BS2, or not, only at the very last moment, at time t_2 (Fig. 7.9), 'after the photon has left BS1', i.e., after t_1. Thus, in Wheeler's words, *one decides whether the photon shall have come by one route, or by both routes, after it has already done its travel.*

A more humongous, astronomical version has also been attributed to Wheeler, where light emitted from a distant star is split into two beams by the gravitational-lensing effect of a stellar object. The light from the star began in the distant past, many light years away. By observing these photons either as an interference pattern on a detector screen, or using telescopes oriented to give *which-path* information, Wheeler contended that *the distant past could be created or altered by our manner of observing it now*. It is of course assumed here that the superposition states associated with the two paths do not get corrupted by random phase factors that would be acquired by the action of other fields, dust particles etc., that may be in the path (stretching over astronomical length and time scales). In real life, an interferometer of astronomical size (unlike a laboratory version of Fig. 7.9) cannot be isolated from the environment.

Niels Bohr [31], in commenting on this very experiment has stated that, in his view, the notion of a path taken by the photon has no meaning, and that 'it is just arguments of this kind which recall the impossibility of subdividing quantum phenomena and reveal the ambiguity in ascribing customary physical attributes to atomic objects'. However, such an answer is insufficient to dispel the difficulties of many physicists as well as philosophers who now find that the past may have no objective existence if Wheeler is correct. Physicists will continue to draw diagrams showing the 'paths' taken by the photons, and will attempt to understand the experiment in a convenient language which is also fully consistent with QM.

We review Wheeler's experiment using Bohmian concepts (BM, Sec. 6.7). Bohmian mechanics of photons is less familiar since the 'wavefunction of the photon' is not well established. However, the discussion in terms of electron waves gives a clear picture of the Bohmian view. In BM a particle always follows just one path, even in cases where a final interference pattern would occur. Here we recollect our discussion of the two-slit experiment within Bohmian mechanics. Thus Wheeler's view that an interference pattern is evidence that the photon has 'come by both routes', is not accepted by BM. On changing the experimental set up at time t_2, the pre-established part of the path ($t < t_2$) remains intact.

Objectively real past history is embodied in the trajectory of the particle. This trajectory of the particle was determined by the initial conditions at the start of the trajectory at time t_0, and the quantum potential $Q(x,t_0)$ acting on the particle at x. When the experimental setup is changed at time t_2 by inserting or removing BS2, after the time t_1 when the particle left the beam splitter BS1, we only change the quantum potential for $t > t_2$. Thus there is absolutely no paradox, and there is an objective past. At the same time, the measurement process continues to have the 'participatory' character in that the experimentalist can choose to measure an interference process or obtain path information. If BS2 is introduced to force an interference, then the trajectories of the particles (for $t > t_2$) would be affected by the new quantum potential, and bunch together as required to generate an interference result [30].

Thus, there is no technical difficulty in holding that the photon always takes one route or the other, but statistically arriving so as to form the interference pattern when needed. Nevertheless, a different view is held by Wheeler, Penrose and others. Penrose claims that 'it seems inescapable that the photon must, in some sense, have actually *traveled* both routes at once.' [160]. It is also stated that 'quantum mechanics gives us a wavefunction to describe the reality of the photon's position, and between the half-silvered mirrors, the photon's wavefunction is a doubly peaked state, the distance between the two peaks being sometimes very considerable'. Since the concept of a photon wavefunction is not well established, let us think of an experiment carried out with electrons. According to the conventional QM-Born interpretation, the probability of finding the particle has a two peaked structure. It does not mean that the mass of the particle is split or distributed between the two paths. Standard quantum mechanics gets at the answer without asking 'which way did the photon go?' type of questions. Hence the views of Wheeler and Penrose seem to differ from conventional QM. In the case of photons, the interpretation may be even more seriously open to question, as photons are always detected only as polaritons, i.e., photon-like localized excitations near matter. In our view, Penrose's, or Wheeler's conclusions cannot be supported, and inconsistent with, e.g., Clauser's experiment [48]. The original *gedanken* experiment soon became technically feasible, and many experiments, of increasing precision as well as sophistication [115], have been carried out, confirming the wave-particle duality predicted by QM. However, whether we actually changed the past as implied by Wheeler is not testable, as Bohr had already fore-warned.

The wave-particle duality in the classical domain is quite familiar, for example in crystallography, where a crystal has a real-space crystal structure (position specification), while its reciprocal-space structure (momentum-space specification) is its 'Bravais lattice'. However, unlike in the quantum domain, the two descriptions seem to co-exist. For macroscopic length and momentum scales, the quantum uncertainties are extremely small since the thermal de Broglie wavelengths are in the micro-femtometer range (see Eq. 7.56).

7.7.6 *The quantum Zeno (Watchdog) paradox*

Zeno was a contemporary of Socrates who claimed that motion and change are illusions. Thus, according to Zeno, a flying arrow is actually at rest at any instant, and yet the 'sum' of these rests is the content of its motion. Several paradoxes of classical philosophy are ascribed to Zeno. In Zeno's arrow paradox, by looking at the arrow at every instant, we conclude that the arrow is immobile. Some may claim that this paradox did not see a genuine solution until the arrival of the quantum theory which denied a position to a moving particle.

Beskow and Nilsson argued in 1967 that frequent position measurements, implicitly carried out by a bubble chamber on an unstable particle, could prevent its decay. The *quantum Zeno Paradox* [149], was the name given to the effect of frequent observations on a quantum state, forcing the state to persist in that state indefinitely. This is also known as the 'Watchdog effect', or the 'watched-pot effect'. The latter name is based on the saying 'A watched pot never boils', as used by, e.g., the Victorian novelist Elizabeth Gaskell, author of *Cranford*.

The quantum Zeno effect may at first sight look like a riddle or paradox. However, it is completely consistent with QM, and also with many everyday experiences. Thus we may consider a water tap which is 'motion activated'. When the hand is under the tap but not moving, no water flows. We may label this the $|0\rangle$ state. When the hand is moved as in a washing action, water begins to flow for a short duration of time τ. We may denote the flow state by $|1\rangle$. If we constantly move the hand, sufficiently frequently, then we can maintain the system in the flow state $|1\rangle$ indefinitely. Similarly, we may activate the state $|0\rangle$ and maintain it there by the operation of 'keeping the hand still'. This is a 'real-life example' of a Zeno effect involving the 'participation of the experimentalist' in the outcome.

The quantum Zeno effect is also an example of the participation of the experimentalist in the outcome of a quantum process. It is often presented in recent writings with a thick coat of metaphysics. In the simplest version of the 'Zeno effect', instead of acting continuously on a quantum system with some perturbation for a long time $t = T$, we act on it many times N, for short time intervals $\Delta t = T/N$. The long-time evolution of the system under the perturbation is found to be proportional to t (liner in t), while the short time evolution is quadratic in Δt. The short-time procedure allows the system to evolve during ΔT and then it is checked, or *measured*, to see how much of the initial state is still left. This is also known as *projecting* the time-evolved state back to the initial state. Since the quadratic value of $\Delta t \ll 1$ is even smaller than Δt, every application of the perturbation, however frequent, is less effective in forcing the quantum state to undergo a transition, and hence the system stays in the initial state. The repeated application of the 'measurement', followed by the perturbation is the continual act of watching, and the Zeno effect is the inhibition of the transition of the quantum state to a different state, normally expected as the effect of the perturbation.

The theory of the Zeno effect can be compared with the usual Fermi 'golden rule' that discusses the rate of transition of a quantum state $|i\rangle$, to some final state $|f\rangle$ under some interaction $\hat{H}_{int}$. The interaction acts continuously, and the golden-rule transition rate $R_g = 1/\tau_g$ is given by

$$R_g = 1/\tau_g = \sum_{i,f} |\langle i|\hat{H}_{int}|f\rangle|^2 D(i,f)\,. \tag{7.57}$$

In the above equation, $D(i,f)$ is a joint density of states associated with the transition $i \to f$. Most discussions of the Zeno effect are confined to a single initial state, e.g., $i = 0$, and a single final sate, e.g., an excited state $f = 1$. The joint-density of states is taken to be unity. The Fermi Golden-rule rate is usually found to lead to a life-time τ_g such that the rate of decay from the initial state is exponential, with τ_g the half-life. If the number of atoms in the state i is n_i, then

$$n_i(t) = n_i(0)\exp(-t/\tau_g) \;\simeq\; n_i(t)/n_i(0) = 1 - t/\tau_g + \cdots \tag{7.58}$$

On the other hand, the time evolution of the state $i = 0$, i.e., ψ_0, is given by the action of $U(t) = \exp(-iHt)$ on ψ_0, where $H = H_0 + H_{int}$. We have (with $\hbar = 1$),

$$H_0\psi_0 = E_0\psi_0, \quad \psi(t) = e^{-iHt}\psi_0, \tag{7.59}$$

$$p(t) = |\langle\psi_0|\psi_t\rangle|^2. \tag{7.60}$$

Clearly, $p(t)$ is the probability that the system is in the state ψ_0 after a time t. On expanding the time evolution operator $U(t)$, it is easy to show that:

$$p(t) = 1 - t^2/\tau_z^2 + \cdots, \quad \text{where } \tau_z^{-2} = \langle\psi_0|H^2|\psi_0\rangle - \langle\psi_0|H|\psi_0\rangle^2. \tag{7.61}$$

Here τ_z is the Zeno time, or the lifetime at very short time scales where the low-order expansion of $U(t)$ holds. This is quadratic in t, and should be compared with the linear t behaviour of Eq. 7.58, found in exponential decay. If the interaction is such that $\langle 0|\hat{H}_{int}|0\rangle = 0$, then τ_z^{-2} reduces to the term $|\langle 0|\hat{H}_{int}|1\rangle|^2$ found in the Fermi golden rule but the time dependence is quadratic. Thus short-time behaviour shows much less decay of the initial state, i.e., a higher 'survival probability'. This can be presented in a dramatic manner if we consider N short-time applications of $\hat{H}_{int}$ lasting $\Delta t = T/N$ seconds per 'shot', followed by projection to $|0\rangle$, i.e., taking a measurement of the survival fraction $p(t)$. Then, taking the $N \to \infty$ limit shows that the system remains in the initial state, with $p(t) = 1$. Thus the survival probability $P_N(t)$ after N measurements is:

$$p_N(t) = \{p(\Delta t)\}^n = p(T/N)^N = \left[1 - (t/N\tau_z)^2\right]_{N\to\infty} = 1. \tag{7.62}$$

Instead of letting the system in $|0\rangle$ to evolve freely under the interaction, we frequently intervene and 're-prepare' the system and retain it in $|0\rangle$, and the free evolution is halted. The phenomenon has been demonstrated in a variety of quantum systems (as reviewed in [71]), and it is now an actively researched area due to its possible importance in quantum computers and in controlling decoherence.

It is of course clear that the above discussion is highly deficient in many aspects. The validity and convergence of the perturbation expansion have to be examined. The clear problem with $\Delta E\Delta t \sim \hbar$ as $\Delta t \to 0$ and $\Delta E \to \infty$ requires treating a bigger range of energy states of the time evolving systems. In fact, an 'anti-Zeno effect', where the watched state 'evaporates' into other nearby quantum states is common for multi-level systems (see the review [71]).

Much confusion has arisen by taking the anthropocentric language used in standard measurement theory too literally. The quantum Zeno effect itself has been claimed to be the process used by 'the stream of human consciousness' to cognitively interact with the material world and disrupt its deterministic time evolution (Chapter 6, Henry Stapp [202]). Stapp talks of 'quantum states of the brain'.

Rigorously stated, Stapp holds that such states can only exist *via* a participating mind (that is, some one should be observing your mind for your mind to have such states, unless it can be boot- strapped as in Descartes' *cogito ergo sum*). In grappling with the question of how a 'person's effort could influence his or her physical action', Stapp calls upon the quantum Zeno effect for his rescue. He is forced to do this because the 'only dynamical freedom offered by the quantum formalism in this situation is the freedom to perform at a selected time, some process 1 action'. By process 1, Stapp means the 'basic probing action that partitions a potential continuum of physically described possibilities into a (countable) set of empirically recognizable alternative possibilities'. This is roughly what von Neumann called the 'gedankische innerlaben', and deals with the choice made by the human observer. Thus Stappp's quantum model of the brain, invoked to free himself from the alleged clockwork determinism of classical physics, finds itself totally restricted by the unitary dynamics of QM. Only the Zeno effect, and the *literal* invoking of the hypothesis of wavefunction collapse allow him any freedom. The Zeno effect occurs only if there are quantum states, and only if the 'measurement' intervals (presumably these are acts of volition) are short enough to beat the usual linear decay. In reality, if quantized energy states exist in the brain, an anti-Zeno effect is more likely because of the large densities of states found in 'macroscopic' systems. The 'stream of consciousness' which couples to the physical brain remains totally mysterious and un-elucidated. Given the quasi-continuous energy states expected in a system like the brain, and the temperature of the brain, Stapp's discussion becomes, in our view, inapplicable to real biological brains.

Chapter 8

Many Particle Systems and the Classical Limit

In this chapter we study the bridge between elementary quantum systems and the more complex many-particle systems of every-day reality. The quantum to classical connection is examined. Interacting many-particle quantum systems are found to be accurately describable by simple classical models containing non-local potentials. Density-functional theory is reviewed. The many-particle wavefunction can be subsumed within density functionals or classical distribution functions of the many-body system. Purely quantum as well as hybrid systems at finite T are described seamlessly. We examine explicit decoherence models added to QM, and find them to be unnecessary.

8.1 The 'real world' of many-particle systems

Textbooks on quantum mechanics begin with treatments of an electron in a box, a hydrogen atom, and such 'simple' few-body systems. Similarly, discussions on the nature of quantum reality often focus on individual electrons, photons, two-level superpositions, or systems with perhaps two electrons or spins in some state of entanglement (see Chs. 6 and 7). Alternatively, the discussion goes to another extreme where the wavefunction of the Universe is considered (Sec. 6.10.4). Here too, usually just 'one object' is treated in, say, the WKB approximation. However, the real world contains many particles and their interacting fields.

In this chapter we are concerned with interacting many-particle systems typical of terrestrial bulk matter. A 100-electron atom is already an unstable system in the world of nuclei. Much of the universe is made up of lighter atoms, mainly hydrogen and helium, while terrestrial matter contains C, Si, O, N, P, etc., and the common metals. The number of atoms contained in an everyday sample of matter (e.g., a gram mole) would be at the scale of the Avogadro number, $N_{A\nu} \approx 6 \times 10^{23}$. That is, if we use the word billion to refer to a thousand million, then we are talking of particle numbers of the order of billion-billion-millions. Clearly we have changed our vantage position in the hierarchy of energy scales and length scales. None of the hall marks of QM, e.g., 'spooky action at a distance', interference of matter waves etc., are seen at the length scales of everyday reality.

The unit of quantization $\hbar$ is not zero, and the transition from quantum to classical behavior is a topic of continuing investigation. Should we treat everything using the quantum theory since it is the over-arching theory, or construct a *hybrid theory* where one part is treated by QM, while the rest is treated by classical theory? Systems with a mass exceeding $\sim$ 500 amu have de Broglie wavelengths smaller than a Bohr radius even at liquid-He temperatures, and they will behave as classical particles (see Eq. 7.56). Hence treating the temperature as an intrinsic part of the quantum theory becomes important in hybrid theories.

In this chapter we address how the quantum theory merges into the familiar classical description of the world. We do this by examining the following:

(1) QM is applied without modification to atoms, molecules, solids and such objects of everyday reality using the methods of quantum materials science. Here we use the Schrödinger equation, or possibly the Dirac equation, with a semi-classical or quantum description of light. Non-local potentials e.g., the Hartree-Fock potential, are standard fare in materials simulations. Furthermore,, effective classical potentials that capture the quantum behavior can be constructed; these when used in *classical* simulations (e.g., in molecular dynamics) reproduce the observed quantum behavior, if non-local potentials are admitted. These suggest that *classical reality emerges automatically from QM as the length scales change, and that no special steps are needed.*

    David Bohm recasts the Schrödinger equation into a non-linear Newtonian equation of motion, and arrives at conclusions similar to those of the materials-physics community.

(2) Density-functional theory (DFT) offers a description of quantum systems based on the one-body density distribution $n(\vec{r})$ that emerges to take the place of the many-particle wavefunction. All the properties at $T = 0$ or at finite-T become functionals of $n(\vec{r})$. The latter is an experimentally accessible quantity applicable to quantum systems, classical systems, or hybrids at any temperature.

(3) We briefly review some of the theories which set out to explicitly ensure the emergence of classical behavior in quantum systems when the energy and length scales are suitably modified [189]. Theories of the sort given by Ghiradi, Rimini and Weber (GRW), or those of Gell-Mann and Hartle, fall in to one group as they invoke some form of direct decoherence. A related class of theories (see Bub, [37]) calls itself the 'new orthodoxy', and invokes decoherence by attaching 'ancilla' to the system under study and averaging over the ancilla variables to obtain equations containing decoherence effects. The other class of theories invokes an active role for the consciousness of the observer [202], leading to theories of varying degrees of subjectivity.

The world of every day reality is the regime of cold matter, shown in the lower left corner of the phase diagram of nuclear-matter, Fig 8.1. When a large number of electrons and positively charged nuclei share the same region of space and time, they condense into atoms if their total internal energy (E) per particle is low enough to establish a common temperature T which is sufficiently low.

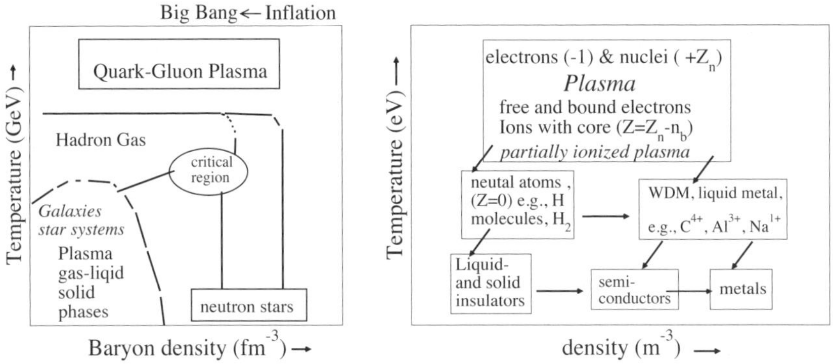

(a) phase diagram, nuclear matter (b) phase diagram, cold matter

Fig. 8.1 (a) The quark-gluon plasma of primordial matter cools to produce galaxies, stars, planets etc., and very dense objects like neutron stars (Sec. 5.4.1). Terrestrial matter is in a tiny region near the origin, i.e., in the gas-liquid-solid region. This is shown in more detail in panel (b). The phase diagram of cold matter. WDM is 'warm-dense matter', where warm is $T < T_F$, and T_F is the Fermi temperature. The diagram is not to scale; n.b., femtometers^{-3} in (a) and meters^{-3} in (b).

The temperature T of a system of classical particles, if measured in energy units is proportional to the average kinetic energy (KE) of the particles. In a quantum system (e.g., a gas of electrons), the average kinetic energy is *not* directly related to the temperature. Thus the average KE of electrons at $T = 0$ is not zero, but 3/5 of the Fermi Energy E_F of the system. The Fermi energy of a piece of Aluminum may correspond to 12 eV (i.e., $\sim$ 140,000 K) while its temperature may be 300K. The temperature T of a quantum system has to be calculated from, say, the equation of state (EOS) of the system, or determined from the heat bath coupled to the system. The EOS relates density, pressure and temperature, or other equivalent thermodynamic variables. There is no operator in QM whose meanvalue is T. It can be considered as a Lagrange multiplier ensuring the conservation of energy in a system. In thermodynamics, temperature appears as the energy derivative of entropy.

If the energy E is very high, then the electrons and ions do not combine and the particles exist as a fully or partially ionized gas of charged particles. Such a system is known as a *plasma*. Sometimes the ions may condense to form a warm-dense state of matter (WDM) which is like a molten, compressed metal. This may cool to a nearly rigid lattice, and the electrons become a common pool of charged carriers held trapped by the attractive potential of the lattice of ions. Such a system is a solid *metal*. Depending on the temperature of the system, (or equivalently, on the magnitude of E), the bare nuclei of charge Z_n usually acquire a few bound electrons, n_b. They are known as the 'core electrons'. Hence the effective charge Z of the ions becomes $Z = Z_n - n_b$. If enough bound electrons are acquired by the nucleus, Z can become zero and we have neutral atoms.

These neutral atoms join up with other atoms, to share their electrons forming entangled singlet states, i.e., chemical bonds of the type discussed in Sec. 7.3.1. In energetic (i.e., excited) situations, the interactions may occur via triplet states, when chemists talk of *free radicals*. Free radicals correspond to un-bonded electrons with unpaired spins. The upper, non-bonding curve in the black board behind Paul Dirac in Fig. 7.4 contains two H-atoms with un-bonded electrons.

Protons join with electrons to form H atoms, and they in turn form stable H_2 molecules via covalent singlet bonding. The amount of protons, electrons, H-atoms and H_2 molecules present in a system at any moment is determined by a minimization of the energy of the system using the density matrix which contains statistical factors corresponding to each eigenstate of the system. The density matrix determines the free energy F of the system. It is equally accurate to say that the density distribution $n(r)$ of a system is such that it minimizes the total free energy. Such considerations lead to chemical equilibria. The chemical balance equations are usually written as:

$$\mathrm{H}^+ + e^- \rightleftharpoons \mathrm{H}, \tag{8.1}$$

$$\mathrm{H} + \mathrm{H} \rightleftharpoons \mathrm{H}_2 . \tag{8.2}$$

The *Saha equation* provides a very simplified treatment of the ionization equilibrium given in Eq. 8.1. The interactions among atoms and molecules can be presented as inter-atomic or inter-molecular potentials. They generate further condensation of atoms into molecules, molecular or ionic liquids and solids. These low-energy interactions, associated with the valance forces between atoms, electrostatic interactions, and residual forces like 'van der Wall forces' (see sect. 6.4.1), produce the world of chemistry and biology. The interactions are precisely those of electric charges obeying quantum mechanics.

Quantum chemistry is the subject which attempts to describe chemical phenomena entirely using quantum mechanics. The *ab initio* calculation of molecular properties, treating a molecule as a collection of electrons and nuclei is now a well understood problem for which *off-the-shelf* computer codes are available [81]. These codes solve the Schrödinger equation of the molecule, and provide a many-electron wavefunction. Accurate calculations for systems containing hundreds of electrons are now a routine matter. Similar computer codes exist, for solids, liquids, and partially-ionized hot plasmas. The calculated values for individual properties, or thermodynamic properties (e.g., equations of states) and linear transport properties are usually in good agreement with experiment. If there is a disagreement, that is not indicative of a failure of QM or statistical mechanics. Instead, it points to some shortcoming in the mathematical approximations used in solving the many-electron Schrödinger equation.

8.2 Many-electron systems

Most discussions which focus on interpretations of QM deal with a few simple particles with two levels, and their manipulation by an observer. When many particles are included in the system under study (SUS), even if they are 'non-interacting', there are nevertheless special quantum correlations which come into being, due to the quantum nature of the particles (i.e., Bosons, Fermions). This was discussed to some extent in the context of the Bell-CHSH inequalities for two quantum particles. Mermin [141] has explicitly discussed three correlated quantum particles. In this section we look at the 'real world' limit typical of materials science. The particle number in a piece of solid Aluminum is in the Avogadro-number regime ($\sim 10^{23}$/mole). The sort of properties that we are concerned with, in everyday reality, are simply found to be proportional to the mass or extension (volume) of the sample under study. Thus we discuss the properties of the system in terms of the density ρ which is the number of particles per unit volume of the material. In the following we use $\rho(\vec{r})$ for the nuclear-density (or ion density) distribution, and $n(\vec{r})$ for the electron density distribution. In examining density-functional theory we find that the density distribution $\rho(\vec{r})$ or $n(\vec{r})$ of non-uniform systems, known as the one-body density is a property which takes over the job of the wavefunction for many-particle systems.

Keeping in mind that quantum systems are holistic in all sorts of non-intuitive ways, we work with the total number of electrons N_e, and the total number of nucleons N_{nu} in a volume Ω, selected by the experimentalist. Usually the quantum mechanics is done at zero temperature, and an isolated system is considered. However, this restriction can be relaxed in a variety of ways once the excitation spectrum of the fully isolated system (a 'reduction') is determined by a full quantum mechanical calculation. The Hamiltonian H was already introduced in Sec. 6.4 as the operator which corresponds to the total energy of a quantum system. This H usually contains the 'non-interacting' or 'one-body' part H_0 and the interacting part H_{int}. Thus

$$H = H_0 + H_{int}\,. \tag{8.3}$$

For instance, in the model of a trapped electron studied in Sec. 6.4, H_0 was simply the kinetic energy operator (inside the square-well potential). In the case of the harmonic oscillator, H_0 included a parabolic potential which depended *only on the position of the particle under study*. Thus, H_0 contains the kinetic energy operator and one-body potentials which are entirely local. The potential which acts on the electrons is sometimes known as the 'external potential', although it could actually be generated internally. In the case of atoms, the positively-charged nuclei, being at the very least three orders of magnitude (i.e., 10^3) heavier than electrons, can be regarded as being at rest (to a first approximation), and providing the 'external potential' which traps the electrons.

The idea of treating the classical system of ions as the source of an 'external potential' is used in a very simple but surprisingly useful model of condensed matter. Thus various versions of the *jellium model* of charged particles (electrons) held in place by heavier positive particles (ions) are useful in condensed-matter physics, plasma physics, nuclear physics and astrophysics.

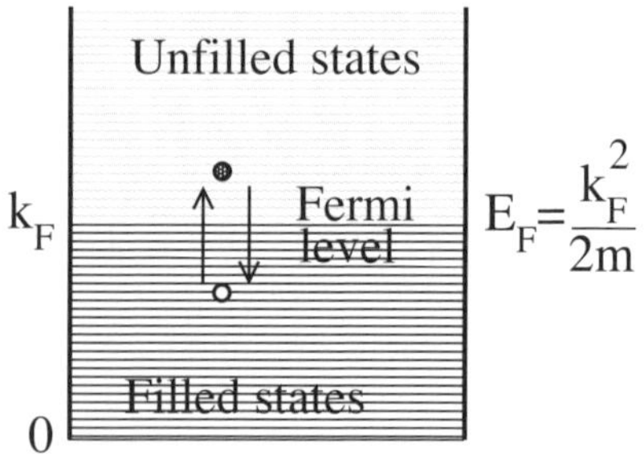

Fig. 8.2 Electrons occupy momentum states k up to the Fermi momentum k_F, at the Fermi energy E_F. Electrons can be excited to $k > k_F$ states leaving a 'hole' in the 'Fermi sea' of filled states. The hole and the particle (above the Fermi level) form a 'particle-hole' pair. Superposition states of particle-hole pairs screen the Coulomb interactions among electrons, converting them to weakly interacting Landau 'quasiparticles'. The 3-dimensional Fermi sea is a sphere, and the Fermi level is the *Fermi surface* of the sphere.

8.2.1 *The jellium model*

The so-called 'jellium' model of metals assumes that the nuclei plus their core electrons merely provide a uniform static positive charge exactly sufficient to neutralize the charge of the electrons. Thus, such a piece of metal (or such a plasma) is electrically neutral, but contain a 'gas' or 'liquid' of electrons that move about inside the metal. In the case of a liquid, the time-averaged ion distribution may be regarded as fixed and used to define its spatial extension. The mobile particles of this liquid are the electrons. Their distribution among the various momentum states (or energy states) is governed by Fermi-Dirac statistics. That is, each level, characterized by a momentum $\vec{k}$ and spin index σ can have only one electron, or no electrons at any instant. Lev Landau (1908–1968), the great founding figure of Soviet theoretical physics, argued that even when the Coulomb interactions among the electrons (contained in the H_{int} part of the Hamiltonian) are taken into account, the electrons behaved like virtually non-interacting 'quasi-particles' constituting a 'Landau Fermi liquid'.

Electrons in many metals (e.g., sodium, aluminum) behave like non-interacting particles even though they have strong Coulomb and exchange interactions among them. Wigner drew attention to this in the 1930s. Landau suggested that electrons manifest themselves as 'quasi-particles' with very diminished or 'screened' interactions among them. Electrons occupy all the momentum states up to a maximum value k_F sufficient to contain the N electrons with any state occupied by not more than two electrons (up, and down spin). Today we know that this process involves the formation of a superposition state of many 'particle-hole' pair-states that are entangled with the electrons (see Fig. 8.2). The resulting many-body quantum field has excitations which are the Landau quasiparticles, known also as 'dressed electrons'. They are essentially very weakly interacting Fermions. These superpositions are very stable, suggesting that there is no decoherence of such superpositions (Sec. 8.5).

The jellium model simplifies the quantum mechanics of a large ensemble of electrons held trapped inside a metal (or a plasma) by the positive charges of the ions. The point-like positive charges are smeared out to give a uniform charge density. For instance, a piece of Al consists of a nearly rigid periodically positioned (cubic) lattice of Al^{3+} ions, with three 'free' electrons per ion released into the common electron liquid that pervades the whole sample of the metal. In the jellium model of solid Aluminum or Aluminum plasma (below temperatures $T < 20$ eV), at normal density (~ 2.7 g/cc), we have a uniform, static positive charge in which electrons move freely. The density of these electrons is the same as the density of free electrons in a piece of Al at normal density. The word 'free' is used here in the sense that these electron can move about within the confines of the metal (or the plasma), and their interactions are weakened due to 'screening'.

Using the atomic weight in grams of Aluminum, the normal density ρ_0, and the Avogadro number which is the number of Al ions contained in an atomic weight, it is easy to determine the average electron density (i.e., the number of electrons per unit volume, n_e) in Aluminum. Rather than using centimeters or meters, it is more appropriate to use atomic units, (a.u., see Sec. 6.25) where the Bohr radius of the H atom is taken as the unit of distance. Then it is found for Al that there are $n_e = 0.0268$ electrons per unit of atomic volume. Since the positively-charged ion distribution is assumed to be uniform, the electron density is also uniform. A very interesting measure of this uniform density is the radius of the sphere which contains, on average, just one electron. This is known as the *Wigner-Seitz radius*, and denoted by r_s. Thus,

$$\text{in 3-D} \;\; r_s = \left[\frac{3}{4\pi n_e}\right]^{1/3}, \;\; r_s = \left[\frac{1}{\pi n}\right]^{1/2} \;\; \text{in 2-D}. \tag{8.4}$$

Thus the value of r_s in Al at normal density and temperature is ~ 2.1 a.u. [11]. This is ~ 0.1 of a nanometer (10^{-9}m), while the r_s of matter in a white dwarf is below the femtometer (10^{-15}m) range. The r_s sets a scale which controls the correlation lengths of entanglement in interacting many-electron systems.

8.2.2 *Non-interacting electrons*

We consider a gas of non-interacting electrons. Their Fermionic nature forces them into superposition states with built-in non-local correlations. We look at the wavefunction of entangled electrons, and calculate the pair-distribution function (PDF) of the N-particle system. We already displayed this PDF in Fig. 7.1. Such N-electron entangled states are typical of electrons found in metals. Such superpositions involve quantum correlations, and are totally stable.

Let us now consider a jellium model of N non-interacting electrons contained in a volume Ω such that the uniform electron density n_e is N/Ω. In fact, we may assume that both N and Ω are very large, while the density n_e, or equivalently the r_s parameter is held fixed by the experimentalist. The external potential $Vext(\vec{r})$ acting on any electron is zero except at the boundary of Ω which is very large. Below we write $V_{ext}(\vec{r})$ as $V(\vec{r})$ for brevity. Hence the wavefunctions of the electrons are taken as simple plane waves, occupying momentum states $\vec{k}$ corresponding to the energies $\varepsilon(k) = (\hbar k)^2/(2m_e)$. The values of $\hbar$ and the electron mass m_e are unity in atomic units. If there is no external magnetic field present, the up-spin electrons have the same energy as the down-spin electrons, and hence the energies $\varepsilon(k)$ do not depend on the spin state. At the risk of some sloppiness in notation, we use the form $|1\rangle$, or sometimes $|\phi\rangle$ or $|k\rangle$ to briefly indicate the one-particle wavefunction $\phi(\vec{r}_1)$ itself. Let us also note that each electron has some spin state σ which may be 'up' (u) or 'down' (d), corresponding to $\pm\hbar/2$. Thus, when we write $|1\rangle$, it is implicitly $|1,\sigma_1\rangle$. Hence, the Schrödinger Equation, eigenvalues ε_k, and eigenstates $\phi_k(\vec{r})$, denoted by $|k\rangle$ can be given as follows:

$$H_0|k\rangle = \varepsilon_k|k\rangle, \quad \varepsilon(k) = k^2/2, \tag{8.5}$$

$$|k\rangle = \phi_k(\vec{r}) = \Omega^{-1/2}e^{-i\vec{k}\cdot\vec{r}}. \tag{8.6}$$

A uniform one-body density $n(\vec{r}) = n$ results from $V(\vec{r})$ being zero. If the external potential $V(\vec{r})$ were non-zero, $n(\vec{r})$ becomes non-uniform, and we need to solve the many-body Schrödinger's equation, using the boundary conditions set by $V(\vec{r})$, to get $n(1,2,\ldots) = |\psi(1,2,\ldots)|^2$, where r_1, r_2 etc. are abbreviated to 1,2, etc. Then all except one of the r_i are integrated to get the one-body density $n(\vec{r})$ (see Eq. (8.18)).

The one-electron solutions $|k\rangle$, or 'orbitals', obtained from Eq. 8.5 have to be combined to form the many-electron wavefunction $\psi = |1,2,3,\ldots,N\rangle$ of the total system. Since electrons are Fermions, the wavefunction must change sign if any two electrons are exchanged by swapping their quantum numbers, as required by the Pauli exclusion principle. Thus, for two electrons at $\vec{r}_1$ and $\vec{r}_2$, in the states ϕ_1 and ϕ_2, the correct wavefunction is the entangled state:

$$|1,2\rangle = (1/\sqrt{2})\left[\phi_1(\vec{r}_1)\phi_2(\vec{r}_2) - \phi_1(\vec{r}_2)\phi_2(\vec{r}_1)\right]. \tag{8.7}$$

It is easily verified that this can be written as a 2×2 determinant:

$$|1,2\rangle = \frac{1}{\sqrt{2}}\begin{vmatrix} \phi_1(\vec{r}_1) & \phi_1(\vec{r}_2) \\ \phi_2(\vec{r}_1) & \phi_2(\vec{r}_2) \end{vmatrix}. \tag{8.8}$$

This is already the form of the wavefunction that we came across in Sec. 7.1. The determinantal form can be used for the entangled state of any number of electrons. In general, the N-particle fully-entangled wavefunction for non-interacting electrons is the $N \times N$ *Slater determinant*:

$$|1,2,\ldots,N\rangle = \frac{1}{\sqrt{(N!)}}\begin{vmatrix} \phi_1(\vec{r}_1) & \phi_1(\vec{r}_2) & \cdots & \phi_1(\vec{r}_N) \\ \phi_2(\vec{r}_1) & \phi_2(\vec{r}_2) & \cdots & \phi_2(\vec{r}_N) \\ \cdots & \cdots & \cdots & \cdots \\ \phi_N(\vec{r}_1) & \phi_N(\vec{r}_2) & \cdots & \phi_N(\vec{r}_N) \end{vmatrix}. \tag{8.9}$$

We should now explicitly take account of the spin-state of the electrons. If all the electrons are in the same spin state, e.g., if they are all in the state σ_1, we have a fully spin-polarized electron gas. Then a factor $\sigma_1(1)\sigma_1(2)\sigma_1(3)\cdots\sigma_1(N)$ can be pulled out of the Slater determinant, and the antisymmetric product is entirely made up of the space parts of the electron wavefunctions.

In the two-electron case, if the electrons have the same spin we simply pull out the factor $\sigma_1(1)\sigma_1(2)$ from Eq. 8.7 and we are left with just the space part:

$$\psi(1,2) = \frac{1}{\Omega\sqrt{2}}\left[e^{i\vec{k}_1\cdot\vec{r}_1}e^{i\vec{k}_2\cdot\vec{r}_2} - e^{i\vec{k}_1\cdot\vec{r}_2}e^{i\vec{k}_2\cdot\vec{r}_1}\right] \tag{8.10}$$

$$= \frac{\sqrt{2}}{\Omega}e^{(i\pi/2+i\vec{K}\cdot\vec{R})}\sin(\vec{k}\cdot\vec{r}/2). \tag{8.11}$$

In the last line we have introduced the relative coordinates and momenta $\vec{r} = \vec{r}_1 - \vec{r}_2, \vec{k} = \vec{k}_1 - \vec{k}_2$, and the center of mass coordinate $\vec{R} = (\vec{r}_1 + \vec{r}_2)/2$, for the pair of electrons which have become entangled due to the Pauli exclusion principle. The Fermi statistics excludes the second electron from the neighborhood of the first, as they cannot coexist in the same spatial region without violating the exclusion principle. It is also seen that the Bohm quantum potential replaces the relative kinetic energy of the two particles, just as in the case of a particle trapped in a quantum well (Sec. 6.4). That is, in the Bohm picture, the 'relative particle' is at rest.

That is, even though the particles are non-interacting, the entanglement (due to the exclusion principle) has produced extensive space correlations. If all the particles are in the same spin state (spin-polarized system), an effective repulsive correlation comes into being. On the other hand, if the spins are anti-parallel, (unpolarized system) no such spatial correlation is produced. The spin-polarization of a system can be defined in terms of the parameter ζ

$$\zeta = \frac{n_u - n_d}{n}, \quad n = n_u + n_d \,. \tag{8.12}$$

The spatial correlations due to the Pauli principle are a maximum for $\zeta = 1$ and zero for $\zeta = 0$.

We have discussed these non-local correlations for particles obeying Fermi statistics. Similar correlations arise for Bose particles (e.g., photons), where the total wavefunction, instead of being antisymmetric, assumes a symmetric form. 'Photon bunching' is a consequence of such correlations. The Bose-Einstein condensation is a result of Bose particles all existing in the same quantum state as there is no exclusion principle prohibiting co-occupation. These effects are non-local and instantaneous, in the sense that they are built into the many-body time-independent wavefunction satisfying the Schrödinger equation. These are precisely the correlations that were termed 'spooky action' by Einstein. However, these instantaneous correlations cannot be used to transmit information faster than light, and hence do not violate special relativity (see Ch. 7 of [28]).

The many-electron wavefunction $\psi(1,2,\ldots,N)$, and its square $|\psi|^2$ for a laboratory sample of metal is an unwieldy object which depends on a large number of variables. On the other hand, most experimental observables (e.g., kinetic energy, dipole moment, total energy) depend on just a few variables. Hence it is enough to use the particle density, i.e., the one-body density $n(\vec{r}_1)$ for determining the average value of the dipole moment or the kinetic energy as they are observables of one-body operators. The total energy involves the Coulomb potential which is a two-body operator. Hence we need the two-body density $\langle n(r_1)n(r_2)\rangle$, obtained by integrating over all the variables in $\psi(1,2,\ldots,N)$, except for $r_1 = (\vec{r}_1, \sigma_1)$ and $r_2 = (\vec{r}_2, \sigma_2)$. The one-body density distribution plays a special role in density functional descriptions of matter.

The two-body density can be scaled by the average density n to give the probability of finding an electron at $\vec{r}_2$ given an electron at $\vec{r}_1$. In fact, taking $\vec{r}_1$ to be the origin of the coordinates, and since the system is spherically symmetric, we drop vector notation and use the radial $r = |\vec{r}_2|$ to discuss the pair-distribution function $g_{12}(r)$ for any pair of electrons in the system containing N particles. Such a pair distribution function (PDF) for a spin-polarized electron fluid was shown in Fig. 7.1. Starting from the square of the N-particle determinantal wavefunction, i.e., Eq. 8.9 and integrating over all but two of the space and spin variables, the zero-temperature PDFs $g^0_{\sigma,\sigma'}(r)$, where σ can be up (u), or down (d) are given by

$$g^0_{ud}(r) = 1\,, \tag{8.13}$$

$$g^0_{uu}(r) = g^0_{dd} = \left[1 - \{3j_1(rk_F)/(rk_F)\}^2\right]\,. \tag{8.14}$$

(The large-r behavior is normalized to unity, whereas if $\zeta = 0$, it is common to normalize the spin-resolved functions to 1/2). Here $j_1(x)$ is the spherical Bessel function, while its argument x contains $k_F = 1/(\alpha r_s), \alpha = 0.52106$. This is the Fermi momentum defined in Eq. 8.16 and is the highest momentum state occupied by an electron at $T = 0$. The probability of finding an electron at r, given that there is an electron at the origin of coordinates is affected by their mutual Coulomb interaction. This is not included in the above calculation, and hence the superfix 'zero' in g^0.

A most important characteristic of this PDF is that it is a universal function which scales with r_s, due to the argument $x = rk_F = r/\alpha r_s$. That is, in *laboratory-scale many-particle systems, the 'spooky-action' correlations scale down to the size of* r_s, *i.e., a fraction of a nanometer* and hence not perceivable. However, if two electrons are separated by a lunar distance $R = 2r_s$, the spooky action adjusts to the new length scale. Hence these effects *breach* the form of local realism required by Einstein *et al.*, in the EPR paradox (Sec. 7.3.7).

In summary, these correlations arising from basic Fermi or Bose statistics have some important characteristics:

(a) Their range (i.e., non-locality) is controlled by the density of particles in the system. That is, for laboratory-scale systems their range becomes comparable to the Wigner-Seitz radius r_s.
(b) The correlation effects can be mimicked by equivalent classical potentials, as shown by Lado [123] in the 1960s (see Fig. 7.1).
(c) They can also be expressed as generalization of Bohm quantum potentials which are independent of the strength of the quantum field (wavefunction) and prevail at all length scales. These potentials have the character of 'information'.

Thus, the restriction of non-local correlations to length scales of the order of the Wigner-Seitz radius of the particles in the system, and the possibility that the PDFs can be mimicked by classical potentials establish in detail that the quantum mechanical description of normal material systems merges seamlessly with the classical system, as the the particle numbers per unit volume are increased. How ever, this does not exclude us from *engineering* large scale superpositions or entanglements. Hence we have SQUIDS, fullerene cat states (Sec. 7.7.4.1) etc., where macroscopic quantum correlations are made to persist.

8.2.2.1 *Electron-electron interactions.*

In the above discussion we have only considered the non-interacting part of the Hamiltonian, viz., H_0. It now remains to examine the effect of interactions, viz., H_{int} on the quantum mechanics of the system. Generally speaking, the most common contribution to the interaction part of the Hamiltonian comes from the Coulomb potentials that occur among electrons and nuclei in matter.

The interaction with the positive background is written as H_{b-e}. We have:

$$H = H_0 + H_{int} = \sum_{i=1}^{N} \frac{-\nabla_i^2}{2} + \sum_{i<j}^{N} \frac{e_i e_j}{|\vec{r}_i - \vec{r}_j|} + H_{b-e}. \tag{8.15}$$

Here we have explicitly written e_i and e_j for the electron charge which is -1 in atomic units. The Coulomb interaction term involves the two-body form $1/|\vec{r}_i - \vec{r}_j|$. There is no 'singularity' arising from $|\vec{r}_i - \vec{r}_j|$ going to zero as such terms are eliminated by counter terms in the interaction with the uniform static positive background used in the jellium model (this is the only role played by H_{e-b}). The non-interacting part of the Hamiltonian H_0 is the free kinetic energy of the system. That is, we add electrons to the momentum and spin states $(\vec{k}, \sigma)$, starting from $k = 0$ up to the highest occupied value $|k_F|$ (See Fig. 8.2), with energy E_F. Thus, summing over all levels $k = 0, \ldots, \vec{k}_F$ we get a total of N electrons, with each level occupied by two electrons (spins up, down) if the system is spin-unpolarized ($\zeta = 0$). Then, using atomic units where $\hbar = 1$, $|e| = 1$, $m_e = 1$, we have [11]:

$$E_F = \frac{k_F^2}{2} = \frac{1}{2}(3\pi^2 n_e)^{2/3}, \; k_F = \frac{(9\pi/4)^{1/3}}{r_s}. \tag{8.16}$$

Here n_e is the number density, i.e., the number of electrons per atomic unit of volume. The Fermi energy E_F is a measure of the kinetic energy (KE) of the system. The KE is the mean value of the quantum mechanical operator $\sum_i \hat{p}_i^2/2$ where $p_i = -i\hbar\nabla_i$. Thus, unlike the Coulomb potential, the KE operator truly acts on the wavefunction in a non-classical way. The Fermi energy in 3-D scales as $1/r_s^2$, while the interaction term (the Coulomb potential), scales as $1/r_s$. Thus the ratio of the potential energy to the kinetic energy, Γ, obtained here may be contrasted with the classical limit, Γ_c, where the kinetic energy is the temperature T in energy units.

$$\Gamma = \langle H_{int} \rangle / \langle H_0 \rangle = r_s; \;\; \Gamma_c = 1/(r_s T). \tag{8.17}$$

Γ is known as the 'coupling constant' of the system. If we remain at $T = 0$, the Coulomb interactions (a classical interaction) increases in strength (it varies as $1/r_s$) when the density is decreased, relative to the quantum part of the Hamiltonian which is the kinetic energy term which varies as $1/r^2$. That is, as materials systems become strongly correlated, (strongly coupled), they become more classical.

On the other hand, if an attempt were made to calculate the many-body wavefunction of such an interacting system, then the simple $N \times N$ Slater determinant, with one basis function per electron becomes woefully inadequate. That is, a basis set of N functions for N electrons needs to be greatly extended to $N_b \gg N$. Then, all states with the label i such that $i > N$ are nominally empty. When the interactions are switched on, an electron occupying a given eigenstate i can get excited to any arbitrary empty state $j > N$, creating a hole (an empty state) in i, and an excited electron in j (see Fig. 8.2). That is, new distributions, or configurations of electrons arise. In addition to the simple Slater determinant discussed before, we now need to construct all possible ${}^{N_b}C_N$ Slater $N \times N$ determinants out of all the N_b basis functions to include all possible excitations, or at least the most important ones. Thus the many-body wavefunction becomes a linear combination of all possible Slater determinants. Such a wavefunction is known as a 'configuration-interaction' (CI) wavefunction. The resolution of the N-electron problem in CI increases in complexity extremely rapidly (in a non-polynomial manner), and exact diagonalizations become impossible except for problems involving a handful of electrons. However, clever approximations, as well as methods based on density-functional theory (discussed below) provide the means to deal with larger systems.

Thus, the computational tools of quantum chemistry enable us to see that quantum behavior seamlessly becomes classical behavior as we consider increasingly realistic systems with normal densities where r_s remains at nanometer length scales. The 'spooky action' effects are restricted to such length scales, and manifest mainly as familiar chemical bonding!

8.3 Density functional theory

A remarkable theorem which suggests the possibility of a classical view of a wide class of quantum systems (and perhaps all many-particle systems) was proved by Hohenberg and Kohn in 1964 [105]. This theorem displaces the many-body wavefunction ψ and brings forward the humble single-particle density distribution as the key quantity that determines all the properties of a quantum system at zero or finite temperatures.

The key quantity is the one-body density distribution $n(\vec{r})$; it arises from the quantum-mechanical interaction of the particles with each other and the external potential $V_{ext}(\vec{r})$. The external potential may be, for instance, that of a quantum well (see Fig. 6.6) where it is zero inside the well, or the attractive potential $(Z/|\vec{r}-\vec{R}|)$ due to an ion of charge Z at $\vec{R}$, or the potential due to the positive ions $(\Sigma_i Z_i/|\vec{r}-\vec{R}_i|)$ in a metal or a plasma.

The Hohenberg-Kohn theorem asserts that all properties of an inhomogeneous electron system are known if the one-body density distribution $n(\vec{r})$ is known. This has been extended in many directions by Walter Kohn (Fig. 8.3) and collaborators. However, there has been a great hesitation to formally do away with the wavefunction. Nevertheless, this is the inescapable conclusion, at least for static properties of quantum systems even at finite-T (cf., David Mermin). A formulation which uses the time-dependent density $n(\vec{r},t)$ has been formulated in terms of the principle of least action. This theorem, if true, implies that even time-dependent phenomena can be treated without the wavefunction.

The energy E of a quantum system is a functional of the one-particle density $n(\vec{r})$, and E takes its minimum value when $n(\vec{r})$ is the true ground-state density. At finite T, we use the Helmholtz free energy F instead of the total energy E. Thus a new approach to the computation of quantum systems, known as density-functional theory (DFT) becomes possible. Of course, if ψ is known, the one-body density distribution is obtained from it by integrating over all particle coordinates except one. Thus, writing 1,2, etc., to mean $\vec{r}_1, \sigma_1, \vec{r}_2, \sigma_2$, where σ is a spin etc., we have:

$$n(\vec{r}) = \int |\psi(1,2,3,\ldots,N)|^2 d2d3\cdots dN\,. \tag{8.18}$$

Calculating an accurate N-body wavefunction becomes extremely difficult as N increases, and various clever approximation schemes, based on mean-field theories and corrections to them are used. Mean-field approximations simplify the many-body problem by replacing it by a set of one-body problems. That is, any electron at $\vec{r}$ feels only an averaged-out 'mean-field' potential $V_{mf}(\vec{r})$ acting locally at $\vec{r}$. A simple Schrödinger equation for just one electron moving in the mean potential is solved. Douglas Hartree used just such methods in solving for the wavefunctions of many electron atoms, even before the advent of digital computers. In such theories, the many-electron wavefunction involving many determinants (used in the configuration interaction method) is replace by a single-determinant (or just a few symmetry-adapted forms) and the problem becomes computationally very tractable.

The power of density-functional theory is that it is *not* a mean-field theory. It is an exact theory, and provides, at least in principle, the possibility of treating the many-body problem as a set of effective one-body problems. The original theorem asserts that the exact ground state density distribution can be calculated, without using the Schrödinger equation, via a variational principle involving only the density distribution. Thus, for a system of electrons interacting via the Coulomb potential, and interacting with an external potential $V_{ext}(r)$, the true ground-state density $n(r)$ and the energy E are obtained by minimizing the energy functional

$$E[n] = T_{ke}[n] + W_{cou}[n] + \int dr^3 V_{ext}(r)n(r)\,. \tag{8.19}$$

Fig. 8.3 Walter Kohn, born in 1923 began physics as a teenage-refugee in Canada. After studying under Schwinger he moved to condensed-matter physics. The Hohenberg-Kohn theorem (1964) and generalizations imply that the quantum physics of a many-particle system in an external potential can be completely specified via its one-particle density distribution. Photo, 1994, by the author.

Here we use the notation $f[n]$ to indicate a functional of $[n]$. The Coulomb energy $W_{cou}[n]$ has a known form. The kinetic energy functional $T_{ke}[n]$ is a universal functional (i.e., independent of the external potential), but its exact form is unknown. At finite-temperatures, the corresponding kinetic free-energy functional is needed, and the functional minimization gives the interacting equilibrium free energy $F[n]$ and the density $n(\vec{r})$ at the temperature T.

The theory of 'hybrid systems' is a complex issue in standard quantum mechanics [104]. Furthermore, incorporating finite temperatures involves doubling all quantum operators to define a *thermofield dynamics*, as in Umezawa's method [103], or using the doubled-contour techniques of Martin, Schwinger and Keldysh. However, 'hybrid' finite-T systems are equally easily and seamlessly treated in DFT. Given a 'hybrid system' containing classical particles (e.g., ions) with a density distribution $\rho(\vec{r})$, and quantum particles with a density distribution $n(\vec{r})$, the free energy is now a functional of both densities, viz., $F[\rho(\vec{r}),n(\vec{r})]$, and the treatment is at finite-T.

The corresponding coupled Euler-Lagrange variational equations of the hybrid system are:

$$\frac{\delta F[n(\vec{r}),\rho(\vec{r})]}{\delta n(\vec{r})}=0; \quad \frac{\delta F[n(\vec{r}),\rho(\vec{r})]}{\delta \rho(\vec{r})}=0. \tag{8.20}$$

The first of these, involving a functional derivative with respect to the electron-density can be written as a Kohn-Sham equation (see Eq. 8.21), while the second equation is a classical functional derivative for the ion density. The latter leads to a Boltzmann-type distribution for the ion density $\rho(\vec{r})$ in terms of an effective classical Kohn-Sham potential (see Eq. 8.25).

The Kohn-Sham equation side-steps our ignorance of the kinetic energy functional by recasting the electronic Euler-Lagrange equation as a one-body effective Schrödinger equation.

The ion distribution $\rho(\vec{r})$ is essentially fixed in solids. Hence it is treated as the generator of an 'external potential' and further approximated by a uniform distribution ρ. This is the 'jellium model' that we already discussed. Amazingly, it even works for hot plasmas if simple modifications are carried out [59]. Hence we will simply regard ρ as a uniform density selected to just ensure the charge neutrality of the system. Then we need not solve coupled equations for $n(\vec{r})$ and $\rho(\vec{r})$.

In DFT, each particle moves in an effective potential usually known as the Kohn-Sham potential, $V_{KS}(\vec{r})$. This potential is a *functional* of the density distribution $n(\vec{r})$ and hence involves information from the whole system. That is, although V_{KS} is a one-body potential, it is in general non-local and contextual, as required by the non-local nature of the quantum theory. Hence it is often written as $V_{KS}([n(\vec{r})])$. Here we note that the Bohm quantum potentials are also functionals of the density $n(\vec{r})$ via the wavefunction. The essentially non-electrostatic many-body components of the Kohn-Sham potential are lumped into what is known as the exchange-correlation potential $V_{xc}(\vec{r})$. In effect, the two-body correlations contained in the PDF of the system are incorporated in $V_{xc}(\vec{r})$.

There exists a vast literature on constructing the exchange and correlation ('xc') part V_{xc} of the Kohn-Sham potentials for quantum calculations of electronic systems. Remarkably enough, one of the most successful such schemes treats $V_{xc}(\vec{r})$ as a *local property* of the density distribution $n(\vec{r})$, and is known as 'the local density approximation' (LDA). The LDA works well when the densities under study are typical every day materials. It works very poorly for very dilute systems of the sort considered by foundations-of-physics theorists, i.e., systems containing only a few particles. Although DFT is exact even in such limits, a convenient and accurate non-local xc-functional valid there is lacking, although gradient expansions have been constructed.

On the other hand, an exact expression valid at all temperatures and spin polarizations exists for $V_{Xc}(r)$ as the density-functional derivative of the exchange-correlation free energy $F_{xc}([n])$. The latter is the energy of the system beyond the mean-field (e.g., Hartree-Fock) energy. That is, F_{xc} can be expressed as a coupling-constant integration over the PDF. If we consider a jellium model of electrons in a uniform neutralizing positive background, then we can calculate $g(\lambda, r)$ for Coulomb systems, with the interaction λ/r, for *lambda* in the range 0–1, where λ is known as the 'coupling constant'. The value of F_{xc} can be expressed (Eq. 9.5) as an integration over λ and r [61]. Thus we see that DFT has replace the need for a knowledge of the PDF by a need for $F_{xc}[n]$ or $V_{xc}[n]$.

All these DFT results show that the classical description emerges naturally from the quantum description as we move to laboratory length scales. The quantum effects remain at the quantum length scales of the Wigner-Seitz radius, while the laboratory length scales are much larger. *The measured results are averages over suitable distribution functions that can be calculated from the one-body density distribution.*

8.3.1 *The Kohn-Sham method for quantum systems*

Density functional theory provides a means of replacing the interacting quantum many-body problem with an effectively non-interacting, exact, quantum one-body problem. The corresponding effective Schrödinger equation is known as the Kohn-Sham equation [121].

The effective one-body potential is the Kohn-Sham potential, $V_{KS}(\vec{r})$. This includes the external potential, the electrostatic potentials, and an *exchange-correlation potential* V_{xc} which mimics many-body effects as a one-body potential dependent on $n(\vec{r})$. At finite T, there is a corresponding temperature-dependent form of V_{xc} [162].

The Kohn-Sham equation is of the form:

$$\{-\nabla^2/2m + V_{KS}(\vec{r}) + V_{ext}(\vec{r})\}\xi_j(\vec{r}) = \varepsilon_j \xi_j(\vec{r}), \tag{8.21}$$

$$V_{KS}(\vec{r}) = V_{cou}(\vec{r}) + V_{xc}(\vec{r}). \tag{8.22}$$

This equation is solved iteratively, starting from a trial density distribution which determines the trial V_{KS} at that step, to obtain a set of Kohn-Sham eigenfunctions $\xi_j(\vec{r})$ and eigenvalues ε_j. At each iteration, a new estimate of $n(\vec{r})$ is obtained and fed back into the equations. The all important density distribution $n(\vec{r})$ is given by:

$$n(\vec{r}) = \sum_i f(\varepsilon_i, T)|\xi_i(\vec{r})|^2, \tag{8.23}$$

$$f(\varepsilon_i, T)) = 1/\{1+\exp(\varepsilon_i - \mu)/T\}. \tag{8.24}$$

Here $f(\varepsilon_i, T)$ is the Fermi distribution at the temperature T, and μ is the chemical potential. At $T = 0$ this is just unity for occupied states, and zero for unoccupied states. Then μ simply becomes the Fermi energy E_F, i.e., the energy of the highest occupied level. The index i includes summations over spin states as well. The Kohn-Sham eigenfunctions are merely calculational intermediates. In fact, the original Hohenberg-Kohn theorem permits, at least in principle, a determination of the optimal $n(\vec{r})$ by a direct energy minimization if the Hohenberg-Kohn kinetic energy functional $T_{ke}[n]$ were known.

8.4 Classical maps for quantum systems

The Hohenberg-Kohn and Mermin (HKM) variational principle allows us to determine properties of finite-T many-particle systems without using a wavefunction. It is interesting to look at the HKM theory in the classical limit of extremely high temperatures, and see how it modifies towards lower temperatures. Such a study enables us to directly construct approximations to the pair-distribution functions of electronic systems at arbitrary temperatures. Such PDFs allow us to predict at least the static properties and linear transport properties of quantum systems without appeal to wavefunctions, using purely classical descriptions, together with some non-local effective potentials.

Well-established answers are available for uniform systems. In order to treat a uniform system as a Kohn-Sham problem, it is necessary to consider a charge (an electron) at the origin of coordinates, and study the inhomogeneous electron distribution $n(\vec{r})$ around the origin. The presence of a charge at the origin will modify the distribution, but it will 'heal' sufficiently far away, due to screening effects by other electrons, and the density profile $n(\vec{r})$ will tend to the average density $n_e = n$ of electrons in the fluid. In the following we assume radial symmetry applicable to a fluid or a jellium model and use r instead of $\vec{r}$. In the classical limit of high temperatures (i.e., $T \gg E_F$), the profile $n(r)$, and the pair-distribution function $g(r)$ are simply classical Maxwell-Boltzmann distributions given by

$$n(r) = n_e \exp(-U(r)/T), \tag{8.25}$$

$$g(r) = \exp(-U(r)/T). \tag{8.26}$$

An electron at a radial distance r feels the potential $U(r)$ due to all the other particles in the system. This involves the direct Coulomb interaction $(1/r)$ with the electron at the origin (we use atomic units), plus the mean-field potential V_{mf} of all the other electrons in the electron gas, as well as the 'many-body' effect $V_{mb}[n]$ of correlations beyond the mean-field approximation. The latter is a non-local functional of $n(r)$ as it depends on the PDF.

$$U(r) = 1/r + V_{mf}(r) + V_{mb}(r). \tag{8.27}$$

All such effects, including the mean-field terms are included in various integral equations for classical fluids. A very successful integral equation that can be used here is known as the modified *hyper-netted-chain* (M-HNC) integral equation. This is essentially the hyper-netted-chain equation inclusive of bridge-correction terms which are non-reducible many-body terms not included in the chain sum.

Clearly, at high temperatures, the Kohn-Sham density, Eq. 8.23 evolves to become the Boltzmann form, Eq. 8.25. The Kohn-Sham potential becomes the so-called 'potential of mean-force' contained in the Boltzmann exponent. This, being purely classical, does not contain exchange effects, but contains the electrostatic terms and all the correlation effects.

8.4.1 *Quantum electron distributions from classical methods*

One can explicitly examine the quantum to classical transition by an examination of the system at $T = 0$ and its evolution to the classical limit using a finite-T calculation of the PDFs. The $T = 0$ electron gas is entirely quantum mechanical. Let us, for simplicity, consider a fully spin-polarized system where all the electrons are in the 'up' spin state. Then the non-interacting pair-distribution function $g^0(r)$ for an electron fluid at the density n_e, corresponding to $r_s = \{3/(4\pi n_e)\}^{1/3}$ can be calculated from a Slater determinant of plane waves [137]. At any temperature T, we have:

$$g^0(r) = 1 - F_T^2(r), \tag{8.28}$$

$$F_T(r) = (6\pi^2/k_F^3)\int f(k)\frac{sin(kr)}{r}\frac{kdk}{2\pi^2}. \tag{8.29}$$

Here k_F is the Fermi momentum and $f(k)$ is the Fermi distribution given in equation (8.24), at the energy $k^2/2$ (atomic units) while the temperature T is in energy units.

The correlations contained in $g^0(r)$ are entirely due to the quantum statistics of the particles. A classical potential $\beta P(r)$ which will reproduce $g^0(r)$ exactly can be constructed following Lado [123], by inverting the HNC-integral equation for $g^0(r)$. This potential is known as the *Pauli exclusion potential*. In fact, as a first guess one sees that

$$\beta P'(r) \approx -\log\{g^0(r)\}. \tag{8.30}$$

This P' is really the potential of mean force corresponding to $g^0(r)$. It contains the Pauli potential, as well as its mean-field term and many-body effects. These can be removed by properly inverting the *HNC* equation, and the actual $\beta P(r/r_s)$ is displayed in Fig. 7.1. It is a scale-independent potential and has the same strength for all densities if r/r_s is used as the variable.

The logarithmic form of Eq. 8.30 strongly suggests its kinship with the concepts of information and quantum entropy (Sec. 7.5.3). Here we cannot separate the β from the $P(r)$ as it is the *product* that can be determined from the known $g^0(r)$. That is, the 'β' refers to an internal inverse temperature which is not the physical temperature. Indeed, the physical temperature can be zero and yet $\beta P(r)$ is well defined. Hence $1/\beta$ can be used as the measure of an internal 'quantum temperature' T_q. This prompts one to consider a classical Coulomb fluid at T_q where the particles interact with each other via the total pair-potential $\beta V(r) = \beta\{P(r) + V_{cou}(r)\}$ where $V_{cou}(r)$ is the appropriate Coulomb interaction. It turns out that the quantum fluid can be very accurately 'mapped' via such a classical fluid using the pair-potential $V(r)$ and a quantum temperature T_q whose physical meaning could be related to density-functional concepts. The classically calculated PDFs given by Perrot and Dharma-wardana [61] are found to be in very good agreement with the quantum Monte Carlo PDFs at $T = 0$, in spite of the short-comings in the treatment of the bridge corrections that go into the modified HNC equation. This method, based on an approximate classical map of the quantum problem, and using the HNC equations, is known as the classical-map HNC, or CHNC technique.

These results show that the special correlations found in quantum systems can occur perfectly well within a classical scheme, if we consider appropriate non-local pair-potentials. The non-local correlations of Bohm quantum potentials are no different from the correlations produced by the Pauli potential. Robert Laughlin's classical map for fractional quantum-Hall fluids also illustrates the possibility of representing quantum systems by equivalent classical plasma models at finite temperatures.

The existence of classical models which can accurately mimic quantum systems — i.e., classical maps of quantum systems — is indeed the main message of Bohm's quantum mechanics where the wavefunctions of the Schrödinger equation is used to generate quantum potentials that modify the kinetic energy. The latter also arises from the non-local operator $-i\hbar\nabla$ in the quantum formulation. The classical maps given in this section use the concepts of DFT, e.g., the one-body or two-body densities rather than ψ, in generating classical representations.

8.5 The quantum to classical transition and 'decoherence'

While we have argued that DFT and Bohmian mechanics work seamlessly between quantum and classical finite-T phenomena, conventional QM leave the issue of the transition to the classical limit in obscurity. John Bell [19] writes:

'There is a fundamental ambiguity in quantum mechanics, in that nobody knows exactly what it says about any particular situation, for nobody knows exactly where the boundary between the wavy quantum world and the world of particular events is located.'

The consternation expressed by John Bell is misplaced. It is a result of his neglect of the role of temperature in 'real-world' quantum mechanics. At $T = 0$, all systems are quantum mechanical and there is no quantum-classical boundary. At finite temperatures, the thermal de Broglie wavelength tells us where to pose the boundary between the 'wavy quantum world' and our world. In Sec. 7.7.3, Eq. 7.56 we calculated the thermal de Broglie wavelength λ_c of a cat whose mass is one kilogram. We found that even at $T = 1$ Kelvin, the cat's wavelength is about a millionth of a femtometer — i.e., a millionth of the size of a proton. So, superpositions involving the atoms of a cat are not possible unless it is compressed beyond the densities prevailing on white dwarfs and such stellar objects.

Schrödinger formulated the cat paradox to show that QM naively applied to classical systems (treated as elementary quantum systems, e.g., two-level systems), leads to absurdities. Von Neumann's formulation of quantum measurement as a collapse of the wavefunction has been another source of confusion. Although it provides a mathematically convenient description, it is by no means necessary. In Sec. 7.5.1 we treated measurement as a process involving sampling of a large many-particle system in the thermodynamic limit (i.e., $N/V \to$ the density n as $N \to \infty$). Samples can be drawn from such systems where the correlation lengths are much smaller than the system size, while this is impossible if the whole system is just a pair of entangled spins. The world of particular events mentioned by John Bell is not the two-spin world of text-book theory. The real world 'many-particle systems' can be sampled. The measurement destroys the sample, but the main many-particle system is unaffected. Consequently, unambiguous observer-independent results are obtained in physics experiments.

Nevertheless, many theorists have produced decoherence mechanisms, e.g., using quantum gravity, to hunt out Schrördinger's cats that may lurk in the 'real world'. These were discussed in Secs. 7.5.3–7.5.4. In this section we re-examine the theory of Ghirardi, Rimini and Weber (GRW), known also as a 'dynamical-reduction model', or the 'spontaneous localization' model [85].

8.5.1 *GRW spontaneous localization theory*

The essential ingredients of GRW are:

(i) introduction of a non-linear and stochastic modification of Schrödinger's equation without changing the microscopic dynamics of the standard theory at nanometric and shorter length scales;

(ii) for longer, macroscopic length scales, a consistent derivation of the reduction of the wave packet which in the standard theory would lead to superpositions of macroscopically different states.

GRW assume that the many-electron wavefunction $\psi(1,2,\ldots,N)$ where we write 1,2 etc., for $\vec{r}_1,\vec{r}_2$ for brevity, evolves according to the time-dependent Schrödinger equation. However, there is additional dynamics, which requires that the evolution of the wavefunctions is controlled by a decay constant $\lambda = N/\tau$ where N is the number of particles and τ is a new universal constant chosen such that its effect is negligible for few particle systems at microscopic length scales. Thus for instance, the j-th particle in ψ is modified at random times (e.g., with a mean frequency of 10^{-16} sec^{-1}) as follows:

$$\psi_{grw}(1,2,\cdots j\cdots,N) = \int d\vec{r} g(\vec{r}-\vec{r}_j)\psi((1,2,\ldots,N), \tag{8.31}$$

$$g(\vec{r}-\vec{r}_j) = A\exp\{-|\vec{r}-\vec{r}_j|^2/r_0^2\}. \tag{8.32}$$

Here the particle coordinate $\vec{r}_j$ is assumed to be weighted by a Gaussian g which has a decay length of r_0, and A is a suitable normalization constant. Thus, after a time τ/N, each particle gets narrowed down to a range of about r_0. Hence, interference effects will become negligible for distances significantly bigger than r_0. Thus, particles in macro-scale many-body systems will be localized to distances within an order of magnitude of r_0, while few-particle systems would show normal quantum behaviour.

The GRW *ansatz* will certainly ensure that paradoxes of the cat type will not happen as all superpositions would die out for systems with large numbers of particles. Similarly, in coupling an apparatus to a quantum system, the macroscopic length scales are reached and classical behavior is obtained, with the pointer reading uniquely selecting one of the states instead of a supposition. However, we showed that already at 1 Kelvin, the de Broglie wavelength of a 1 kg object is in the micro-femtometer range, i.e., far smaller than the r_0 of GRW.

The addition of spontaneous random localization centers to the system implies that in solving the Schrödinger equation (SE), the effect of these centers has to be taken into account as boundary conditions on the solution. Thus the GRW mechanism is a prescription for including a specific type of boundary conditions in the SE. Equivalently, this is a way of modifying the quantum potentials of Bohm theory to ensure localization. This has been explicitly demonstrated by Beddingham as a 'hidden-variable interpretation of localization theory' [16], using a 'continuous collapse' GRW model.

The GRW theory runs counter to the stability of macroscopic samples of superfluids, quantum-Hall fluids, or certain types of macromolecules where the bonding involves a macroscopic superposition. A large graphene sheet or a π-conjugated macromolecule is perfectly stable. The π-bonded electron system is made up of superpositions of sp^2-electron pairs (see Fig. 7.5). While each 'double bond' is nanometric, the conjugated system of double bonds is a macroscopic superposition, and these should crumble under the GRW mechanism. No such thing is observed in nature. Standard quantum chemistry results for these large molecules give excellent agreement with experiment. On the other hand, if these superpositions are destroyed by 'spontaneous localization', then a large conjugated hydrocarbon should rapidly generate free-radical centers at the GRW localization points and combine with atmospheric oxygen. Instead, such organic materials are very stable, and well known even in astrochemistry (Ch. 10).

Some writings of Ghirardi and collaborators, and their critics, have begun to approach the ideas that may be related to density-functional theory, but without shedding the stochastic modification of the Schrödinger equation. These approaches (cf., Roderiche Tumulka) are known as the 'mass-density ontology' approach. However, as already remarked, these theories continue to ignore thermal fields which set de Broglie wavelengths in micro-femtometers, and hence already far smaller than the *ad hoc* r_0 localization lengths envisaged by GRW.

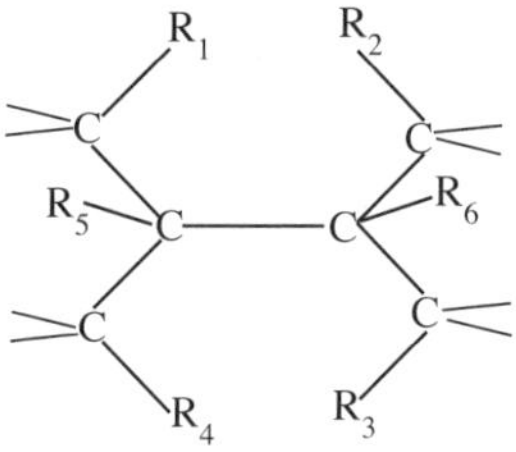

Fig. 8.4 The side groups R_1, R_2 etc., interfere sterically, and hinder the free rotation about the C-C bonds. The groups R_i may contain various groups like aromatic rings, sugar or peptide side chains. They introduce electrostatic interactions, H-bonds, van der Wall interactions, as well as non-bonded hard-sphere interactions, within the molecule as well as with the solvent water molecules. Myriads of such interactions exist in a large protein molecule.

8.6 Emergence of indeterminism in complex systems

Our study of the nature of many-particle systems, including extreme quantum systems like Landau Fermi liquids, showed that a valid classical representation for particle correlations in such systems exists, even when the particles are entangled [61]. In other words, their thermodynamics and low-energy dynamics can be treated using classical mechanics and a set of effective potentials which mimic the electrostatic as well as exchange-correlation effects *evaluated at some finite temperature*. Simulations of properties of complex macromolecules, e.g., proteins, polysaccharides etc., are always treated within such schemes using molecular dynamics. However, even the classical problem of computing the behavior of a sufficiently complex system can become an indeterminate problem. Turbulent motion, collisions of classical 'Sinai billiards' etc., leading to chaotic motion are discussed in various sections of this book (e.g., Sec. 9.2.1).

A more familiar problem is the computation of the realistic trajectories of flipped coins and their final out-comes — i.e., a fully deterministic series of events which finally emerges as a set of 'random' events beyond our prediction. That is, we report the momentum and the initial starting point of the coin flip as the input I, and get Burbaki the mathematician to calculate the outcome as 'head', or 'tail'. We engage Burbaka, another mathematician, and give her the details of the initial state as $I+\varepsilon$ where ε is an infinitesimal change. Burbaka's prediction is completely different from that of Burbaki, for arbitrarily small, different values of ε ! Poincaré was the first person to realize that even though we have deterministic laws, the outcome may for all purposes be indeterminate and non-computable.

Hence the length of Cleopatra's nose could be more important to the course of history than the power of a Roman army! However, leaving such matters to historians and Hegelians, we consider a macromolecule of biological interest. It turns out that here too, we have an indeterminate problem.

Most molecules of biological interest contain -C-C-, -C-N- and other covalent single-bond components where the atoms C, N, carry additional side groups (see Fig. 8.4) which we may indicate as R_1, R_2 etc.

Usually, free rotation is possible around the covalent-bond axis if R_1, $R_2, \ldots$, are small atoms like H, F etc. However, R_i could itself be a benzene ring, an aliphatic chain or some complex molecular group, where by rotations are significantly sterically hindered or entirely blocked. The different substituents generate *steric interactions* which define the three dimensional geometry of the whole molecule. This geometry describes the *conformation* of the molecule. Thus we may have the bond configuration -C(R_1R_2)-C(R_3R_4)- where we need to define in detail the conformation of the atoms in $R_1, \ldots, R_4$ with respect to each other as well as to the whole carbon chain. The problem of calculating the conformation of a molecule with n atoms now involves, in the lowest binary-interaction model, taking account of $n \times (n-1)$ interactions modeled as classical potentials. This is of course enormously simpler than the n electron quantum problem where the wavefunction is an an expansion in Slater determinants in a basis set of N functions, with $N > n$, leading to at least a non-polynomial problem of $N!$ complexity. Even if we restrict the quantum problem to a single-determinantal calculation at zero temperature, we have a problem whose complexity grows as N^4 in the electron-electron interaction matrix elements.

However, the $n \times (n-1)$ binary-interaction problem in protein folding turns out to be totally inadequate. Unlike quantum processes among elementary particles (where only binary potentials play a part), classical effective potentials involve three-body, four-body and multi-body cluster interactions. We have traded-in the simple linearity of the quantum description for a highly non-linear classical description where the interactions involve a holistic complexity. These multi-center classical potentials have to be formulated at finite temperatures. The problem gets confounded by a gamut of competing energy scales as well as times scales. The effective classical potentials contain very weak van der Waals potentials as well as very strong potentials that can change the bonding structure among isomeric forms or between tautomeric forms.

Real proteins are large molecules that carry a dynamic solvation sheath of water molecules highly sensitive to the salinity of the ambient water. Liquid water is a very complicated tetrahedral infinite network of bonds which are continually making and breaking at the temperature (300 K $\sim$25 meV) of biological interest. One may neglect the medium altogether and make a drastic approximation, or include it in some simple electrostatic approximation or in a dielectric approximation (i.e., the medium is modeled as merely weakening the Coulomb interactions via a dielectric constant). Such approaches are woefully *inadequate* for most purposes. For instance, an important factor in protein folding is uncharged non-polar groups of particular amino acids that interact and avoid water. Repelled by solvent water, 'hydrophobic' non-polar groups attract each other by van der Waals forces and move into the protein interior. As a result, intra-protein hydrophobic pockets are formed. These pockets are $\sim$ 0.4 nm^3 in size and cannot be neglected. They contain side groups of non-polar (but polarizable) amino acids such as leucine, isoleucine, phenylalanine, tryptophan, tyrosine and valine. Though small, hydrophobic pockets may be critical to protein function. These hydrophobic pockets may change shape and define two stable conformations of the protein, having very different biological activity. Tubulins are proteins having such pockets. The pocket in the photopsin proteins is big enough to hold a light-sensitive molecule like retinal in its *cis*-conformation, and too small to hold retinal in its *trans*-conformation (see Fig. 12.3). These details of rodhopsin conformations turn out to be crucial to the mechanism of vision (see Sec. 12.3.1). Theoretical models cannot neglect them.

The degree of salinity of the water is critically significant to many protein functions and defining the protein conformation. The cations and anions congregate at the polar centers of the protein, and form electrostatic double layers. Furthermore, since some of the protein groups may take a very long time to come to their optimal conformations, their action may be occurring at some *non-equilibrium conformation* unconnected with the energy optimized 'final' conformation. The 'final optimal conformation' itself may be a meaningless concept for a system with an energy surface having a multiplicity of pockets of low energy which are nearly equal within the range of the thermal energy k_BT. Hence, even with simplifications, the general problem turns out to be intractable, although much headway has been made, for instance, in explaining specific issues, e.g., the action of enzymes (e.g., [42]).

The problems of protein folding, or coin tosses, illustrate how indeterminism arises in a sufficiently complex system or process governed by deterministic laws of great simplicity. It is not mere complexity that has led to this property. In addition to dynamical indeterminacy arising from intense sensitivity to the specification of initial conditions, there can be static indeterminacy in the energy landscape itself. This combines with any dynamics contained in the system, including thermally induced movements. Such indeterminacy may be shown by any system having a ground state energy surface with many many nearly equivalent pockets of minima in the parameter space which defines the system. If the energy differences in the barriers between these pockets are of the order of the thermal energy $k_B T$, then we have a macromolecule whose configuration at any given interval of time is indeterminate.

Clearly then, if we cannot predict the detailed dynamical behavior of a single protein, or a coin toss, we are even worse off when dealing with even the simplest cells of biology. Thus indeterminism has arisen from determinism. Individualized behavior has emerged from the indistinguishable behavior of quantum particles like electrons or atoms. Since the indistinguishability of members of a set is cardinal to many aspects of the second law of thermodynamics, (Sec. 9.2.4) here we see how a subsystem (i.e., a part of a bigger system) which has become sufficiently complex could defy thermodynamics in its subdynamics.

8.7 Conclusion

Our study of quantum mechanical systems containing a few particles led us to the conclusion that the manifest, observed reality involved the observer's participation in selecting *one* outcome from a complementarity of *potentialities*. The observer's role is in setting up the boundary conditions and details of the whole system inclusive of the measuring apparatus. However, many-particle systems (macroscopic matter) can be studied using suitable samples of matter from the main system. Thus the participatory role is restricted to the manner used for making samples. If samples are made in the same way, observer-independent results can be obtained. The laws of QM establishes special contextual, non-local, instantaneous correlations within the sample being studied. This is true even if the sample system were of large macroscopic dimensions as long as the sample was held isolated from external decoherent perturbations. These special correlations could be modeled as arising from quantum potentials that are non-local except in some trivial cases.

The Pauli exclusion potential is an example of a quantum potential associated with particle entanglement. The action of such potentials depend on characteristic length scales such as the Wigner-Seitz radius of an electron. In few particle systems, such length scales are of the same magnitude as the system size itself. However, the length-scales associated with the quantum potentials in 'bulk matter' have shrunk to nano-scale dimensions. That is, the 'spooky' quantum effects are no longer a direct and evident part of our manifest reality. In any case, the boundary between the 'wavy quantum world' and our world is set by the size of the thermal de Broglie wavelength. Its value for macroscopic objects is in the micro-femtometer range even at liquid He temperatures. Consequently, no cat states of macroscopic objects exist in nature. Hence it is not necessary to construct *ad hoc* theories of decoherence designed to avoid macroscopic Schrödinger cat sates.

Furthermore, the description in terms of the many-body wavefunction ψ could be replaced by a description in terms of the one-body density distribution $n(\vec{r})$, as established in density-functional theory. The one-body distribution and other bulk properties may be regarded as emergent properties at the new macroscopic length scale. The manifest reality of bulk matter is the usual reality of classical physics. In many systems, this deterministic classical physics leads to a manifest indeterminism as well as the emergence of 'individualized behavior'.

Chapter 9

Energy, Entropy and Emergent Properties

We review the emergence of indeterminism and irreversibility from the deterministic time-invertible laws of physics. Thermodynamics, entropy, information, statistical mechanics, ergodicty, mixing, Poincaré cycles, etc., are discussed. Dynamic response, the emergence of collective modes, phase transitions and order parameters are presented. Maxwell's demon, Gibbs paradox and other 'paradoxes' of statistical physics are examined.

9.1 Thermodynamics

The taming of fire by early man immediately put the *Homo Sapien* far ahead of all other competing creatures. Prometheus, the Greek God of fire, and *Agni*, the Vedic God of fire are the mythic figures revered for man's mastery over fire. This mastery is perhaps mockingly incomplete, and *global warming* is part of an unfolding Greek tragedy that surely began before Prometheus. The way out of such a tragedy hinges on our understanding of energy and its transformations. Thermodynamics, the science which provides the 'no-go' legislation concerning energy and its transfer between equilibrium systems is also a key arbiter of our everyday reality. The theoretical physicist Elliott Lieb names thermodynamics as the greatest scientific achievement of the 19th century, and claims classical thermodynamics as a perfect physical theory [133]. Indeed, like the quantum theory, thermodynamics does not have exceptions.

Although the mechanical energy of animals, wind and water had been harnessed, the relationship among the various forms of energy, and in particular heat, was very unclear. Mechanical energy was called *vis viva* in Latin, or the *living force* in English. The caloric theory of heat which held sway into the early 19th century claimed that heat was *caloric*, a fluid-like substance that existed in every object. 'Caloric' increased when the particles were heated, and decreased when cooled. The ideal gas laws ('$PV = nRT$') could be derived from this hypothesis, e.g., as shown by Pierre-Simon Laplace (1749–1827). On the other hand, the prediction of the increase of mass with increase of caloric on heating was not observed. Perhaps, like the human soul, caloric was a weightless fluid? Other experimental results began to loom against the caloric theory. Thus Count Rumford demonstrated that mechanical friction generated seemingly limitless heat (caloric). These experiments, and the interest in steam engines at the dawn of the industrial revolution in the 19th century led to the formulation of the first two laws of thermodynamics. The kinetic theory, favoured by Daniel Bernouilli (1700-82) even before the rise of the Caloric theory, regained the attention of scientists.

9.1.1 *The laws of thermodynamics*

In 1847 James P. Joule published a paper entitled 'On Matter, Living Force, and Heat' where *the principle of conservation of energy* was enunciated. The German Scientist von Mayer had already come to similar conclusions. Joule's experiments were inspired by earlier work by Count Rumford, and led him to conclude that heat is not a substance but a form of energy.

> 'Experiment has shown that whenever *living force* is apparently destroyed or absorbed, heat is produced. The most frequent way in which living force is thus converted into heat is by means of friction'.

Joule had determined the amount of heat equivalent to the converted kinetic energy. His work established the first 'no-go' law of thermodynamics, which states the 'impossibility of creating a perpetual (i.e., runs forever) machine. Energy can be converted from one form to another, but it cannot be created *ex nihilo*.

Leonardo da Vinci's drawings of possible machines show that he had wondered about perpetual machines. Einstein's $E = mc^2$ does not invalidate the first law of thermodynamics. Instead, it identifies matter also as a capsuled form of energy. Although the nature of the first law is thoroughly understood and unquestioned, belief in the possibility of perpetual machines, automobiles which run on water, electromagnetic devices with more power output than the power input abound even today. The energy machine of Joesph Newman was one of the most famous and led to litigation between Newman and the US Patent office in the 1980s. The case, *Newman v. Quigg* (the patent Commissioner) is cited as case-law giving the patent office authority to reject claims of perpetual-motion inventors. Present day proponents no longer talk of perpetual machines. They talk of 'over-unity devices', 'zero-point-energy machines' etc.

If the first law is justly famous for outlawing perpetual machines, the second law of thermodynamics outlaws other kinds of machines, and establishes an arrow of time. The second law has proved to be a rich gold mine of ideas linking heat engines, information, order-disorder, black holes and even social phenomena. The second law brings in the important concept of entropy and the impossibility of extracting the full energy difference that exist between two bodies at two different temperatures T_1 and T_2. The law was first stated by Sadi Carnot in 1824, at a time when even the first law of thermodynamics was not understood, in a work entitled *reflections on the motive power of fire*. Carnot used the *caloric* concepts of the time and argued that the production of motive power in a steam engine was not 'consumption of caloric', but resulted from transportation from a hot body to a cold body. He also formulated the efficiency of the process.

In modern language, Carnot showed that even the most efficient, ideal heat engine (now known as a Carnot engine) is not 100% efficient. A fraction T_1/T_2 of its energy is spent as waste heat in transforming heat to mechanical energy or other useful forms of energy. Thus only the fraction $1 - T_1/T_2$ is 'useful energy' or *free energy*. By the time Rudolf Clausius restated Carnot's ideas in 1850s, thermodynamics was a vibrant science. Clausius systematized Carnot's ideas and restated the 'no-go' form of the second law as "heat can not pass from a cold body to a warm body by itself, without some other simultaneous change, connected therewith". The 'wasted energy' that we cannot get away with has been compared with the 'corruption' and 'inefficiency' inherent even in social transactions.

Clausius introduced the concept of entropy as measure of the 'wasted energy' in getting work out of heat.

Using the symbol S for entropy, the change in entropy δS, when a small quantity of heat δQ is added or subtracted reversibly at the temperature T of the system is given by $\delta S = \delta Q/T$. Entropy was soon recognized as a thermodynamic-state property like energy, volume or temperature. Since the added heat changes the internal energy E, $\delta Q = \delta E$, it implies that

$$\partial S/\partial E = 1/T. \tag{9.1}$$

Hence when the internal energy is conserved, entropy and temperature are conjugate properties. Furthermore, if F is the 'free energy' available for useful work in a system, then

$$F = E - TS \tag{9.2}$$

where TS is the 'unavailable' energy associated with the entropy component.

Willard Gibbs, James Clerk Maxwell and Ludwig Boltzmann were the giants who carried forward thermodynamics and also founded statistical physics. They gave an interpretation of thermodynamics in terms of microscopic atomic processes, at a time when many scientists did not even believe in atoms. The energy wasted in converting heat to work is understood within statistical physics as the increased disorderly thermal energy of atoms or molecules in their final states. Hence, the entropy of Clausius is a measure of the disorder arising from the energy transfer. Thus *the second law of thermodynamics* can be stated in terms of the ideal efficiency that one can expect from heat engines, or by stating that the total entropy of a system has to remain constant or increase with time. Entropy does not decrease on its own, just as heat does not flow from a cold body to a hot body on its own. We return to a discussion of entropy in Sec. 9.2.4.

Given a set of pictures of a set of events, if the entropy is found to increase along a certain sequence, then that sequence occurred in precisely the forward direction of time! The free energy of the system decreases, and *disorder increases* along the direction of the arrow of time. However, a set of pictures showing a seed growing up into a tree shows a sequence where an increasingly ordered structure is being formed. The seed, and the tree that grows are actually in direct interaction with the environment and the photon field of the sun (solar radiation). Hence the definition of the system under observation (SUS) is crucially important in discussions of the second law. The photons bring energy into the SUS. In effect, the second law states that the change of entropy as a result of irreversible phenomena inside the SUS is always positive. This does not prevent entropy from decreasing at some *part of the system*, provided that entropy increases at another part to compensate this loss. Further, the second law applies to large numbers of particles. As stated succinctly by Maxwell in his *Theory of Heat*, 'the second law is drawn from our experience of bodies consisting of an immense number of molecules ... it is continually being violated ... in any sufficiently small group of molecules ... As the number ... is increased ... the probability of a measurable variation ... may be regarded as practically an impossibility'.

The work of Leo Szilard (1929), Claude Shannon (1948), Rolf Landauer (1961) and Charles Bennett (1973) has led to a new, widened formulation of the second law where the increase of entropy is understood in terms of loss of information. Thermodynamics, viewed as relating properties like P, V, T, S, free energy F etc., does not directly connect with information. This needs the statistical formulations of the second law and understanding entropy (see Sec. 9.2.4).

Temperature is a statistical quantity independent of the direction of motion of the particles, while useful energy changes depend on energy and momentum transfer among particles at the collisional level. Statistical physics uses classical mechanics or where appropriate, quantum mechanics, to derive thermodynamics from the molecular or atomic dynamics in bulk matter. As we noted in the previous chapter, quantum mechanics naturally leads to classical mechanics at larger length scales, e.g., within a Bohmian or density-functional setting. The wavefunction can be dispensed with, and the particle distribution function $n(\vec{r})$ can be used rigorously for the calculation of thermodynamic functions. A practical way of calculating the two-body correlation functions is to use classical maps or simulations of many-body systems, using effective potentials that have been constructed from density functional or bench-mark QMC calculations. These provide practical calculational approaches within the framework of statistical mechanics. Statistical mechanics allows us to relate the properties of microscopic physics to the measured properties of the macroscopic world.

9.2 Statistical mechanics

The basic question confronted by statistical mechanics is the following: the value of any microscopic property, say, the number density, at some spacial location $\vec{x}$ is a highly fluctuating property, changing stochastically from moment to moment, as the individual atoms enter the location, suffer collisions with other atoms, and leave the region. *How do we relate such a microscopic property (e.g., kinetic energy density) dependent on all the particles in the system, calculated form our basic laws of physics, to a 'well behaved', repeatedly measurable, observer-independent bulk property (e.g., temperature) of the system?* This brings to mind the measurement problem in traditional formulations of quantum mechanics!

Newton's equations can be applied to a system of N particles whose dynamical momenta ($\vec{p} = p_1, p_2 \cdots$) and dynamical coordinates ($\vec{q} = q_1, q_2, \cdots$) define a property $a(\vec{p}, \vec{q}, t)$. If it were a quantum system, in principle we proceed as follows. We specify the dynamical coordinates ($\vec{q}$) at some given instant t_0 and solve the many-body Schrödinger equation for $\psi(t)$ subject to the relevant boundary conditions and calculate the mean value $\langle\psi|\hat{a}(-i\hbar\nabla, \vec{q})|\psi\rangle$. How do we calculate the corresponding macroscopic, physically observed property $A(\vec{x}, t)$, where $\vec{x}$ is a space coordinate in the laboratory, and t is the time? The physical property $A(\vec{x}, t)$ may be an energy density $\varepsilon(\vec{x}, t)$ or a particle density $n(\vec{x}, t)$ at the space location $\vec{x}$, at time t, and is a continuous function of these variables — i.e., it is *a field*, limited only by the boundary conditions defining the 'walls' of the container or the simulation box.

On the other hand, the microscopic quantities, e.g., $n(p_1, p_2, \cdots, q_1, q_2, \cdots)$ are functions of the *microstates*, viz., $(\vec{p}, \vec{q})$, of the system. They reflect the discrete atomic nature of matter, and widely fluctuate among different spatial locations, and time instants, as particles collide and move away from a given region.

9.2.1 *Ergodicity and mixing*

Boltzmann attempted to answer the basic question of statistical mechanics (stated in the previous section) using the *ergodic principle*. He used the idea that an experimental measurement takes a time which is very long over atomic time scales. Hence a time-averaging of the dynamical effects occurs.

The N-particle system corresponds to $6N$ dynamical variables, or degrees of freedom $\vec{p} = p_1, p_2 \cdots \vec{q} = q_1, q_2, \cdots$. Thus we can define a $6N$-dimensional *phase space* and mark the dynamical state $(\vec{p}, \vec{q})$ as a point, called the representative point. As the particles suffer collisions with each other, and are acted on by the potentials included in the Hamiltonian $H(\vec{p}, \vec{q})$ of the system, the values of $\vec{p}(t), \vec{q}(t)$ will change. This change will appear as a trajectory of the point $(\vec{p}, \vec{q})$ in phase space. Boltzmann conjectured that in many-particle systems, as in bulk matter, the point representing the dynamical state of the system will wonder all over the phase space, essentially sampling every location in phase space. This was Boltzmann's *ergodic hypothesis*. Thus, he argued that a time-average over the system is essentially an average over the phase-space of the system.

Boltzmann's approach is not directly used as such today (see Sec. 9.2.2). However, investigations of the motion of the representative point have led to important dynamical insights regarding dynamical instabilities, sensitivity to initial states, and dynamical *irreversibility*. If we consider a system executing harmonic oscillations (e.g., a pendulum), the trajectory will be a closed orbit periodic in time. Such a system is clearly *not* ergodic. If we consider a phase point subject to two simple harmonic oscillations, we obtain periodic orbits known as Lissajous's figures if the the ratio of the two frequencies ω_1 and ω_2 is a rational number. The Lissajous figures correspond to periodic orbits of the phase point on the surface of a torus in phase space. However, if the frequency ratio is irrational, periodicity is destroyed and the phase point wonders around in phase phase, without ever returning to the initial state, filling up the surface of a torus in phase space. If dynamical coupling among the oscillators is considered, as was done by Kolmogorov, Arnold and Moser, highly unstable trajectories arise, and the phase point does not even stay on the Tori, but wonders all over the phase space, essentially filling the phase space. Such a situation becomes similar to what Boltzmann envisaged.

Consider a $6N$-dimensional volume element $d\vec{p}d\vec{q}$ in the phase space around the representative point, and follow the motion of the volume element. Given periodic motion, the volume element remains undistorted, and the trajectory is a closed loop. Henri Poincaré in his *Traité de Mécanique Céleste* examined the question of two trajectories which begin very close to each other, and asked if they will always remain close, or diverge wildly. This issue came up as a problem associated with the dynamical stability of planetary motion. Poincaré found that when interactions among planets are included as perturbations, and if the orbital frequencies of the planets satisfied certain resonance conditions, then the trajectories become very sensitive to initial conditions and became 'chaotic'. Very slight differences in initial conditions were sufficient to create widely different trajectories. A very weak radio signal, when brought into resonance by turning the tuner knob can be made strongly audible. That is, very small perturbations can have very large effects when resonances are encountered. Poncaré's discovery is similar in principle, but applies to the dynamical evolution of the trajectories in phase space. This line of research was extended by von Neumann, Berkhoff and Hopf in the 1930s. It provided support to Boltzmann's ergodic hypothesis, and launched the study of chaotic dynamics.

Another line of development was the construction of a formal basis for the idea of *mixing* in phase space. As already noted, the volume element of periodic trajectories retain their shape. and sweeps a finite fraction of the phase space. On the other hand, we may have ergodic motion where the representative volume element is moderately distorted, while its measure (i.e., 'volume') is held invariant. The representative point never returns to its original location, and the volume element eventually samples the full phase space. A third possibility is when the volume element becomes violently distorted, developing branches, filaments and threads that fill the whole volume, while still retaining its measure invariant. This is the case of *mixing*, originally discussed by Willard Gibbs in terms of a drop of ink inexorably and irreversibly spreading in water. To understand the nature of mixing, we consider at time zero ($t = 0$) two regions of phase space, A and B, both of finite measure $\mu_0(A)$ and $\mu_0(B)$. We keep the region B fixed and let A evolve in time. Then if there is mixing, A will evolve to intersect B and and spread over it uniformly. Also, A will spread over the whole of phase space, and hence any part of B will look locally the same as any locality in phase space. *Mixing implies ergodicity but ergodicity does not imply mixing.*

In 1962 Sinai proved an important, surprising result. He showed that a system of N hard spheres enclosed within hard-wall boundaries was a mixing system, even for N as small as 2. This destroyed the long-held belief that ergodicity is a property of large N systems only. However, since pair-potentials between atoms tend to have a hard-wall like feature for close encounters, ergodicity and mixing may be very common phenomena in bulk matter. Under certain conditions, what appears as mixing may imply a type of hidden or implicate order, as discussed by David Bohm [28] where an ink drop in a viscous fluid is mixed in a uniform manner, and then shown to be reversible. However, the work leading from Poincaré to Sinai, and subsequent developments show that irreversible mixing is a common dynamical feature of most systems, and essentially opens the key to understanding irreversibility and the arrow of time. The exponential sensitivity to initial conditions is the hall mark of these chaotic systems (see review in Cvitanović *et al.* [45]), as already surmised (but not proved) by Willard Gibbs.

9.2.2 *Ensembles as the bridge to macroscopic properties*

The approach to statistical mechanics advocated by Boltzmann, can be summarized as follows. Given a time independent Hamiltonian, the energy of the system is constant (1st law), and this can be used to somewhat simplify the problem. For such a system, Boltzmann held that macroscopic properties were long-time averages of microscopic properties occurring at short time scales.

A macroscopic physical observable is the time average of a microscopic quantity over a sufficiently long time. The ergodic principle is used to replace this time average by a statistical average taken over an *ensemble* uniformly distributed over an energy surface. Thus, the microscopic property $a(\vec{p},\vec{q})$ is averaged over the probability $W(\vec{p},\vec{q})$ of finding the system in the microstate $(\vec{p},\vec{q})$, to yield the observed physical property A. In other words, the measure of the domain in phase space is the *probability density* of the representative point being in the volume element.

$$A = \langle a \rangle = \int a(\vec{p},\vec{q})W(\vec{p},\vec{q})d\vec{p}d\vec{q}. \tag{9.3}$$

The integral is taken over the whole of phase space, and the dynamical problem is now shifted to the determination of the probability weight function W, while a serves to specify the dynamical property to be measured.

In view of the work of Poincaré, Birkhoff, Sinai and others, the ergodic principle may be considered valid except for special elementary systems which are of measure zero in most naturally occurring systems. However, the above approach to macroscopic dynamical quantities *can be applied only for properties which are independent of time*. Hence Boltzmann's approach may be satisfactory for static thermodynamic properties, and also for linear transport properties which are directly expressible in terms of correlation functions of the equilibrium system. However, this approach is insufficient for hydrodynamics, electrodynamics and other branches of physics where time dependence is of basic importance. Thus Lev Landau and others have suggested that Eq. 9.3 should be considered the *cornerstone* of statistical mechanics, without an appeal to ergodicity etc. The observed bulk properties are the *ensemble averages* of the microscopic quantities. The macroscopic physical property A can now be a time-evolving quantity, with the weight function W averaging over $a(\vec{p},\vec{q},t)$. In effect, if we are unable to specify exactly the initial conditions of a dynamical system, an infinitesimal error would lead to wildly wrong estimates of the microscopic quantity a. Hence we cannot predict the outcome of a given experiment at the microscopic level. However, we can predict definitively the most probable experimental outcome if we consider a large number of essentially identical repetitions of the experiment. In repeating the experiments, we only set macroscopic boundary conditions to within experimentally realizable tolerances, and here we do not have exponential sensitivity to these conditions. This possibility depends not only on the mixing property of the representative point in phase space, but also on the fact that we are dealing with large numbers of particles where the concept of the thermodynamic limit, viz, $N \to \infty$, $A/N \to$ (finite value) holds valid.

Our discussions on statistical mechanics show that our knowledge of bulk physical properties rests on a probability weight function, just as in quantum mechanics. If the de Broglie-Bohm formulation (Sec. 6.7) of the Schrödinger equation or density-functional theory is used, we have a seamless formulation of quantum and classical observables within the same formulation of the probability weight function of a many-body system. The averaging over the weight function W is equivalent to taking the trace over the quantum mechanical density matrix.

In the following we discuss practical calculations where the probability weight function $W(\vec{p},\vec{q})$ which depend on the $6N$ variables of the N-particle many-body system is reduced to a few-body distribution function. In the quantum problem, the system is a mixture populating the N-body eigenstates $\Psi_i(\vec{q},t)$ of the N-particle Schrödinger equation. This can be described by a density matrix which can be reduced to the one-body or two-body density matrix. In the end we mostly need the one-body or two-body distribution functions, as most physical properties do not call for more complex distribution functions.

9.2.3 *Thermodynamic properties and particle-correlations*

In our every day reality we are dealing with many-particle systems at normal densities and temperatures, and the properties under study are known as 'bulk' properties. These are a class of emergent properties of many-particle systems. We seek to connect them with the quantum properties of the elementary constituents, i.e., electrons and nuclei that go to form these systems, by averaging over a suitable probability weight function, as outlined in the discussion on statistical mechanics. Some of these bulk properties are the state variables of thermodynamics. The energy (E) and mass (M) are known as 'extensive' properties. They can be specified as, e.g., energy per unit volume, or mass per unit volume. The temperature T, and the pressure P, are examples of 'intensive' variables. The amount of heat energy Q absorbed or released when the system undergoes a phase transition at a temperature T defines the associated entropy change as $\Delta S = Q/T$. The free energy per unit volume, $F = E - TS$, and the chemical potential μ complete the list of one well known set of thermodynamic variables that describe bulk matter.

These properties are not defined if we are dealing with just a pair of quantum or classical particles. They emerge as meaningful average properties of 'bulk' systems. These emergent properties of bulk matter can be calculated from the quantum theory of the elementary constituents. The essential effect of particle interactions, be it classical or quantum, can be incorporated in the effective potentials and the probability distribution functions (e.g., pair distributions) that characterize the system completely. In principle, according to DFT, one can use only the one-body density distribution $n(\vec{r})$ and an exchange-correlation potential $V_{xc}(\vec{r})$ for all calculations. However, it is usually more transparent to formulate the problem in terms of the one-body and two-body distributions. Then the exchange-correlation potential is (at least in principle) directly constructed *in situ* from the fundamental interactions and the pair-distribution functions, in a fully non-local manner. Such non-locality is in fact a hall mark of many-particle quantum systems.

The pair-distribution function (PDF), denoted by $g(\vec{r}_1,\vec{r}_2)$ gives the probability of finding a particle at $\vec{r}_2$, given that there already exists a particle at $\vec{r}_1$. It can also be directly calculated from the many-electron wavefunction which is explicitly available for systems containing a few particles. However, in dealing with many-electron systems, the density-functional or classical-map hyper-netted-chain (CHNC) or molecular-dynamics type approaches provide computationally practicable methods (Sec. 8.4.1). The essential interaction in ordinary matter is the Coulomb potential acting among the nuclei and electrons (Fig. 9.1). The statistical effects (e.g., Fermi statistics) impose additional effective correlations which appear already at the level of the non-interacting pair-distribution function $g^0(\vec{r}_1,\vec{r}_2)$. These correlations can be regarded as arising from additional pair-potentials. Thus the effects of Fermi statistics can be mapped into a spin-dependent effective interaction known as the Pauli-exclusion potential $P(\vec{r}_1,\vec{r}_2)\delta(\sigma_1,\sigma_2)$, as shown by Lado [123] (see Fig. 7.1). This was further extended and utilized by subsequent workers [61] to yield a full calculation of the fully interacting PDF. Correlation functions involving three-particles and more are not directly needed as all the potentials which enter into a description in terms of electrons and nuclei involve only two-body potentials.

The equation of state (EOS) of a material provides the relation among P, V, and T for a given mass of matter, and characterizes the thermodynamic phases of the substance. For instance, $PV = nRT$ is the EOS for n gram moles of an ideal classical gas, with $R = N_{Av}k_B$ being the gas constant. It is simply the Boltzmann constant k_B for an Avogadro number of particles. If T is in energy units, k_B is unity. The ideal gas always remains a gas, and does not show any phase transitions. However, if the gas particles have interactions, liquid and solid phases appear.

Quantum statistical mechanics enables us to calculate the EOS and all thermodynamic properties of a bulk piece of matter from 'first principles'. In practice, because of the intractability of the N-body problem, such direct calculations are

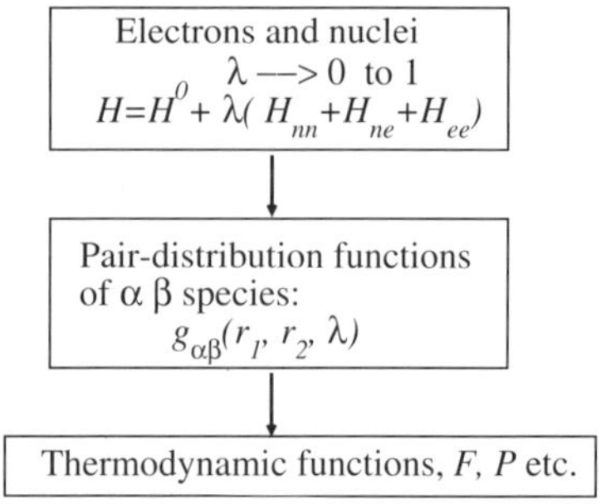

Fig. 9.1 The calculation of the thermodynamics of a system of electrons and nuclei, via a quantum Hamiltonian with Coulomb interactions. The PDFs for all species are evaluated as a function of the coupling constant λ using DFT, CHNC and other methods, avoiding a direct calculation of the many-body wavefunction. An integration over the coupling constant λ from 0 to 1 provides the free energy etc. The finite-T PDFs and Kubo formulae give linear transport properties. Bound-states, molecule formation and other details lead to complications.

feasible only if the coupling constant Γ of Eq. 8.17 is smaller than unity. This measures the strength of the interactions against the kinetic energy. At zero temperature, the kinetic energy is of the order of the Fermi energy, while at high temperatures, it becomes the thermal energy T. We noted that this coupling constant is essentially r_s for jellium at zero temperature. If $r_s < 1$, the interactions can be treated as perturbations and the wavefunctions, density matrices etc., or Greens functions can be evaluated. However, when $\Gamma > 1$, reliable results become sparse. 'Brute-force' computer simulations for small but significant numbers of particles can be carried out using Quantum Monte Carlo (QMC) methods, and they provide useful bench marks for testing approximate methods. For instance, consider the gas of electrons found confined to the 2-dimensional interface between a Si surface and a SiO_2 layer as found in a Si-metal-oxide field effect transistor (MOSFET). Its properties for $r_s < 1$ can be calculated reliably. If $r_s > 35$ in 2-D, the electrons crystallizes into a two-dimensional hexagonal array known as a Wigner crystal. The properties of such crystals are also tractable because a problematic part of the Hamiltonian, viz., the kinetic energy operator, is no longer relevant as the electrons have localized themselves in space, as in the lowest state of a harmonic oscillator. However, perturbation theory methods are not applicable for treating the intermediate densities when the coupling strengths are such that $1 < r_s < 35$ in 2-D. QMC simulations as well as analytical investigations are needed to make progress in such strongly-coupled regimes of many-body systems.

A particularly successful practical calculational method has been provided by DFT where a calculation of the many-body wavefunction is not attempted. Density-functional theory, as well as CHNC-type classical maps of quantum many-body systems, allow us to calculate the quantum-mechanical density distributions in a given system, and hence the thermodynamic functions (Fig. 9.1) via coupling-constant integrations.

For simplicity, if we consider a jellium-like electron gas, the only interaction is the electron-electron Coulomb potential. This is written with a coupling constant λ and the total Hamiltonian is written as $H(\lambda)$.

$$V(r_{12},\lambda) = \lambda/|\vec{r}_2 - \vec{r}_1|, \;\; H = H^0 + \lambda H_{ee}\,. \tag{9.4}$$

Thus, when λ=0, the system is a gas of non-interacting electrons with the PDFs $g^0(\vec{r}_1,\vec{r}_2)$. When $\lambda = 1$ we have a fully interacting system and the PDF evolves into $g(\vec{r}_1,\vec{r}_2)$. In calculating thermodynamic functions, we determine $g(\vec{r}_1,\vec{r}_2,\lambda)$ as a function of λ and carry out a coupling-constant integration over λ from 0 to unity. This enables us to evaluate the 'excess values' of typical thermodynamic functions like the free energy. The excess value is the correction to be added to the non-interacting free energy to obtain the full free energy. In fact, from a DFT point of view, the excess free energy per particle is simply the exchange-correlation correction per particle, f_{xc}. This has to be added to the non-interacting free energy per particle $f^0 = F^0/N$, to obtain the full free energy per particle, $f = F/N$. Writing $n = N/V$ for the total electron density, and $x_\alpha = N_\alpha/N$ for the composition fraction of electrons with spin α, it can be shown that

$$f_{xc} = \int_0^1 d\lambda \frac{n}{2} \int_0^\infty \frac{4\pi r^2 dr}{r} \sum_{\alpha,\beta} x_\alpha x_\beta \{g_{\alpha\beta}(r,\lambda) - 1\}\,. \tag{9.5}$$

Once the Helmholz free energy f has been evaluated as a function of the temperature, a variety of other thermodynamic functions can be easily evaluated. We denote the internal energy per particle by $\varepsilon = E/N$ and the excess internal energy by ε_{xc}, and similarly for other state functions. Then, writing $\beta = 1/T$ (i.e., T is in energy units, and $k_B = 1$),

$$\varepsilon_{xc} = \partial\{\beta f_{xc}\}/\partial\beta\,, \tag{9.6}$$

$$\mu_{xc} = \partial\{n f_{xc}\}/\partial n\,, \tag{9.7}$$

$$P_{xc}/n = f_{xc} - \mu_{xc}\,, \tag{9.8}$$

$$s_{xc} = \beta(\varepsilon_{xc} - f_{xc})\,. \tag{9.9}$$

Thus we can calculate the excess entropy s_{xc} per particle that has to be added to the non-interacting entropy to obtain the full interacting entropy of the system.

A detailed DFT-type 'first-principles calculation' of the EOS of Al from solid to plasma conditions, as well as a review of existing methods can be found in Perrot *et al.* [163]. Issues arising from bound-state formation and multi-ionization states etc., are treated in detail there.

We conclude this section by noting that the thermodynamic properties of a bulk sample of matter, treated as a mixture of classical ions and quantum electrons (a hybrid system), can be determined unambiguously by calculating all the relevant pair-distribution functions as a function of the coupling constant.

9.2.4 *Entropy and statistical mechanics*

The second law of thermodynamics is an experimental fact. It requires that the entropy of the world increases with time. If the internal energy of a system changes by ΔE, the free energy change ΔF available for useful work is $\Delta E - T\Delta S$. Thus we see that the difference in the internal energy and the available (useful) energy involves ΔS. This suggests that the nature of entropy has some connection to the internal states of the system undergoing change. This connection comes out most clearly when the nature of entropy is examined using statistical mechanics. The increase of entropy as we go from one equilibrium state to another is related to the dissipation of energy, and linked to the notion of *irreversibility*. A process is said to be *irreversible* if it is impossible for the micro-states of the system to return to the initial state that it had prior to the process.

As already discussed in Sec. 9.2, Boltzmann was led to consider the microscopic states in a system of particles, in his attempts to deal with a number of paradoxes (Sec. 9.4) in thermodynamics. The microstate of a simple classical particle can be specified by giving its three-position coordinates $\vec{r} = x, y, z$, and three momenta $\vec{p} = p_x, p_y, p_z$. In dealing with quantum particles both momenta and position cannot be specified at the same time, but the $\psi(\vec{q},t)$ or $\psi(\vec{p},t)$ is needed. Thus, a single classical particle requires the specification of six variables (or six 'degrees of freedom'). Hence, as discussed in Sec. 9.2, the N-particle system can be characterized by $6N$ variables which define its microstate. In all these equations, we deal with the probability w_i of finding the system in the microstate i, and this is sometimes regarded as expressing our lack of knowledge (or *ignorance*) of the detailed microstate of the system. In quantum mechanics, if the microstates are denoted by some eigenstates $|i\rangle$, and if the system is a mixture of all the states totaling N_s, occurring with probabilities $w_i = |c_i|^2$, then the state vector $|\phi\rangle$, of the system is:

$$|\phi\rangle = \sum_{i=1}^{N_s} c_i|i\rangle\,. \tag{9.10}$$

Of course, Boltzmann, writing in the latter half of the 19th century, did not use a quantum representation in terms of eigenstates. He used the language of classical microstates. In practice, the number of accessible microstates N_s is a very large number even for systems close to their ground state at absolute zero. In order to take account of the possible distribution of the system in all possible microstates (i.e., in phase space), Boltzmann considered each state to have a probability or "weight" w_i, and considered the sum

$$S = -k\sum_{i=1}^{N_s} w_i \log w_i + C\,, \tag{9.11}$$

$$\text{if } w_i = 1/N_s,\ \ S = k\log W + C, \quad W = N_s\,. \tag{9.12}$$

Here k and C are constants yet to be specified. If the number of microstates is N_s, the 'space of states' available to the system W is simply N_s. If each state is

Fig. 9.2 Boltzmann was far ahead of his peers in statistical physics, kinetic theory, acceptance of the atomic hypothesis, etc. He suffered from mental illness, and hanged himself in September, 1906, aged 62. He had wanted his entropy formula engraved on his tomb, located in the cemetery in Vienna. His pioneering ideas acquired recognition soon after his death.

equally likely, $w_i = 1/N_s$. Hence the second equation, viz., Eq. (9.12) holds for an ensemble where all states are equally likely. Such an ensemble is known as a *micro-canonical ensemble*. One may regard this as an assumption of statistical mechanics, i.e., in equilibrium a closed system has an equal *a priori* probability to be in any of its accessible states. This latter equation, Eq. (9.12), was the form first derived by Boltzmann. It is the famous Boltzmann entropy equation engraved on Boltzmann's tombstone (Fig. 9.2). The form Eq. 9.11 is not restricted to the micro-canonical ensemble, and was first discussed by Gibbs. How a form of the $k \log W$ equation is used in a practical calculation is illustrated in Sec. 9.4.1.

Using arguments based on gas laws and heat engines, it is possible to relate S to the internal energy and free energy of thermodynamics. In particular, it is possible to calculate the distribution functions of the ideal gas (Maxwell distribution) and calculate the entropy from first principles, as was indeed done by Boltzmann in 1872. Such analyses show that the constant k is simply the Boltzmann constant k_B which occurs in the ideal gas law $PV = k_B N_{Av} T$ for one gram-mole of gas containing the Avogadro number of atoms. If the temperature T is measured in energy units, then $k_B = 1$ and hence Boltzmann's entropy equation is sometimes written as $S = \log W$. If the system is in contact with a large heat bath held at the temperature T, the probabilities $w_i = \exp(-\varepsilon_i/T)$, where ε_i are the energy eigenvalues of the states $|i\rangle$. This corresponds to a *canonical ensemble*.

The volume element $d\vec{q}$ in QM cannot be made arbitrarily small as in classical mechanics since the uncertainty principle does not allow us to localize a particle into a cell smaller than of the order of h^3 in the three space dimensions. This leads to a constant correction term S_0 of the order of $\log(1/h^3)$ that was not contained in Boltzmann's result (cf., third law of thermodynamics), although he had envisaged the possibility of a constant C in Eq. (9.12). We may also note that since the volume element in phase space cannot be arbitrarily reduced, an automatic *coarse graining* is introduced into the distribution functions by QM.

One may associate such elementary volume elements $d\vec{q}$ with the atoms of an ideal gas, and calculate the entropy of the system from the number of microstates available to the gas molecules. On the other hand, if we arbitrarily label every other molecule by some means, this labeling allows us to distinguish one set form the other unlabeled set, and the ability to distinguish objects leads to extra configurations (e.g., by permuting them), and hence additional contributions to the entropy occur. However, the labeling process was only a tagging for our use and not supposed to change the thermodynamics of the system. Thus there seems to be a paradox, now known as the *Gibbs paradox*. The statistical physics definition of entropy even for an ideal gas needs a solution to the Gibbs paradox. This is discussed in Sec. 9.4. The entropy of a system grows logarithmically with the number of particles. Consider a particle with just two possible micro-states (e.g., two spin states in the absence of a magnetic field). Using natural units ($k_B = 1$) and Boltzmann's formula for the entropy, its entropy is $\log 2$. For a gram mole of such particles we have $2^{N_{Av}}$ states and hence $S = N_{Av} \log 2$, i.e., a very large number. If we consider a box with all the particles in one side (i.e, similar to just one spin state), the N_{Av} particles have $1^{N_{Av}} = 1$ state and hence zero entropy, compared to what happens if we allow both halves of a box (analogous to two spin states) to be occupied.

Thus, although the entropy law is probabilistic, the high-entropy state is so highly probable that it is a certainty when dealing with macroscopic systems with N_{Av} particles. Similarly, there is just no likelihood that all the gas particles in a box would, by chance, find themselves on one half of the containing box.

9.2.5 *Entropy and information*

Already in the 1920s, Leo Szilard and Leon Brillouin had considered if measurement, i.e., collecting information, had something to do with entropy. They were interested in the so called *Maxwell's demon paradox* that we discuss in Sec. 9.4.4. It may at first seem surprising that 'information' should have something to do with 'entropy'. Information in the everyday world is what someone tells or 'informs' us about. Generally speaking such information can be broken down to a series of 'yes' or 'no' answers which correspond to zeros or ones, as in a sequence of binary numbers or *bits*. If Alice, a recipient of information, were to consider all the possible statements (bits of information), totaling N_s that Bob may convey to her, and if she assigns a set of probabilities w_i for being 'yes', then it is clear that Boltzmann's Eq. (9.12) can be used to define an entropy S_{inf} to characterize Alice's lack of exact information regarding what Bob is going to say. Once Bob gives the answers, some of the w_i will become 1, and others will become 0, giving us a new total S, normally adding to zero if we have complete information. Thus S_{inf} is indeed a measure of our lack of knowledge, *ignorance*, or entropy of information.

Claude Shannon realized that the above method of quantifying ignorance, i.e., measuring the entropy of information, is in fact the only correct formulation, since any measure of S_{inf} must satisfy certain characteristics. Thus, if we divide our information into parts, then the entropy of the total must be just the sum of the parts. If the information is received as a branching process, the answer to each successive branch or sub-question provides some further refinement of the probabilities. Then we should be able to write the total information gained as the weighted sum of the information gained at each branch point.

Instead of using the natural logarithm to the base e, as in Boltzmann's entropy equation, it is customary to use the base 2, when defining information entropy, S_{inf}. However, choosing the base of the logarithm is equivalent to using different units, and hence merely modifying the value of the constant k in the formula for the Boltzmann-Gibbs form for the entropy S by a factor of ~ 0.693. As we already noted, if temperature is expressed in energy units, $k = 1$ in Boltzmann's entropy formula, and it can be directly used in any situation. Since we wish to use a 'bit' as the unit of information, we have:

$$S_{inf} = -\sum_i p_i \log_2 p_i = -(\log_e 2) \sum_i p_i \log p_i = -0.693S. \tag{9.13}$$

To illustrate the working of the above formula, consider the number of micro-states in a register or word made up of 8 bits, i.e., $N_s = 2^8$. If no special constraints are in play, all these possible states (i.e., 0 or 1 labels) are equally probable. Hence $p_i = 1/256$. Then S_{inf} can be evaluated from the above formula and comes out as 8 bits.

Rolf Landauer argued that all information is a part of 'physical reality", and considered in detail the heat generation in computing and in information processing [126]. The work of Landauer and Bennett [21] led to important conclusions. It was shown that a measurement (which can be considered as making a copy of existing information) does *not* cost energy and does not create entropy. In contrast, erasure of information creates entropy. Each bit (binary unit) of erased information increases the entropy of the world by $k \log 2$. To quote Landauer

> 'Note that our argument here does not necessarily depend upon connections, frequently made in other writings, between entropy and information. We simply think of each bit as being located in a physical system, with perhaps a great many degrees of freedom, in addition to the relevant one. However, for each possible physical state which will be interpreted as a ZERO, there is a very similar possible physical state in which the physical system represents a ONE. Hence a system which is in a ONE state has only half as many physical states available to it as a system which can be in a ONE or ZERO state.'

Thus, Landauer identifies computation as simply another process subject to the second law, and the constant k in the entropy equation is exactly the same as in the theory of heat engines!

9.2.6 *Non-equilibrium thermodynamics?*

The second law is a probabilistic law which is statistically a certainty. Then, given a low-entropy initial state arising from initial cosmological conditions, the existence of a direction in time (that of increasing entropy), and the existence of irreversibility are no longer mysterious. However, Ilya Prigogine and associates have not been satisfied with this. They, in what is sometimes known as the *Brussels-Austin program* were a prominent group in the latter half of the 20th century. They claimed that 'when we go beyond the conventional formulations of (statistical mechanics) in terms of Hilbert spaces,...the statistical formulation leads to new solutions which... introduce probability as the basic dynamical object and break time symmetry". Prigogine and associates claim that the use of an extended Hilbert space, known as a 'rigged Hilbert space', and the ideas of chaos and non-integrability in the sense of Poincaré, when applied to the thermodynamic limit lead to irreversibility.

Derivations of the efficiency of heat engines (e.g., the Carnot engine) assume that the processes (of compression of gases along isotherms etc.) are carried out reversibly. In effect, the second law is usually interpreted to be valid essentially at thermodynamic equilibrium where processes are carried out infinitely slowly. What happens if one is confronted with systems far from equilibrium? It appears that the hope of the Brussels-Austin program was to formulate a theory of 'thermodynamics' of such non-equilibrium systems. In this they were encouraged by the earlier work of Onsager who established a number of *reciprocity relations* connecting entropy production and thermodynamic fluxes, and generalized thermodynamic forces. However, it can be shown that the microscopic basis of the Onsager reciprocity relations is nothing but the basic laws of mechanics, and the principle of detailed balance (i.e., microscopic reversibility). Similarly, the work of Källen and Welton, Nyquist, Green and Kubo had established close relations between the dissipation of energy (e.g., when an electric current flows in a metal) and the density fluctuations in the system as reflected in the distribution functions. This is sometimes called the *fluctuation-dissipation theorem*, and asserts that linear transport coefficients (e.g., the conductivity) could be expressed in terms of suitable correlation functions of the equilibrium system. Such results are valid for very weak perturbations acting on the system in equilibrium.

However, the Brussels-Austin group was looking for a treatment of truly out-of equilibrium systems. Prigogine [174], in his book, *From Being to Becoming* claimed that 'the usual formulation of classical (or quantum) mechanics has become 'embedded' in a larger theoretical structure which also allows the description of irreversible processes'. The Brussels-Austin program expected to make a direct, rigorous mathematical connection between deterministic, trajectory dynamics and a non-deterministic Markov type dynamics, and to establish a 'non-equilibrium thermodynamics'. Many interesting results for various discrete maps (e.g., Baker's transformation) have been obtained by them. Similarly, non-equilibrium stationary-state chemical systems arising from the simultaneous interplay of several rate processes (non-equilibrium self-organization) have also been established on a firm footing by the Austin-Brussels group. However, the more fundamental programs of linking deterministic dynamics with non-equilibrium processes, as envisaged by Prigogine and his associates, are yet to be realized. In fact, many workers are now skeptical of the validity of this program.

New developments in computer simulations of particles interacting with each other, and subject to interactions with thermostats (the environment) have led to an improved understanding of irreversible non-equilibrium behaviour. This type of computer-simulation based investigations are known as *Nosé mechanics* [108], and provide a formal link between non-equilibrium molecular-dynamics simulations and irreversible processes in realistic experimental situations. For systems close to equilibrium, Nosé dynamics reproduces the Green-Kubo results familiar from liner-response theory. For the purpose of the present discussion we may regard Nosé dynamics as Newtonian dynamics with the inclusion of friction-like terms which are dynamically reversible (i.e., preserves time — symmetry). Irreversibility arises because steady states which could evolve to violate the second law and (reverse the arrow of time) span only a negligible volume of phase space (that of a 'zero-volume attractor'). Although brief fluctuations violating the second law are possible, a long-term violation would require an infinitely long simulation. A small number of theorists have claimed that these ideas are complementary to, and consistent with the ideas of the Brussels-Austin program.

9.2.7 *Quantum mechanics and irreversibility*

We have developed our discussion in the previous sections with some emphasis on classical mechanics. This is mainly because our observational world is governed by the large-scale limit of quantum mechanics, i.e., classical mechanics. However, it is instructive to go back and look at *irreversibility* starting from an elementary quantum mechanical model where we can explicitly spell out the assumptions we make as we introduce 'damping' or 'irreversibility".

Let us consider a system of particles in a large cubic box, of side L, and let the one-particle density distribution be $\rho(\vec{x})$. We can, at least in principle, calculate the eigenstates of the system (say, with periodic boundary conditions), and obtain $\phi_j(\vec{x})$ and energies ε_j. Then we can write the density as a sum over the eigenstates, as follows:

$$\rho(\vec{x}) = \sum_{jj'} \langle j|\tilde{\rho}|j'\rangle A_j^{+} A_{j'} \tag{9.14}$$

where $\tilde{\rho}$ is the density operator, and A_j^{+}, $A_{j'}$ indicate operators of the 'bra' and 'ket' states $\langle j|$ and $|j'\rangle$ (these are also called creation and annihilation operators). Since the ϕ_j are the exact solutions of the Schrödinger equation, the time dependence of the the density is also given by the time-dependent Schrödinger equation. This can be written in the present notation as:

$$\rho(\vec{x},t) = \sum_{jj'} e^{i(\varepsilon_{j'}-\varepsilon_j)t} \langle j|\tilde{\rho}|j'\rangle A_j^{+} A_{j'}\,. \tag{9.15}$$

Since the cubic-box length L is finite, the energy-level spacing between different ε_j would be finite (except for some accidental degeneracies). Hence each term in the sum given in Eq. 9.15 is periodic in time. As we discussed in the case of two coupled pendula, if the numbers ε_j are commensurate (i.e, if their ratios are rational numbers), then the time dependence $\rho(\vec{x},t)$ is strictly periodic. This means, after a given time T_p, the system recovers its original density distribution. On the other hand, in the more general case, the theory of periodic functions can be used to show that, if the energy levels are discrete, then $\rho(\vec{x},t)$ is periodic in the sense that, given any small difference Δ, there exists a time T_p such that the systems comes back to the neighborhood such that $|\rho(\vec{x},T_p)-\rho(\vec{x},0)|<\Delta$. The time T_p is the *cycle time* of the system, and is similar to the Poincaré cycle time encountered in classical dynamical systems. If the box size L is made increasingly large, then T_p becomes very very large. Nevertheless, if we wait long enough, the system will return to its initial conditions. Thus the dynamical system is reversible.

For real world systems with 10^{23} particles, the cycle time T_p is many many orders of magnitude larger than the estimated age of the universe. That is, it is infinite in terms of laboratory time scales of interest to us. In fact, there are several distinct aspects associated with the emergence of irreversibility in statistical mechanical calculations. These are:

(a) The time dependence is calculated with the volume of the quantization box tending to infinity, while the density distribution is held constant (this is known as the thermodynamic limit). The $t\to\infty$ limit is taken *after* the thermodynamic limit.

(b) An ensemble of all possible states consistent with the density distribution $n(\vec{x})$ is considered and an ensemble average is taken. This process includes the constraint that the ensemble satisfies additional criteria as required by various conservation laws and symmetries obeyed by the system.

(c) A third aspect of irreversibility is associated with the possibility that, even with a quantization box of finite size (i.e., $L\to\infty$ is not yet taken), all information that strikes the internal surfaces of the box are not reflected back. Some information is lost to the outside world. This can happen when we assume that a particle (or an oscillator) is coupled to an external field, e.g., an electromagnetic field.

An electromagnetic field is a system with an infinite number of degrees of freedom. The oscillator acquires a damping due to the reaction of the radiation field, both classically and quantum mechanically. Thus, even without taking the $L\to\infty$, we bring in irreversibility when we admit the existence of electromagnetic fields in nature. If we consider the one particle propagators (Green's functions) of the system, the simple poles of the discrete system acquire imaginary parts (and possibly branch cuts) when coupled to the field.

The above discussion once again shows how irreversibility is connected with the coupling of a system under observation (our external world), usually containing a large number of particles, to considerations involving essentially infinite numbers of modes, be they described as ensembles, fields, ancillary variables etc., that need to be averaged over in making an observation. The coupling to the field allows the loss of information from the system to the outside. The effect is identical to the issue of 'decoherence' discussed in regard to quantum systems.

9.3 Dynamic response of a many-particle system

Discussions of many-particle systems consider very simplified systems where the particles are non interacting, or possibly interacting via very short-ranged forces similar to those of hard spheres. Even then we saw that many systems are extremely rich, yielding a variety of unexpected properties like mixing and chaotic dynamics. Furthermore, QM, with its Fermi or Bose statistics, imposes highly non-local correlations. However, we saw that these can be represented by non-local pair-interactions, as far as particle distribution functions are concerned (see Fig. 7.1). The *many-electron wavefunction* could be dispensed with, and replaced by a DFT description which uses the *one-particle density distribution* $n(\vec{x})$. When interactions among particles are included, another level of complexity begin to emerge. The single-particle excitations of the system couple together to form new, collective modes, while the vestiges of the one-particle modes that survive acquire a damped character. The system become highly non-local and contextual when viewed from the point of view of the original one-particle modes. On the other hand, the description using the new collective modes turns out to be similar to that of a set of independent particles. A new kind of locality, based on new length scales arises.

Well-known examples of this are: (a) the strongly interacting electrons in metals are replaced by weakly interacting quasi-particles in Landau Fermi-liquid theory; (b) Bohm and Pines showed that the ground-state energy of a system of interacting electrons can be be re-expressed as the zero-point oscillation energy of a set of weakly-interacting collective modes known as plasmons; (c) the replacement of strongly interacting electrons in the fractional quantum-Hall effect by virtually non-interacting 'composite Fermions'.

In this section we briefly examine an example of a simple interacting system and chart the emergence of new collective modes that replace the highly correlated, non-local single-particle modes at the shorter length scales.

Our knowledge of an external object is based on observing its response to a probe. We may shine a light on the system and the light that returns from it is its 'optical response'. This optical response, i.e., the reflected and transmitted light can be described by the 'frequency dependent refractive index' $\eta(\omega)$. We may throw a projectile at a system, and the system would react by shaking or oscillating under the impact. The projectile itself would hurtle away from it with modified energy and momentum. Rutherford's study of atomic nuclei using α-particles as projectiles, or the 1969 SLAC experiments using electrons to probe the quark-structure of protons, were two classic experiments of this type. The response of a system reveals the 'constituents', or more correctly, the excitation modes of the system. Most such properties can be related to a response function known as the *dynamic structure factor* $S(\vec{k},\omega)$ of a given system. Here, $S(\vec{k},\omega)$ is a measure of the probability of transferring a momentum $\hbar k$ and an energy $\hbar\omega$ to the system using an energetic probe. Thus $S(\vec{k},\omega)$ can be considered as a mirror or map of the 'ring tones' of the system. If the system is going to respond strongly at a frequency ω, then there would be a sharp peak in the dynamic structure factor at that frequency. Thus, the emergence of new, more extended collective modes in a system, replacing the interacting modes of the shorter length scales, can be detected from the response of the system to an external perturbation. In the following we consider a very simple but instructive example of how collective modes emerge in interacting systems, replacing the elementary modes of the original system.

9.3.1 *Emergent modes of many-particle systems*

Let us consider N charged particles distributed in a box of volume $V = L^3$ such that the particle density is $n = N/V$, together with a neutralizing background. This is our simple jellium model (Sec. 8.2.1) of a charged fluid. It could also represent a simple metal like Na or Al, where the charged fluid is the 'electron gas' moving about inside the metal. The ion substructure is essentially static and retains the electrons via electrostatic forces.

The electrons interact by the Coulomb repulsion, and attempt to keep away from each other, with a mean energy e^2/r_s, where r_s, the Wigner-Seitz radius, is a measure of the mean distance between electrons. On the other hand, the kinetic energy makes the electrons move about rapidly, colliding with each other. If we take a *time-averaged picture* of such a relatively high-density electron gas, we have a uniform distribution of electrons (with their motions averaged out) in the 'cage' made up of atomic nuclei which are very heavy and immobile. In the jellium picture of the metal, the ions are simply assumed to be an impassive 'background charge' which does not come into the play, except to ensure overall charge neutrality.

The uniform density n of charged particles can be disturbed by applying an external charge to it briefly. This is similar to creating a density disturbance in a liquid 'by sticking a finger' in it. The displaced charge density in some locality is like a little density bump δn. This extra density will create restorative forces that will force it back towards its original distribution where $\delta n = 0$. However, the restoring effect will make it overshoot the original uniform distribution and recreate the δn bump a short time later, just as with a pendulum bob which oscillates around the mean equilibrium position. Thus the charge fluid will set up charge-density oscillations that damp out after some time. It turns out that the oscillations have a well defined frequency known as the plasma frequency ω_p. The charge-density oscillations are plasma oscillations, and their energy quanta are known as 'plasmons'. They are cooperative (collective) motions of large numbers of charged particles in step with each other. The interacting charged liquid behaves like a collection of plasmons, with energy $\hbar\omega_p$, and some damping γ. The single-electron dynamics, with kinetic energies of their individual motions have become subsumed in the plasma modes. A new set of weakly-interacting 'normal modes' of the system has emerged from the the interacting N-electron system.

We came across such oscillatory motion in our study of a charged particle in a parabolic potential (Sec. 6.4), where the ground state energy was the 'zero-point' energy. Thus we should expect that the total energy of the system of interacting charged particles is nothing but the zero-point energy of our collection of plasmons, as first shown by Bohm and Pines in 1952. Hence the new collective description provides a new perspective on properties that we had understood and accounted for previously using our 'lower-level' (i.e., more microscopic) variables of individual electrons.

Even the response functions of the system, written in terms of the single-particle modes of the original system, can be re-written using the new modes. We develop the density response function of the system using simple arguments. Let us use atomic units where $\hbar = 1$ and the mass m_e of the electrons is unity. Similarly, the absolute value of the charge $|e| = 1$. Let us consider an electron gas, with energy states $\varepsilon_k = k^2/2$, occupied up to the Fermi level E_F (see Fig.8.2). Consider an electron in the state k_1 promoted to a level k_2 by applying a disturbance to it. The promoting of particles to higher energies is possible only to the extent that the initial level k_1 is occupied, and to the extent that the final level k_2 is unoccupied. The average occupations of a level k at a given temperature $T = 1/\beta$ is given by the Fermi-Dirac function

$$n(k) = 1/(1+e^{\beta(\varepsilon_k-\mu)}). \tag{9.16}$$

Thus, the occupation of the state k_1 is $n(k_1)$, while the extent of un-occupation of k_2 is $1-n(k_2)$. Thus the probability of the transition $k_1 \to k_2$ is proportional to $n(k_1)\{1-n(k_2)\}$. As soon as particles begin to occupy the level k_2, the reverse transition $k_2 \to k_1$ also becomes possible. Hence the net rate of disturbing the system is proportional to

$$n(k_1)\{1-n(k_2)\} - n(k_2)\{1-n(k_1)\}$$

i.e., $n(k_1) - n(k_2)$. If these transitions are provoked by a probe with energy ω, the energy change $\Delta E = \omega + \varepsilon_{k_1} - \varepsilon_{k_2}$ corresponds to a time $\Delta t = 1/\Delta E$ by the uncertainty principle. Hence the transitions $k_1 \to k_2$ can occur during a time duration Δt. The total set of transitions which correspond to the disturbance of the system by the probe is in fact the *response* of the system. Denoting it by $\chi(\vec{q},\omega)$, and writing $\vec{k}_1$ as $\vec{k}$, and $\vec{k}_2 = \vec{k}+\vec{q}$, we have

$$\chi(\vec{q},\omega) = \sum_k \frac{n(\vec{k}) - n(\vec{k}+\vec{q})}{\omega + \varepsilon_k - \varepsilon_{k+q} + i\gamma}. \tag{9.17}$$

An infinitesimal damping γ has been added to regularize the denominator. If we consider the fact that all electronic transitions involve an electromagnetic field, then adding a damping factor is fully consistent with the existence of a natural width to every transition (see section 6.10.3). The quantity $\chi(\vec{q},\omega)$ is known as the Lindhard density response function of the system. It can be derived rigorously using time-dependent perturbation theory. Here we have used the uncertainty principle.

The response is a collective process involving all the single-particle excitations among the single-particle states k. The remarkable thing is, it can be shown that this is actually equivalent to a response function $\chi_p(\vec{q},\omega)$ written entirely in terms of the plasmon excitations of the system.

$$\chi_p(\vec{q},\omega) = 2\omega_p/\{\omega^2 - \omega_p^2(q)\}, \tag{9.18}$$

$$\omega_p^2 = 4\pi n e^2/m, \quad \omega_p(q) = \omega_p + f(q). \tag{9.19}$$

Here $\omega_p(q)$ is the actual plasmon energy which is weakly dependent on the momentum transfer $\vec{q}$ via the function $f(q)$ whose form does not concern us, except that $f(q) = 0$ for $q \to 0$. The $\omega_p(q)$ may be considered to contain a real part and an 'imaginary part' which corresponds to the damping of the plasmon modes. The total strength of the response is measured by what is known as the f-sum rule. The plasmon description is sufficiently complete to almost satisfy the f-sum rule.

If we consider a solid or a liquid with charge-neutral particles, the corresponding density excitations are simply sound waves. The quantized particles corresponding to sound waves are known as *phonons*. Collections of atomic-sized individual magnets, when formed into an interacting many-body state produce ferromagnetism, anti-ferromagnetism and other types of magnetic behaviour. Their response can be described in terms of magnetic waves, or *magnons*. Condensed matter physicists and plasma physicists are very familiar with a variety of such collective modes which emerge from the under-lying atomic interactions. The physics of such systems can be discussed using these new modes, leaving aside the previous, more microscopic description involving quantum eigenstates etc. That is, a new description of the system (in terms of new modes) that subsumes the more microscopic length scales has emerged in the many-particle system.

9.3.2 *Phase transitions*

The onset of new collective excitations are entirely a result of interactions. If we think of a phonon as a dynamic cooperative oscillation of a set of atoms, defining a wavelength for these oscillations, there would also be critical situations where the system could 'freeze' in some ordered state which mimics the oscillation of the dynamic mode. That is, the interactions may be strong enough to create a transition of the system to a new phase. Thus, when a liquid is cooled, the kinetic energy decreases and the potential energy can overcome the atomic motion and induce freezing. This is the onset of the liquid $\to$ solid phase transition. The onset of the phase transitions can be determined by simpler approaches using the equilibrium energies of the various competing phases (e.g., solid phase and the liquid phase). The lower-energy phase will be the observed phase.

Since these collective effects are a result of inter-particle interactions, it has been very fruitful to study them using simple mathematical idealizations or 'models' where the particles are regarded as static objects positioned on a lattice, (as in a solid), and the interaction is taken to be a simple constant J acting between nearest neighbour particles. Each particle is assumed to be capable of being in one of two states. These may for example be the singlet and triplet states of a molecule. The two states, i.e., a two-level system, may be simply modeled as an *up-spin* state $|\uparrow\rangle$ and a down-spin state $|\downarrow\rangle$. More simply, we may associate a classical spin S^z pointing 'up', or 'down' along the z-axis, and think of the two states as 0 and 1. This leads to a model of interacting classical spins, known as the *Ising model*. This is defined by the Hamiltonian:

$$H = \sum_{[ij]} J S_i^z S_j^z + Bg \sum_j S_j^z \,. \tag{9.20}$$

Here the notation $[ij]$ means that nearest-neighbour pairs are to be summed, and the indices i, j run over the lattice of atoms. The interaction strength J is known as the nearest neighbour *exchange interaction*. This name arises from the fact that the triplet state may exchange electrons and stabilize itself relative to the singlet state, thus favouring a 'magnetic ground state' if J were negative. The last term describes the interaction of the spins with an external magnetic field B, and the weight factor g is called the Landé g-factor. We may assume that the external field B is negligibly small, but non-zero. That is, the presence of $B \sim 0$ is useful as it fixes a definite direction to the z-axis, breaking the rotational symmetry of the system. The simplest case is that of a one-dimensional chain of spins, interacting with each other by the nearest-neighbour interaction J. Such a chain would contain up-spins and down-spins in some random fashion. The net number of up spins in the equilibrium state of the many-spin system is defined by

$$M = \sum_i S_i^z = (\text{sum of} \uparrow \text{spins}) - (\text{sum of} \downarrow \text{spins}). \tag{9.21}$$

This is the net magnetization of the system. A one-dimensional chain of such spins can be shown to have no net magnetization at any temperature. That is, at finite temperature T, the free energy $F = E - TS$ of the system, where E is the internal energy, and S is the entropy, is lowest for a random-spin system. The entropy term TS wins over the energy term. Thus the spins are randomly oriented. However, the situation is very different in a two-dimensional or three-dimensional lattice of spins. Now there are more interactions, and the energy term E wins over the entropy term at sufficiently low temperatures.

Lars Onsager, in a brilliant break through in statistical mechanics, provided in 1948 an exact analytical theory of the energy and magnetization of the 2-D Ising model. There he showed that the spins which are randomly oriented at high temperatures, spontaneously order themselves into a ferromagnetic state on cooling below a critical temperature T_c. Below this temperature, all the spins are oriented pointing up. In fact, Onsager's T_c for the Ising-model has its analogue in the real world of magnetism, known as the Néel Temperature T_N. When a magnetic material is heated above its Néel temperature it looses its magnetism. Magnetism may be thought of as an *emergent property* of the whole system, brought about by the cooperative action of all the individual spins ordering in a definite manner. The magnetic moment is the *order parameter* of the system. The critical temperature (in energy units) is the energy parameter which defines the onset of the ordered phase.

The situation can be more complex in real solids. Instead of randomizing all spins, some *domains* of spins may reverse their spins as a group. The domain size is associated with the spin-spin correlation length. The spins of a given domain are oriented parallel to each other. There is a transition region between domains, known as the *domain wall*, where the spin direction changes (Fig. 9.3). The net magnetization of the whole solid is obtained by taking the vector sum of the different-domain spins. The domain-wall region itself provides new emergent many-particle structures which contribute to the entropy term TS in the total free energy. The formation of domain walls in an ensemble of spins has many thermodynamic similarities to the formation of cells and cell walls in lipid-water systems essential to the origin of life (see Sec. 10.4.2.1).

A quantum mechanical model of magnetism has also been formulated by extending the Ising model where each lattice site (say j) is occupied by an atom whose state is represented by a spin *operator* $\hat{S}_j$ rather than a classical spin. The resulting model is described by the Heisenberg

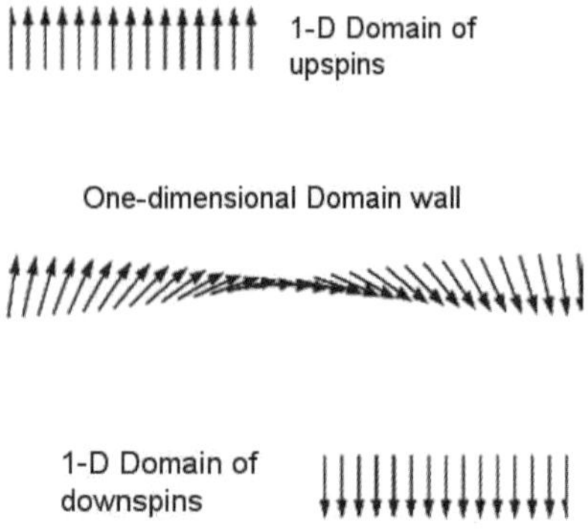

Fig. 9.3 We use a one-dimensional chain of spins to depict two spin-orinted domains and the transition region known as the 'domain wall'. A chain of 1-D spins has no net magnetization as the net effect of different domains cancels out. However, in 2 and 3 dimensions, magnetic materials below their Néel temperature show net magnetism as there are more domains with up-spins, than those with down spins.

Hamiltonian operator for the magnetic system:

$$H = \sum_{[ij]} J\hat{S}_i \cdot \hat{S}_j + Bg\sum_j \hat{S}_j^z. \tag{9.22}$$

Unlike the 2-D Ising model, the Heisenberg model is not amenable to an analytical solution. The usual procedure is to calculate the density matrix of the system, or equivalently, suitable Green's functions. The mean magnetization is evaluated as the trace of the total-spin operator over the density matrix.

Thus Onsager's work on the 2-D model of classical spins (Ising model) opened the way for a deep and accurate theory of phase transitions where an order parameter appears, or disappears as the temperature T or the strength of the interactions J is modified.

The Ising model can also be used to approximately simulate phase transitions like the solid $\rightarrow$ liquid melting transition. The solid is the ordered phase analogous to the magnetized phase. The process of melting involves loss of long-range crystalline order, with the liquid phase having only short-range order. The liquid $\rightarrow$ gas transition (evaporation) is a process where the short-range order of the liquid is lost and the molecules of the gas move about in random fashion. The tendency for a system to go into an ordered phase (or not) depends on the strength of the correlations between the particles, as opposed to their kinetic energy which allows particles to overcome potential energy of interactions. The ratio of the potential energy to the kinetic energy is often used (see Sec. 8.2.2.1) as a measure of this, and it is known as the coupling constant, denoted by Γ. That is, $\Gamma = J/T$ for the Ising model. We had already come across the coupling constant in electron fluids in the quantum regime. There the kinetic energy is typically the Fermi energy E_F. In classical systems, the kinetic energy is very simply the thermal energy T.

Lev Landau had given a general theory of phase transitions, based on a mean-field estimate of the potential on a given particle (cf. Sec. 8.4.1. This *mean-field theory* of phase transitions, as well as the extension to the Landau-Ginsberg form had elucidated many key concepts like various types of order parameters which correspond to different emergent properties of interacting systems. While Landau theory gave a first approximation to these, Onsager's exact theory enabled one to delimit the strengths and weaknesses of Landau's mean-field theory. Modern scaling theories, and computer simulations (e.g., molecular dynamics or quantum Monte Carlo methods) provide accurate results even for systems where no analytical theory could be envisaged.

Thus we may conclude that many-particle systems, due to their inter-particle interactions begin to behave like *fields*. These fields spawn new collective modes which emerge as new modes at larger length scales associated with the extent of their *order parameters*. The shorter length scales are almost completely subsumed by the new emergent modes with the new length scales. The deep-lying shorter length scales become evident only if energetic probes consistent with those modes are used. They become the 'hidden variables' of the higher-scale theory.

9.4 Paradoxes proposed within statistical physics

A number of paradoxes with titles like 'Laplace's demon', 'Maxwell's demon', 'Loschmidt's paradox' etc., are well known. These focus on the seemingly rigidly deterministic nature of the laws of physics, as well as on their microscopic time reversibility. If every event in the world could be reduced to mechanical motions defined by Newton's equations, then their motion is perfectly reversible. That is, the periodic motion of a pendulum may be used to run the clock forward, or equally validly in the backward direction. Even in quantum mechanics, the wavefunction which describes a system evolves backwards or forwards in time, in a perfectly deterministic manner, with the normalization of the amplitude completely conserved (unitary evolution, see Sec. 6.21). Such deterministic evolution is disrupted in QM only in the act of measurement — i.e., outside intervention.

These questions contain their counterparts in the basic questions posed by philosophers since time immemorial. The ancient philosophical questions are often about the nature of free will, determinism, and *karma*, These were also the questions posed by the Greeks in terms of first causes and final causes. Such questions become invariably linked with the nature and direction of the flow of time. The rise of physics and thermodynamics enabled one to get away from semantics and ask such question in a more precise manner. The new questions re-phrase the old questions in terms of the possibility or impossibility of building heat engines which refute the laws of thermodynamics.

The Gibbs paradox is in a different class, as it focuses on the essential indistinguishability of 'elementary particles' while the macroscopic world is made up of distinguishable objects. This also touches on information and its association with the second law. The preceding sections have largely dealt with the issues raised in these paradoxes. In the following we restate those issues directly focusing on the paradoxes as stated in the literature.

9.4.1 *Gibbs paradox*

In everyday life we easily distinguish one marble from another, and in discussing the permutations among n marbles, this characteristic is respected. In using the statistical concepts of entropy (cf., Eq. 9.11) Gibbs came across a difficulty when the concept of distinguishability of microscopic particles is retained at the basic level. If you have n particles, and N boxes to put them in, different ways of distributing them in the boxes correspond to the 'microstates' of the system that go into Boltzmann's expression (Eq. 9.11) for the entropy of the system.

If the particles are indistinguishable, then the total number of different arrangements gets reduced by factorial n. Gibbs poses the issue in the form of the 'mixing paradox' where the assumption of distinguishable particles is shown to contradict the second law. In QM particles are indistinguishable and can occupy eigenstates having quantum numbers to label the eigenstates, but not the particles. Hence the paradox is automatically resolved in quantum statistical mechanics.

However, QM is not needed at the thermodynamical level. There we distinguish between thermodynamic states, but we are not concerned with individual molecular positions and momenta. In the monumental work *Heterogeneous Equilibrium* Gibbs in 1875 resolves the thermodynamic form of the paradox directly, by noting that distinguishability and indistinguishability (applied to thermodynamic states and not to microstates) provides the key to the problem.

A quantity of matter equal to the molecular weight is called a gram mole. Thus one gram mole of H_2 is very nearly 2 grams, while a gram mole of O_2 is $\sim$32 grams. Consider one gram mole ($V_a = V_m$ = 22.4 liters) of a gas a, and one gram mole ($V_b = V_m$) of gas b. Both gases are at the standard temperature (273 K) T and pressure P of one atmosphere. The gases are in two chambers separated by a diaphragm. The diaphragm is then removed. Now the total number of moles is 2 with a volume $2V_m$; the fractional compositions are $f_a = f_b = f$ = 1/2. Since the volume occupied by the gas is proportional to the available number of configurations that the gas molecules can occupy, we can use the volume for W in Boltzmann's entropy equation. Hence we can easily calculate the change of entropy due to mixing of the two gases a and b. We denote the Avogadro number i.e., the number of individual molecules in one gram mole by n_0,. Then Boltzmann's constant k in Eq. 9.11 multiplied by n_0, i.e., the gas constant $R = k_B n_0$ is the relevant constant for one gram mole.

$$\Delta S = S_{mix} - S_{ini} = 2R\log 2V_0 - [R\log V_a + R\log V_b] \tag{9.23}$$

$$= R\log 2. \tag{9.24}$$

Surprisingly, the result does not depend on the types of gases used. If the distinguishability between a and b is $p = |a-b|$, it seems to hold even in the limit $p \to \varepsilon$! When identical gases ($\varepsilon = 0$) mix, there should be no increase in entropy. This is the thermodynamic form of the Gibbs paradox.

What do we mean by 'the entropy increased by $R \log 2$', on mixing the two unlike gasses? It means that if we attempted to separate the gases to their *original state*, then we need to do work that changes external bodies in the environment (e.g., run a heat engine).

Although statistical physics deals with microstates, thermodynamics deals with observable changes in macrostates. Each macrostate is compatible with a large set of microstates. When we 'return' the mixture to 'its original state' all molecules of type a are returned to the chamber a with any momentum compatible with the temperature. However, each molecule is *not returned to its exact original position and momentum*. In fact, if we reinstate the diaphragm, or a semi-permeable diaphragm while the system is in the mixed state, we find that the thermodynamic functions are exactly the same as prior to mixing. So there is no entropy of mixing based on the identity of the gases. E. T. Jaynes in his 1992 discussion of the Gibbs paradox in *Entropy and Bayesian Methods* says 'The second law that Clausius gave us was not a statement about any property of microstates'.

On the other hand, the assumption of the knowledge of the identity of the gas molecules is additional *information*, as discussed in terms of the ideas of Sec. 9.2.5. We can assign 0 and 1 to the properties 'like' and 'unlike'. If such information were treated as an experimentally accessible thermodynamic property, then *the unmixed, labeled state was associated with more information*. 'Entropy of mixing' is just the erasure or loss of such information. The experimentalist has to decide on the thermodynamic variables and degrees of freedom chosen to characterize the system under study, and the second law will hold within them. If extra properties are arbitrarily included in the entropy calculation, but not included in defining the thermodynamic-state properties, then one may find apparent contradictions with the second law. If the identity of the n molecules at the statistical level cannot be included, then this feature has to be included by dividing the total available microstates by $n!$ in calculating thermodynamic functions. This is done in the *Sakur-Tetrode* equation for the entropy of a classical ideal gas.

9.4.2 *Determinism and the demon of Laplace*

The application of Newtonian mechanics to celestial bodies had been very successfully carried out in the century which followed Newton. New machines and steam engines had ushered in the industrial revolution. Ernst Hoffmann (1776–1822) had written *Der Sandmann* which presented a young lady by the name of 'Olympia' who is revealed to be a pure automaton. Olympia was in fact the precursor of *Coppelia, Nut-cracker* and other works which represented life-like individuals who are revealed to be puppets. Thus, while an intensely Christian, capricious world of angels, spirits and witches dominated popular thinking, the 'age of reason' had created an intellectual class which strongly believed in determinism, law and order.

Marquis Pierre Simon de Laplace (1749–1827) was the most profound mathematical physicist of his era, and set landmarks in Newtonian Astronomy that hold even to this day. He did not actually mention a 'Demon' of Laplace, but defined the concept of determinism as follows.

> 'We may regard the present state of the universe as the effect of its past and the cause of its future. An intellect which at any given moment knew all of the forces that animate nature and the mutual positions of the beings that compose it, if this intellect were vast enough to submit the data to analysis, could condense into a single formula the movement of the greatest bodies of the universe and that of the lightest atom; for such an intellect nothing could be uncertain and the future just like the past would be present before its eyes'.

The 'intellect', referred to in the above passage from Laplace's *Essai philosophique sur les probabilités* (1814), has been renamed 'Laplace's Demon', in analogy with 'Maxwell's demon', by modern writers. Victorian writers invariably considered the 'intellect' to be God, although Laplace had explicitly stated that he had 'no need for that hypothesis', when Napoleon had asked him why the Creator is not mentioned in Laplace's *Exposition du système du monde*.

The above form of determinism is difficult to defend even within the physics of the 18th century. Laplace himself noted the difficulty that if the intellect was itself a part of the universe, then it cannot *choose* to 'submit the data to analysis' etc. In fact, Laplace's demon can only analyze *a selected part of the universe*, within his gaze, while he himself stands outside the system under study. A similar problem was faced by St Augustine. According to St. Augustine, (*Confessions, Bk. XI*) God created Space (Universe) and Time while being eternal and non-spatial. The past, present and future are directly in God's mind.

Poincaré's work on the three-body problem at the end of the 19th century led to the recognition that dynamical systems can be infinitely sensitive to even the slightest error in the specification of the initial conditions. Even the solar system, previously thought to be the paragon of deterministic mechanics, is known to be a chaotic dynamical system as some moons of planets, and many asteroids follow chaotic orbits. That is, in practice, precise prediction is impossible. Prediction is possible for some special situations, within times consistent with various specific time scales. This chaotic property of dynamics is not the exception but the rule. In our discussion on ergodicity and mixing we noted that any N-body system, where N is sufficiently large, can show mixing and irreversibility. For hard-sphere billiard balls, $N > 2$ ensures chaotic dynamics (see Sec. 8.6).

Quantum mechanics, being based on *linear* differential equations, does not directly admit chaos. That is, strictly speaking there is no such thing as 'quantum chaos'. The wavefunction evolves in a deterministic and unitary manner (Eq. 6.21). However, the wavefunction is only a means of calculating the probabilities of experimental outcomes. Furthermore, the Schrödinger equation can be rewritten in terms of trajectories and quantum potentials as in Bohm. Hence the sensitivity to initial conditions reappears, although the degree of sensitivity is another issue that needs more specialized discussions.

Information is irretrievably lost at the boundaries of the system under observation, or the boundary of our event horizon set by the velocity of light, as well as by the expansion of the universe. Loss of information leads to increased entropy, irreversibility and predictive impossibility. Even a single particle, e.g., a charged particle, when coupled to its own electromagnetic field, creates damping and irreversibility. The conformation of a macromolecule like a protein, made up of such atoms is totally beyond numerical prediction. Thus modern physics recognizes the intrinsic many-body or *field-like* character of nature. It has moved away from the deterministic physics of Laplace. The determinism of physical law as applied to elementary systems, is found to be rich enough to spawn indeterminism when applied to 'realistic' systems at biological length scales.

9.4.3 *Loschmidt's reversibility paradox*

Boltzmann had introduced the hypothesis of 'molecular chaos' in his formulation of statistical mechanics, and the kinetic theory of gases. This discussion was at the heart of Boltzmann's approach to the second law of thermodynamics, where every spontaneous process leads to an increase of entropy.

In 1876 Johann Loschmidt proposed the 'Reversibility Paradox'. He argued that since all the laws of dynamics used by Boltzmann are reversible with time, there cannot be true irreversibility in a system obeying such laws. That is, in effect Loschmidt applied Laplace's demon in reverse to demolish Boltzmann's dynamical arguments for the second law. Loschmidt's time-reversal arguments violate Boltzmann's hypothesis of molecular chaos. All particles which have just had a collision will collide again under time reversal and re-reversal, and thus their velocities are correlated in a regular manner, rather than being 'chaotic'. Loschmidt claimed that *there are just as many initial conditions which lead at least for a short time to a decrease in the entropy of the system as there are initial conditions leading to an increase in the entropy*.

In the real world we never observe a decrease in entropy for large isolated systems. In N-body systems, if N is small, one can observe a decrease of entropy in the form of statistical fluctuations, e.g. density fluctuations, local pressure and temperature fluctuations, whose probabilities scale as $1/\sqrt{N}$. Let us number all the state of a system and put the numbers in to a hat, and drew them at random. Then we would almost always draw a number tagged to an equilibrium — or quasi equilibrium — state. The *transition from non-equilibrium to equilibrium thus corresponds to a transition from exceptionally rare non equilibrium-states to extremely probable states*. This is the gist of the statistical interpretation of the second law of thermodynamics given by Boltzmann to the issues raised by Loschmidt.

We discussed the issue of irreversibility in the context of Nosé dynamics in Sec. 9.2.6. While a set of dynamical equations may be *invertible* in the sense that we can replace the t by $-t$ and get a meaningful evolution in reversed time, the probability of such an eventuality was found be be controlled by a phase space typical of a zero-volume attractor, and hence was unlikely to be observed unless infinitely long simulation times were considered. Similarly, in our discussion of irreversibility in the context of quantum mechanics (Sec. 9.2.7) we concluded that for N-particle systems with sufficiently large N, the cycle times, or equivalently the Poincaré period T_p could be assumed to be infinitely long compared to even the age of the universe. That is, infinitely long cycle times of dynamics of many-body systems is sufficient to ensure irreversibility. Thus it is unnecessary to invoke the arrow of time, and cosmological initial conditions (e.g., the big bang) in explaining irreversibility of many-body systems.

9.4.4 *Maxwell's demon*

Maxwell's demon (Max-D) is a microscopic being who could sit at a trap door in a chamber of gas and sort out the fast moving gas atoms to one side of the trap door, and the slow moving molecules to the other side. Thus, the demon could make one end of the chamber hot, and the other end cold. Clerk Maxwell introduced the demon in 1867 to show that the second law is only valid in a statistical sense. In this sense there is no paradox.

If the system were just one object and not a statistical mixture of particles, the concept of temperature does not apply to such a body.

If a Max-D which is not subject to the laws of physics exists, then indeed the second law can be flouted. One can always postulate 'angels' or 'spirits' that can flout some natural law and force a metaphysical exit. So we consider the case of a Max-D subject to the laws of physics. One may well ask if a small micro-chip could be build, which is 'sensitive enough' to distinguish between hot and cold atoms. Such questions have led to a clear identification that erasing information involves an increase of entropy (Sec. 9.2.5). Max-D has to monitor each atom as it approaches the trap door and collect information. Thus, the sorting out of the hot atoms from the cold atoms by Max-D requires it to erase records in its memory to make room for uploading new information. Thus the demon has to erase information continually, i.e., create entropy. Charles Bennett [21], following on Landauer's analysis, established that Max-D creates at least as much entropy as it decreases by its Maxwell-demoniacal actions. This is in effect a proof that a perpetual-motion machine of the second kind cannot be made even in the form of a computer microchip at the molecular scale.

9.4.5 *The paradox of spontaneous ordering in nature*

All spontaneous processes occur with a release of free energy. The free energy F of a system at a temperature T consists of two parts, since $F = E - TS$, where E is the internal energy, while S is the entropy. If the energy and temperature remains constant, then a spontaneous process must increase the entropy S. This is in fact what is stated in the second law of thermodynamics. Thus, although all spontaneous processes should occur with an increase of entropy, i.e., decrease of order, many natural processes seem to contradict this. Some examples are:

(1) Liquids are more disordered than crystalline solids. A supersaturated or supercooled liquid spontaneously crystallizes when a seed crystal or a nucleation center is added. Similarly, a supersaturated vapour in a Cloud chamber condenses and produces a vapour track when seeded by ionized particles from radioactive sources.
(2) A collection of magnetic atoms (particles with spin) spontaneously arrange themselves in an ordered array, with all magnets pointing in the same direction. At higher temperatures, domains of spins and domain walls separate out from larger regions of oriented spins (see Sec. 9.3.2), somewhat like new cells replicating from mature cells.
(3) Chlorophyll in plants converts carbon dioxide and water into highly complex ordered structures (sugars), when sunlight is available. Photosynthesis is seemingly an entropy-lowering paradoxical process where chemical products with high free energy are produced.
(4) A single bacterium placed in a dish containing some inorganic salts and sugars in water, can multiply into a colony containing millions of bacteria. Such processes occur even in the dark. Each bacterium is a highly organized chemical factory and hence they are low-entropy objects. At first sight, this seems paradoxically contrary to the second law.

(5) Darwinian evolution itself seems to imply the spontaneous formation of increasingly complex, highly ordered sophisticated structures with new, emergent, information-gathering low-entropic properties.

The examples involving crystallization, the growth of a bacterial colony, photosynthesis, and Darwinian evolution are all consistent with the second law of thermodynamics. They are all 'driven' processes. In the case of photosynthesis, it is the solar-energy input which drives the photosynthetic reaction. The total free energy of the initial state involves not only the free energy of the chemical constituents in the photosynthetic system (a plant leaf), but also the free energy of the photons (solar light) falling on the leaf. When such an energy balance is taken into account, then there is a net entropy increase, as required by the second law.

When a bacterial colony grows from a single bacterium, the free energy driving the process comes from oxidizing sugars which are converted into CO_2 and H_2O as more bacteria are formed. The net effect is an crease of entropy. If the colony grows very fast, then the corresponding entropy production is proportionately fast and the process is correspondingly more 'energy inefficient'.

All forms of life may all be regarded as *natural heat engines* which have evolved to dissipate the free energy of solar radiation falling on the earths surface, as well as the internal heat of the earth itself. Thus Darwinian evolution itself may be regarded as a complex dissipative process encompassing the terrestrial biosphere. Seen from the information point of view, living organisms are complex *adaptive* systems that record and use information. This creates entropy. Life's purpose, seen purely from the thermodynamic point of view, is the degradation of solar energy. Thus it is entirely similar to dissipative structures like hurricanes, ocean currents and convective processes, weathering of mountains and rocks, volcanic action and other natural processes which degrade solar energy. Intelligent(?), conscious organisms are capable of dissipating energy even more actively than it would be otherwise. They have contributed to global warming and rapidly burnt fossils that took millennia to deposit. We study complex adaptive systems (like living cells) in the subsequent chapters of this book.

Bibliography

[1] Afshar, S. S.: Proc. SPIE **5866**, 229–244 (2005) `http://arxiv.org/abs/quant-ph/0701027`

[2] Y. Aharonov, Y., Albert, D. Z. and Vaidman, L.: Phys. Rev. Lett. **60**, 1351 (1988).

[3] Aharonov Yakir and Rohrlich D., Quantum Paradoxes: Quantum Theory for the Perplexed (Wiley-VCH Verlag, Weinheim (2005).

[4] Aitchison, I. J. R. and Hey, A. J. G.: Gauge Theories in Particle Physics (Adam Hilger, Bristol, 1989)

[5] Albert, D. Z., Aharonov, Y., D'Amato, S.: Phys. Rev. Lett. **54**, 5 (1985).

[6] Altepeter, J. B., Jeffrey, E. R., Kwiat, P. G., Opt. Express **13**, 8951 (2005).

[7] Anderson, P. W.: Science **177**, 393 (1972)

[8] Anderson, J. A.: An Introduction to Neural Networks (MIT Press, Cambridge, Mass., 1995).

[9] Arndt, M., Nairz, O., Vos-Andreae, J., Keller, C., van der Zouw, G., and Zellinger, A. Nature **401**, 680 (1999).

[10] Ashlab, S. and Nori, Franco, arXiv:0907.4823v1 [quant-ph] `http://arxiv.org/abs/0907.4823`

[11] Ashcroft, N. W. and Mermin, D.: Solid State Physics (Saunders College, Philadelphia, 1976).

[12] Aspect, A., Grangier, P. and Roger, G., Phys. Rev. Lett. **49**, 91 (1982).

[13] Atkatz, David: Am. J. Phys. **62**, 619 (1994).

[14] Alpher, R. A., Bethe, H., Gamow, G.: Phys. Rev. **73**, 803 (1948).

[15] Barrow, John D. and Tipler, F. J.: The Anthropic Principle (Oxford University Press, Oxford, 1986).

[16] Bedingham, D. J.: J. Phys. A: Math. Theor. **44**, 275303 (2011).

[17] Begelman, M.: Science, **300**, 1898 (2003).

[18] Bell, J. S.: Speakable and Unspeakable in Quantum Mechanics (Cambridge University Press, Cambridge, 1987).

[19] Bell, J. S.: Themes in Contemporary Physics II, ed. Deser, S. and Finkelstein, R. J. (World Scientific, Singapore, 1989).

[20] Berman, Marshall: *Intelligent Design; the new creationism threatens all of science and society*. American Physical Society News, **14**, October (2005), `http://www.cesame-nm.org/index.php?name=Sections&req=viewarticle&artid=35&page=1`

[21] Bennette, C. H.: *The Thermodynamics of Computation — A Review*, International Journal of Theoretical Physics, **21**, 905 (1982).

[22] Bennette, C. H. and Brassard, G.: *Proceedings of IEEE International Conference on Computers, Systems and Signal Processing*, Bangalore, India, 175 (1984).

[23] Bennett, C. H., Brassard, B., Crépeau, C., Jozsa, R., Peres, A., Wootters, W. K.: Phys. Rev. Lett. **70**, 1895 (1993).

[24] Bialynicki-Birula Iwo: Progress in Optics, **36**, 245 (1996), `http://xxx.lanl.gov/abs/quant-ph/0508202`

[25] Bohm, David: Quantum Theory (Prentice-Hall, Englewood Cliffs, N. J. (1951).

[26] Bohm, David: Phys. Rev. **85**, 166 (1952); **89**, 1040 (1953).

[27] Bohm, David: Wholeness and the Implicate Order (Routledge, New York, 1980).

[28] Bohm, David and Hiley, B. J.: The undivided universe: An ontological interpretaion of quatum theory (Routledge, New York, 1993).

[29] Bohm, David and Hiley, B. J., Phys. Rep. **144**, 323 (1987).

[30] Bohm, D. J., Dewdney C. and Hiley, B. J., Nature **315**, 294 (1985).

[31] Bohr, N., in [226].

[32] Bohr, N.: Atomic Physics and Human Knowledge (Science Editions, New York, 1961).

[33] Bohr, N.: Albert Einstein, Philosopher-Scientist, Ed. P. A. Schillp; The Libraray of Living Philosophers, Evanston, Il. (1949). `http://www.emr.hibu.no/lars/eng/schlipp/Default.html`

[34] S. F. Boys, Rev. Mod. Phys. **32**, 300 (1960).

[35] Brock, T. D. and Freeze: H. Thermus aquaticus, a Nonsporulating Extreme Thermophile J. Bact. vol. **98**, 289 (1969).

[36] Bjorkern, J. D. and Drell, S. C.: Relativistic Quantum Mechanics (Mc-Graw Hill, New York, 1965).

[37] Bub, J.: Interpreting the Quantum World (Cambridge University Press, Cambridge, 1997).

[38] Bray, D.: Wetware (Yale University Press, New Haven, 2009).

[39] Bromberger, Sylvain: Why-Questions, in Mind and Cosmos, Ed. R. Colodny, (University of Pittsburgh Press, Pittsburgh, 1966).

[40] Adán Cabello: Phys. Rev. A **27**, 012113 (2005).

[41] Callender C. and Weingard, R.: Philosophy of Science, **63**, 470 (1996).

[42] Cairns-Smith, A. G.: Genetic Takeover and the Mineral Origins of Life (Cambridge University Press, New York, 1985).

[43] Carter, Brendon, in Ed. Longair, M. S.: Large Number Coincidences and the Anthropic Principle. In Confrontation of Cosmology Theory and Astronomical Data (Dordrecht, Reidel, 1974), pp. 291–298.

[44] C. Caves, C. A. Fuchs and R. Schack, Stud. Hist. Phil. Mod. Phys. **38** 255 (2007).

[45] Cvitanovic P., et al., Classical and Quantum Chaos: A Cyclist Treatise, `http://www.nbi.dk/ChaosBook/`

[46] Chanadrasekhar, S.: Monthly Notices of the Royal Astronomical Society, **95**, 207 (1936).

[47] Clauser, J. F., Horne, M. A., Shimony, A. and Holt, R. A., Phys. Rev. Lett. **23**, 880–884 (1969).

[48] Clauser, J. F.: Phys. Rev. D **9**, 853 (1974).

[49] Clifton R., Bub J. and Halverson H.: Foundations of Physics **33**, 1561 (2003).

[50] Cushing, James T.: Philosophical Concepts in Physics (Cambridge University Press, Cambridge, 1998).

[51] Dunlap, J. C., Laros, J. J., Decoursey P., Eds.: Chronobiology: Biological Timekeeping (Sinauer, Massachussets, 2004).

[52] Davis, Paul: About Time: Einstein's Unfinished Revolution (Powell Books, Portland, 1995).

[53] Darwin, Charles: The Origin of the Species. Murray (Harvard Facsimile Edition), London (1859).

[54] Dawkins, Richard: The Selfish Gene (Oxford University Press, Oxford, 1976), 2nd Edn. 1989.

[55] Dennett, Daniel C.: Darwin's dangerous idea: Evolution and the Meaning of Life (Simon and Schuster, New York, 1995).

[56] Deutsch, David: The Fabric of Reality: The Science of Parallel Universes and its Implications (Penguin, New York, 1998).

[57] DeWitt, Bryce: Physics Today **58**, 32 (2005).

[58] DeWitt, B. S. and Graham, N. The Many-worlds Interpretation of Quantum Mechanics (Princeton U. P., Princeton, 1973).

[59] Dharma-Wardana, M. W. C., in *Density Functional Theory*, Eds. E. K. U. Gross and R. M. Dreizler (Plenum Press, London, 1995)

[60] Dharma-Wardana, M. W. C., Physical Review B **75**, 075427 (2007).

[61] Dharma-Wardana, M. W. C. and Perrot, F.: Phys. Rev. Lett. 84, 959 (2000)

[62] Dharma-wardana, M. W. C., MacDonald, A. H., Lockwood, D. J. Baribeau, J.-M.., Houghton D. C.: Phys. Rev. Lett. **58**, 1761 (1987).

[63] Dennis Dieks and Sander Lam: Am. J. Phys. **76**, 838 (2008).

[64] Home, Dipankar: Phys. Rev. A **68**, 042105 (2003).

[65] Dirac P. A. M.: Principles of Quantum Mechanics (Oxford University Press, Oxford, 1947).

[66] Dürr, D., Goldstein, S. and Zanghi, N. in *Bohmian Mechanics and Quantum Theory: An Appraisal*, Eds. Cushing, J. T., Fine, A. and Goldstein, S. Kluwer Academic Press, New York (1995), `http://arxiv.org/abs/quant-ph/9511016`

[67] A. Einstein in Scientific papers presented to Max Born (Festschrift), 33, Hafner, New York (1953); D. Bohm in *op. cit*, pp. 13–19.

[68] Ehlers, J., Rindler, W and Penrose R.: Energy Conservation as the Basis of Relativistic Mechanics II, Am. J. Phys. **35**, 995 (1965).

[69] Eldredge, Nileas and Gould, Stephen Jay: Punctuated equilibria: an alternative to phyletic gradualism, in Schopf, T. J. M., ed.: Models in Paleobiology, Freeman Cooper, San Francisco (1972), pp. 82–115.

[70] Einstein, A., Podolsky, B., Rosen, N.: Phys. Rev. **47**, 777 (1935).

[71] Facchi, P. and S. Pascazio, S.: J. Phys. A **41** (2008) 493001, arXiv:0903.3297.

[72] Feyerabend Paul: Against Method, 1975, 2nd Edn. (Verso, London, 1988).

[73] Feynman, Richard P.: Feynman's Thesis: A New Approach to Quantum Theory (1942) (World Scientific, Singapore, 2005).

[74] Feynman, R.: The Feynman Lectures on Physics (Addison-Wesley, Reading, MA, 1964).

[75] Feynman, Richard P.: Character of Physical Law (B.B.C. London, 1965 and Random House, New York, 1994).
[76] Forrest, Barbara and Gross, Paul R.: Creationist's Trojan Horse. The Wedge of Intelligent Design (Oxford University Press, Oxford, 2004).
[77] Fuchs, C. A.: Quantum Theory — reconstruction of Foundations. Ed. krenikov A. Växjö University Press, Växjö, Sweden (2002); Christopher Fuchs, `http://arxiv.org/abs/1003.3509`, downloaded August 2012.
[78] Galileo Galilei: Discourse on Two New Sciences, Discorsi e dimonstrazioni matematiche intorno a due nuove science (Elsevier, Leiden, 1638).
[79] Galileo Galilei: Dialogue concerning two world systems, Dialogo di Galileo Galilei, Linceo, dovo ne I congressi di quttro giornate si discorre sopre: due massimi sistemi delmondo, tolemaico e copernicano (1632). An English translation is available at `http://webexhibits.org/calenders/year-text-Galileo.html`
[80] Galileo Galilei: Discoveries and Opinions of Galileo, Trans. Stillman Drake, Doubleday, New York (1957).
[81] Gaussian computational codes for quantum chemistry (Quantum Chemistry Program Exchange, Program No. 237, 1970) `http://www.gaussian.com/`.
[82] Geim, A. K and Novoselov K. S., Nature Materials, **6**, 183 (2007).
[83] Gell-mann, Murray, The Quark and the Jaguar (Freeman and Co., New York, 1994).
[84] Geymonat Ludovico: Galilée, translation to French by Francoise-Marie Rosset, p. 171 Laffont, Paris (1968).
[85] C. Ghirardi, A. Rimini and T. Weber, "Unified dynamics for microscopic and macroscopic systems," Phys. Rev. D **34**, 470 (1986).
[86] Gill P., Ivanov P L., Kimpton C., Piercy R., Benson N., Tully G; Evett I., Hagelberg E., Sullivan K: Nature Genet. **6**, 130 (1994), see also, Science **306**, 407 (2004).
[87] Godfrey-Smith, P.: An introduction to the philosophy of science; Theory and Reality (University of Chicago Press, Chicago, 2003).
[88] Goodman, Nelson: Fact, Fiction & Forecast (Harvard University Press, Cambridge, MA, 1955).
[89] Susantha Goonatilake: Toward a Global Science (Indiana University Press, Bloomington, 1998).
[90] S. J. Gould and R. Lewontin, The Spandrels of San Marco and the Panglossian Paradigm: A critique of the adaptationist program, Proc. Roy. Soc. (London) B **205** 581 (1979).
[91] Green, D. A. and Stephenson, F. R.: arXiv:astro-ph/0301603v1, `http://arxiv.org/abs/astro-ph/0301603` downloaded April 2010.
[92] Grunbaum, B. and Shephard, G. C.: Tilings and Patterns (Freeman, London, 1987).
[93] Alan H. Guth: The Inflationary Universe: The quest for a new theory of cosmic origins (Addision-Wesley, Reading, MA, 1997).
[94] Haig, D.: *The social gene*, Eds. Krebs, J. R. & Davies, N. B., Behavioural Ecology: an Evolutionary Approach (Blackwell, London, 1997).
[95] Hannestead, S. and Mortsell, E.: Journal of Cosmology and Astro-Particle Physics, **9**, 1 (2004).
[96] Hardy, L.: e-print quant-ph/0111068 (2001).
[97] Haroche, S., Raimond, J.: Exploring the Quantum: Atoms, Cavities, and Photons (Oxford University Press, Oxford, 2006).

[98] Harris, Sam: The End of Faith Norton, New York (2004).
[99] Haugland, J.: Artificial Intellegence, The Very Idea (MIT Press, Cambridge, MA, 1985).
[100] Hawking, Stephen: A Brief History of Time (Bantam, New York, 1988).
[101] Heyak, F. A.: The use of knowledge in Society, Am. Econ. Rev. **35**(4), 519 (1945).
[102] Hempel, Carl G.: Aspects of scientific explanation and other essays, (Free Press, New York, 1965).
[103] Henning, P. A.: Physics Reports **253**, 235 (1995).
[104] Heslot, A.: Phys. Rev. D **31**, 1341 (1985).
[105] Hohenberg, P. and Kohn, W.: Phys. Rev. **136B**, 864 (1964).
[106] P. Holland: Phys. Rev. A **60**, 4326 (1999).
[107] Holland, P.: Quantum Theory of Motion (Cambridge University Press, Cambridge, 1993).
[108] Holian, B. L., Hoover, W. G. and Posch, H. A.: Phys. Rev. Lett. **59**, 10 (1987.)
[109] Horgan, J.: Profile: Paul Karl Feyerabend: The Worst Enemy of Science, Scientific American, **269**, 16 (1993).
[110] Horgan, J.: The End of Science: Facing the Limits of Science in the Twilight of the Scientific Age (Broadway Books, New York, 1999).
[111] Hull, David L.: Science as a Process; An evolutionary account of the social and conceptual development of science (Chicago University Press, Chicago, 1988).
[112] Huxley, T. H.: *On the Physical Basis of Life, Lay Sermons, Addresses and Reviews* (D. Appleton and Co., New York, 1890), pp. 120–146.
[113] Ingoli Fransesco: Letter entitled: De Situ et quiete Terre contra Copernici systema disputatio (The position and immobility of the earth, defended against the Copernican position) 1616.
[114] Jackson, M. W.: Spectrum of Belief, Joseph von Fraunhofer and the Craft of Precision Optics (MIT, Cambridge, 2000).
[115] V. Jacques, E. Wu, F. Grosshands, F. Treussart, P. Grangier, A. Aspect and J.-F. Roch, Experimental realization of Wheelers delayed-choice Gedanken experiment, arXiv:quant-ph/0610241V1, Science **315**, 966 (2007).
[116] Jayatilleke, K. N.: Early Buddhist Theory of Knowledge, Sec. 91, Allen & Unwin, London (1963), quotes *Sutra-kritange*, Vol. II.
[117] Jaynes, E. T.: in *Maximum Entropy and Bayesian Methods*, Eds., Erickson, G., Neudorfer, P., and Smith, C. R., Kluwer, Dordrecht (1992)
[118] Jeannerod, M.: The Cognitive Neuroscience of Action (Wiley-Blackwell, Hoboken, NJ, 1997).
[119] Jeffrey, R.: Subjective Probability: The Real Thing (Cambridge University Press, Cambridge, 2004).
[120] Stuart A. Kauffmann: Origins of Order, Self-Organization and Selection in Evolution (Oxford University Press, Oxford, 1993).
[121] Kohn, W. and Sham, L. J.: Phys. Rev. **140**, A1133 (1965).
[122] Kuhn Thomas: The Structure of Scientific Revolutions, 3rd Edn. (University of Chicago Press, Chicago, 1992), 1st Edn. (1962).
[123] Lado, F., J. Chem. Phys. **47**, 5369 (1967).
[124] Lamb Jr. W. E., Appl. Phys. B **57**, 27 (1989).

[125] LAMBDA: The Legacy Archive for Microwave Background Data Analysis (LAMBDA) data center, `http://lambda.gsfc.nasa.gov/`.
[126] Landauer, R.: "Dissipation and Heat Generation in the Computing Process", IBM J. Res. Dev. **5**, 183 (1961).
[127] Latour, Bruno: Science in action: How to follow scientists and engineers through society, (Harvard University Press, Cambridge, 1987).
[128] Latour, Bruno and Woolgar, Stephen: Laboratory life: The construction of scientific facts (Sage Publications, Beverly Hills, CA, 1979).
[129] Lederman Leon M. and Hill C. T.: Symmetry and the Beautiful Universe (Prometheus Books, Amherst, NY, 2004).
[130] Leavens, C. R.: Solid State Com. **76**, 253 (1990).
[131] Leavens, C. R. and Aers, G. C.: Scanning Tunneling Microscopy III, eds., Wiesendanger R. and Guntherodt, H. J. (Springer, New York, 1993).
[132] Leshno, M., Lin, V. Y., Pinkus, A., WSchocken, S.: Neural networks **6**, 861 (1993), `http://citeseerx.ist.psu.edu/viewdoc/summary?doi=10.1.1.145.6041`
[133] Lieb, E. H. and Yngvason, J.: Phys.Rept. **310**, 1 (1999), `http://arxiv.org/abs/cond-mat/9708200`
[134] Linde Andrei. D.: `http://arxiv.org/abs/hep-th/0503195` and `0503203`.
[135] Lundeen, J. S., Steinberg, A. M., Phys. Rev. Lett. **102**, 020404 (2009).
[136] MacKenzie, Donald A., Statistics in Britain, 1865–1930: The Social Construction of Scientific Knowledge (Edinburgh University Press, Edinburgh, 1981).
[137] Mahan, G. D.: Many-Particle Physics (Plenum, New York, 1981).
[138] Discourses of the Buddha, Medium compendium (Majjima Nikaya), discourse 63 (sutta 63). `http://bodhimonastery.org/courses/MN/MN_course.html` downloaded April 2012.
[139] Mather, J. C., Fixen, D. J., Shafer, R. A., Mosier, C., Wilkinson, D. T.: Astrophys. J. **512**, 511 (1999).
[140] Mayr, Ernst: How to carry out the adaptationist program, Am. Naturalist **121**, 324 (1983).
[141] Mermin, D. N.: Rev. Mod. Phys. **65**, 803 (1993).
[142] Mermin, D. N., Journal of Philosophy **78**, 397 (1981).
[143] Mermin, D. N., Boojums all the Way Through, (Cambridge University Press, Cambridge, 1990).
[144] Mermin, D. N.: In praise of measurement, arXiv:quant-ph/0612216v1 (2006).
[145] Metropolis N., Rosenbluth A. W., Rosenbluth M. N., Teller A. H. and Teller E., J. Chem. Phys. **21**, 1087 (1953).
[146] Metzinger, Thomas: Being No One: The Self-Model Theory of Subjectivity (MIT Press, Boston, MA, 2003).
[147] Metzger, W: Gesetze des Sehens, Waldemar Kramer, Frankfurt (1975).
[148] W. H. Newton-Smith: The Rationality of Science (Taylor and Francis, London, 2007).
[149] Misra B. and Sudarshan E.: J. math. Phys. **18**, 756 (1977).
[150] Moore, G. E.: Principia Ethica (Cambridge University Press, Cambridge, 1903).

[151] Murphy, N. and Ellis, G.: On the moral nature of the Universe: Cosmology, Theology and Ethics (Fortress Press, Minneapolis, 1996).

[152] Nyanaponika Thera: Heart Of Buddhist Meditation, Weiser, Main (1965) and Buddhist Publication Soc., Kandy (1954).

[153] Paley, William: Natural Theology (1802), Eds. Eddy Matthew D. and Knight, David M. (Oxford University Press, 2006).

[154] Parrot, Stephan, Quantum-Physics arXiv:0909.0295v2, `http://arxiv.org/abs/0909.0295`.

[155] Patchkovaskii, S., Zhao, Z., Brabec, T., and Villeneuve, D. M.: Phys. Rev. Lett. **97**, 4923 (2006).

[156] Pauli, Wolfgang: Handbook der Physik, Vol. 24 (Springer, Berlin, 1933).

[157] Linus Pauling: Nature of the Chemical Bond, 3rd Edn. (Cornell University Press, Ithaca, 1960).

[158] Peat, David F.: Infinite Potential, The Life and Times of David Bohm (Addison-Wesley, New York, 1997).

[159] Peat, D.: A Flickering Reality: Cinema and the Nature of Reality (Pari Publishing, Pari, Italy, 2012).

[160] Penrose, Roger: The Emperor's New Mind (Oxford University Press, Oxford, 1989).

[161] Penrose, R.: *On gravity's role in quantum state reduction*, Gen. Relativ. Grav. **28**, 581 (1996).

[162] Perrot, F. and Dharma-Wardana, M. W. C.: Phys. Rev. A **30**, 2619 (1984).

[163] Perrot, F. and Dharma-Wardana, M. W. C.: Phys. Rev. E **52**, 2920 (1995).

[164] C. S. Pittendrigh: Annu. Rev. Physiol. **55**, 17 (1993).

[165] Poincaré, Henri: Science et Méthode. Flammarion, Paris (1897), `http://www.ac-nancy-metz.fr/enseign/philo/textesph/Scienceetmethode.pdf`

[166] Popper, Karl: The Logic of Scientific Discovery. Hutchinson, London (1959) — Translation of Logik der Forschung, Springer, Vienna (1935).

[167] Popper, Karl: Conjectures and Refutations, The Growth of Scientific Knowledge (Routeleg, London, 1963).

[168] Popper, Karl: Quantum Theory and the Schism in Physics, (Bowman and Littlefield, New Jersey, 1982).

[169] Pospelov, Maxim and Romalis, R.: Physics Today **57**, 40 (2004).

[170] Post, S. G.: Altruism and Altruistic Love: Science, Philosophy and Religion in Dialogue (Oxford University Press, Oxford, 2002).

[171] Premack, D. and Woodruff, G.: Behavioural and Brain Sciences, **4**, 515 (1978).

[172] Premack, D. and Premack, A.: Original Intelligence, The Architecture of the Human Mind (McGraw-Hill, New York, 2003).

[173] Price, H. H.: Hibbert Journal, Vol. II, 472 (1955).

[174] Prigogine, I.: From Being to Becoming: Time and Complexity in the Physical Sciences (Freeman, New York, 1980).

[175] Psillos, Stathis: Scientific Realism: How science tracks truth (Routledge, New York, 1999).

[176] Pusey, M. F., Barrett, J. and Rudloph, T.: Nature Physics **8**, 476 (2012).

[177] Quine, Willard V.: Two Dogmas of Empiricism, in From a Logical Point of View (Harvard University Press, Cambridge, 1953).

[178] Readhead, M.: Incompleteness, Nonlocality and Realism (Clarendon Press, Oxford, 1987).

[179] Riess, A. G. *et al.*, "Tests of the accelerating Universe ...", Astrophys. J. **369**, 62 (2000).

[180] Rosenberg, Alex: Philosophy of Science; a contemporary introduction. Routleg, New York (2000).

[181] Rosenblatt F., Psychological Review, **65**, 386 (1958).

[182] Rosenblum B. and Kuttner, F.: Quantum Enigma: Physics encounters consciousness, Oxford University Press, Oxford (2006).

[183] Strogatz, S. H.: SYNC, the Emerging Science of Spontaneous Order. Hyperion, New York (2003).

[184] Russell, Bertrand: History of Western Philosophy, Ch. XXIV Alen & Unwin, London (1961).

[185] Ryle, Gilbert: The Concept of Mind. University of Chicago Press, Chicago (1949).

[186] Shannon, C. E.: Bell System Technical Journal **27**, 379 (1948), `http://cm.bell-labs.com/cm/ms/what/shannonday/paper.html`

[187] Schrödinger, Erwin.: Annalen der Physik **79**, 361–376 and 489–527 (1926).

[188] Schrödinger, Erwin. Proc. Cambridge Philosophical Soc. **31**, 555 (1935); ibid. **32**, 446 (1936).

[189] Schlosshauer, M. Decoherence and the quantum-to-classical transition (Springer Verlag, Heidelberg, 2007).

[190] Searle, John: *Minds, brains and Programs*. Behavioral and Brain Sciences **1**, 417 (1980).

[191] Serres, Michel: La Naissance de la Physique dans le texte de Lucrèce, les Editions de Minuit, Paris (1977).

[192] Shanks, Niall: God, the Devil and Darwin, A Critique of Intelligent Design Theory. (Oxford University Press, Oxford, 2004).

[193] Shapin, Steven: The Scientific Life: A Moral History of a Late Modern Vocation. University of Chicago, Chicago (2008).

[194] Shapin, Steven: The Scientific Revolution. University of Chicago, Chicago (1996).

[195] Shapin, Steven and Schaffer, Simon: Leviathan and the air-pump: Hobbes, Boyle and the experimemntal life, Princeton University Press, Princeton NJ (1985).

[196] Shultz, Bernard F.: A first course in general relativity, Cambridge University Press, Cambridge (1985).

[197] Simon, Herbert: Proceedings of the American Philosophical Society **106**, 467 (1962).

[198] Singer, Tania: Science **303**, 1121 (2004).

[199] Sokal, Allan: Transgressing the Boundaries: Toward a Transformative Hermeneutics of Quantum Gravity, Social Text, **14**, 217 (1996).

[200] Bacciagaluppi, G., and Antony Valentini: Quantum theory at the crossroads: Reconsidering the 1927 Solvay Conference, Cambridge Univ. Press (2009). `http://science.videosift.com/video/Solvay-Conference-1927-Einstein-Schrodinger-Bohr;`

[201] Stapp, H.: Mind, Matter and Quantum Mechanics, Springer Verlag, Berlin (2004).

[202] Stapp, H.: Mindful Universe, Quantum Mechanics and the Participating Observer, Springer Verlag, Berlin (2007).

[203] Stcherbatsky, T. H.: Buddhist Logic, Vol I, 1st Ed. 1926, reprinted Barnasidas Publishers, Delhi, India (2000).
[204] Taubes, Garry: Bad Science: The Short Life and Weird Times of Cold Fusion, Random House, New York (1993).
[205] Nassim Nicholas Taleb: The Black Swan. The Impact of the Highly Improbable, Random House, New York (2007).
[206] Tegmark, M. *et al.*,: Phys. Rev. D 69, 103501 (2004).
[207] Tuszynski, J., The emerging physics of consciousness, Ed. J. Tuszynski Springer Verlag, Berlin (2006).
[208] Unruh, W. G. and Opat, G. I., Am. J. Phys. **47**, 743 (1979).
[209] Unruh, W. G.: Am. J. Phys. **49**, 589 (1981).
[210] Utiyama, R. Rhys. Rev. **101** 1597 (1956).
[211] van Kampen, N. G.: Physica A **153**, 97 (1988).
[212] van Kampen, N. G.: Am. J. Phys. **76**, 989 (2008).
[213] Volovik, Grigory E.: The Universe in a Helium Droplet, Oxford University Press, Oxford (2003).
[214] von Neumann, J. Mathematische Grundlagen der Quantenmechanik, Springer Verlag, Berlin (1932).
[215] Warneken, Felix: Science, **311**, 1301 (2006).
[216] Widrow, B. and Lehr, M. A.: Proc. IEEE **78**, 1415 (1990).
[217] Weber, Max: Wissenshaft als Beruf, Munich (1918) English version: "Science as a vocation", in From Max Weber: Essays in Sociology, pp. 129–156, Oxford University Press, Oxford (1946), or `http://www.zephoria.org/alterity/archives/2004/11/max\_weber\_scien.html`
[218] Weinberg, Steven: Phys. Rev. Lett. **59**, 2607 (1987).
[219] Weinberg, Steven: The first three minutes: A modern view of the origin of the universe, Basic Books, New York (1977).
[220] Weinberg, Steven: Dreams of a Final Theory. Pantheon, New York (1992).
[221] Weinberg, S.: Conceptual Foundations of Quantum Field Theory, edited by Tian Yu Cao, p. 242. Cambridge U. P., Cambridge (1999).
[222] R. S. Westfall: Never at Rest — A biography of Isaac Newton, Cambridge University Press, Cambridge (1980).
[223] Wheeler, J. A., *The 'Past' and the 'Delayed-Choice' Double-Slit Experiment*, pp. 9–48, Ed., Marlow A. R., Mathematical Foundations of Quantum Mechanics, Academic Press, New York (1978).
[224] Wheeler John A.: A Journey into Gravity and Spacetime. Freeman, New York (1992).
[225] Wheeler John A.: Recent Thinking about the nature of the Physical World: It from Bit. Frontiers in Cosmic Physics: Symposium in memory of Sere Alexander Korff. Eds. R. B. Mendell and A. I. Mincer, New York Academy of Sciences, New York (1992).
[226] Wheeler, J. A. and Zurek W. H., Eds. Quantum Theory and Measurement. Princeton University Press, Princeton (1983).
[227] Wigner, Eugene P.: *The Unreasonable Effectiveness of Mathematics in the Natural Sciences*, Communications in Pure and Applied Mathematics **13**, No. I, John Wiley & Sons, New York (1960).

[228] Wigner, Eugene P.: Remarks on the Mind-Body question, reprinted in [226] and available on-line:
`http://www.phys.uu.nl/igg/jos/foundQM/wigner.pdf`

[229] Willis, T.: Cerebri Anatome, cui accessit Nervorum description et usus, see The Anatomy of the Brain and Nerves, Ed. Feindel, T., McGill University Press, Montreal (1965).

[230] Witten, Edward, `http://www.sns.ias.edu/~witten/papers/string.pdf`

[231] *Philosphische Untersukungen*, English Trs. G. E. M. Anscombe: Philosophical Investigations, McMillan, New York (1953).

[232] Woese, C. R., Kandler, O. and Wheelis, M. L.: Towards natural systems of organisms: Proposal for the domains Archaea, Bacteria, and Eucarya, Proc. Natl. Acad. Sci. USA **86**, 4576 (1990).

[233] Wooters, W. K. and Zurek, W. H.: *Physics Today*, p. 76, February (2009).

[234] Zellinger, A. e-print quant-ph/0212084 (2002).

Part II.
Complex Systems and Consciousness

Chapter 10

Bio-Molecules, the Sub-Slime of Astrochemistry

The early earth was made up of hot inorganic materials strongly out of equilibrium, and subject to radiation fields. We review astrochemistry, the nature of the early-earth and its atmosphere, scenarios for the formation of prebiotic molecules and more complex genetic machinery in 'Darwin's warm pond'. Extra-terrestrial, atmospheric-oceanic, and hydro-thermal-vent theories of pioneer organisms are considered in some detail.

10.1 Introduction

The universe as we know it today, made up of nebulae, galaxies, stars etc., is predominantly a vast mass of hydrogen ionized into the plasma state of protons and electrons. Some of the cooler regions of this universe contain tiny, almost negligible amounts of matter that we refer to as organic molecules. These are molecules containing carbon atoms, often bound to ubiquitous hydrogen atoms, and the much rarer nitrogen, oxygen and other familiar elements of biological matter. An even more inconceivably minuscule fraction of such organic molecules is found as biological matter, almost like a little bit of slime that we may see floating on the surface of a vast ocean. Indeed, bio-molecules, and life itself can be dismissed as an inconceivably tiny smear of inconsequential slime appearing in the vast scheme of degradation of the energy unleashed in the big bang. However, *we are part of the slime* and its origins are of direct concern to us. A brief popular account of 'life-ascending' from pre-biotic matter may be found in Lane [55].

The previous chapters of this book examined the nature of the laws that govern the universe, to the extent accessible to our measuring instruments and theorizing capabilities. We know how quarks assemble into nucleons, and how nucleons and electrons assemble into atoms and molecules. These processes occur in 'containers' known as stars which are powered by nuclear energy. Thus, stars themselves are 'cells' which metabolize and degrade energy. The degraded energy, at lower energy scales, is still large enough to sustain chemical reactions. Fermentation is such a process where large sugar molecules are broken down releasing energy. Anaerobic processes (i.e., processes which occur in the absence of air or oxygen) would have been favoured in the early earth. Once an oxygen mantle was formed, newer aerobic mechanisms would have become increasingly important. Such chemical degradation of energy, both anaerobic and aerobic, can also occur by more complex pathways involving the formation of 'chemical' containers which can self-assemble, replicate and re-consume the chemical energy. Living cells are just such objects. The transformation of simple inorganic matter into more complex pre-biotic matter, and the transformation of prebiotic matter into living matter (*biopoiesis*) involve a multitude of very complex and subtle chemical pathways. This is in fact the story of the evolution of life form of inorganic matter [90]. Although many scientifically plausible schemes can be formulated, a complete 'turn-key' laboratory demonstration that satisfies *everyone* is another issue. However, the successful implantation of a digitally determined genome sequences into a bacterium i.e., the synthesis of a computer designed organism by the Craig Venter Institute, involving one million nucleotides already demonstrates the current capabilities of synthetic biology [40]. Even if a complete turn-key demonstration, or several such demonstrations based directly on pre-biotic material were presented, the actual scenarios that prevailed in primordial times would remain a matter of conjecture.

Nevertheless, our knowledge of the basic laws of chemistry is complete. Our knowledge of the physio-chemical conditions prevailing in the early Earth have increased tremendously. We even have Archaean fossil evidence to consider, according to Schopf [85], that a "true consensus for life's existence" dates it to more than some 3500 Million years ago. William Schopf had studied a string of rock capsules and fossil stromatolites in Shark Bay, Australia, that he initially claimed to be fossils of ancient cyanobacteria. This view has come under intense attack by Martin Brassier and others, but we may use a figure like $3 \pm 0.5 \times 10^9$ years as a possibility, since paleontological estimates continue to attract bitter controversy. The uncertain dating of fossil evidence does not affect the very plausible scenarios available for the formation of pre-biotic molecules and for biopoiesis itself. In this chapter we examine how astro-chemistry and terrestrial chemistry evolve into living matter — slime from one point of view, and sublime from another point of view.

10.2 Chemical elements and molecules in the cosmos

The Universe is estimated to be about 13.7–14 billion years old. As the universe expanded from the Big Bang, cooling and decreasing in density, only the lightest elements, e.g., Hydrogen, Helium, and traces of Deuterium, Tritium, and possibly Li and Be would have been formed. Heavier elements were formed by nucleosynthesis only after stars were formed. Carbon, the element most important for life itself is produced by the triple-α-process involving three α-particles. Oxygen is formed by an additional step.

$$3\,^4\mathrm{He} \to {}^{12}\mathrm{C}, \quad \text{and} \quad {}^{12}\mathrm{C} + {}^4\mathrm{He} \to {}^{16}\mathrm{O} \tag{10.1}$$

The nucleus with the highest binding energy is ^{56}Fe. Its accumulation as an end product of nucleosynthesis in the stellar core destabilizes the star and initiates stellar collapse. Just as some fruits burst and spew out seeds, stellar explosions spew out gas clouds containing the chemical elements with the nuclear charge upto that of iron ($Z \leq 26$) into space. Supernova (see Sec. 2.6) explosions send out clouds of heavier (higher Z) elements into interstellar space. Such clouds again re-condense forming proto-stars and planets. Hence we may think of stars as little factories which burn up hydrogen etc., and seed the inter-stellar medium with heavier elements. The abundances of elements [6] in the cosmos relative to hydrogen is shown in Fig. 10.1. This clearly shows that the chemistry of space is mainly that of H, C, N, and Oxygen. The interstellar medium constitutes $\sim$10% of the mass of the galaxy. Interstellar clouds, stellar envelopes, comets, asteroids etc., provide the crucible for the formation of molecules from these elements. Molecular hydrogen (H_2) is by far the most abundant molecule in these clouds. CO is the most abundant carbon-containing species, with the CO/H_2 ratio being of the order of 10^{-4}. Spectra of highly excited CO molecules emanating from the most distant quasar currently known (with a redshift of 6.42) have been detected. This indicates that CO has been present in the universe at least since approximately 800 millions years after the Big Bang.

In cold clouds, dust particles and carbon grains adsorb ice mantles of water [100] and other compounds [30] like CO_2, CO, CH_3OH, with smaller amounts of CH_4, NH_3, H_2CO, and HCOOH. Carbon-containing molecules (i.e., organic molecules) are observed in our galaxy and in distant ones with similar abundances and composition. Observations of galaxies with high redshifts probe the early universe and show that organic molecules were already present and that dust-grains were already in existence. Given the dimensions of our universe, the widely distributed observations of aliphatic and aromatic molecules, as well as carbon-bearing ices (such as CO-, graphite-, or fullerene-containing ices), provide clear evidence that chemical processes have proceeded in similar ways in

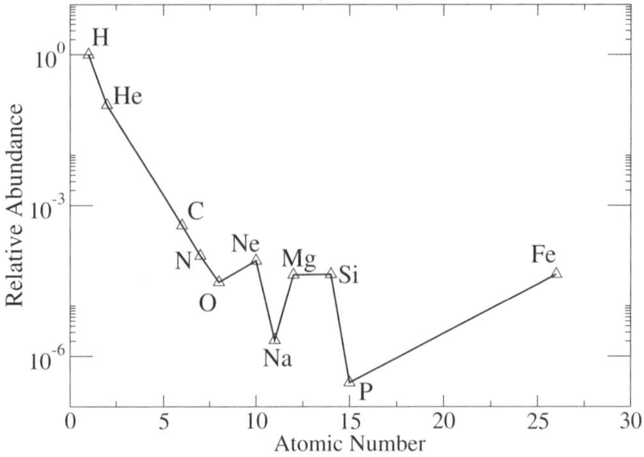

Fig. 10.1 Cosmic abundances of the elements relative to hydrogen (logarithmic vertical axis).

diverse parts of the Universe. The prebiotic molecules in diverse regions are thus quite likely to be similar. The conditions for the formation of sufficiently complex C-based molecules is determined by the strength of the C-C covalent bond; carbon chemistry is likely to be the preferred basis of biological molecules everywhere in the Universe.

The detection of more than 170 extra-solar planets and numerous proto-planetary disks by 2010 show that planets are quite common in the Universe. On the other hand, the Earth may have provided a rare combination of conditions suitable for biological processes. A moderate surface temperature, presence of abundant water, a stable orbit, suitable inclination of the axis, and a dynamic atmosphere and interior structure are features favouring life on the Earth. Even this view is being modified rapidly, as more and more 'earth-like' planets are being continually discovered.

10.3 Envisioning the early Earth

The solar system was probably created from the nebular debris produced by a red-giant star which exploded in a supernova, roughly five billion, i.e., 5×10^9 years ago, referred to as 5 Giga-years ago (Gya). As the cloud of debris collapsed under its gravitational field, it compressed and heated its center forming a proto-star which formed the solar embryo. The gravitational forces began to produce an inward moving, rotating mass of gas which naturally rearranged into the shape of an accretion disk extending away from the equator of the rapidly growing sphere of the sun. The matter in such an accretion disk radiates away its energy and begins to cool. Depending on various details, the proto-star may end up as a double star or a single star, the latter being probably the case with our sun.

The process may be thought of as *a natural engine* where gravitational and electromagnetic energies are converted into mechanical energy. The spiraling in and compression of charged particles into the central core create a body heavy and hot enough to sustain thermonuclear conversion of hydrogen to helium, powering the nascent sun.

The dust particles in the accretion disk aggregate to form larger rocks and asteroids. The detailed mechanism may involve instabilities that lead to the formation of rings that collapse on themselves, forming proto-planets. In principle, such processes have similarities to laboratory condensation processes and hydrodynamic instabilities. However, astronomical processes are driven by enormously different energy scales, caused by the gravitational fields, radiation fields and strong magnetic fields that exist in the stellar accretion disk where matter is present in the form of ions and electrons (plasma).

The initial slow aggregation can become a fast 'runaway process' which strongly favours further growth of the larger aggregates. Detailed modeling of planetary formation is consistent with the observed features of the sun and its planets in our solar system, although many assumptions remain. Mercury, Venus, Earth and Mars turn out to be dense, rocky planets, while Jupiter, Saturn, Uranus and Neptune turn out to be 'gaseous' planets with deep atmospheres composed primarily of hydrogen, helium, and oceans of methane.

The earth itself may be $\sim$4.5 Gy old, while the other planets would have very similar ages. Planets, asteroids, comets, dust clouds and other objects would already have a pre-formed stock of elementary molecules (e.g., CO, CN, H_2O, NH_3), radicals (CH, NH, CH_3 etc.) and ions (CH^+, NH_2^+, OH^- etc.). These would be contained in the atmospheres and mantles of the early planetary objects [95]. Thus conditions and materials suitable for the formation of more complex chemical building blocks of biological matter could have existed in the Earth, Mars, and some other parts of the solar system. Examples of such molecules are the simple amino acids, nucleotides, fatty acids etc., needed to make more complex biological molecules like RNA, DNA, and proteins found in living cells. The blanket name *prebiotic matter* is used for such simple organic compounds that are believed to be necessary precursors of biological molecules.

The living matter that we know of are those of terrestrial matter, and microbial matter found in comets and asteroids. Hence, much of our discussion would be based on envisioning the conditions that may have existed in the early earth during its first few billion years of existence. The nature of the terrestrial atmosphere and the oceans in the Hadean (4.5–3.8 Gya) and Archean (3.8–2.5 Gya) aeons are directly of interest, although we have little or no explicit information of that formative epoch. In fact, we have more definite information about the first few moments of the 'Big Bang', than about planetary events much closer to us in space and time! We can only use our knowledge of physics and chemistry, as well as observational planetary science and geophysics to guide us here.

10.3.1 *A hydrogen-rich atmosphere for the early Earth*

The Hadean aeon is associated with the formation of the earth, its oceans and the atmosphere. This was a period dominated by frequent meteoric showers as well as intense radiation from the sun falling on an earth without a protective cover of Ozone. The consensus among planetary scientists in the latter half of the 20th century was that the early Earth had a low concentration of hydrogen containing molecules like methane and ammonia, and that the earth's mantle was more reduced than its current oxidation state. However, more recent studies suggest that the earth's mantle has been in a similar state of oxidation during the past 4 billion years [24]. Similarly, current views [93] favour a significantly hydrogen-rich early-Earth atmosphere in which organic molecules could be produced more efficiently than previously envisaged.

The amounts of hydrogen, carbon, nitrogen and such key elements in the atmosphere, oceans and the crust of the early earth determined the prebiotic organic compounds that could have paved the way for life itself. Biologically important molecules can be efficiently formed in a highly reducing atmosphere, i.e., an atmosphere rich in methane (CH_4), and/or ammonia (NH_3). They form efficiently in a weakly reducing atmosphere where the main issue is the ratio of hydrogen to carbon in the atmosphere.

The total hydrogen concentration (i.e., the hydrogen in the form of methane, ammonia, water, etc., and hydrogen) in the atmosphere is set by the balance between the volcanic outgassing rate of hydrogen and its escape to space. The present-day volcanic hydrogen outgassing rate is about 1.8×10^{10} hydrogen molecules per sq. cm per sec. (see [93], footnote 7). Because of higher heat flow in the past, the overall outgassing rate of hydrogen and other gases might have been some five times greater on ancient Earth. Oxygen is dominant in the present-day atmosphere at the level of the *exobase* (this is the boundary beyond which rapidly out-moving molecules may escape without collision). The current *exospheric temperature* T_e, i.e., the temperature at the exobase, is believed to be high (1000 K to 2500 K) because of the efficient absorption of solar-UV radiation by oxygen.

The exospheric temperature T_e determines the kinetic-energy distribution which is a bell-shaped curve (Maxwell distribution, see Sec. 6.2.1). The tail of the distribution contains the high-velocity molecules having speeds exceeding the gravitational escape velocity $V_e = GM/r$. Here r is the radial distance from the center of the Earth, G is Newton's gravitational constant and M is the mass of the Earth. All such molecules do not escape as they may suffer collisions or move in the wrong direction. The escape of molecules, accounting for collisions etc., is known as *Jeans escape*, named after the British astronomer Sir James Jeans. It can be shown that the probability of collisions for molecules at the exobase is of the order of $1/e$, where $e \sim 2.718$ is the base of natural logarithms. Thus energetic molecules above the exobase are very likely to escape without hindrance.

If the exospheric temperature T_e on the early Earth were what it is today, Jeans escape of hydrogen from the atmosphere would have been efficient, and the diffusion of hydrogen through the background gases to the *homopause* level (i.e., up to the height where there is efficient molecular mixing in the atmosphere, without separation based on mass ratios) would have been the limiting process. However, the early Earth's atmosphere was *anoxic* (i.e., deprived of oxygen) [33] and probably contained substantial amounts of CO_2, similar to present-day Venus or Mars. Carbon dioxide absorbs Ultra-Violet (UV) radiation, but unlike O_2, it can re-radiate energy back to space and keep T_e low. The CO_2-rich venusian and martian exobases have temperatures of 275 K and 350 K, respectively. Jeans escape of hydrogen would have been strongly curbed in the anoxic early-Earth atmosphere with low T_e. Thus hydrogen could have been a major component of the early-Earth atmosphere. Tian et al. [93] modeled the escape behaviour of such an atmosphere, taking account of the much higher extreme-UV radiation prevalent in the Hadean and Archean aeons. They found that a hydrogen-rich early Earth atmosphere could be maintained, even if the outgassing rates were as modest as those consistent with assuming that the oxidation state of Earth's mantle 3.9 Gya were the same as it is today.

Thus Tian et al. state [93] that 'On the basis of our new model of hydrodynamic hydrogen escape, we conclude that diffusion-limited escape theory does not apply to a hydrogen-rich early-Earth atmosphere. Rather, the escape of hydrogen was energy limited. Hydrogen mixing ratios greater than 30% could have been maintained in the atmosphere of prebiotic Earth without either invoking huge volcanic hydrogen outgassing rates or assuming a reduced mantle. *The efficient production of organics in a hydrogen-rich early-Earth atmosphere would have led to an organic soup in the oceans and ponds of the early Earth*. The world oceans could have been the birthplace of life'.

An early Earth with an atmosphere high in hydrogen concentration has important consequences for the origin and evolution of life. Production of prebiotic organic compounds within the Earth's atmosphere (endogenous sources), such as production by lightning or photochemistry, becomes a major possibility in a reducing atmosphere. On the other hand, exogenous sources such as the production of prebiotic matter in hydrothermal systems, deep-ocean crevices, and the disposition of extra-terrestrial prebiotic matter, or even living matter [48] via meteorites and comets become important alternatives in an atmosphere poorer in hydrogen [16]. However, the above discussion in terms of a hydrogen-rich atmosphere brings us back to the early Urey-Miller model that we discuss in Sec. 10.4.2.

10.4 The formation of prebiotic molecules

The nature of the prebiotic molecules that would be relevant to early life would depend on the earliest forms of life that we may envisage. These forms need not be based on life that is most abundant today. If prebiotic molecules arose as a result

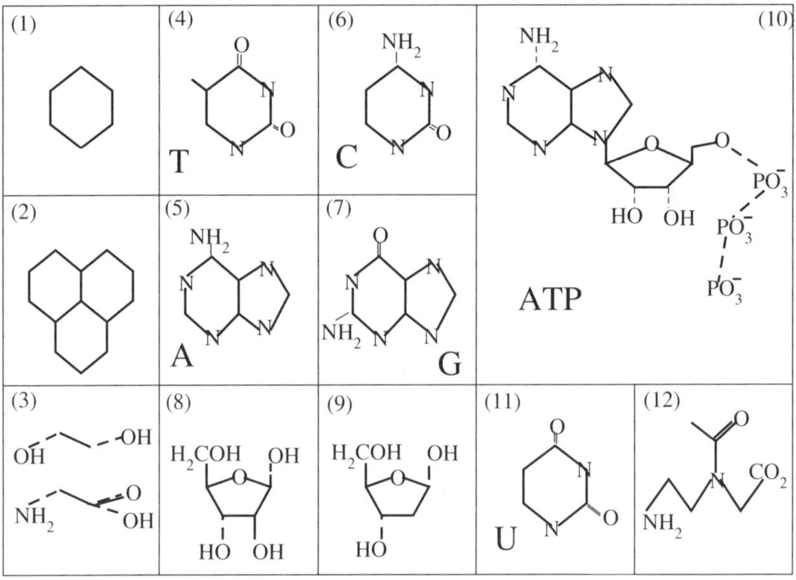

Fig. 10.2 Prebiotic molecules at various levels of complexity. (1) benzene (2) a tricylic example of a poly-aromatic-compound (3) Glycol (di-alcohol) and Alanine (amino acid). Panels (4)-(7) display the four fundamental base pairs. Thiamine (T) and Adenine (A), Cytosine (C), and Guanine (G). These bases join with sugars, and also pair via hydrogen bonds (*base pairs*). Sugars are molecules which contain OH groups and aldehyde (CHO) groups which have formed into a ring shape as shown in (8). This sugar contains five carbon atoms, and is known as *ribose*. (9) deoxyribose has one OH group less. A base attached to a ribose or deoxyribose sugar is known as a *nucleoside*. The nucleoside derived from Adenine is known as adenosine. Nucleosides join with phosphate groups to form *nucleotides*. These can form long chain structures as found in RNA and DNA, where a phosphate group and a sugar group alternate along the chain. A nucleoside can react with phosphoric acid to pick up to three phosphate groups. Thus ATP is adenosine triphosphate (ATP), shown in panel (10). Uracil, shown in (11) is a base found in RNA. Instead of using sugars to form nucleotides, peptide-acids like the molecule shown in (12) can be used to form polymeric peptide-nucleic acid (PNA) chains.

of delivery of organic molecules from outer space, then we may envisage initial biological molecules which have some relation to the predominant chemistry of cosmic dust. Since polycyclic aromatic compounds (fused benzene rings, precursors of graphite) are predominant in meteorites and other inter-planetary debris, Ehrenfreund [31] and others have considered the possibility of life initiating in an 'aromatic world'. Simple molecules like benzene (Fig. 10.2) and polycyclic aromatics, as well as heterocyclic compounds (i.e., aromatic compounds containing other elements like nitrogen, besides carbon), are considered as possible pre-biotic molecules ushering in the 'aromatic world'. Within this picture, it is necessary to explain how these pre-biotic aromatics proceed to become replicating molecules, and how the present DNA world of cells and organisms evolve.

On the other hand, if the early Earth's atmosphere is taken to be the source of most of the prebiotics, then a different picture, more in line with traditional mainstream thinking emerges. The formation of key prebiotic molecules such as amino acids from simple mixtures of elementary C, H, N, O compounds like methane, ammonia, water, nitrogen, carbon monoxide, and hydrogen was demonstrated in the 1950s onwards. When molecules react to form more stable products, the final-state free energy is lower than the initial-state free energy. However, usually an *activation energy* has to be supplied to make the reaction go forward. For instance, a mixture of H_2 molecules and O_2 molecules at normal temperatures do not spontaneously become water because an *energy barrier* separates the initial state from the final state. However, if the mixture is heated, sparked with electric discharges or compressed by a shock wave, the exothermic formation of water will begin as some molecules become energetic enough to overcome the energy barrier. They will generate enough heat to initiate an explosively rapid formation of water.

The Swedish chemist Arrhenius showed that the rate k of chemical reactions depends exponentially on the activation energy and the temperature T of the system, viz.,

$$k \propto Ae^{-\Delta E/T} . \tag{10.2}$$

Thus a very modest increase in temperature can have a large effect on the rate of a reaction. Here A is a prefactor which can be shown to be a measure of the average number of collision (or 'attempts') between the reactant molecules that are needed for the reaction to occur irrespective of the activation energy. The activation energy ΔE is a quantum-mechanical quantity which depends on the energy differences between the bonds of the reactants and those of the transition state. Thus quantum mechanics controls the processes at their microscopic length scales.

Hence we can envisage that prebiotic molecules would have formed in the atmosphere of the early Earth as well as in other favourable stellar environments, via lightening (electric discharges), or interaction with energetic photons which generate chemically active radicals and reactive species (photolysis) which react further. In electric discharges, highly energetic electrons as well as ions are created and they collide with simple molecules in the air and break up their bonds, forming free radicals and active species which recombine to more complex molecules. Ionizing particles present in cosmic rays (e.g., mesons) can also disrupt H_2, CO and other atmospheric molecules and convert them into active species which recombine into prebiotic molecules. Compression waves and shocks can also initiate reactions as they provide the energy necessary to break stable bonds and generate chemically active species. Of course, many of these processes can also act in reverse. However, the free energy balance is definitely in favour of the formation of more complex molecules from simpler ones. The amines and other prebiotics formed in the Archean atmosphere would come together in the oceans to form more complex nucleosides (containing heterocyclic bases), nucleotides, i.e., phosphates of nucleosides (see Fig. 10.2).

The big puzzle was how these molecules assembled themselves into larger, self-replicating molecules that went on to form the earliest living cells. These reactions need a catalyst (i.e., an enzyme), and a template to guide the chemical process towards the desired products. RNA molecules are in fact able to perform these two functions, while DNA is not known to have any enzymatic activity. Thomas Cech [14] of Colorado University, and Sidney Altman [3] of Yale University demonstrated in 1983 that Ribozyme, a type of RNA molecule, was capable of enzymatic activity. They were awarded the Nobel prize for this work, and gave rise to the concept of an early-Earth biology of an 'RNA world' [41, 73].

A third possible model of early-Earth biology has emerged from the discovery of deep-ocean hydrothermal vents in 1977. Over 200 'black-smoker' vents have been discovered in the Pacific, Atlantic and Indian oceans, some of them as tall as sky-scrapers. They pump out broiling metal sulphides and compressed volcanic gases at ~ 700 K into the ocean above them. Hydrothermal circulation would have existed early in the Earth's history. It is likely that vents were present in the Archean oceans. Organic compounds are very likely to be produced during passage through the temperature gradient of the 350°C vent waters to the 0°C ambient ocean waters under high pressure [18]. Polymerization of the organic compounds formed, followed by their self-organization, has also been proposed to take place in these environments. Hydrothermal vents are also the nursery envisaged by Wächtershäuser [96] who has proposed a prebiotic 'iron-sulphur world' based on the interaction of hydrogen sulphide (H_2S) and Ferrous sulphide (FeS). Hydrogen and Iron pyrite (FeS_2) are formed with the release of energy which can be used for further chemical processing involving the formation of prebiotic molecules like methane, amines etc.

Geological processes convert various minerals (e.g., iron silicates like Fayalite, magnesium silicates like Forsterite) into other minerals like Magnetite (Fe_3O_4), or minerals containing more silicon and water (Sepentinite). The formation of such water-bearing rocks is known as *serpentinization*. They are important in the theory of prebiotic matter since serpentinization can generate hydrogen which may reduce CO_2 into methane and other prebiotic molecules. These models are important for precursors of anaerobic life.

Serpentinization is believed to be operative in a second type of hydrothermal vent, quite different from the black smokers. They were discovered in 2000 by scientists aboard the submersible *Atlantis* at the *Lost City*, a vent system near the Mid-Atlantic Ridge. This type of vent is not volcanic, and reactions of sea water and the magma are not directly invoked. The sea water reacts with the rock of the upper mantle (such as olivine) that has been freshly upturned by tectonic-plate activity. The high-pressure water and the rock form hydroxides, just as Calcium oxide (quick lime) forms slaked lime (Calcium hydroxide). The newly formed hydroxide rock, closely allied to the mineral serpentine, expands, cracks, reabsorbs water, and gets into tectonic fissures to release superheated vapour. The reaction of sea water with mantle-rock releases vast amounts of heat, hydrogen, and hydrogenated products (pre-biotic molecules) like methane, ammonia and hydrogen sulphide held as sulphides of the alkaline earths that are pumped up into the ocean forming *alkaline vents*. Recalling some ideas that even predated the discovery of alkaline vents, M. J. Russell, W. Martin and others have proposed that the porous, cellular rock structures in alkaline vents could have acted as incubators of life on earth [60].

The view that biological processes do not extend below bedrock has now given away to the realization that life exists up to perhaps three kilometers under ground [108]. The weight of such subterranean life may far exceed the weight of bio-forms above ground. The prebiotic forms that produced such life forms, or their biochemical and evolutionary pathways are not yet known.

Thus a variety of processes are available for the formation of different prebiotics and possibly yielding several start-up incubators for early-Earth biology. We examine some of these in greater detail below.

10.4.1 *Prebiotics from extra-terrestrial sources*

The stellar dust clouds and nebular debris that existed prior to the formation of the earth are rich in many organic molecules [2]. Some 150 such molecules have been identified by 2010. Polycyclic aromatic compounds which are structural precursors of graphite have been detected [30] in meteors, asteroids and interstellar clouds.

There is little doubt that the early Earth would have received some pre-formed molecules, and possibly also some living cells [48, 101] from the inter-planetary medium. It is difficult to reliably estimate the organic material delivered to the early Earth by comets, since the impact record is inadequate. The delivery of organic compounds by inter-planetary dust particles (IPDPs) is better understood, being about 10^7 kg/year currently, and perhaps, $\sim 10^9$ kg/year (i.e., some 100–200 times more intense) at 4 Gya. Interestingly, prebiotic molecules (as used in life on Earth) observed in the interstellar medium have negligible abundances compared with aromatic moieties. The chemical composition of comets and asteroids would provide us with a specific cross section of the type of molecules

Table 10.1 Organic materials in the Murchison meteorite.

Compound group	Parts/million
macro-molecular carbon	1.5×10^4
amides	72
carboxylic acids	370
CO_2	106
amino acids	60
hydrocarbons	30–60
Ketones + aldehydes	25–30
sugars	25
urea	25
alcohols	11
amines	8
CH_4	0.14
N-bases	0.1
CO	0.06

that may have been delivered to earth from space. As seen from Table 10.1, most of the organic material found on the famous Murchison meteorite is in the form of macro-molecular organics (carbonaceous matter, polycyclic aromatics), while amino acids, sugars and other key prebiotic molecules are also found in smaller amounts [86]. A significant number of amino acids found on such meteorites are not found on earth.

The macromolecular carbon compounds which are found in greater abundance in comets and meteorites may perhaps be degraded into biochemically useful smaller compounds once they are delivered to earth. However, if the early earth had a hydrogen-rich atmosphere [93], the rate of atmospheric production of prebiotics turns out to be possibly two orders of magnitude greater than the delivery of organic compound from outer space or the synthesis of organic compounds in hydrothermal systems at 3.8 Gya.

The *panspermia* suggestions of Arrhenius in the 19th century, those of Hoyle [48] and Wickramasinghe [101] in more recent times, go more strongly for extra-terrestrial life sources, and not just pre-biotics. That is, these authors suggest that living microbes arrived on earth from space, and continue to do so from time to time, nourishing and modifying the evolutionary process, introduce diseases, epidemics etc. There seems to be evidence of microbial remnants in meteorites, comets, and other extra-terrestrial bodies that arrive on the earth. However, even such matter needs pre-biotic nutrients to take off. Furthermore, pushing the problem of the origin of life to an extra-terrestrial planet still leaves open the intriguing question of how life originated in the first place.

If life originated in space, those living cells need not have the diurnal and seasonal cycles of the earth imprinted in their genetic code. However, the living cells found on earth have a Circadian clock as well as diurnal and seasonal cycles consistent with their formation on earth (Ref. I-[164]). Of course, one may argue that these came by the adaptation of living cells to terrestrial conditions. However, if microbes with a genetic machinery consistent with the day and season of an extra-terrestrial world land on earth and survive, they should retain some genetic tell-tale signs in their DNA that would show evidence for that. Hence, investigations of genetic signatures of Circadian clocks could be of interest in testing claims of extra-terrestrial origins of life.

10.4.2 *Prebiotics from electric discharges and photolysis*

Miller [65] and Urey et al. [66], working in the 1950s at the University of Chicago applied electrical sparks to various gas mixtures to simulate the possible effect of lightening in the Archean atmosphere. They showed that small amounts of amino acids and other prebiotic molecules were formed in their experiments. The formation of glycine and other amino acids, simple sugars and lipids were demonstrated. This mechanism becomes very attractive for hydrogen-rich atmospheres. Recent scientific opinion has began to favour such a hydrogen-rich early-earth atmosphere.

The formation of prebiotic organic compounds in an atmosphere of CO_2 or CO by electric discharge is almost as efficient as that in an atmosphere of CH_4, when the hydrogen/Carbon ratio is unity or more. Tian et al. [93] give a conservative estimate of the production of 100 kg/year of amino acids by electric discharge under such conditions. Although the early Earth's atmosphere may have been dominated by CO_2 immediately after the heavy bombardment period, as continents formed on early Earth (Hadean aeon), the atmospheric CO_2 concentration would decline because of weathering, and the H_2/C ratio would become suitable for efficient formation of prebiotic organic compounds through electric discharge. Formation of prebiotic organic compounds by electric discharge at the rate of 100 kg/year in a hydrogen-rich early atmosphere would have created an ocean with a steady-state amino acid concentration of about $C_{amin} = 10^{-6}$ mols/liter, assuming an ocean volume of 1.4×10^{21} liters, and a mean amino-acid molecular weight of 100. This C_{amin} is orders of magnitude greater than the amino acid concentration estimated for a hydrogen-poor early-Earth atmosphere. However, this amino acid concentration is highly uncertain because neither the production rate nor the destruction rate is well known. In addition, organic foams may have formed at the ocean surface, leading to higher concentrations of organic compounds than in the bulk sea

Table 10.2 Synthesis of some prebiotic molecules, and structures, from basic molecules under simulated 'early-earth' conditions.

Compounds	Details	Year, Ref.
HCOOH, HCHO	Garrison, ... Calvin *et al.*, H_2O, CO_2, accelerated ions	1951, [39]
Formaldehyde, formic acid, HCN, and amino acids	Miller, Urey, using H_2O, CO_2, N_2	1953, [65], [66]
Adenine Adenosine triphosphate	Ponnamperuma *et al.*, irradiation of CH_4, NH_3, and H_2O, phosphate	1963, [77] 1965, [78]
Nucleic acids	Sanchez, Orgel *et al.*	1966, [84]
Urea, an important catalyst, and guanidine	Lohrmann, Orgel	1972, [57]
Amino acids, polypeptides	Flores, Ponnamperuma	1972, [35]
Phospholipids	Hargreaves *et al.*	1977, [46]
Fossils, proteinoid microspheres,	Young (1932), Reviews: Schopf, Strother *et al.*,	1932, [91, 85]
Plasma-membrane self assembly	Deamer and Fleischaker's book	1994, [22]
Peptide nucleic acids	Nelson *et al.*	2000, [70]
Cytocine, uracil etc.,	Cleaves, Nelson, Miller, using prebiotic eutectic mixtures	2006, [17]

water. Such foam bubbles may in fact have trapped prebiotic matter and acted as the precursors of living cells. Another issue is the seepage of prebiotic matter into the oceans from the deep mantle of the earth where geophysical heat and compression are available to drive the chemical reactions forward.

Organics also can be formed through photolysis of methane by UV-radiation (Lyman-α photons). These UV-photons generate radicals like CH, CH_2, CH_3 and related charged radicals which undergo subsequent polymerization and reactions with other molecules or radicals, forming prebiotics. The rate of such photo-processes, and also the formation of HCN in atmospheres of N_2, CO_2, and CH_4 are critically dependent on the CH_4/CO_2 ratio. When the hydrogen concentration in early-Earth atmosphere simulations change to a hydrogen-rich (30%) model, the rate of production of hydrocarbons by photolysis is found to increase from $< 10^7$ kg/year to 10^{10} kg/year, i.e., by three orders of magnitude (10^3). On this basis, it has been argued that the atmospheric production rate of organics through UV photolysis would have been *orders of magnitude greater than* the rate of either the synthesis of organic compounds in hydrothermal systems or the delivery of organic compounds to early Earth from space.

10.4.2.1 **Darwin's warm pond**

Although molecules suitable for forming living matter could be formed under prebiotic conditions, reverse reactions are always possible, especially under high temperatures. Also, the presence of steam in hot atmospheres or hydrothermal vents leads to hydrolysis, reversing reactions which are favourable to the formation of desirable complex molecules. Hence some low-temperature protective situations may help to concentrate useful complex molecules. Miller, Orgel *et al.*, showed that cytosine, formed under prebiotic conditions could be naturally formed in useful concentrations under freezing conditions [17].

A reducing-atmosphere scenario is very favourable to the formation of suitable prebiotic molecules, and this scenario is supported by the work of Tian et al. [93]. Reducing conditions are also found in hydrothermal vents. Hence, looking at Table 10.2, one would conclude that given a suitable protective 'container' or incubator that would enhance the reactants coming together and forming complex molecules, the emergence of living matter over geological times becomes a very reasonable hypothesis. If we consider a 'container' about the size of $\sim$100 microns (1μ = 1000 nm) in diameter, it is big enough to contain millions of molecules, and hence many of our ideas from statistical physics remain valid. If prebiotic molecules and their more complex products could be held together in 'incubator-containers' for sufficiently long periods, then the probability of the emergence of such an organism would increase by leaps and bounds. Clay templates (see Subsection 3.2.1) proposed by Cairnes-Smith, zeolite pores, sediment layers, colloidal suspensions, mineral substrates, porous-rock cells in hydrothermal vents, and even froth balls can be entertained as possible precursors or incubators for the first living cells. A small amine molecule would cross such a container or 'hatchery' in a fraction of a second, get reflected by the wall of the container, and return to its original vicinity often enough, unlike in an open medium. Hence the probability that two molecules would remain within the range of influence of each other greatly increases. This ensures that chemical bonding would occur if the activation energy for the reaction could be mitigated by the presence of a suitable catalyst or energy source (see Eq. 10.2, and Figs. 10.3–10.5).

Such a container which would be a prototype of living cells could be called a *protocell* environment. The nature of this 'container' is still a hot topic of investigation. Darwin had imagined a 'warm little pond', where prebiotic molecules stay together when chemical evolution happens. In Darwin's terms, life would be generated *spontaneously, as a natural process*, without any intervention of mystical forces or prior input of living matter. This is the process of *abiogenesis*. Although Darwin cautiously avoided discussing the abiogenetic origin of life in

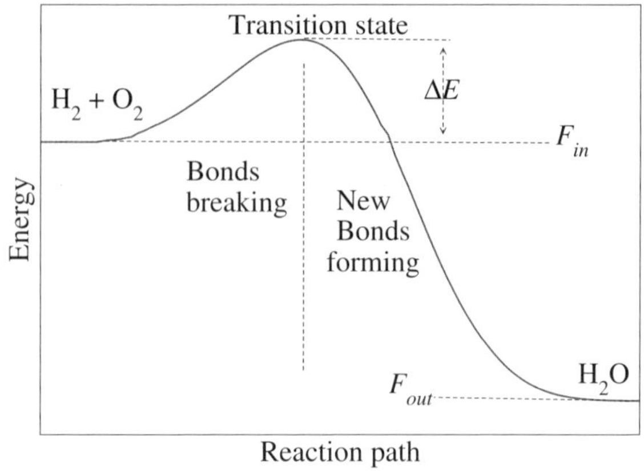

Fig. 10.3 A mixture of molecules, e.g., a mixture of H_2 and O_2 at room temperature, is quite stable, and has a free energy F_{in}. However, the final state where the gases have combined to form water, H_2O, has a much lower free energy F_{out}. Nevertheless the reaction does not take place because of the *energy barrier* ΔE which is also known as the 'activation energy' of the reaction. If a catalyst were present (e.g., platinum powder), the reaction occurs smoothly and easily as the energy barrier is removed. When the catalyst is present, the reaction path occurs on the surface of the catalyst, between adsorbed gas molecules. Alternatively, the molecules have to be physically excited using light, shock waves or heat. The figure is not to scale. The formation of one mole of water from H_2+0.5O_2 is accompanied by the release of $\sim$230 kJ of energy, and corresponds to $F_{in} - F_{out}$.

The Origin of Species, his views favouring abiogenesis are clear from his correspondence with colleagues like Richard Hooker [75].

Holding the prebiotics till complex molecules are formed is only a first step. They have to stay together in protocells for them to further evolve and form an organized entity, i.e., an organism. Such an organism must evolve suitable molecules, i.e., the genetic material, capable of replication and evolution. DNA is an efficient replicator molecule but does not catalyze these reactions which have to overcome an energy barrier (in the absence of a catalyst). Proteins make efficient enzymes which are the needed catalysts, but they cannot replicate. Fortunately, the ribozymes discovered by Thomas R. Cech and Sydney Altman in 1980, which are essentially RNA structures, could replicate themselves. Thus there is no chicken and egg problem, as ribozymes can be both! They could also catalyze the formation of specific proteins which could produce suitable catalysts. Even more significantly, a protein enzyme known as *reverse transcriptase* which could copy RNA into DNA exists. If the primitive proteins and the ribozymes in the prebiotic soup could synthesize reverse transcriptase, the formation of organisms with cell machinery close to those of modern organisms would occur. Instead of having full RNA molecules and transcriptase, catalytic activity of di-nucleotides and

Fig. 10.4 Charles Darwin stands on the same intellectual firmament as Galileo. However, unlike Galileo, Darwin took great pains to avoid controversy and confrontation. He delayed the publication of his ideas, already well formed by 1838. Darwin went public when Alfred Wallace proposed to publish similar ideas on evolution. The *Origin of the Species* (1859) was conceived as a book for the educated layman. The picture is an adaptation from a satirical rendering by the society magazine *Vanity Fair*, 1871.

other simple molecules have also been noted. However, these concatenated processes would not occur unless the reactants are held together and prevented from diffusing apart.

An old but still very attractive idea going back to extensions of Darwin's warm pond uses the properties of bubbles having lipid layers with water-lipid-water interfaces. Organic molecules in the prebiotic oceans would have been like primordial oil slicks being dispersed by wind and ocean currents. These may get driven to shorelines where foam as well as oil slicks gather. The foam is made up of millions of little bubbles, each enveloping a bit of prebiotic liquid inside an oil film. Soap bubbles are somewhat similar fatty-acid layers trapping a bit of moist air. Phospholipids are fat molecules that would occur in prebiotic systems. They carry phosphate groups and show a strong tendency to form 'bubbles', or vesicles trapping a bit of fluid, creating a container for the trapped fluid. Vesicles or bubbles made up of lipid membranes are known as 'liposomes', where the Greek *somes* means 'bodies'. Such lipid layers are most stable as bilayers of molecules (see Fig. 10.5). The bilayer consists of an outer layer with the phosphate groups pointing out of the bubble, while the inner layer is made up of molecules with the phosphate group pointing into the bubble. The phosphate end of the lipid molecule is attractive to water (the hydrophilic end). It is thus arranged to be immersed in the watery environment of the inside as well as the outside of the liposome, conferring great stability to the lipid bilayer and the liposome. For the moment we use the descriptive name 'lipid-bilayer bubble' (LBB) to denote these liposomes. The liquid trapped in the liposome is called 'liposol', 'cytosol', or cytoplasm. The word stem 'cyto' in Greek denotes a 'hollow' region.

One may note that the interactions between the lipid molecules, and the water molecules, have led to the emergence of a new complex adaptive system, i.e., in the present case, a cellular membrane, and a liposome. Objects like liposome are precursors to living cells. Living cells contain a membrane or wall (as well as a lipid bilayer, sometimes called a plasma membrane) which separates them from the ambient environment. The content of the cell, which contains various complex structures, is also called the *protoplasm*. Thus we see that the prebiotic liquid trapped in the liposome is the precursor to the protoplasm.

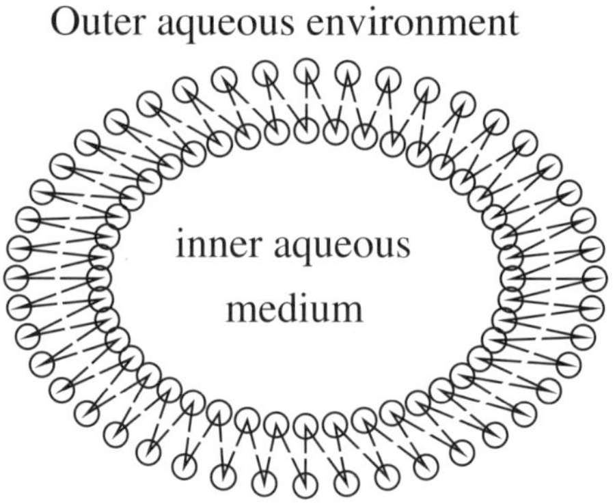

Fig. 10.5 Lipid molecules, represented with a circular head (containing hydrophilic functional groups) and two leg-like lines (one or two hydrophobic aliphatic chains) form a stable spheroidal vesicle in lowering the free energy of the system by excluding the hydrophobic tails while including the hydrophilic heads in aqueous regions. The intermolecular interaction between lipid molecules is the origin of surface tension in the bubble which attempts to minimize surface energy by moving to a spherical shape. Thus energetic reasons lead to the emergence of cell-like objects from the prebiotic chemical mixtures. See also Fig. 10.6.

Fatty acids, alcohols and amines contain a hydrophobic chain of carbon atoms (called the *tail*), and hydrophilic functional groups like COOH, OH and NH_2, or the phosphate group PO_4^{--} called the *head* of the molecule. In work largely associated with Irvine Langmuir, the 1932 Nobel laureate in chemistry, a whole branch of surface chemistry and surfactant technology based on the study of molecular layers of fatty acids formed at oil-water interfaces have emerged. The physical chemistry of Langmuir films can be taken over for quantitative theories of liposomes. The simplest case is an oil-water interface populated by fatty acid molecules. A typical aliphatic acid would have the formula R−COOH where *R* denotes a chain of carbon atoms carrying a suitable number of hydrogen atoms on each carbon. Thus if R is CH_3-, the aliphatic acid is simply acetic acid. If R = C_3H_7-, i.e., $CH_3-CH_2-CH_2-$ the corresponding acid, C_3H_7COOH is butaric acid. The fatty-acid molecules arrange themselves with their 'heads', i.e., the COOH end sticking into the water, while the 'tail', i.e., aliphatic chain R sticks into the oily region. If the concentration of acid molecules is small, there is a random distribution of molecules distributed in the 2-dimensional plane defined by the oil-water interface. However, as the number of molecules increases, they begin to interact with each other and an ordering process begins to emerge. Such ordering processes are similar to the condensation of gases into liquids, and the freezing of liquids to solids, when long-range order appears in a system where there was disorder (see section 9.3.2). This is an example of the emergence of a new collective property, not found in an individual element, but found only in the whole system.

The inclusion of some defects (e.g., mis-arrangements, vacant spots, inclusions of other molecules) into such ordered structures is favoured by the entropy term in the total free energy of the membrane (see Sec. 3.2.1). The possibility of such defects energetically favours the formation of membrane channels and the inclusion of other structures associated with cell membranes.

Millions of lipid-bilayer bubbles (LBBs) would be formed due to winds, waves, tides and other ocean processes along shore lines and inlets of the ancient earth. The LBBs are like millions of tiny 'test tubes' containing slightly different

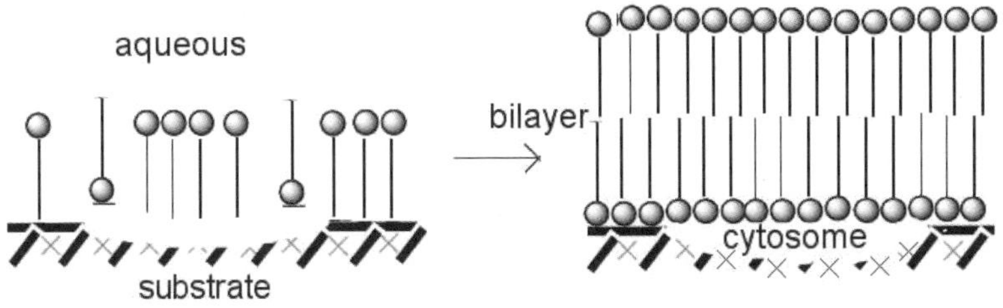

Fig. 10.6 Lipid molecules arrange on the substrate pointing their hydrophilic end (shaded circle) towards the aqueous region, and the hydrophobic carbon chains away from the aqueous region. Two lipid molecules are mis-oriented, but when enough molecules accumulate, a molecular rearrangement to a bilayer occurs, and produces a prototype of a cell membrane separating the chemically active fluid from the aqueous region outside. See also Fig. 10.5.

samples of the prebiotic mixture of chemicals. Most of these LBBs would evaporate or be broken down. However, many LBBs could collect and endure in favourable circumstances, e.g., under an oil slick, or in undisturbed nooks and crannies. Some of these LBBs may include chemicals which interact to produce other chemicals which may render the bilayers more stable. Such stabilized LBBs would last even longer, and their internal chemical evolution would produce substantially more complex products. The formation of molecules containing one heavy atom, two, three and many are listed in Table 10.2. The dusty early-earth atmosphere would have rained many minerals into the soup, and hence catalytic activity could exist in some cells, favouring the formation of important organic precursors like pyruvates, aldehydes, and even the energy-carrier molecule adinosine triphosphate (ATP).

When such chemically evolved LBBs burst, the cell fluid returns to the prebiotic aqueous medium, and its complex molecules serve to enrich it further. Thus, a new generation of subsequently formed LBBs may already begin with these more complex molecules, enhancing the likelihood of the evolution of complex systems like ribozymes. This may be thought of as a rudimentary process of evolution occurring at the LBB stage. When millions and millions of years as well as huge numbers of LBBs are considered, small probabilities for such processes turn into near certainties. It is believed that the first cells may have formed by such processes on earth some 3.5 billion years ago.

10.4.3 *Hydrothermal vents and geological sources*

The common living cells depend on respiration, a process similar to fermentation, for converting nutrients to energy by slowly 'burning' sugars to give CO_2, water and other waste matter, while using up oxygen. However, many 'anaerobic' life forms exist at the bottom of the ocean, deep underground, and in the gut of animals. Anaerobic chemistry may have been the only source of energy available for complex, life-like structures in the early hydrogen-rich oxygen-deficient period of the earth's history, when terrestrial surface conditions were extreme. Active under-sea volcanoes venting out magma material appeared as soon as liquid water accumulated on the Earth more than 4.2 billion years ago. A variety of chemical processes become possible in such energy-driven chemically rich systems. Vents of a different kind occur due to tectonic-plate activity. Thus the theories of prebiotic matter and emergence of life from geological sources and vents have become very topical [60].

Furthermore, the environments at vent sites might be similar to the more extreme conditions on other planets.

The movement of tectonic plates produces earthquakes, mountain ranges and volcanoes. When these occur in parts of the earth's crust covered by oceans, hot lava can come into contact with water and create under-water geysers which develop into chimney-shaped vents, known as *hydrothermal vents*. They were discovered only in 1977 although their existence should come as no surprise. These vents produce significant outputs of CO_2, H_2S, or dissolved metals under nominally acidic conditions, at temperatures in excess of 300°C, and are known as 'volcanic black-smoker vents'. The 'lost-city hydrothermal fields' are another class of vents, discovered in 2000 in the mid-Atlantic ocean. These are located on a highly faulted submarine mountain range of $\sim$20 km extension, and rising to a height of $\sim$4000 m above the sea floor. The lost-city vents differ from the black-smoker vents as the former release methane and hydrogen into the ocean at 40–90°C, under nominally alkaline conditions, but do not produce significant amounts of CO_2, H_2S, or metal sulphides [53]. On contact with sea water they deposit chimney structures up to even 6o m high, made up of calcite, aragonite and brucite ($Mg(OH)_2$). Investigations of these calcite vent sites, which are now known to be abundant, are changing our views, not only about how life can arise and thrive on our planet, but perhaps also on other planets.

When ocean-water comes in contact with hot lava, a hot leach of minerals and water superheated to as much as 400°C is formed (in black-smoke hydrothermals) and bursts through vents and cracks in the seafloor. The rising jet-like column of water deposits minerals and lava matter as it meets the colder layers of water near its freezing point. The deposited matter rapidly forms into a typical chimney-like structure. Some of the vents contain iron sulphides, hydrogen sulphide and associated chemical products containing sulphur.

Table 10.3 Some aqueous-media reactions relevant to prebiotic chemistry in hydrothermal vents.

Reaction	Details
Methanogenesis	$H_2 + CO_2 \to CH_4 + H_2O$
Sulfaneogenesis	$S + H_2 \to H_2S$
'Knall gas' reaction, Aerobic	$1/2\ O_2 + H_2 \to H_2O$
Sulphate reduction	$SO_4^{--} + H^+ + 4H_2 \to HS^- + 4H_2O$
Nitrogen fixation	$N_2 + 3H_2 \to 2NH_3$
Hydrogen Cyanide formation	$CH_4 + NH_3 \to HCN + H_2O$ $CO + NH_3 \to HCN + H_2O$
Iron-sulphide (pyrite) processes Refs. [97], [47]	$N_2 + 3FeS + 3H_2S \to 3FeS_2 + 2NH_3$ $CO_2 + 3FeS + 4H_2S \to CH_3SH + 3FeS_2 + 2H_2O$
serpentinization Ref. [69]	$(Mg, Ca)_2SiO_4 + H_2O \to (Mg, Ca)(OH)_2 + (Mg, Ca)_2SiO_3$ $(Fe, Ni)_2SiO_4 + H_2O \to (Fe, Ni)(OH)_2 + (Fe, Ni)_2SiO_3$

A hydrothermal vent is a natural thermal engine where the heat energy of the magma is converted into mechanical energy, while also setting up strong gradients in the chemical composition of dissolved and suspended matter. The prebiotic chemicals that can develop within such an ambiance can be quite different from those produced by atmospheric electric discharges or extra-terrestrial environments. These systems have energies sufficient to drive reactions even in the absence of catalysts. The microbial (and more advanced) organisms found today in these extreme environments [53] are indeed quite different to what is found in more familiar situations. They may point to very different initial forms of life, or adaptations from more familiar forms. The prebiotic chemicals formed in these vents involve reduction reactions caused by hot H_2 which is found in magma gases, and a product of the reduction of H_2O at high temperatures inside the earth's crust. The reaction processes that become possible in hydrothermal processes are summarized in Table 10.3 and describes the so-called 'iron-sulphide' world.

Wächtershäuser [96, 97] and collaborators have attempted to reconstruct a plausible metabolic mechanism for a 'pioneer organism' which could have arisen in *acidic* hydrothermal vents, within the so-called 'ion-sulphide world'. Wächtershäuser's scheme begins by establishing a number of relevant steps involving CO, S and (Fe,Ne) catalysts:

(i) Thio-acetic acid (CH_3COSH), where one oxygen of the COOH group is replaced by S, is made from thio-methanol (CH_3SH) and CO, with NiS or (Ni, Fe)S as a catalyst.

(ii) Lipids of the form R-CHA-COOH, where the A attached to the α-position could be H, NH_2, OH; more complex groups are formed with KCN and CO in the presence of Ni^{2+}, and $(Ca,Mg)(OH)_2$ buffer. Aliphatic acids, glycol aldehydes (a simple sugar), thioesters etc., are also formed.

(iii) α-Amino acid activation and peptide formation were found when α-amino acids were reacted with CO and H_2S (or CH_3SH) under alkaline conditions. These reactions are the basis of a proposed *peptide cycle* run by the energy supplied by converting CO to CO_2. Here H_2S, (Ni,Fe)-sulphides, COS, i.e., carbon mono-oxy sulphide, and a urea derivative act as intermediates. One weakness here is that black-smoker vents are believed to be acidic.

(iv) Polyphosphates, pyrophosphates, phosphoric oxide emitted by volcanic vents act as phosphorylating agents for the amino acids and peptides.

(v) A medium flow region with a temperature below the critical temperature of water, with a stationary solid structure, a mobile gas phase, and a mobile liquid water phase with dissolved volcanic exhalations are assumed to exist in hydrothermal vents. The formation of an organic superstructure, on serpentinized material containing (Fe, Ni) catalysts, (similar to the growth of stalactites or stalagmites), is envisaged in such medium flow regions.

(vi) Some ligands with several functional groups which attach on the metal catalyst atoms may boost the catalytic action of the metal centers, thus creating an autocatalytic ligand-feedback mechanism for the growth of a superstructure. Further, Wächtershäuser *et al.* propose a 'double feedback' mechanism, somewhat like in co-enzymes which catalyse the synthesis of the catalyst as well as the superstructure.

(vii) As noted in step ii, lipids are formed from CO and CN^- in these metal-catalysed reactions. These lipids are envisaged to form a lipid layer at the surface of the superstructure and the outer aqueous hydrothermal medium. This is known as *lipophilization*. Once there are enough lipid molecules in the lipid layer, a surface rearrangement to a bilayer is energetically favoured, and kinetically supported by the dynamic nature of the hydrothermal environment. Once the chemically active fluid become sustained by its own autocatalytic processes, it can detach from the substrate and the bilayer can completely envelope the protocell (cytosome), producing a single-cell object or vesicle. This has been called a *chemo-autotrophic pioneer organism*.

(viii) The laws of chemistry. taken in the context of the chemicals contained in the pioneer organism, should also yield the genetic machinery which enables the cell to replicate. Wächtershäuser, Nielson [71] and their collaborators have proposed that chemo-autotrophically produced polypeptides containing cyclic groups (made by condensation of urea and glycol, i.e., hydrantoin) , would have played the role of 'nucleic acids' in the earliest stages. These were then followed later on by the (phosphorylated-sugar based) nucleic acids found in current life forms.

Thus a black-smoker vent based 'recipe' for the origin of life is available, and further research is needed to tighten or vitiate the various steps in the scheme. The discovery of the alkaline, calcite hydrothermals in the year 2000 have prompted great interest in a different model. Instead of forming near-vertical black chimneys with a single violent emission mouth, calcite hydrothermals are complex labyrinthyine objects, with many porous compartments, channels, and orifices. The hot, compressed alkaline fluid seeps out of these multiply structured vents and into the water at the bottom of the ocean.

Molecular CH_4 is produced from the hydrothermal transformation of olivine-rich up-turned earth layers. The olivine hydrolysis process is known as serpentinization and probably occurs in the 450–550 K temperature range, at the ~ 0.03 gigapascal range of pressures. Martin and Russell [61] propose the following scheme:

(i) Geochemistry of the hydrothermal vent provides H, HS^-, CH_4, H_2CO, NH_3, CH_3SH and other simple pre-biotic molecules in a hot, reduced, alkaline solution near 373K, and pH $\sim$10.

(ii) Colloidal FeS deposited as structured FeS and NiS precipitates are formed on the labyrinthine cellular structure of the vents.

(iii) A redox gradient of about 1/2 volt across the walls of the FeS deposits is established which provides an energy source, and also provides catalytic action together with NiS.
(iv) Martin and Russell point out that this compartmentalized, reactant-retaining, energy-supplied system of monomeric building blocks could be plausibly shown to reach *the RNA world* of Gilbert [41], using the already available chemical and conceptual tools.

We have presented sceanarios for the origin of life in hydrothermals of two kinds, warm Darwinian ponds, froth balls, porous structures etc., with the early-earth hydrogen-rich atmosphere generating pre-biotics via photolysis and related Muller-Urey mechanism.

10.5 Conclusion

In this chapter we have begun from Astrophysics and prebiotic chemistry. and examined several scenarios for the spontaneous assembly of cell-like bodies, i.e., protocells. These protocells need the basic chemical compounds, as well as more complex constituents like ribozymes, peptide-nucleic acids, sugars, lipids and other components that form the machinery of life. The interactions among these constituent molecules, and with the aqueous environment, substrate matter etc., set the stage for the emergence of complex adaptive systems containing molecular orderings and processes not found at the lower length scales. The chemistry of the RNA-world complements with the extreme conditions that prevail in the hydrothermal vents and provide robust scenarios for the origin of living cells on Earth. Furthermore, the several possible scenarios discussed here can potentially lead to essentially the same tree of life that we associate with the prokaryotes.

Chapter 11

The Cell as the Basic Unit of Life

We discuss the structure and primary functions of the simplest cells which are the basic units of life. The exchange of matter and information between the cell and the environment, production of energy, replication, sex, signaling, 'awareness' of cells, 'artificial life', and 'cells as computers', etc., are considered.

11.1 The nature of cells

The development of science has been controlled by the consolidation of knowledge at a given length scale, together with exploration of new length scales using new tools (see Sec. 2.4.1). It was the discovery of the microscope in the 17th century that led to the study of *cells*. Robert Hooke in 1665 saw a regular, repeating honey-comb like structure in sliced pieces of cork viewed under a microscope, and coined the name 'cell'. Almost two centuries later, with a wealth of observations not available to the world before the advent of the microscope, botanists and zoologists like Henri Dutrochet, Matthias Schleiden, and Theodore Schwann suggested that *the cell*, a microscopically small bag of slime, is in fact the basic unit of all life. Cells may be smaller than a micron, i.e., 1000 nm in size and here we note that a human hair is about eight microns (8×10^{-6}m) in diameter. Some cells (e.g., the neuron cells of the giant sea slug) are big enough to be visible to the naked eye, and even some species of *amoeba* may be one mm in diameter.

What did life mean to the ancients, and how have those ideas evolved up to now? Many people even today believe that there is a fundamental difference between living matter and inanimate matter. Living matter, originally called *organic matter* was claimed to possess 'spirit', a vital force or *vis vitalis*. Living organisms were claimed to act with a purpose, i.e., *teleologically*, while inanimate matter was generally not considered to move with any purpose. Theologians held

Table 11.1 Approximate atomic and molecular compositions typical of simple living cells.

Element	atoms per 1000 Oxygen atoms	**Molecular species**	molecules per DNA molecule	%Wet weight†
O, C, H	1000, 4000, 6000	DNA	1	1
N, P, K, S, K	200, 20, 10, 10	RNA	~4,000	6
Mg, Na, Ca, Cl	10,10, 8, 4, 2	Protein	~400,000	15
Co, Fe, Ni	0.5, 0.3, 0.2	Lipid	~4,000,000	2
Mn, Zn, Cu	~0.08 each	Polysacch.	~10,000	3

Adapted from: [56, 98, 99].
†Water is 70–75% of the cell weight., and ~99% of the *number* of molecules.

that humans are very special compared to other living beings. Even today, opposition to research using human stem cells, and opposition to birth control reflects this line of thinking. Theologians claimed that every embryo was bequeathed a soul by God at conception. This troubled St. Augustine enormously as conception out of wedlock seemed to have the complicity of the Divine spirit. Indian thinkers too believed that higher animals possessed an *Ātma* or soul. Even in the more subtle doctrines of Buddhist thinkers, which denied the existence of a soul, i.e., (*anatta*), some form of 'consciousness' or *vignana* different from the body was introduced (part I-[116]). This 'consciousness' departs from the dead body and enters into an embryo, in a manner consistent with a legacy of past deeds or volitions, i.e., the *karma* of the dying creature. Thus the Indian thinkers allowed for the possibility of humans being reborn into animals and *vise versa*. The empiricist philosophers among the ancient Greeks, or among the ancient Indians were a negligible minor voice compared to the enormous power of religion and mysticism. The modern view, that all life as well as 'consciousness' are emergent physical properties which are completely described by the laws of physics and chemistry, came into acceptance only very gradually (see the related discussion in Sec. 1.4.1).

It was the discovery of the microscope which truly opened the eyes of a world tired of medieval metaphysics. Anton van Leeuwenhoek (1632–1723) was the first to see bacteria, yeast cells, blood corpuscles, and the teaming mass of life in a bit of saliva or in a drop of pond water. He was able to make microscopes with magnifications reaching a factor of 270. Anyone witnessing the writhing mass of life in a drop of water is amazed by the intricately controlled movements of living cells, superimposed on the effect of Brownian motion. It was soon confirmed that the movement originated internally in the cells, and not purely due to external forces, surface tension effects etc. Living cells take up nutrients (which may be other, simpler cells) and expel waste matter. Cells which take up toxic matter recognize it and expel the harmful material. Living cells procreate by cell division, passing on their genetic information to the newly replicated cells. Living cells are self-regulating, and may adjust to challenging environmental conditions by entering metastable defensive states. When many cells work together, as in an organism, they are capable of receiving and sending signals, which may be electrical or chemical, among each other. Some biologists view this sensitivity to the environment as a very rudimentary form of 'consciousness'. Most others reserve the name 'consciousness' for a class of emergent properties found in groups of specially adapted cells known as neurons.

We argued in Chapter 10 that living cells evolved from prebiotic matter.This is re-confirmed by the chemistry of living cells. Leaving aside matters of detail, or possibly some anaerobic deep-sea vent-life forms, the chemical compositions of living cells, be it archaea, bacteria, or plant or animal cells (see Fig. 2.7), all turn out to be chemically similar in terms of the pre-biotic building units that have gone into them. Table 11.1 gives the atomic composition, and also the composition in terms of important molecular species found in cells. The row marked O, C, H, shows that in a typical DNA molecule there is enough oxygen to make 1000 water molecules leaving 4000 C and 4000 H atoms. The second footnote suggests that since 99% are water molecules (which was 1000), the remaining 1%, i.e., 10 molecular fragments, are made with 4000 C atoms, and 4000 H toms. That is, much of the DNA is made up of molecules where there is just one C per H atom. In fact the base pairs (labeled A,T,C,G,U in Fig. 10.2) found in DNA and RNA satisfy this requirement.

Simple cells have just one chromosome, containing one DNA molecule. The chemical composition of the building blocks of the cells is in accord with the prebiotic chemistry of the early-earth atmosphere. The presence of Co, Fe and Ni is suggestive of hydrothermal-vent origins for terrestrial life. However, the early earth atmosphere would have had enough volcanic and formative unsettled mineral debris. Hence atmospheric prebiotic models can also accommodate the presence of such metals. The chemistry of present-day cells need not be a good guide to the early cells, although fossil records support it.

11.1.1 *The main 'purpose' of the cell*

We have already remarked that living matter offers new, ever more intricate channels for the degradation of the torrent of solar energy that falls incessantly on the earth, as well as the geophysical energy stored inside the earth. Ocean currents, winds, rain, forest fires – in fact almost all terrestrial events are caused by the energy of the sun. Evaporation, formation of clouds, transportation by winds, condensation as rain, transportation by rivers, and re-evaporation provide a *cyclic* process for energy degradation. By energy degradation we mean the conversion of 'free energy' into random thermal energy which is incapable of doing useful work (see Chapter 9).

We saw in Chapter 10 that the energy degradation processes present in the atmosphere, in the oceans and in hydrothermal vents produced pre-biotic molecules, sugars etc. They too contain useful energy, even if it be minuscule in comparison to the fury of stellar and planetary processes. Thus, living matter is simply a set of heat engines which have evolved to do work within a certain low-energy niche. Hence it is not surprising that living matter is found confined to within a few kilometers of the earth's surface and in oceans. That is, essentially where prebiotic matter is likely to be found, somewhat like slime forming on a wet sandstone. Given that cells are the basic unit of living matter, it already suffices to study the simplest of cells, i.e., prokaryotes (bacteria+archaea), depicted in Fig. 11.1, to appreciate the basic 'business' of cells. The name 'prokaryote' means 'before the nucleus', in that these cells, unlike the more advanced eukaryotes, do not have a feature (an organelle) known as the nucleus. These cells have evolved into complex processors equipped with sensors, motor ability and signaling capabilities to carry out their task of energy degradation. They have evolved into eukaryote cells which are more elaborate than the prokaryotes in having specialized organelles inside the cell. They have also evolved by many cells grouping together cooperatively, to form multicellular organisms, or, in some cases colonies which are better adapted to their environment, and are today capable of even transforming the environment in a significant way.

All bacteria have essentially the same simple building plan, consisting mainly of three parts. These are (i) the cell membrane which is usually rigid, (ii) protoplasm (or cytoplasm), and (iii) a chromosome and some plasmids. Our knowledge of the components of the cell resulted from successive advances in new tools and techniques like staining procedures, microscopy, centrifuge technology, X-ray and electron diffraction techniques etc. These depend on advances in the physical sciences making an impact in engineering and instrumentation. Ultimately they produce the conditions needed for paradigm shifts in science (see Sec. 2.4). One such example is the technology of centrifugation that made nuclear fissile materials as well as cell components available to science [29].

The cell membrane is a phospholipid layer (see Fig. 10.5), as already discussed in the context of prebiotic liposomes. The lipid bilayer may sometimes be further enclosed in an additional *cell wall* which may be made up of polysaccharide molecules.

The liquid enclosed inside the cell membrane is known as the protoplasm, or the cytoplasm. It is a gel-like fluid containing mostly water. But the small amount

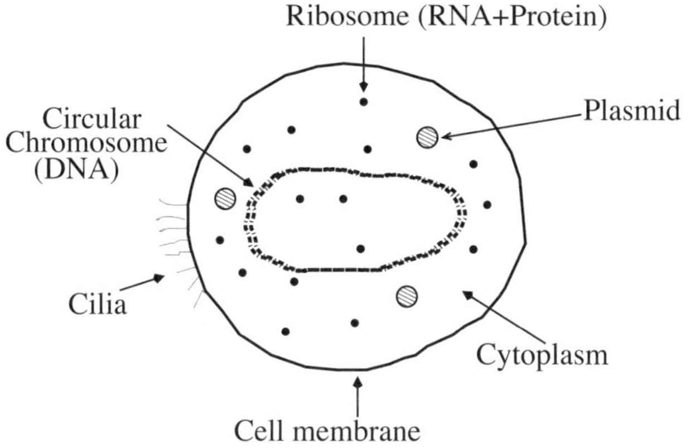

Fig. 11.1 Prokaryotes, commonly called bacteria, all have a simple structure, consisting of a cell membrane enclosing a fluid called the cytoplasm and a single chromosome in the form of a loop made up of molecules. Some bacteria have one or more smaller chromosomes called plasmids. Millions of small granular shaped macro-molecules, known as ribosomes, used for protein synthesis by the cell are found in the cytoplasm, and a few are shown. Ribosomes are made up of RNA and protein. Some bacteria have hair-like objects known as cilia growing on their cell membranes (only a few are shown). Unlike eukaryotes, these simple cells do not have a nucleus and other cellular organelles. The cells may be nearly spherical (coccus), rod shaped (bacillus), or spiral like (spirillum), and may be single, paired or loosely linked like pods on a bean.

of enzymes (protein catalysts), salts, nucleic acids, lipids etc., convert into one of the most amazing concoctions that evolution has produced. It is only the recent advances in biotechnology that have given scientists the capacity to produce synthetic approximations to it. The cytoplasm contains millions of spheroidal bodies called ribosomes. High resolution imaging shows that prokaryote ribosomes are far from spherical, and are made up of some 50 or 60 proteins, and three RNA molecules per ribosome. The RNA molecules catalyze the formation of new proteins [3, 14], while the proteins in the ribosome itself provide a structural scaffolding for the RNA to function.

The cell's chromosome is also contained in the cytoplasm. It holds all the prokaryote genes, collectively known as the cell's *genome*. The prokaryote genome is found to contain, typically, some 2000–4000 genes, with each gene providing the code for a protein. The plasmids, found in many bacteria, are also containers of a few genes. The cell has one copy of its chromosome, and many plasmids. Plasmids play a role as containers of antibodies for fighting toxins (such as antibiotics) and pathogens which may attack the bacterium. In eukaryotic cells there are many more chromosomes, containing very long strands of DNA, specially packed into the small space of the cell nucleus.

All the components of the cell work together in a highly organized fashion, to carry out the functions of the cell that constitute its life. The operation of the cell can be stated in terms of several primary functions. We discuss them below:

11.1.2 *Exchange between the cell and the environment*

A lipid bilayer containing a low concentration of salt, when placed in salty water, will allow the salt ions to flow into it because of the concentration gradient of the salt. This is known as *osmosis*. The cell is able to gather nutrients from its environment in a more sophisticated manner than merely relying passively on such elementary physical processes. It has machinery to create the elementary physical processes to suit its purpose. Its cell membrane has a level of sensitivity such that it can detect another object which may be its *prey* or food, living or nonliving. Then the cell membrane (e.g., that of an amoeba) modifies its shape and engulfs the prey, opens itself sufficiently to absorb it into the cytoplasm. The chemicals in the cell break down the engulfed material, and 'the amoeba *digests* the nutrients'. This function of the cell will be discussed under energy production. The output from these chemical processes, as well as indigestible substances, are the waste material which the cell has to excrete. This too is done by a contortion of the cytoplasm to push out the material from a temporary opening created to expel the waste matter. Further more, the cell membrane has specialized openings or pores controlled by proteins which can change shape and control the access to the cell, as discussed below.

The cell membrane plays a very crucial role in controlling the exchange of material between the interior of the cell and the external environment. It safely holds the enzymes, the ribosomes, the DNA and other cytoplasmic materials which the cell needs for its own function, and allows in or lets out only selected substances. The cell membrane, made up of the phospholipid bilayer has numerous portals incorporated into it by embedding various proteins and other special molecules. This is not surprising since the lipid layer may be thought of as a layer of oil which can dissolve other molecules in it or hold them on its surface. In fact, a well-known model of the cell membrane is the 'fluid-mosaic model', introduced by Singer and Nicolson in 1972 [87]. In this model, the proteins associated with the membrane are classified as 'peripheral proteins', and intrinsic proteins. The peripheral proteins are mostly resident on the surface of the lipid bilayer. The intrinsic proteins are globular proteins whose hydrophobic part is buried in the lipid bilayer, while the hydrophilic globular parts protrude out of the lipid bilayer. Some of these proteins traverse the entire width of the bilayer (see Figs. 12.2–12.3) and provide channels for transporting ions across the lipid membrane, or embedding sensitive molecules like retinal basic to light sensitivity. Similarly, special proteins embedded in the cell membrane are especially designed to detect and bring in the all-important glucose molecules into the cell. Glucose molecules can slowly diffuse through the lipid membrane, but these *transporter proteins* allow the cell to take up glucose extremely rapidly. The positions of these transporter proteins in the membrane define 'protein channels' for the exchange of material between the cell and the environment. Hence, once a few such molecules got embedded, the evolutionary process selected such cell membranes as the ones most effective for the cell. There are other transfer proteins of the same type (glycoproteins) for other sugars like maltose or lactose. These proteins have further evolved sugar-sensitive feelers that would detect the presence of sugars in the environment, enabling the cell to even move to regions with higher sugar concentrations! Thus the outer surface of the prokaryote cell membrane containing embedded glycoproteins is covered with an enhancing and modulating thicket of sugar molecules known as the *glycocalyx*. Their importance in the cell's sensitivity to the environment will be discussed in a subsequent section.

The proteins in the membrane decide what molecules or prey-organisms are introduced into the cytoplasm. Many living cells have *vacuoles* (empty regions), or create temporary vacuoles by contorting the cytoplasm and the cell membrane. This creates extra surface to facilitate exchange of material between the interior and the exterior. In more advanced cells (e.g., the eukaryotes), the endoplasmic reticulum and the Golgi apparatus participate in this function. Similarly, hair-like extensions of the cell surface, known as *microvilli*, are found within the the glycocalyx, especially in specialized cells. A microvillus is about 1μ (10^{-6}m) long and about 0.08 μ in diameter ($\sim$ 1/16th of a hair). These structures increase the surface of cells by approximately two orders of magnitude (e.g., 600-fold in humans), facilitating absorption of nutrients and secretion. They are used in detecting motion (e.g., sensory cells of the ear), in taste buds and in the olfactory cells of the nose. Thus we see that the *surface to volume ratio* of a cell is a critical physical parameter in determining further evolution of cells and in cell specialization (see Chapter 12).

The transfer of ions like Ca^{++}, Na^{+}, K^{+} in and out of the cell is also done through suitable ion channels in the membranes of eukaryotic cells. These channels are formed by inserting a protein which traverses the cell membrane (see Fig. 12.3) and controls the flow of material through its included volume by changing the protein shape, folded state, and the way they bind to signaling molecules (see Table 12.1). They play an important role in regulating membrane potentials, osmosis, vision and signal transduction in memory. Rudimentary ion channels have been found in prokaryotic cells as well. They have a central role in neurons found in higher organisms.

11.1.3 *Generating energy from nutrients*

The prebiotic oceans had sugar molecules and other substances like amino acids and lipids which could be taken up and further processed within the confines of the lipid bilayers of the early living cells. Glucose is even today the essential nutrient of some specialized cells like neurons. Releasing nutrient energy involves using oxygen to burn sugar, or using anaerobic processes for degradation. The 'burning down' has to be performed at bio-friendly temperatures in a gentle manner.

The energy-rich molecules of the nutrients have to be broken down into water, CO_2 and similar final products of digestion. These are precisely the products of fermentation. Hence we see that 'living' cells are simply microscopic breweries or compost pits which also digest the input material and produce heat, water, CO_2 and waste matter. However, living cells are more complex than compost pits because they can continue to function in an autonomous manner, and also replicate themselves, as discussed below. The food material taken in by the cell is broken down by the enzymes contained in the cytoplasm, and finally synthesizes adenosine triphosphate (ATP) molecules (see Fig. 10.2, panel 10). These are small molecules packed with energy in each phosphate bond. Phosphate bonds are *the universal energy-currency note* of living matter. It is basically like a quantum of energy for biological processes. However, in biology natural selection always looks for alternative pathways that break such rules — there are no strict rules other than optimization! Continuous energy gradients can be set up in biological systems by setting up an electro-chemical potential (as in a battery) across a membrane. Then, pumping charged ions (e.g., Ca^{++}) or protons (H^+, found in acids) can give a continuous energy gradient — i.e., small change. The motion of a solvent (usually water) across a membrane is known as *osmosis*. That is equivalent to moving ions across the membrane (Galilean invariance!). Nevertheless, such ion (proton) transport has been given the special name *chemiosmosis*. If a proton gradient (i.e., basically, a gradient in pH) is set up across a membrane, the flow of protons can drive the synthesis of ATP. Small energy coins are converted to the basic 'energy note' by this process. It is of great importance in many cell processes including photosynthesis.

Guanosine triphosphate (GTP) is a molecule where the adenosine base in ATP is replaced by the purine base known as Gaunosine. This is also used in many cell processes and it is easily converted to ATP; or, the ATP molecule can be used to make GTP when needed. ATP is 'spent' by the cell to supply energy for various activities of the cell. The ATP itself gets changed to adenosine diphosphate (ADP), or even adenosine phosphate. The activities include moving its cytoplasm, ejecting waste matter, moving its cilia, or for ribosomes to make proteins from suitable breakdown products of the nutrients. More complex cells (e.g., in eukaryotes) contain mitochondria and other organelles which are specialized for energy conversion utilizing more advanced forms of 'fermentation'.

In addition to the direct use of glucose-like sugars by prokaryotes, some of them can use an anaerobic pathway (anaerobic fermentation) called glycolysis, which can occur without oxygen. The glucose may be converted to glycogen, a polymer similar to starch for later use; or for the production of ATP. In anaerobic fermentation, one molecule of glucose is broken down to two molecules of pyruvic acid, CH_3-CO-COOH, finally producing two molecules of ATP and other products. The cell's aerobic processes produce many more ATP molecules and this is the essential mechanism of 'cell respiration' analyzed in detail in the *Krebs cycle* (or citric acid cycle), and the *respiratory chain* (also called the electron-transport chain). These metabolic pathways extract energy not only from glucose, but also from polysaccharides, lipids and such nutrients. Here too, pyruvic acid is produced as an intermediate, showing how nature uses essentially the same tools once it evolves them.

Prokaryote respiration occurs in the cytoplasm and on the plasma membrane. Both glycolysis and the Krebs cycle occur in the cytoplasm, while the electron transport-chain and the steps between glycolysis and the Krebs cycle occur on the plasma membrane.

The eukaryote cells (see Fig. 11.2) have special organelles (the mitochondria) which also uses pyruvic acid intermediates, the Krebs cycle and the electron chain for making ATP. The Russian biologist Konstantine Mareschovsky suggested in 1910 that mitochondria were primitive prokaryotes subsequently enslaved by eukaryotic cells and retained for their own energy production. This view was resurrected in the 1970s and advocated by Lynn Margulis and others [59]. Their argument was based on the structural similarity of mitochondria to bacteria. Thus mitochondria have a single loop-like chromosome with a genome structure similar to those of bacteria, synthesize ATP as in prokaryotes, and undergo binary fission, as do bacteria. In this view, the mitochondrian is a bacterium living symbiotically with the eukaryote cell. That is, the evolution of eukaryotes was associated with the early cells acquiring symbiotic bacteria which took up specific tasks. Thus mitochondria and chloroplasts (in plants) are believed to have been acquired, some 2.5 billion years ago, by prokaryotes which retained them in a symbiotic relationship. The nature and size of the mitochondrian has also evolved in the higher animals. The human mitochondrian was shown to have 16,569 base pairs using the

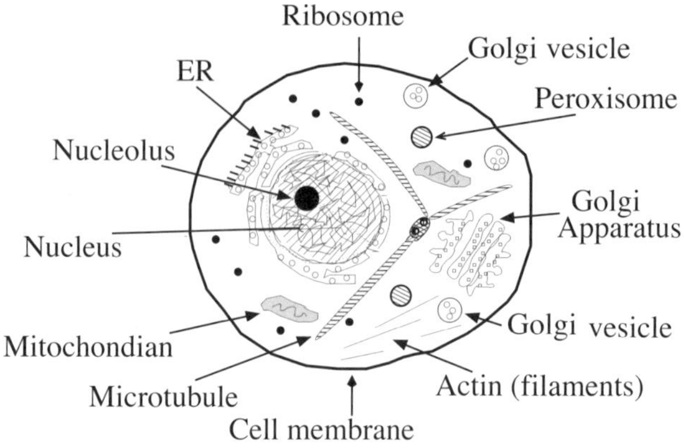

Fig. 11.2 The Eukaryote cell, schematically shown above, has been retained by nature as the basic unit of all living organisms (except the prokaryotic ancestors of the eukaryotes). Not all organelles are shown. The DNA of an eukaryotic cell is now mostly packed into the nucleus. It contains a nucleolus for synthesizing ribosome components. Ribosomes are found in the cytoplasm as well as on the endoplasmic reticulum (ER). The Golgi apparatus and ER contribute to the synthesis and refining of proteins destined mainly for the cell membrane. These are transported by the Golgi vesicles. The many mitochondria convert nutrients to energy-bearing molecules like ATP. The peroxisomes and lysosomes (not shown) recycle cellular waste. The microtubules, attached to the centrosome are crucial to guiding cell division where, unlike in prokaryotes, the DNA which is folded away in the nucleus has to be opened out, replicated and repacked for the two daughter cells. Microtubules also act like 'rail roads' for transporting proteins across the cell. They have also been invoked for exotic explanations of consciousness (see Chapter 13).

sequencing method of Fred Sanger (Sanger protocol), developed at the University of Cambridge, England during the 1970s. The symbiosis of bacteria in the organs of higher animals is well known. Thus humans have millions of bacteria living in the gut, skin, and other parts of the body. The symbiotic bacteria are absolutely necessary for the survival of these organisms.

11.1.4 *Harvesting the sun — photosynthesis*

Within very broad terms, the energy extraction from nutrients cells, be it using mitochondria or otherwise, are elaborations on fermentation that already occurs in the decomposition of non-living matter. The nutrients are molecules formed by plants and algae using the energy of the sun. Photosynthesis is the mechanism used by plants and some prokaryotes to harvest the sun's energy. This energy gets reused by the rest of the animal kingdom. Photosynthesis involves the production of ATP in the first step where one type of light-activated Chlorophyll reacts with water to pull electrons out of it (this step is confusingly called photosynthesis-

II, a historical relic). This process, tantamount to oxidizing water is a chemical wonder because water is in fact a highly oxidized, stable system. Then sugars are made in the next step (photosynthesis-I), where CO_2 is converted into sugar by a second light-activated from of chlorophyll (different from the form involved in photosynthesis-II). This activated chlorophyll uses a molecule named NADPH to reduce the very stable CO_2 molecule and combine it with the the two nascent H atoms from the first stage to from sugar, releasing the oxygen that was originally in the water.

$$6CO_2+6H_2O \rightarrow C_6H_{12}O_6 +3\ O_2$$

This step is also a chemical wonder because a very stable molecule, CO_2 is pushed into reaction. Photosynthesis provides the oxygen that was used up to burn nutrients by cells, back into the atmosphere. This two stage harvesting of light by chlorophyll is known as the Z-mechanism. The cyanobacteria may have been the first form of life to device photosynthesis, adapted from anaerobic forms that uses H_2S instead of water. The sugars produced by photosynthesis are used in the normal fermentation-like cell processes. Photosynthesis is a slow process of getting energy, as one has to wait in the sun to harvest it. Plants do that, while humans and other animals eat plants (and other animals) to get their energy.

Thus plant cells have 'chloroplasts' which absorb sun light and convert them to chemical energy, by converting CO_2 and water into sugar. This photosynthetic process is essentially the reverse process of fermentation, although gathering the solar energy and carrying this out at bio-friendly temperatures is testimony to the amazing effectiveness of the evolutionary process which tinkers away for millennia, selecting an improvement over a less effective product or process.

11.1.5 *Reproduction and replication — cell division*

A sufficiently large soap bubble bursts or it may split into two or more smaller soap bubbles. This process of subdivision is driven partly by energy considerations, and partly by kinetic considerations. The stability of the bubble depends on the ratio of its surface area to its volume. If the bubble radius is r, this ratio, area/volume = $3/r$. The same property holds for a liquid drop, which holds its shape by surface tension. As the liquid drop gets bigger, the surface tension is no longer able to hold the molecules of the drop together, and it becomes energetically more favourable to form two smaller droplets.

All the energy considerations that apply to a liquid drop apply to the simplest of cells as well. In addition, since the entry of nutrients and excretion of waste depend on traversing the cell surface, a large cell having less surface per unit of volume than a small cell, becomes less efficient in exchanging nutrients and waste with the environment. The latter is an example of a kinetic constraint. Hence, once a cell becomes bigger than a certain size, free-energy minimization as well as kinetic constraints will occur to split the cell into two smaller units, essentially in the same way as a water droplet dividing spontaneously into two droplets. This was probably the earliest physical mechanism driving the process of cell reproduction which has now become an immensely elaborate, necessary component of the physiological and cultural aspects of life.

The living cell could increase its surface area while increasing its volume and beat the $3/r$ rule by deviating from the spherical geometry. Thus the prokaryote cell evolved into eukaryote cells, and introduced *vacuoles*, *endoplasmic reticula* (ER) and other *organelles*, increasing the available surface area for a given volume. Another device, produced by evolutionary adaptation was the conversion of loosely associated cells into truly multi-cellular organisms. All the cells are now like organelles inside a bigger membrane called the skin. Now cell division can proceed using well-established mechanisms, and those parts of the multicellular organism undergoing cell division are reproduced or copied *de nouveau*. The organism can now live much longer than the natural life cycle of a single cell which may be as short as 20 minutes, or as long as many days. Thus, the evolution of the multi-cellular organism has paved the way for using cell division (i.e., reproduction at the cellular level) to beat the limiting $3/r$ surface-to-volume ratio, and to prolong the lifetime of the organism, enabling it to evolve and specialize its adaptations to fit its environment even more.

11.1.5.1 *Life cycle of a cell*

Many bacteria live for about twenty minutes and undergo cell division. Thus the life of the cell can be thought of as involving a relatively calm period, followed by the hectic phase where the bacterium splits into two. The 'calm phase' of the life of the cell, known as the *interphase*, is itself treated as three subphases, viz., G_1, S, and G_2, where G stands for 'gap', and S signifies a period of synthesis of DNA. The interphase comes to an end in G_2, and then cell division takes place, at which point the two daughter cells are in G_1, i.e., at the start of the interphase of the new cell cycle.

Prokaryote cells grow continuously during the interphase and the growth proceeds in parallel with the duplication of the chromosome. The duplication of the chromosome requires the separation of the two DNA strands. A number of enzymes are involved in this process. *Helicase* is an enzyme which separates the helical double strand, initiating the process and using up ATP for the energy needed to separate the two strands. Then one strand of the chromosome is attached to the cell membrane at one end of the cell, while the other end is attached to the opposite end of the cell. As the growth of the cell and the DNA replication proceeds, the cell membrane moves inwards (invagination) and eventually divides into two cells. We should compare this quite precise binary division of the cell with the bursting of water droplets and prebiotic liposomes where the two daughter bubbles may not have been identical with each other. While this may mean that one of the daughter bubbles may have been more well adapted to the environment and thus given an evolutionary advantage, the situation is different for more advanced cells. Such cells have enzymes, ribosomes and other valuable components which are too important to lose. Hence sharing them as equally as possible, and creating two well adapted cells is a better strategy.

11.1.5.2 *Sexual reproduction*

The process of cell division in eukaryote cells is much more complex as there are many organelles including the nucleus which have to be duplicated. Furthermore, the chromosomes are well packed with histones and so they have to be opened up, rearranged, replicated and once more repacked for the daughter cells. Eukaryotic cell division, known as *mitosis*, hence involves four stages associated primarily with the behaviour of the nucleus during mitosis. Here, the enzyme helicase once again plays a leading role, together with a whole array of other enzymes. In higher organisms which have specialized cells dedicated to sexual reproduction (e.g., in ovaries and testes), a special process of cell division known as *meiosis* takes place. The objective of this process is to create 'breeder cells' (sperm or ova) containing just half the number of chromosomes of the parent. Such cells are called *haploid cells*. When a sperm and an ovum combine and form a fertilized egg, the resulting cell has the full complement of chromosomes, with half the chromosomes coming from each parent.

The evolutionary advantage of this process is the capacity to sample a bigger gene pool than if all the chromosomes came from one parent. This process is closely related to the operation of an algorithm which samples a landscape of all possibilities by a process of continual bisection, as in a generalization of the Newton-Raphson bisection algorithm for determining the zero of a function. Thus two partners (and not three) is an efficient sampling process in this case. Equivalently, it may be thought of as a simulation process depending on information from two sampling moves prior to a definitive trial step in the evolutionary game.

Although we have presented the above 'algorithmic' explanation for the adoption of a 'bisexual' approach to propagation, asexual propagation where essentially equivalent copies of the mother cell are produced by mitosis is the norm in the world of single cells. As long as they remain unicellular organisms, the asexual mitosis mechanism provides an efficient process of species propagation and hence no evolutionary change is needed except to deal with random mutations. On the other hand, a multicellular organism cannot exactly split all of its cells into doubles and propagate by mitosis. A multicellular organism is capable of evolving specialized groups of cells to do different tasks. Thus such organisms have evolved sexual organs and incorporated a number of advantages that justify the need for evolving the complex machinery of sexuality. The binary sampling process can indeed come up with an individual inheriting two bad genes, e.g., to inherit the sickle-cell disease. However, natural selection eliminates them in the long run. Some of the advantages conferred on the organism and its community via sexuality are as follows. (i) Access to genetic variety via the father *and* the mother. (ii) Such genetic variety is advantages for competition for food and other resources in what Darwin called a 'tangled bank' full of competing species. (iii) Even a well adapted species may face competition from a new species and it needs a mechanism to continually 'run forward' to maintain its lead. This is known as the 'Red Queen' hypothesis, where the Red Queen in Lewis Caroll's *Through the Looking glass* makes Alice run just to stay in the same place. (iv) In asexual reproduction, any changes in the DNA are transferred to the off-spring, even if the changes are not advantageous to it. In bisexual reproduction there arises the possibility of correcting for errors. This 'DNA repair' hypothesis was advanced by Bernstein *et al.* [8] who point out that deleterious mutations can be repaired in germ cells while the cells of the rest of the body (somatic cells) grow old due to copying errors.

Biologists have probably noted the implicit Newton-Raphson type algorithmic character inherent in sexual reproduction. The question of the advantage of sex to the individual versus advantage to the community has to be resolved in deciding on the nature of the fitness function that is being optimized. Hence we see that 'mathematical modeling' of sexuality contain much arbitrariness at present.

On the other hand, sex between a couple is the natural outcome of synaptic communication. The sexual organs of the higher animals seem to be an adaptation of the neuron-synapse pair, where chemical signaling (seminal fluids) are used to communicated between the two mating organisms presenting the pre-synaptic and post-synaptic neuron banks. Thus, sexuality in higher animals can be connected to consciousness and the pleasure principle for motivating species propagation.

11.2 The awareness of the primitive living cell

The sensitivity of the cell to its environment, both outside and *inside* the cell, implies that it has its defence and repair mechanisms. However, the sensitivity of the cell to its inside is nothing but its own *awareness*. We use this term without attaching any 'mysticism' to it, as we observe it and have a basic understanding of the phenomenon. In some sense, our brains are trained to *assume* that anything which moves is sentient, and has a purpose. It has taken us a long journey of re-education to revise these instinctive reactions.

Barbara McClintock studied the response of the Maize genome to various internal stresses and trauma faced by the cell (e.g., from having to deal with chromosomes with broken ends), and went on to discover *transposons*, or genes that can jump about from one location in the genome to another. It was subsequently realized that transposons moving outside cells were essentially what are called *viruses*, i.e., fragments of DNA which have no life of their own, but can relocate into a cell and thrive in it. They can carry resistance to antibiotics by jumping from one cell to another, or into the plasmids in the bacterium itself. Bacteria can stock bits of DNA in plasmids as defences against antibiotics and bacteriophage, i.e., viruses that can attack them. This work was done in the 1940s, before the discovery of the structure of DNA, and its significance was not appreciated at the time. In her 1983 Nobel lecture entitled *The significance of responses of the Genome to challenge*, McClintock states [63] the following:

> The conclusion seems inescapable that cells are able to sense the presence in their nuclei of ruptured ends of chromosomes, and then to activate a mechanism that will bring together and unite these ends, one with another. And this will occur regardless of the initial distance in a telophase nucleus that separated the ruptured ends. The ability of a cell to sense these broken ends, to direct them toward each other, and then to unite them so that the union of the two DNA strands is correctly oriented, is a particularly revealing example of the sensitivity of cells to all that is going on within them. They make wise decisions and act upon them.

Much of the behaviour of a bacterium, as viewed in a microscope can be naively interpreted in terms of the bacterium 'being aware of itself', or 'conscious' in some sense. Bacteria move seeking food, move away from toxic substances like acids, heat, salinity etc. Julius Adler in the 1970s showed the amazing sensitivity of *E. coli* bacteria to many chemicals. It can detect one part in ten million of aspartic acid (α-amino succinic acid) and change its direction to consume it. Even a few molecules of aspartic acid are detected by the glycocalyx on the cell membrane. When the bacterium moves from a low-nutrient region to a high nutrient region, it is responding according to quite well understood physical and chemical processes. One may of course say that the bacteria are 'aware' of the presence of the nutrient. In fact, the bacterium is sensitive to *changes in the amount* of aspartic acid in the medium, since it gets 'habituated' to a given constant concentration of it. This habituation arises from the organic acid setting up an equilibrium concentration of molecules on the cell membrane, and when this remains unchanged, the response of the membrane molecules becomes quiescent. However, any change produces a corresponding change in the cell.

11.2.1 *Sensitivity of the cell to its environment*

Cells interact with the environment mainly to gather nutrients, eject waste, or move away from danger. These are its adaptation to the environment, and its ability to use the environment as part of the cell's energy-degradation strategy. The origin of these sensitivities resides ultimately in simple physical processes which obey the laws of physics. Even atoms and molecules respond to heat, chemical gradients, changes of acidity (pH), hydrodynamic flow etc., via changes in their own kinetic energy, conformation, charge state or changes in the electronic structure. Thus hydrogen bonds break or form in the presence of changes in pH or suitable enzymes. A protein can change its conformation (i.e., the way it is folded, its 3-D form) in response to the presence or absence of specific ions like Ca^{++}, or other proteins. The cell membrane harnesses the sensitivity of these basic chemical processes by equipping itself with suitable proteins and other molecules embedded in the lipid layer. The *glycocalyx* regions associated with such embedded proteins act as sensors, and can communicate with other cells, due to their selective sensitivity to sugars and other substances. If the cell has cilia, they also act as sensors which detect the movement of water, or prey near it. When ions, e.g., H^+, etc., arrive close to a cell membrane, the electrostatic forces affect the membrane and the cytoplasm. Counter ions build up inside the cytoplasm or near special portals, generating electrical membrane potentials. These potentials can be converted to current flow when protons or other ions are allowed to pass through the cell membrane. This electrostatic process is routinely described in a personalized way by saying 'the membrane responds to the ions arriving at its surface'.

The cell membrane is not just a wall separating the cytoplasm from the environment. It is also a very sensitive organ which acts as the eyes, ears, tactile hands and motile legs of the prokaryote. The glycocalyx does the job of being the sense organs of the primitive living cell. The chemical changes which occur at the cell membrane and the associated electrical and chemical signals change the chemical processes inside the cell. These changes are the responses of the cell to the messages from the cell membrane. The responses may involve altered enzyme activity, making new proteins by triggering specific genes (gene expression), and energizing various coiled proteins to push the cytoplasm and the cell membrane to form temporary false legs (pseudopodia, as seen in *amoeba*), or use flagella to move the whole cell, i.e., if needed, run away from an unfavourable environment, or move to a nutrient rich environment. The whole system of chemicals in the cell has evolved in such a way that the water molecules, ions, lipids, nucleic acids, polysaccharides, ribosomes etc., contained in the cytoplasm act cooperatively to respond to the signals sent by the cell membrane.

We see that the 'interpretation' of the signals from sense organs, and the formulation of the response are done by cooperative actions in the cytoplasm which is the 'primitive brain' of the organism. The proteins are the molecular switches or logic gates of the primitive brain. Thus a protein would bind a small molecule, and change its shape. This new protein complex now can function as an enzyme (catalyst) to facilitate (or amplify) a reaction. Here we have a process similar to a transistor, where a small signal, brought in by the small molecule, is amplified in a more complex response. Many such processes are constrained to work in loose but close tandem inside the boundary of the cell membrane.

In a liposome or froth bubble (see Sec. 10.4.2.1), the lipid layer is simply the barrier trapping the cell fluid and separating it from the environment. In contrast, the cell membrane of the primitive cell has to carry out many additional functions. The effect of Darwinian selection, working over millions of years would lead to the specialization of parts of the membrane into the sense organs, hands and feet, and so on. The cytoplasm itself can evolve specialized regions which are more readily equipped to deal with the signals sent by the cell membrane., The glycoproteins embedded in the cell membrane would be a model for developing parts of membrane sensitive to light, i.e., a prototype for a retina.

All this occurs at several levels. Thus prokaryotes evolve into the more complex *eukaryotes*, i.e., single-cell organisms where the cytoplasm has specialized regions for doing special jobs. These are known as *organelles*. The main organelles are the nucleus, the mitochondria, the endoplasmic reticulum (ER), the Golgi apparatus, microtubules, vacuoles. and chloroplasts in plants cells sensitive to light. Fossils of cyanobacteria, i.e., a type of bacteria capable of photosynthesis, are claimed to be some 3 billion years old [85]. This indicates that the evolutionary processes leading to specialization in cells have begun very early in the earth's history.

The *nucleus* of eukaryote cells controls the activity of the cell membrane by determining what openings, known as *protein channels*, are present in the lipid bilayer at a given time. In effect, the nucleus has a say in determining what molecules enter or leave the cell. Similarly, the nucleus has chemical control over all the organelles in the eukaryote cell as well as information flow. The 'words and phrases' of the language used by the nucleus and the organelles for conveying information are the genes possessed by the cell. The nucleus controls information flow by holding a gene pool which is exchanged among the organelles in a centralized manner. Eukaryotes have a much larger genome than prokaryotes. Thus humans have some 30,000 genes, compared to some 3,000 in a bacterial cell. This large number of genes is accommodated by splitting up the single loop-like chromosome found in prokaryotes and instead having many linear chromosomes,. Thus human cells have 46 chromosomes, or 23 pairs, half of these coming from each parent. Each chromosome of such eukaryotic cells is a single long DNA double-strand. Unlike the short DNA molecules of prokaryote cells, these long molecules need to be folded (coiled) and compactly stored in the cell. This is done by coating the DNA with proteins known as histones. The complex of histones and DNA is known as chromatin. The packing of regions of the DNA chain, or exposing genes outwards by controlling the packing are dynamic processes. They are used by the nucleus to control the processes of copying (transcription) and translation (synthesis of the corresponding protein), i.e., to control information flow. Within such dynamic control, the cell makes messaging copies of some of its genes every day. These copies are simply a sequence of base pairs, corresponding to the sequence found in the gene. Such a messenger molecule is called mRNA. The mRNA, once in the cytoplasm, becomes the template for the production of a protein associated with the code implied by the gene (Fig. 11.3). The necessary chemical reactions are catalysed by the enzymes in the cytoplasm. Most of the proteins are then transferred to the ER and the Golgi apparatus, further 'refined' by the addition of sugars, phosphate groups etc. This sugar-coating process is known as *glycosylation*. Such proteins finally end up on the cell surface forming the glycocalyx that we already alluded to. Transportation of proteins and other 'cellular cargo' from one part of the cell to another is done along microtubules which act like 'rail roads'. The microtubules are themselves made of proteins known as tubulins (see Fig. 13.1). The cargo is pushed along the microtubule by specific types of transporter proteins like *kinesin*. The thin (25 nm diameter) but long microtubules provide a primitive structural stiffness to the cell. Hence the network of microtubules is known as the *cytoskeleton*.

Cells also use glycolipids in their 'sensor hardware'. This sensor hardware of the cell is much more developed in eukaryotes, in comparison to the more rudimentary form found in prokaryotes. As prokaryotes do not have ER, Golgi and such organelles, the glycolation is done in the cytoplasm itself.

Another level of evolution, once again driven by the sensitivity of the cell membrane to the environment, is for aggregates of cells to act cooperatively to form multi-cellular organisms. The temporary false limbs that an amoeba has to make by contorting its protoplasm are replaced by actual, permanent

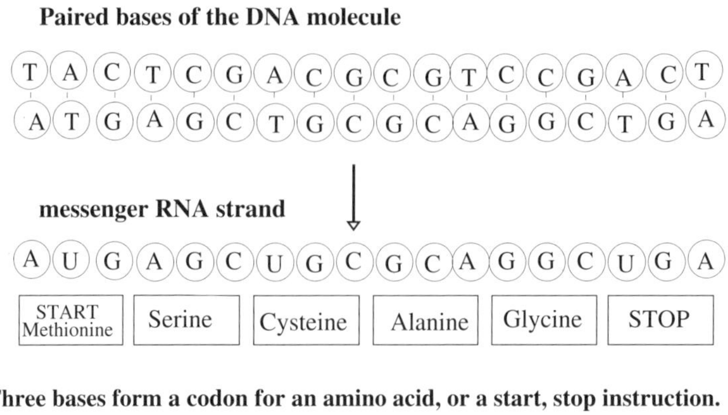

Fig. 11.3 This figure reviews the well known three-base *codon* scheme used in DNA to define an amino acid, and the copying of the information to a messenger RNA molecule. The bases are Adenine (A), Cytosine (C), Guanine (G), Thyamine (T); the last is replaced by Uracil (U) in the RNA molecule. The three bases GCA occurring together in that sequence is the 'codon' which stands for the amino acid Alanine. Similarly, Serine (AGC), Cysteine (TGC $\rightarrow$ UGC), Glycine (GGC), and Methionine (ATG $\rightarrow$ AUG) sequences are shown. Methionine indicates the start of the gene sequence, while the end or 'STOP' is indicated by the three bases TGA which becomes UGA in the mRNA molecule. There is no amino acid corresponding to these three bases occurring together and designating STOP, and the codon has been given the name 'opal'. In bacteria, the ATG codon is read as formylmethionine. The flow of information from the DNA to the mRNA is known as *transcription*. The next step, where the amino acids are assembled into proteins with the aid of tRNA is known as *translation*.

multi-cellular organs like hands and feet. The specialized sensitive regions on the cell membrane become further sensitized and evolve into ears, eyes, noses, mouths, digestive organs and a rectum. Parts of the cytoplasm and the nucleus involved in information processing evolve into neuron cells which form the cells of the brain of the multi-cellular organism.

We have recounted the cascading process of different length scales and energy scales empowering the next stage of longer length scales, smaller energy scales and higher levels of complexity. The sensitivity of molecules to simple physical and chemical processes, translates into the more varied and amplified response of macromolecules to such processes. These, when embedded in a coherent structure within a sensitive lipid membrane, begin to respond to the environment as an individual. The sensitivity of this entity as a whole to the environment is the force behind Darwinian evolution which ultimately generates multi-cellular life.

11.2.2 *The cell as a computer*

The 'awareness' of the cell is an automatic response, similar to that of a thermometer needle to an increase of temperature. The thermometer could also be a complex device which is programmable and responds to many other aspects of the environment around it. In effect, in all such cases, be it a prokaryote or an engineered sensor, we can break down the response process to a series of simple steps governed by the laws of physics or chemistry (which are, in the final analysis, reducible to those of physics). The bacterial motion is simply a result of the chemistry, and the organizational structure of the cell with its cilia (flagella), powered by ATP, which make the cell move. We call it a spontaneous *reflex action*, although the same term is not used for the response of the thermometer. The thermometer is 'an automated control device'. Computer simulations which attempted to mimic bacteria and such life-forms as 'synthetic life' have used the name 'cellular automata' for such systems (see Sec. 11.2.3). In the case of bacteria, and also pre-programed thermometers, there is no learning process which modifies an internal structure of the system during a training period. The bacterium has been programed during its evolutionary history, while the thermometer has been programmed during its installation. The engineering device is made up of a small number of well defined components designed to work more or less faultlessly within set tolerances.

The accepted picture of the bacterial cell is very different from that of a nearly error-free automated device. In the simplest view, the single-cell organism performs like a large number of randomly released balls falling one at a time down a very elaborate pinball machine, and giving a resulting distribution of balls displaying an amazing pattern which could put a da Vinci to shame. The pinball machines themselves have been built, painstakingly slowly, over eons of time, by similar random processes in the pre-biotic stage! As Feynman pointed out in his *Lectures on Computation* [34], even a collection of billiard balls moving against suitably designed baffles can behave as a computer – i.e., a machine capable of executing a number of logical steps and making decisions.

A somewhat more elaborated view of the cell as a battle-planner-type computer i.e., a device which works out survival-strategy based on information collected, has also been proposed I-[38], [45]. The existence of many protein units in the cytoskeleton of the cell is a starting point of these considerations. Microtubules are made up of dimer units of the protein *tubulin* arranged in tubular arrays. tubulins are long chains of amino acids that take up a specific conformation under the competing forces of electrostatics, presence of water, formation of hydrogen bonds, and van der Walls forces. In fact, all proteins behave similarly, and the 'folding of a protein' to a particular conformation is a delicate process which is difficult to compute. Thus, depending on the delicate balance of the net forces acting on the tubulin, the molecule can occupy one of several folded conformations. The tubulins (and many other proteins) can also switch between closely related conformations with only slight changes in energy. That is, they can under suitable conditions act like binary bits which are the primary units of information storage and information processing (see Sec. 7.2.1). Some authors have argued that the array-like structure of tubulins in microtubules enables them to work collectively, and that the cytoskeletons of bacteria have the capacity to do primitive computations [45].

Other possibilities also exist. The exotic correlated states (see Sec. 13.1.3) envisaged by Hameroff and others are not at all necessary. Sol-gel processes involving the cytoskeleton has also been proposed as a means of information processing. The cytoskeleton can already act as a primitive neural network, where some of the protein molecules located in the membranes or at cross connections could act as the nodes of the net. It is precisely from such nodes at the cell membrane of eukaryotes that the cytoskeleton extended to form axons and synapses, generating special cells – neurons – dedicated to information processing (see Sec. 12.4).

What ever the mechanism, such computational capacity would enable organisms like the *paramecium* (an eukaryote belonging to the protista kingdom), that do not have specializations like neurons, to show very complex behavioural strategies.

11.2.3 *Simulations of 'life'*

Science fiction writers have endlessly speculated on strange types of life, different from those on earth. Life systems based on Si instead of C, replacing N and P by other elements like As, or O by S, i.e., elements from the same column of the periodic table, are easily envisaged. The unconfirmed claims for a type of 'As-based life' in 2010 by NASA scientists show that science fiction and science reality may not be too far from each other. Besides the attempt to create life forms based on elements other than C, novel 'synthetic cells' which are still based on C, but use other genome architectures are well within today's capabilities. The eventual goal of such efforts would be to build new organisms with 'engineered genomes' that act in ways that differ from nature. In theory, bacteria may be designed for bio-fuel production or environmental cleanup.

Another, much cheaper line of research into 'alternative life forms' has been via computer simulations. Such studies became popular in the early days of the advent of inexpensive desk computers. If life is a result of evolution where molecules follow some simple rules of chemistry, one may try to simulate this on a computer which can execute millions of steps in a short time. Thus 'artificial life (Alife)', simulations on a computer may attempt to model abstracted features of life as we understand it, and as 'it might possibly be'.

Computer games are the most well known type of simulation which is also most prolific. Many such games embody elementary process of eating up the prey, escaping from predators, growing and even replicating. It seems that inventors of games have watched amoebas through a microscope and used that as their basic game plan for manipulating pixels on a 2-dimensional computer screen! Even the more sophisticated first-person shooting games or massively subscribed on-line multi-player 'virtual-reality' games using fractal techniques, are still elaborations of schemes of attack and survival. Battle-planning strategy, military and stock-market applications are a more serious application of these 'games'. The associated developments in software (e.g., in neural networks) have been important in evolutionary genetics, learning and systems control.

One of the earliest 'games of life' that gained great popularity was the cellular automaton invented by John Conway at Cambridge University, England [38]. The game is played on a grid of squares ('cells') having four side neighbours and four corner neighbours. These cells can be unpopulated (0), or populated (1), using simple rules for populating or unpopulated cells (0s to 1s, or *vice versa*). This at first sight looks like a clever generalization of the 0s and 1s game of tic-tac-toe. However, it goes back to the work of Alan Turing, John von Neumann and others who were interested in formulating machines (specifically, Turing machines) that could 'self-replicate'. John von Neumann developed a self-replicating model, inspired by the mechanism of crystal growth, and named it the 'universal constructor' [112]. The clay theory of the origin of life, introduced by Cairns-Smith in the 1960s, also used a crystal-growth model.

Conway's rules of the game of life can be iteratively applied in discrete time steps, and now it becomes a dynamic process with some similarities to the growth of populations of organisms, but here dependent only on the initial starting configuration. Another important event in this field was the publication of John Wolfram's *A new kind of Life* in 2002, summarizing work begun in the 1980s [109]. The rise in popularity of such games was in many ways linked with the beginning of the marketing of personal computers and electronic information media in that era. These Alife models are in effect new tools that show even to the uninitiated that utterly unexpected complex patterns, behaviours, as well as replication can evolve from simple rules and stark beginnings. However, the processes are completely computable, and attempts to explain consciousness and free will using these finite-element, non-chaotic models should be regarded more as a beacon to our musings rather than an 'explanation' of free will or 'life'.

John von Neumann, in trying to construct a formal demonstration of a robot capable of constructing a working replica of itself found that it was necessary to replicate a whole factory. That needed foundries and other resource infrastructures, sources of raw materials, and in fact a whole supporting environment. This is why he turned to a finite-element cellular-automaton approach of dealing with a few abstract units. The embedding environment was specified by a few discrete rules. In contrast, the living cell cannot exist without the watery aqueous environment around it, or the millions of water molecules which form 99% of the molecules contained inside the cell membrane. The main players of the drama of the cell, listed in Table 11.1 cannot do what they do, without the the unnamed cast of water molecules, ions etc. The movements and scripts of these main players are dictated by the laws of physics, but they are not deterministic. They are incessantly subject to the chaotic, basically non-computable statistical behaviour of the supporting cast of actors in the cell and outside the cell.

Here it is sobering to recollect that even two hard-sphere billiards colliding with each other according to the laws of physics have features of clear irreversibility in their dynamics (see Sec. 9.2.1). The work leading from Poincaré to Sinai, and subsequent developments show that irreversible mixing is a common dynamical feature of most systems, and most certainly for the ensemble of interactions in a living cell. The cell defies determinism and it has acquired an element of 'free will'. It already has a unit of time, and an arrow of time. The diurnal cycle of the cell is encoded in its DNA (part I-[164]). Its dissipative biochemistry provides the arrow, the irreversibility and the indeterminacy typical of life.

Chapter 12

Specialized Cells for Sight, Insight, and Information

Evolutionary pressure drives simple cells to become multi-cellular organisms, with cells adapting to do specialized tasks. The nature of vision, neurons, the molecular basis of memory, phantom limbs, neural nets, information processing etc., are discussed in this chapter.

12.1 Adaptation and survival

Darwin's theory of evolution as originally presented is supported by extensive field observations of nature. More complex technical information comes from physiology, embryology, fossil studies, radio-carbon dating etc. Exacting information and details at the level of molecular organization are provided from sciences like molecular phylogenetics. However, comprehending the theory of evolution does not demand advanced material, or a physics-type foundation of abstract ideas and mathematical techniques outlined in part I of this book. Darwinian concepts are sufficiently natural that such views had been entertained by many thinkers even prior to Darwin, in spite of the mindset of a world dominated by religious dogma.

Darwin's ideas were further elaborated by modern evolutionary biology, fossil-record studies, molecular genetics and other sciences. They are found to hold valid even at smaller length scales than those of our own world. Bacteria, viruses and others systems which *compete for resources* demonstrate how every advantageous adaptation or specialization acquired by such a system (a species) leads to the proliferation of that system. Even in the world of single-cell organisms, the main reason for the formation of colonies of organisms (and multicellular organisms) was to deal with starvation. The amoeba *Dictyostellum discideum*, has been nicknamed 'Dicty' as it has become a popular bench-mark organism for biologists [26]. When Dicty cells runs out of food, they release the chemical cyclic-adinosine monophosphate (cAMP) to call each other to aggregate, forming a composite colony known as a 'slug'. This chemical, cAMP, is retained as *a cell-signaling substance* even in higher organisms. Clearly, working together for 'the good of the many' seems to be an emergent principle arising to optimize the behaviour of organisms faced with scarcity and ruthless individual competition.

At a higher level of specialization, competition for resources means competition for acquiring mates, spouses, subjects, and colonization of lands — i.e., imperialism. Thus it is not surprising that the lessons learnt from 'evolutionary biology' become relevant to all the social sciences as well as all human endeavor. Yet the resistance to Darwinian thought, even today, is found precisely in these last areas of the application of evolutionary theory, partly because we do not know how to *use* it. The *mis-use* of Darwinian theory in the hands of Francis Galton and others with their 'eugenics' movement, followed by Nazi abuses, such as enforced 'racial hygiene', and the extermination of 'undesired' population groups etc., take the blame for this. The phrase 'survival of the fittest' does not imply a value judgment. In reality, Darwinian theory remains value neutral, as is the case with all scientific theories. Evolutionary theory can serve to elucidate social phenomena in a more systematic way than the common non-empirical paradigms used in the social sciences. However, the bad name acquired by the political misuse of Darwinian theory stands in its way.

On the other hand, evolutionary theory applied to biology, and humans via medicine and neuroscience have moved forward vigorously. The American National Science Foundation has even set up the *National Evolutionary Synthesis Center* whose basic mission is to 'help foster a grand synthesis of the biological disciplines through the unifying principle of descent with modification'.

In this chapter we examine a few specialized living cells which are clear adaptations of the original base model of the prokaryote cell and its more advanced edition, the eukaryote cell. Each cell transfers its acquired mutations to its progeny during cell division. If the genomes of the daughter cells provide any evolutionary advantages to the new progeny, they proliferate. Such processes work over billions of years, with zillions of cells undergoing reproduction perhaps every 20 minutes in many prokaryotes. Hence the appearance of specialized cells more adapted to meet the competition is a near certainty.

It turns out that this evolutionary process is a 'tinkering process', or 'bricolage', as noted by François Jacob. He notes that the rise of a new, 'improved' type of cellular structure is not like replacing a candle with an incandescent light bulb. The light bulb involves an entirely new set of scientific and engineering principles (Ohmic heating), making a clean break from the chemical process (burning oil) used in the candle. Thus Jacob says 'Evolution does not produce innovation from scratch. It works on what already exists, either transforming a system to give it a new function, or combining several systems to produce a more complex one'[49]. In this chapter we see how features in the simple prokaryote cell have been adapted and specialized to create amazingly complex structures, by a tinkering process which has taken billions of years to accomplish.

12.2 Multicellular specialization

A single-cell organism has all its DNA in its genome, and where necessary, some components in organelles as well. Cell specialization is best done within a multi-cellular framework where different cells can do different things and yet cooperate for the success of the whole organism. This is done by allowing only a limited number of genes of the genome to 'express' themselves in a given cell. The DNA of a cell is folded and packed with histones so that some genes are exposed, while others are sterically hidden by the folding process. The exposed genes define proteins which are needed for a specific organ. Thus a cell which becomes part of a finger expresses only a select set of proteins, based on the expressed genes. This is called cell differentiation. It enables different parts of the same genome to play major roles in different cells. Thus the round worm *C. elegans* studied by Sydney Brenner and his colleagues in their pioneering work has 959 somatic cells. These represent the major differentiated tissue types including neurons (302 cells), muscles (111 cells), intestine (34 cells) and the epidermis (213 cells) [11].

Long before specializations like intestines, muscle, eyes and limbs could be evolved and incorporated into a multi-cellular organism, groups of cells have to be brought together. This may happen if a warm pond nearly dries out and forces close contact among the prokaryote or eukaryote cells in the water. In fact, the main reason for the emergence of colonies of cells seems to be to deal with starvation, as with the Dicty amoeba. Here we shall call such life in each others proximity, enhanced by some chemical signaling process 'cohabitation'. This colony formation leading to multi-cellular forms is sometimes called the *colonial hypothesis*. Other theories of cell aggregation use ideas like symbiosis leading to multi-cellularization. Multi-cellularity has probably arisen by several paths, in algae, fungi, plants, and animals. Clearly, detailed mechanisms are a subject of detailed research, and links with the emergence of multi-cellular reproduction.

However, we can envisage that simple colonies like those of Dicty gave rise to the simplest multicellular organisms, such as the sponges. These are organized with less than a dozen types of cells. With the progress of time, more and more complex organisms, with cells capable of specialized tasks appeared.

Multicellular functions require coordination and signaling among the cells. Two types of chemicals have evolved from the basic ingredients that the cell already had in the unicellular stage. These are hormones and neurotransmitters. In chemical communication using hormones, a gland releases a chemical message (a hormone) into the blood stream; the hormone molecules reach other cells in distant tissues, and this constitutes the signal. The distant tissue cells react to the hormone. For example, when food is ingested, the sugar level in the blood rises. This rise activates certain cells in the pancreas to release the hormone *insulin*. Insulin carried in the blood stream acts on insulin receptors in muscle cells. This makes the muscle cells absorb the glucose and convert them into glycogen, for later use as an energy reserve. On the other hand, chemical communication using neurotransmitters is specific to transmission of signals between nerve cells (neurons). These are either amino acids or their derivatives such as glutamate, γ-aminobutaric acid (GABA), acetylcholine, ephinephrene, serotonin and dopamine. Signaling using neurotransmitters occurs across neuron cells and involves much shorter distances than with hormones. The cell membrane of the neuron receiving the signal is very close to the neuron which is sending it. Hence the chemical is simply released into the synaptic cleft, i.e., the space between the cell membranes. These processes are discussed in greater detail in Sec. 12.4.

The variety of organs occurring in the body corresponds to the variety of specialized cells. However, the principles behind the specialization are very similar, and use the basic chemicals present in single-cell organisms, over and over again, by tinkering with them in the most intricate of ways, as noted by François Jacob [49]. It is sufficient for our purpose of understanding nature's ways, to look at light sensitive cells and neurons. These lead to thc key functionalities of vision and consciousness in higher organisms.

One type of light-sensitive adaptation is to harvest sunlight as a source of energy, as found in chloroplasts of plant cells. Another type of adaptation is vision, enabling an organism to pursue prey and acquire nutrients for energy. Pursuing prey, or escaping a predator requires organizing information collected by the primitive eye and other sense organs of the cell (see Sec. 11.2.1). Those organisms with specialized cells for information processing exploited their evolutionary advantage and proliferated. The needed specialized cells are known as 'nerve cells' or *neurons*. Neurons use electrical signals and chemicals to process information. Recall that the cell membrane of an amoeba can detect even a few aspartate molecules dissolved in a million other molecules. Neurons also detect the presence of chemicals, and open (or close) their membrane pores to let in (or let out) various types of ions. These transfers of ions create electrical signals which propagate to other banks of neurons and cells which make proteins, ATP and other cell chemicals. These chemicals are used 'to act' according to the incoming electrical impulses. The 'act' is also mediated by further signals which release chemicals using the energy of ATP molecules contained in muscle cells.

In the initial stages of evolution when organisms had only a few neurons, they could be used to hard-wire set behaviours. The set behaviours were transmitted to the off-spring by the inherited DNA. That is, automatic protocols for carrying out a sequence of chemical reactions can be configured via the DNA into the organelles of the cells. These produce the inherited reflexes of an organism.

Cnidarians (e.g., jelly fish) developed a simple nervous system. This was further evolved in clams, snails and other mollusks. The sea slug *Aplysia*, used by Eric Kandel and collaborators [51] for pioneering work in neuroscience, contains some 20,000 neuron cells. When there are many neurons, the number of possible interconnections also increases rapidly, as in a human with ten billion (10^{10}) neurons in the fore-brain. We discuss the bit capacity of the brain in Sec. 12.6.2. By exploiting the extra capacity for handling information, the higher vertebrates evolved the collective-mode behaviour of groups of neurons known as *consciousness*.

However, sensitivity to light probably came much before neurons arrived. Thus *Bacteriorhodopsin* and *proteorhodopsin* are light sensitive molecules found in Archaea and in marine bacteria. These molecules are closely related to rhodopsin, the key light-sensitive molecule found in the human eye. These early precursors of rhodopsin (proteorhodopsins) enable bacterial cells to supplement respiration as a cellular energy source. This ability to withstand oxygen deprivation probably explains why so many ocean bacteria have proteorhodopsin. Just as cell cooperation and aggregation into multi-cellular organisms were provoked by the threat of starvation, turning to solar energy for power was probably a result of deprivation of a necessary resource. Interestingly enough, the currently most prevalent method of harvesting solar energy by cells uses chlorophyll instead of molecules derived from the rhodopsin family. When the rhodopsin family of molecules was finally adopted by the evolutionary process for vision in vertebrates, the amino acid sequences of the proteins have evolved extensively and hence differ significantly from those in bacteriorhodopsins. Chlorophyll itself is a molecule evolved from the bacteriochlorophylls. These are used in bacterial photosynthesis reactions where no oxygen is produced.

Vision is a process where no oxygen is produced, and the energy of the collected photons are used to trigger electrical signals to the brain. Vision is the most poignant link between a 'person' and the external world. It is the subject of much metaphysical speculation too. It is vision that has often been claimed to demonstrate the 'Creator's intelligent design', or the need for God or Gestalt. Hence it is heartening to find that a completely molecular description of vision that is detailed and coherent is available to us.

12.3 Cells adapted for vision

A chromophore is a molecule that can absorb light, usually in a well defined colour range. The absorption of light occurs when the electronic state of a molecule changes by a transition to an excited state. This is a process governed by the quantum theory which ascribes wavefunctions to electronic states. Chemists describe electronic states by the arrangement of bonds, non-bonding electrons, dangling bonds, free radicals etc. A chemical bond is made up of a *pair* of electrons (see Sec. 7.17). Dangling bonds and free radicals are usually single, active electrons sitting on an atom waiting to react with a partner. If a covalent bond is broken, it produces two dangling bonds. Thus free radicals, dangling bonds etc., are quickly converted to more stable molecular states. In addition to the most stable 'ground state', there are many other nearly equivalent quasi-ground states for a large molecule. In the case of molecules where bonding arrangements give rise to nearly identical equilibrium states with different atomic positions, we refer to them as sterio-isomers. Thus the *cis-* and *trans-* isomers of molecules containing different arrangements of substituents at C=C bonds, or at rotationally hindered C–C bonds are well known (see Fig. 12.1). In fact, hindered rotations, multi-center interactions and other issues of energy scales and time scales lead to the intractable problem of the *folded conformation of proteins* (see Sec. 8.6), where by a deterministic system becomes computationally indeterminate.

If the photon is capable of exciting the double bond in a *cis* configuration into a transient single bond with two dangling bonds, the two parts of the molecule can rotate about the transient single bond, and then the dangling bonds can recombine, leaving the molecule in a *trans* configuration. This is exactly what happens in vision related chromophores. The basic light sensitive molecule in these chromophores is known as *retinal*. It's *cis-* and *trans-* isomers are shown in Fig. 12.1. The compact *cis*-isomer is held clasped inside a protein, but the trans isomer does not fit, and unfolds the protein located in the cell membrane (see Fig. 12.2), setting off a series of reactions which produces a charge imbalance between the inside and outside of the cell. This effect, known as *hyperpolarization* produces the electrical impulse which travels out of the cell, as described below.

Clearly, the proteins themselves have evolved hand in hand with the light-sensitive retinal. A number of closely similar proteins known as *opsins*, and

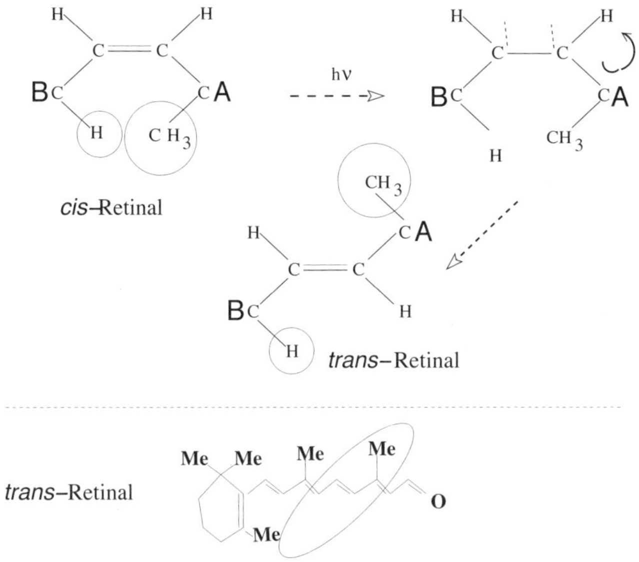

Fig. 12.1 An H atom and a methyl group (CH_3) sterically hindering each other in the *cis*- isomer of a molecule. A and B are chemical groups. The circles around the H-atom and the methyl group are schematic and fluctuate as the molecule vibrates. The double bond prevents the molecule from rotating about it. However, when a photon ($h\nu$) is absorbed, the double bond is broken, and a pair of dangling bonds (freed π electrons), marked as dashed lines pointing up are produced at each carbon atom. Now the molecule is free to rotate, and relieves the steric interaction by moving the H and CH_3 groups to the opposite sides, while re-forming the double bond. The result is the *trans*- isomer of the molecule. The *trans* isomer is bigger than its more compact *cis*-form. The molecule *retinal* is shown in the lower part of the figure, showing the methyl groups as Me. The active double bond and the isomerization region are marked off within the oval. The group A is the aldehyde group $-C(CH_3)CH\text{-}CHO$, while B is a hydrocarbon chain containing two more double bonds and a trimethyl-cyclohexene ring.

photopsins are found in the specialized light-sensitive cells of the eye, known as *rods* and *cones*. The rod shaped cells which are located more towards the periphery of the retina are known as *rods*. They all contain the identical protein opsin, and hence they are monochromatic. The cone-shaped light sensitive cells, found in the more central region of the retina are of three types. Three different proteins (photopsins), sensitive to red, blue and yellow light, found in different cone shaped cells provide trichromatic vision. This is the physiological and biochemical basis of the law of color combinations known to artists since antiquity.

Rod-like cells are monochromatic in that they have just one kind of light-sensitive molecule. They contain opsin molecules positioned among the molecules of the lipid bilayer of the cell membrane. Opsin is made up of 348 amino acids linked together in a winding chain. This is folded into seven column-like

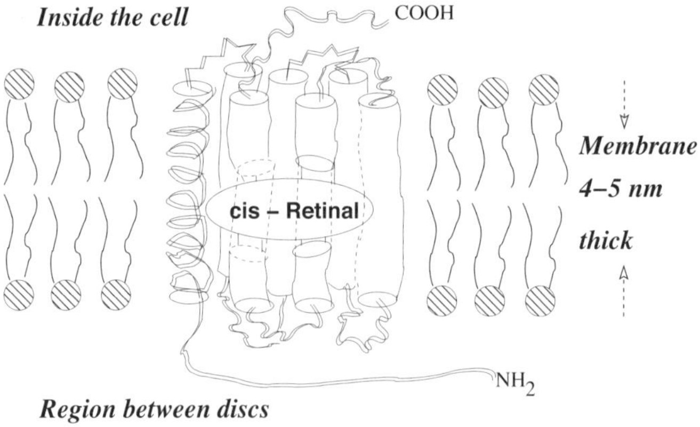

Fig. 12.2 The lipid membrane is traversed by the protein molecule *opsin* which is folded into seven columnar sections, and having an inner cavity sufficient to hold the *cis* form of the retinal molecule. Two columnar parts are cut out to show the retinal region marked as an ellipse.

spirals which are positioned traversing the cell membrane. The opsin helices, seven in number, arrange themselves leaving a pocket-like region which nicely fits the compact shape of the *cis*-retinal molecule (see Fig. 12.3). The opsin part is the 'protein-moiety', while retinal is the 'chromophore-moiety', together forming rhodopsin. Similarly, in the cone-shaped cells of the retina used for color vision, the protein moiety can be one of the three photopsins.

The rod- or cone-shaped ends (pointing into the retina) of photo-receptor cells are separated out from the main cell body. The separated-out rod or cone part has a cell membrane specially arranged like a stack of discs to meet the incoming light repeatedly, as shown in Fig. 12.3. An axon connects the cell body and leads to a synapse (see below) which couples with bipolar-cell dendrites that communicate with ganglion cells. These cells send image-forming as well as non-image visual information to the brain in the form of electrical impulses.

12.3.1 *Visual perception*

In visual perception, information is pre-processed already in the retina, and passes to the intermediary neurons in the lateral geniculate nucleus (LGN, see Fig. 12.4), also known as the lateral geniculate corpus. The electrical impulses go from the LGN to the primary visual cortex (V1, or first visual area), while also receiving inputs from higher regions. This 'feedback' effect from the neurons in the brain is many times bigger than the input from the retina itself! Visual perception is approximately 20% sight-input and 80% mind input. Hence the LGN is probably

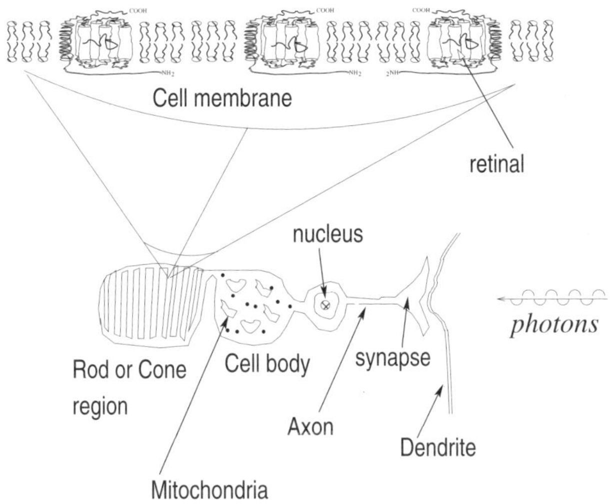

Fig. 12.3 The lipid membrane of the rod-shaped or cone-shaped end of the cells is folded into a stack of 'discs' to maximize the interaction with incoming photons. The membrane is full of many insertions of the protein molecule (opsin or photopsins). It is folded into seven connected helices traversing the membrane, as shown in Fig. 12.2. The *cis*-retinal molecule is held in the columnar-space ('cavity') between the seven helical parts. The absorption of light causes the *cis*-retinal to form *trans*-retinal, and this no longer fits into the cavity of the protein moiety. The latter undergoes transformations which generate an action potential in the cell. This is transmitted to the synapse, to interneurons (intermediary neurons) like bipolar cells, ganglion cells and finally to the brain.

the first point in the visual pathway where the states of the mind begin to greatly influence what we 'see'. So we may recycle the meaning of the word 'insight' to say that sight is generously mixed with 'in-sight' before an image is formed!

That the brain *makes up* a big part of what we see was understood to varying degrees by Isaac Newton, Immanuel Kant, Arthur Schopenhauer, Clerk Maxwell and others. Thus Maxwell wrote 'If colour vision has any laws, it must be something in our nature that determines the form of these laws. The science of colour must therefore be regarded as being essentially mental science' [62]. The point is, the colour of surfaces do not change much even when the spectral distribution (wavelength-composition) of the light they are looked at changes significantly. The perceived 'colour' remains the same, even though the amount and spectral distribution of the reflected light changes. This relative stability of colour under changing conditions of illumination is a capacity of the brain to create a stable, 'corrected' working representation (color) of the surface seen. In effect, the brain has learned how to allow for reflectance variations from surfaces [5]. Patients who have injuries in the V4 area are found to be incapable of this, and may only see

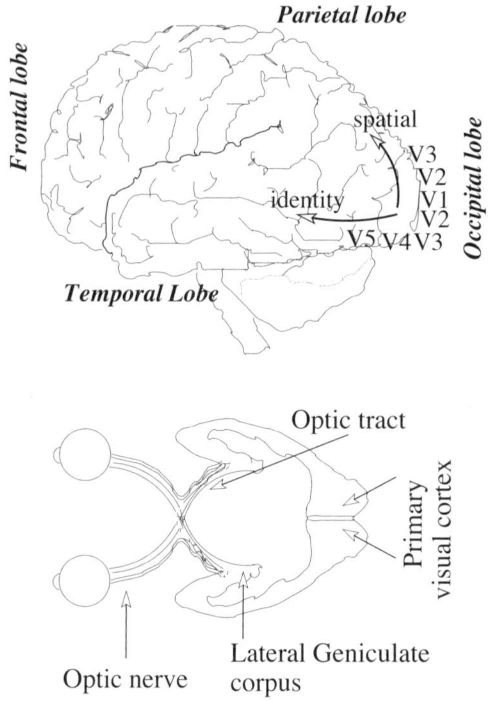

Fig. 12.4 The electrical signals from the retina travel through a bundle of axons known as the optic nerve. They travel to the intermediate neurons in the Lateral Geniculate corpus and to the primary visual cortex which is in the posterior of the brain. Further processing occurs via the neurons in the regions V1, V2-V6. The analysis of *what* (what the object is), is seen is done in the temporal lobe, while the spatial location ('where' the object is) is resolved in the parietal lobe.

shades of grey. It has in fact been proposed by Semir Zeki that the visual cortex of the brain uses a number of micro-algorithms in many, functionally separate processing systems, and that injury to one processing system may not affect others.

The LGN is a synaptic relay (see Fig. 12.4), located in the dorsal (i.e., at the back, as opposed to ventral) part of the thalamus. There is a left LGN, and a right LGN, respecting the bi-hemispheric construction of the brain itself. Its appearance resembles a knee, and hence the name 'geniculate'. Each LGN receives axons from the eye in its own side, as as well as from the other eye. The information is separately processed in parallel in different layers of the LGN.

The visual area V1 is essentially a cellular representation of the retina in the cortex (i.e., a *retinotopic map*) [1]. Other visual areas, known as V2, V3, V4, V5, and V6 take up the information (i.e., electrical impulses and chemical signals). The neurons in the region V4 are responsible for the interpretation of colour.

Hence, if the V4 region is injured due to a tumor or accident, only black and white vision remains. The neurons in the region V5 detect motion, and if this region is damaged, vision becomes a sequence of still images. In fact, the first map of any area in the brain was prepared by a Japanese opthalmologist, Tatsuji Inouye (1909) who correlated areas of depressed vision (visual-field scotomas) with the location of cranial gunshot wounds in survivors of the Russo-Japanese war. A high point in such understanding was achieved by Wade Marshall who began his studies of sensations in Chicago, in the mid 1930s. Information comes to the brain via action potentials traveling along axons. Whether the signal is visual, tactile, etc., is determined purely from what set of nerves the signals are delivered to the brain. Marshall using experiments on monkeys showed that different sensations which carry different types of information end up arranged and stored in different parts of the cerebral cortex. Sensory information is topographically organized in the form of precise maps of the sensory sources such as the retina, the skin, fingers, the tongue or the inner ear. This map was given the name *sensory homunculus* by Wilder Penfield, the Canadian neurosurgeon who worked directly with human patients and advanced our knowledge of the sensory map of the human brain.

As visual information is received via the optic nerve, objects are deconstructed into line segments of different orientations; other aspects of visual perception — depth, form, motion, colour — are separated out and sent via different pathways to the brain. An important step in this separation is the identity (i.e., 'what is seen') and the spatial location ('where') of each bit. When the various bits are brought together and reconstructed in a perception, a large amount of 'feedback' and internally generated material are added as input, as already mentioned with regard to the LGN. The 'what', i.e., the identity of the object seen can be sorted out only when modules of neural networks come into play [88] to resolve the 'binding problem', i.e., the problem of reintegrating all the information (see Sec. 14.2.2). The 'higher level' networks (i.e., neuron nets more up-stream from the sensory signals) propose 'suggestions', and these suggestions are compared with the incoming inputs, to see if a suggestion 'fits' with what is seen. Thus, the higher-level module may suggest that the image seen could be of a Greek vase, or that of two faces, as in Fig. 1.1. Other more specific suggestions would be possible if the data were more specific. One of the suggestions is adopted to identify the incoming sensory stream, and this identification can change dynamically. These types of analyzes are done in conjunction with the temporal lobe, while the spatial aspects are determined in the parietal lobe.

These hierarchical links among a primary cortex, and secondary cortices are found in regard to all sensory and motor cortices. The secondary units from these brain regions are linked by axons and synapses to little-understood *associative*

Fig. 12.5 Ramón y Cajal (1852–1934) was a visionary Spanish scientist who, in the latter part of the 19th century established the basics of neuron science and laid down many concepts that are still found to be valid. Cajal was honoured with the Nobel prize for medicine in 1906.

networks where all signals from different sources are assembled. It should be noted that this view of the integration of information ('binding problem', see Sec. 14.2.2) is not established, as some 'master assembly area' or 'integrating neural network' has not been identified. What ever the details, they lead to an integrated multi-sensory representation of the 'external world'. However, possibly more than 80% of the inputs to this perceived 'external world' are appropriate constructions arising from the neural-net templates of the brain itself.

12.4 Neurons: Nature's telecom units

In simple organisms, e.g., prokaryotes, the cell membrane has developed a glycocalyx which is sensitive to the outer world. The cell membrane has protein molecules traversing through them, and these also served as controllable portals for the intake or egress of ions and other substances. It is precisely these same capabilities, slowly specialized and evolved over the ages, that led to parts of the cell membrane extending to become axons and synapses (Fig. 12.6), while the cell body continued to carry out the biochemical processes of producing energy, synthesizing chemicals etc., as needed by the cell.

Devices that receive, process and output electrical inputs are familiar to us today because electrical appliances and computers are so common. Nevertheless, our initial knowledge of neurons comes from the extremely prescient contributions of the Spanish anatomist Santiago Ramón y Cajal (1852-1934). He recognized, already in the late 19th century, many of the properties of neurons that we know in greater depth today (see Fig. 12.5).

As shown in Fig. 12.6, neurons have four basic parts. They are (i) cell body, sometimes called the *soma*, (ii) a number of branching thread-like extensions known as 'dendrites', (iii) an axon which is a wire-like extension of the cell membrane, (iv) a set of axon terminals, known as 'presynaptic terminals' which are button-like objects that approach closely a soma, or the dendrite of another neuron. These four types of components are found in all neuron cells of living organisms.

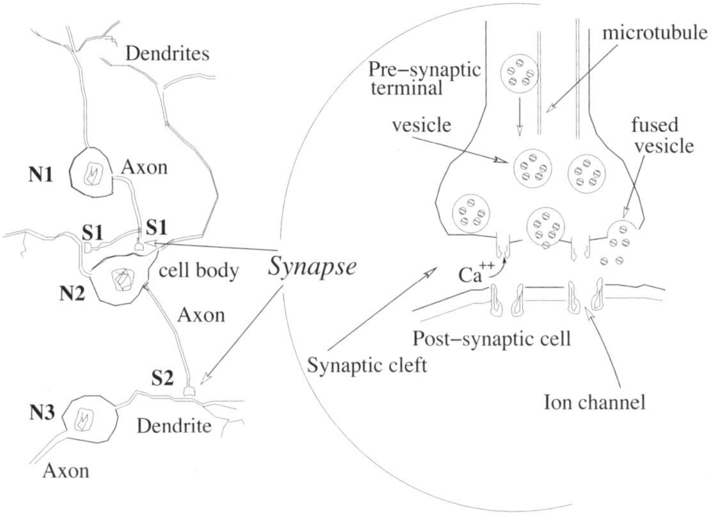

Fig. 12.6 The neurons N1, N2, and N3 as well as their axons and dendrites are shown. The coupling regions between neurons are known as *synapses*. The axon of N1 takes signals to the two synapses S1, where the neuron N2 receives the incoming signal. Then the signal goes along the axon to the synapse N2, where it meets a dendrite of the neuron N3. The axon of N3 relays the signal further. On the right an enlarged view of a synapse (coupling region) is shown. The button-like presynaptic terminal contains little sacks (vesicles) of molecules (e.g. glutamate) which are the neurotransmitters. The vesicles are brought along the axon via the microtubules which act like a 'rail'. The electrical signal received along the axon opens ion-channels and allows Ca^{++} ions to flow in, which causes the vesicles to move to the cell membrane and fuse with it, releasing (exocytosis) the neurotransmitter molecules (e.g., glutamate, see Tables 12.1, 12.2) into the inter-neuron region. This region is known as the 'synaptic cleft'. The molecules, as well as associated ions enter the post-synaptic cell and changes the membrane potential of the receiving cell. This membrane potential becomes an action potential which goes along its axon to the next synapse. Ion channels vary according to the type of ions that can pass through them, and the mechanism used by the cell to open or close them.

The dendrites often appear as tree-like complex branching structures that extend from the cell body. They, together with the membrane of the soma, form the input or receptive area of the incoming signals to the neuron. The axon, a thin tube, is the conduit of the output signal. Axons may be as short as a fraction of a millimeter, or extend to several meters in length. The axon terminates in several fine branches, each of which carries a presynaptic terminal. They are usually positioned very close to a dendrite or cell membrane of another neuron. Today we know that there are other cells, known as 'glial cells' which play a supporting role to neurons, electrically insulating one neuron from another, holding them in place, and also playing a role in the supply of nutrients, control of pathogens etc. Neurons also differ from other cells in that they do not divide to form new cells.

This led to the belief that we are born with a store of some two billion neurons and that we slowly lose them as we grow old. However, more recent studies on *neuroplasticity* show that cortical structures can grow, repair and modify neurons, not only in infants, but also in adults.

12.4.1 *Communication among neurons*

The special features associated with neurons, i.e., axons, dendrites, and synapses are the cellular appendages especially evolved for signaling. The signals used by the neurons are made up of electrical impulses traveling along axons to synapses, and chemical signals across the synaptic cleft, i.e., the small space between the pre-synaptic terminal of the sender-cell membrane and the target cell membrane. The synaptic cleft is about 20 nm. (2×10^{-8} meters) wide. In comparison, we may note that a hair is some 400 times thicker than a synaptic cleft. In Fig. 12.6 we show a part of a neural network (see also Sec. 2.2.5) consisting of three neurons.

A famous and long standing disagreement between Ramón y Cajal and Camillo Golgi, co-Nobelists, was whether the neuron network was an interconnected cytoplasm, or whether each neuron was an independent unit, with the cell membranes spatially separated at the synapses. Golgi claimed that the cytoplasms intermingled, creating a continuous thread of liquid networking among the neuron cells. Cajal in contrast proposed the *neuron doctrine*. The main tenets of the neuron doctrine were: (i) that neurons are fundamental and distinct units of the brain, (ii) the *ionic hypothesis* claiming that ion-migration processes are responsible for the membrane potentials which drive signals in one direction, and (iii) that chemicals are involved in signal transfer across the inter-neural spaces. It was just these inter-neural spaces (synaptic clefts) that Golgi refused to accept. It was only in the 1950s, with the advent of electron microscopy, that Sandford Palay and George Palade showed that Cajal's intuition was dead on.

The electrical impulses used by the neurons are called *action potentials*. The study of action potentials began in earnest with the work of Bernstein, a student of Helmholtz, circa 1906. These signals travel from the neuron to the axons and presynaptic terminals, at speeds ranging from a few meters/sec to $\sim$100 meters/sec, i.e., $\sim$270 miles/hour. They are 'all-or-none' signals somewhat like ones and zeros used in a binary-bit of information. But the chemical signal used at the synapse is a graded one, depending on the amount of chemicals released, the opening of the ion channels and receptors taking up the chemicals. This latter feature turns out to be very important for *memory*.

It should be remembered that channels in cell membranes are structures in the lipid bilayer mediated by large molecules (e.g, amino acid chains) which traverse the membrane. The most common ion channels in cell membranes are Na^+ or Ca^{++} channels. A channel may be opened up (activated) by a change in the membrane potential. Such channels are said to be voltage-gated. Other types of channels are activated by specific molecules (ligand molecules) that bind to the channel. They are known as ligand-gated channels. When the ligand molecule binds to the protein molecule at the channel, the protein changes its shape, opening up the channel. Nature has re-adapted some of the chemicals available in the prebiotic medium as neurotransmitters and messenger molecules (see Tables 12.1, 12.2) for use at the synapses to send signals as well as to modulate signals.

When Na^+ channels are closed, the ions cannot enter the cell. The salt concentration in the cytoplasm (i.e., inside the cell) is smaller than in the exterior. Thus the interior is negatively charged relative to the exterior (which is taken as the reference zero potential). This voltage difference across the cell membrane is known as the *resting potential* which is about -65 millivolts (-65 mV). When the intercellular region changes due to the presence of ions or molecules released into the synaptic cleft, the membrane potential changes. A change of the membrane potential from -65 mV to -75 mV is called *hyperpolarization*, while a decrease in the magnitude, e.g., from -65 mV to -55 mV is called *depolarization*. Once the cell is depolarized, it is able to take up ions and generate an action potential. Hence depolarisation is an excitatory condition. Similarly, hyperpolarization prevents the generation of action potentials, and inhibits the cell. The movement of Na^+ into the cell (via Na^+-ion channels), leads to a rapid increase in the potential, followed by movement of some K^+ ions (via potassium channels) into the cell which drops the potential down. This is similar to the charging and discharging of a capacitor. The outer and inner surfaces of the cell membrane hold layers of charge, just as in the plates of a capacitor. This process of charging and discharging generates a 'spike-shaped' or pulse-like action potential across the cell membrane. The spike shaped potential, carefully measured by Lord (Edgar) Adrian is about 100–120 mV. It travels along the neuron at 1–100 m/sec, tarrying for about 1–10 milliseconds at any location on the axon. Allen Hodgkin and Andrew Huxley pioneered ($\sim$1939) excellent quantitative theories of these electrochemical processes, supporting them with well-conceived experiments that used the millimeter-sized axons of giant squids to stick electrodes directly into neurons. Hodgkin and Huxley shared the 1963 Nobel prize in Medicine, for their work which established the details of the 'ionic hypothesis' regarding the flow of cell currents.

The strength of the action potential, i.e., the 'height' of the spike, is preserved as the signal travels 'down' the axon, i.e., towards the synapse, due to continual strengthening by the cell membrane, by transfer of ions across ions channels in the axon. In fact, as shown by Hodgkin, the action potential is strong enough to jump

Table 12.1 Some neurotransmitters and related molecules — I. See Sec. 12.5 and Table 12.2.

Substance	**Details**
Acetylcholine	A neurotransmitter formed in the *nucleus basalis Meynert* and distributed throughout the brain. It is believed to suppress excitatory synaptic connections. Lack of acetylcholine could lead to excessive connections as in Alzheimer's disease.
Glutamate	Used by sensory neurons for fast chemical messaging. Glutamate binds to ion channels (ionotropic receptors), opens ion channels and allows flow of ions into the postsynaptic cell. Glutamate also activates NMDA ionotropic receptors. It is an ubiquitous, fast (milliseconds), 'first type' of messenger, important in *short-term memory* etc. identified 1950s
GABA, γ-amino butyric acid	'Brain has its own Valium'. Mostly an inhibitory neurotransmitter. Some 20-30% cortical neurons are inhibitory.
Glycine, simple amino acid	Glycine is an inhibitory neurotransmitter active in the spinal cord, brainstem, and retina. When glycine receptors are activated, chloride enters the neuron via ionotropic receptors, generating an inhibitory (hyperpolarizing) potential.
Norephinephrine	This neurotransmitter, works with adenyl cyclase, Ca^{++}-calmodulin complex, cAMP and PKA to generate more neurotransmitters. Relevant to memory processes like *long-term potentiation* (LTP) via neurons in the hippocampus. Ephinephrine (adrenaline) is an important hormone.
Dopamine	Used by neurons in the mid-brain; the lack of dopamine is associated with Parkinson's disease. High dopamine transmission has been proposed in theories of psychosis and schizophrenia.
NMDA, N-methy-d-aspartate	Associated with a class of ionotropic channels normally blocked by Mg^{++} ions. These receptors come into play when a conditioned reflex (CS) and an unconditioned reflex (US) are properly paired, as in associative memory (*classical learning*). LTP also involves a Ca^{++} influx, post-synaptic NMDA receptors as well as a non-NMDA receptor. The maintenance of LTP requires presynaptic activation via NO molecules.
Serotonin or 5HT (5-hydroxy-tryptamine)	Works by activation of the enzyme adenyl cyclase, cAMP and PKA. Serotonin binds to specific receptors in the cell membrane. Essential to *short-term memory* of sensitization to a stimulus, and to other types of memory. Any excess serotonin at a synapse is re-absorbed by the cell. Prozac and other drugs inhibit the re-uptake and make the already produced serotonin available to the cell for a longer time.

over an anesthetized segment of an axon and trigger an action potential further upstream. We are reminded of a signal traveling in a fiber-optic cable which is continually replenished by repeaters and amplifiers which correct for the decay of the signal. It is somewhat in the same manner that a visual sensation, a motor movement or moment of memory is sent along an axon from one part of the body to another! Indeed, if ion channels in neuron membranes fail to work, various neurological diseases (epilepsy etc.) arise. Modern understanding and treatment of these illnesses are directly based on this basic scientific knowledge.

Table 12.2 Some neurotransmitters and related molecules-II. See Sec. 12.5 and Table 12.1.

Substance	Details
Adenyl cyclase enzyme	Also named adenylyl cyclase. This enzyme generates cAMP from ATP molecules. It binds to the Ca-calmodulin-adenyl complex. Serotonin acts more easily with the Ca-calmodulin-adenyl complex, producing more cAMP, and more PKA. Hence more transmitter is released. Adenyl cyclase plays an essential role in *associative memory*, LTP etc.
cAMP; cyclic adenosin monophosphate	The first of the 'second-messenger' molecules. When neurotransmitters act on receptors which are not ion channels (metabotropic receptors) and bind, second-messenger signaling molecules inside the cell are activated. cAMP works by activating PKA. It is important in biological energy transformations as well as memory. Second-messengers can initiate a cascading set of reactions in the cell. Their action persists for longer times than via ionotropic receptors. cAMP molecules are used even in eukaryotes for cell signaling and through out evolution. It is not unique to memory, synaptic plasticity etc.
PKA, Protein kinase A	A phosphorylating agent, found localized in pre-synaptic terminals and cell membranes. It closes down K^+ channels, increases Ca^{++} influx, modifying action potentials. This is a protein which can bind to cAMP and change shape, releasing catalytic molecules which phosphorylate proteins. Under repeated training processes, there is enough time for PKA to relocate to the nucleus of the neurons and activate genetic processes involving sustained growth in the neuron. This is basic to *long-term memory*.
MAP kinase; mitogen-activated protein kinase	Another second-messenger system, acts together with cAMP. Important in memory processes.
Calmodulin, a protein	Some of the Ca^{++} ions which flow into a presynaptic terminal under the influence of an action potential bind to calmodulin and then to adenyl cyclase.
NO nitric oxide gas	This is a retrograde messenger molecule that carries information back to several presynaptic termini by diffusion from the post synaptic cells. The enzyme NO-synthase forms NO from *l*-argenine. Nitrogen oxide is believed to be critical for maintaining LTP.
CREB-1, 2; acronym for *c*AMP-*R*esponse-*E*lement-*B*inding protein	The PKA phosphorylates CREB-1 and, acting on DNA, switches the genes needed for growth of anatomical features associated with long term memory. CREB-2 is controlled by MAP kinase, and can be used to inhibit CREB-1.
CPEB cytoplasmic-polyadenylation-element-binding protein.	A prion-like protein that assumes the 'dominant', self-perpetuating form when subject to serotonin. The dominant CPEB form activates m-RNA and engages protein synthesis needed for anatomical changes in *long-term memory*.

LTP: long-term potentiation.

12.5 Molecular mechanisms of memory

The brain is a specialized organ full of neurons and their connections. What happens in these neurons when we hear, see or touch some thing? How do we learn about it and transfer the knowledge to our 'memory'? How do we recall it? Philosophers had speculated on memory from antiquity, and connected it with various theories of consciousness, due to the lack of a corrective hand from experimental testing of their ideas. While experimental scientists had looked for suitable models, it was only in the late 19th century that microscopy, staining techniques and basic knowledge in physiology converged to enable a very prescient individual like Cajal to make some clever and fruitful suggestions.

Psychologists had proposed that there are several types of memory. Freud and others proposed conscious and subconscious processes. Today, based on modern experimental work, two main types of memory are recognized. One of them is known as *non-declarative memory*, or 'implicit memory'. It is sometimes known by the name 'memory without record', or reflex memory. Thus our knowing of *how* to ride a bicycle, or serve a tennis ball are skills that we perform without having to 'recall' the needed actions 'blow by blow'. We have learned the skills, and they are in our memory. The other type of memory is known as *declarative memory*, or 'explicit memory', 'episodic memory', 'reflective memory', or 'memory with record'. This is the memory of events, of people, faces and facts that we can recall. This type of memory was known as the memory of *what* as opposed to the memory of *how*, by Gilbert Ryle. We can also divide any type of memory into 'short-term' (or 'immediate') memory and 'long term' memory. Short term memory is well known to become long term memory by repetitive training. Training can also be associative, as in unconditioned reflexes (UR) becoming conditioned reflexes (CR), if stimuli are received together, within short overlapping spans of time.

Wilder Penfield, a neurosurgeon working in Montreal (~1938) identified specific sites in the brain related to language comprehension, speech etc. His work suggested that memory may be connected with neurons of the temporal lobe. The earlier view was that the brain acted as a mass, with no discriminating parts, with memory being proportional to the amount of tissue involved. This featureless 'mass-action' model could not be sustained. Brenda Milner, working in the Penfield Institute commenced her now world famous experiments (~1957) on an amnesic patient known only as H. M. The studies on H. M. enabled Milner and her colleagues to establish the following.

(i) Injuries to medial temporal lobe structures, including the hippocampus, separated the immediate memory from long-term memory. Such patients were unable to lay down new memories, although they could recall events prior to the injury.
(ii) Although these injuries to the temporal lobe (Fig. 12.4) affected memory profoundly, they had no effect on the patient's perception and the intellect.

(iii) The importance of the hippocampus etc., for memory disproved the previously held ideas of 'mass-action' and non-specificity of memory location.
(iv) The experiments of Milner *et al.* established experimentally the distinction between declarative and non-declarative memory. For instance, although H. M. could not lay down declarative memory (i.e., remember any events), he could acquire skills (i.e., form non-declarative memory).

One of the skills he learnt was to trace between two outlines of a star while viewing it in the mirror. He improved his skill (i.e., formed a memory) at each day of practicing the skill, but failed to recollect his having done it before.

Taking his cue from the use of large neurons of the giant squid by Hodgkin *et al.*, Eric Kandel and his associates launched (~1960 s) a program of research which elucidated the molecular mechanisms of memory [89]. Thus the sea slug *Aplysia* was introduced as one of the new experimental animals of physiological research. A fascinating and highly readable account of this field has been given by Kandel in his autobiographical work *In search of memory* [51]. Many workers from many disciplines joined in, armed with new tools from DNA research, genetically modified mice, fruit flies, neural-network theory etc., to create, already by the end of the 20th century, a remarkable body of knowledge that applies to invertebrates as well as to humans.

12.5.1 *Habituation and sensitization*

Non-declarative memory is essentially the learning of skills. Hence this was one of the earliest topics studied using invertebrates like *Aplysia*. If the siphon of the sea slug is touched, it immediately contracts it and also withdraws the gill to protect these organs. However, if the touchings were repeated, it gets habituated to the touching stimulus, and does not show the withdrawal reaction. The nature of *habituation* was one of the earliest memory processes elucidated at the synaptic and molecular level.

The neuron connections for the gill-withdrawal reflex were set in place during the early development of the sea slug. When the siphon is touched, the sensory neurons on the siphon excite interneurons (inter-mediating neurons). Their signals as well as the initial input of the sensory neurons combine and the action potential reaches the motor neurons of the gill, and causes the withdrawal of the gill. If the stimulus (touching) is repeated, habituation occurs. Then, when the sensory neuron is stimulated (by touch), the resulting action potential is found to have become very small. Hence the synaptic connection, whose strength depends on the input action potential, also becomes very weak. At the molecular level, this was reflected in a decrease in the *number* of vesicles of neurotransmitter molecules (Table 12.1) released at the synapse. Electron microscopy showed that short-term habituation did not alter the number or size of the active zones in the presynaptic terminals. Even the total number of vesicles did not decrease, but only the number that was docked for release, or ready to fuse with the cell membrane had been reduced. The effect of long-term habituation could also be studied using the same experimental techniques. While short-term memory involved changes in available neurotransmitter vesicles in the pre-synapses, long-term habituation involved a reduction in the number of synaptic connections. Thus the synaptic plasticity involves both modifications in the amount of chemicals spewed out in signaling, as well as changes in the shape and structure of the synapses. Also, the same basic 'hardware' is used for both types of memory.

What molecular mechanisms lead to changes in vesicle availability or modification of synaptic structure? Such questions were elucidated by a study of more complex memory processes. The opposite of habituation is sensitization. For instance, if the sea slug had been sensitized by an electric shock, its reaction to a touch is a very quick, strong withdrawal. A single shock to the tail sensitizes the sea slug to any touching of the siphon and lasts for a short time. However, if the electric shock were repeated four or five times, a long-term sensitization lasting several days is produced. A shock to the tail excites sensory neurons that activate interneurons that couple with the siphon skin and the gill-withdrawal circuit. The intermediate neurons (which use serotonin and other 'modulatory neurotransmitters') act to increase the number of available neurotransmitter (glutamate) vesicles released at the motor neurons to activate gill withdrawal.

Neurotransmitter molecules work by binding to receptor molecules in the membrane of the post-synaptic cell. We have already encountered the opsin molecule which has a space for the retinal molecule. The receptor molecules of interest here have a structure which gives rise to a passage or channel through the membrane. The passage opens up when a neurotransmitter like glutamate binds to it, allowing ions to pass through. These are channels associated with *ionotropic receptors*, first described by Bernard Katz and colleagues in the 1950s. These use a commonly encountered type of synaptic action which occurs fast and lasts a few milliseconds. Another type of receptor found in cell membranes does not have ion channels. When a neurotransmitter binds to them, the ends of the receptor which are inside the cell act to modify the concentration of molecules inside the cell. They act using *second messengers*, or intracellular messengers. They are called second messengers as the transmitter molecule which binds to the receptor on the outer surface of the cell membrane is regarded as the 'first messenger', or 'extracellular' messenger (e.g., glutamate or serotonin). Since these receptors influence the metabolic pathways of the target cell, they are known as *metabotropic receptors*, first discussed by Earl Sutherland, Paul Greengard and their colleagues, almost a decade after the discovery of ionotropic channels.

The first example of a second messenger turned out to be cyclic adenosine monophosphate (cAMP), a molecule well known to be a a part of the primordial arsenal of all living cells including bacteria (see Table 12.1). It is also involved in the Krebs cycle fundamental to cell respiration. Metabotropic receptors, when activated by a first messenger, use the enzyme adenyl cyclase to make cAMP. The latter is a versatile agent affecting many cell processes by activating the ubiquitous phosphorylating agent known as PKA, i.e., protein kinase A. These linked processes enable a second messenger to amplify the signal received at the synapse by orders of magnitude, and put the cell into a distinct metabolic state. Kandel and colleagues were able to establish experimentally that the second-messenger mechanism, with serotonin as the first messenger, was responsible for short-term memory by sensitization.

12.5.2 *Classical conditioning*

Habituation and sensitization involve memory regarding a single stimulus. A more complex form of learning and memory involving the association of two stimuli is known as *classical conditioning*.

Everyone has read about Ivan Pavlov's dogs who begin to salivate at the sight of the attendant who feeds the dogs, even before any food is presented. The attendant is initially a neutral (ineffective) stimulus which becomes the *conditioned* stimulus (CS), while the food is the spontaneously effective *unconditioned* stimulus (US). Normally, the appearance of food produces salivation, while the attendant's appearance does not. But after repeated association of the attendant and the food, they become paired, and the sight of the attendant is enough to produce salivation (conditioned response CR). Thus, the dogs have learnt to associate the CS and the US. Detailed experiments with snails, slugs, flies, mice, rabbits and humans show that there is an optimal time interval between a CS and a US that allows the formation of an association between the CS and the US. The optimal effect is obtained when the CS precedes the US by a short interval ($\sim$0.2 to one or two seconds), but terminate together. If the US precedes the CS, no conditioning results. The sense of 'cause and effect' that is ingrained in us may be closely associated with the molecular mechanisms of conditioning.

The process of behavioural conditioning requires that the CS and the US must excite the same sensory neurons one after the other, and within a short time interval. preferably with a slight overlap. When classical learning has occurred, the modulatory neurons (interneurons) excite the sensory neurons with the correct timing. That is, the CS (e.g., a touch to the syphon of the sea slug in Kandel *et al.*) comes first, and the US (electric shock to the tail) comes just after, leading to the conditioned response (CR), i.e., gill withdrawal in the experiments on the sea slug. Experiments circa 1980 showed that when the CS and the US are correctly paired, the sensory neurons have a greatly enhanced transmitter release than with mere sensitization. The molecular mechanism behind this occurs partly in the presynaptic stage, and partly in the post-synaptic stage.

The arrival of an action potential in the presynaptic terminals leads to an influx of Ca^{++} ions. Their usual action is to release transmitter vesicles. In addition, Ca^{++} binds to a protein known as Calmodulin (Clm) to form Ca-Clm which in turn binds to the enzyme adenyl cyclase (Adc) present at the cell membrane, and needed to form cAMP from the ATP molecules present in the cytoplasm. However, the Ca-Clm-Adc is sensitive to the presence of serotonin. Thus, if serotonin is pumped in just after the formation of Ca-Clm-Adc, more cAMP and more PKA are formed, generating highly enhanced glutamate-transmitter release. Thus the classical learning mechanism at the presynaptic stage involves the timed arrival of serotonin to act on the Ca-Calmodulin-Adc complex.

An additional contribution to classical conditioning comes from the post-synaptic cells. The glutamate arriving at its receptors may bind to the usual ionotropic receptors, or to a class of receptors known as NMDA (N-methy-D-aspartate). This receptor is normally blocked by Mg^{++} ions. However, if the CS and US are correctly paired temporally, then the resulting action potentials act to release the Mg^{++} ions, opening up the NMDA receptors, and allowing Ca^{++} to flow through. These ions produce small molecules like nitric oxide (NO) and carbon monoxide (CO) which diffuse back to the presynaptic region and stimulate the further release of neurotransmitter molecules. The NO molecule is a *retrograde messenger* that returns to the presynaptic region to create a feedback effect.

Classical learning or conditioning can be a short-term memory effect, or it may persist for days and weeks. In long-term memory, the changes in synaptic strengths last longer. In addition, the actual numbers of active synapses also change. In long-term habituation, the number of presynaptic connections among sensory neurons and their target neurons (say, motor neurons) decreases, while in sensitization new connections are formed and persist as long as the memory is retained. Thus, long-term memory involves anatomical changes, which arise during repetitive training. The periods of repetitive training are found to be accompanied by the synthesis of new proteins and even prion-like molecules which generate the new anatomical changes.

12.5.3 *Declarative memory*

Eric Kandel, summarizing some of his studies on memory stated [51] that 'in its simplest forms, learning selects among a large repertoire of pre-existing connections and alters the strength of a subset of those connections'. In the previous sections we have seen how this works in detail in regard to several forms of non-declarative memory or 'reflex memory'.

Declarative memory or 'reflective memory', unlike 'reflex memory', is not an involuntary execution of a set of learned skills. It involves the ability to 'remember' episodes, faces, songs etc., from the past. However, memory does not work like a recorder or camera which faithfully store events for later viewing. Instead, remembering (i.e., retrieval) involves a continually new synthesis of a whole story from available bits and pieces of memory. The retrieval process is not passive. The memory is by rehearsal and retrieval, as well by the acquisition of new 'suggestive' material that fit into the memory.

As we noted in regard to Brenda Milner's studies (see Sec. 12.5) of H. M., studies of *amnesia* have been profoundly important in understanding the nature of memory. Any type of damage to the medial temporal lobe or associated areas like the amygdala and the hippocampus may produce amnesia. Another feature of declarative memory is that injury to the hippocampal region not only inhibits new learning, but also impairs some of the memories acquired prior to the injury. This phenomenon is known as *retrograde amnesia*. In Alzheimer's disease, degeneration first appears in the medial temporal lobe. An amnesic patient like H. M. can hold immediate information, but cannot convert it into a persistent memory. This 'forgetfulness' is a condition independent of other cognitive functions or problems like, say, mental depression. However, the experiments that began in the 1950s with Milner's work, as well as later experiments on animals have convincingly proved that *memory is a separable and distinguishable function of the brain* in much the same way that digestion is a function of the digestive organs.

Animal studies took advantage of the fact that declarative memory is not merely 'conscious recollection', but involves other aspects. For example, the ability to match recently encountered shapes, or recognizing and distinguishing between different objects, remembering spatial details of mazes, recognizing odours etc., have been used in animal studies. Indeed, the pioneering studies on *place cells* in the hippocampus were carried out on mice [105]. Here it was found that an animal's sense of location in space was coded in neurons of the hippocampus. Such neurons are called 'place cells'. Groups of cells, and not just one neuron, are involved in mapping the location, as an animal visits an environment and becomes familiar with it, via visual, tactile and other clues transferred to its memory. It is precisely the memory in these place cells that we use when we drive around using 'familiar spatial clues'. Of course, as we discuss below, such information is gradually transferred from the hippocampus to the cortex for long-term memory.

Specific genes in an animal's DNA can be 'knocked out', or included, to produce genetically modified animal lines which are incapable or capable of synthesizing a given protein. Hence it became possible to use mice, rabbits etc., to study memory at the level of molecular genetics as well.

Unlike non-declarative memory, the context of the sensory information both in time and space has to be available in declarative memory. This type of memory is the cohesive element which defines a person's sense of individuality and his place in the world as perceived by him. The molecular mechanisms behind declarative memory have to involve links and clues which associate different bits of information and build the context. The first understanding of the nature of this associative process came from the work of Norwegian researchers Tom Bliss and Terje Lømo in 1973. They showed that if a brief strong high-frequency electrical signal were applied to a neural pathway in the hippocampus which is also receiving a weak action potential, an increase in the synaptic strength occurred, and lasted for hours or even days. The weak action potential by itself would have produced little or no synaptic strengthening. This increased synaptic strength (facilitated by the burst of high-frequency signal) could last for many weeks if the process were repeated even a few times. This type of strengthening of memory came to be known as *long-term potentiation*, LTP, or 'long-term facilitation', and provided a first working model of memory retention. In fact, Bliss and Collingridge [10] in their well-known review article state the following:

Long-term potentiation of synaptic transmission in the hippocampus is the primary experimental model for investigating the synaptic basis of learning and memory in vertebrates. The best understood form of long-term potentiation is induced by the activation of the N-methyl-D-aspartate receptor complex. This subtype of glutamate receptor endows long-term potentiation with Hebbian characteristics, and allows electrical events at the postsynaptic membrane to be transduced into chemical signals which, in turn, are thought to activate both pre- and postsynaptic mechanisms to generate a persistent increase in synaptic strength.

The LTP idea led to further studies of the neuronal processes in the hippocampus which was well appreciated to be important for long-term declarative memory. Furthermore, the neural pathways of the hippocampus were rather well known, and enabled researchers to activate neurons via different pathways. These experiments showed that in LTP, connections between neurons that share information are strengthened in processes which are active at the same time. Such time-associative processes had already been studied in the context of Pavlovian conditioning (see Sec. 12.5.2) in regard to non-declarative memory. Neuron cells from sliced sections of the hippocampus were studied (as was done with the California sea slug), and the molecular mechanisms of LTP were established. If only a simple action potential were at the synapse, the usual glutamate process at an ion channel occurred. However, if the facilitating signal were present, it acted on the NMDA receptors opening them by expelling the Mg^{++} ions which usually block them in the absence of excitation. Thus, as already discussed in the context of the conditioned reflex, Ca^{++} ions rush into the cell via the NMDA channels and activate the Ca-Calmodulin dependent adenyl-cyclase pathway. This generates increased cAMP, and increased PKA which activates protein pathways. The facilitating neurotransmitter signal in the Sea slug *Aplysia* was serotonin liberated by an intermediate neuron. In the case of the hippocampus in humans, mice and other vertebrates, the interneurons use norephinephrine as the excitatory signal.

Another surprising development was the evidence that the postsynaptic cells subject to LTP could generate nitrogen oxide (NO) molecules. These small molecules could simply diffuse out of the cell, and some would diffuse back to the presynaptic cell. There it would have the effect of increasing the amount of liberated neurotransmitter. Thus NO, and possibly the molecule CO have been proposed as *retrograde-signalling molecules* as they carry a feedback signal in the direction opposite to the flow of action potentials. This feedback messaging enables the synaptic strengthening by LTP to be maintained for a longer time.

Given that the chemical mechanisms of LTP and classical learning of skills are so similar, it is of some interest to examine the actual differences in the way they are implemented in the brain. From the experiments of Brenda Milner on the patient H. M., we learn that the learning of skills (non-declarative memory) does not involve the hippocampus, but requires repetition of the skill to be acquired. The information is believed to be directly processed in the synapses of the cortex, with the synaptic strengths increased by repeated training sessions. On the other hand, how do we acquire long-term declarative memory, seemingly without training — sometimes with just one striking incident? It appears that the hippocampus plays a very interesting intermediary role here. It picks up the information by LTP and then repeatedly *downloads* the information to the cortex, where each such downloading becomes a training session!

12.6 Neurons and information processing

Neurons do not process energy, but they use energy to process *information*. We are familiar with computers where information processing units are distinct from memory units. One type of memory used in computers is to convert the information into bits (yes or no signals), and store them as magnetized (or demagnetized) locations known as *addresses* on a matrix of ferrite. The computer uses the stored memory and processes information in a central processor unit (CPU), by very rapid shuffling of binary units electronically, and in a serial manner. Parallel processing of information is achieved only by having a set of CPU's working in parallel. Computers do not tolerate any errors what so ever. Hence computers can carry out error-free arithmetic operations (e.g., multiplying two very large numbers) with extreme rapidity. In contrast, the brain is very slow and inaccurate. The brain is not a standard computer or calculational device, but a very sophisticated device for *pattern recognition*. The brain is said to function by 'parallel distributed processing' of information, as opposed to pixel by pixel processing. Neural networks (see Sec. 2.2.5) also attempt to emulate this idea, since all inputs are treated by all the nodes of the net.

The brain seems to use each bit of information as its own address, and the processing is also done right there. According to von Neumann [111], the reliability of a neuron, as compared to a silicon chip is many *orders of magnitude* smaller. If a single memory chip in a computer were to fail, the computer itself would fail. In contrast, neurons fail to work quite often. This does not seem to matter because neurons work in a neural network involving a very large number of neurons. Such networks can function even if many nodes (neurons) fail to work. Thus there may be only a minor reduction in the cognitive abilities of people who have very limited amounts of cortical matter compared to normal people [54] who have billions of neurons. The popular literature is full of unconfirmed and sensational claims about people with little or no brain matter 'living normal lives'.

A study of the neuron networks in the brain showed that nerve cells are connected in specific ways. A given neuron will always connect with certain other sets of neurons, excluding others. In fact, modern molecular genetics has clarified that the basic 'wiring-diagram' of the connections is coded in the DNA of an organism. It is determined by the genes that are selected for expression in the genome of the organism. However, although the neural connections are preset by 'heredity' and remain essentially fixed, the synaptic strengths and connections change during learning and memory formation. In previous sections we discussed the molecular mechanisms underlying such synaptic plasticity.

In fact, more recent studies have shown that the concept of synaptic plasticity has to be extended to include *neuroplasticity*. The view that the neural system in the brain is essentially established at birth, and that 'only some minor developments could occur after that', was the traditional view. This has undergone revision since the last decades of the 20th century [54]. Although earlier animal studies had suggested the existence of neuroplaticity, the work of Michael Merzenich and colleagues provided more definitive evidence in the case of humans (for a discussion, see [88, 64]. Merzenich *et al.* studied the response of patients to artificial inner-ear implants. Although initially the patients only heard rumbling noises, within a year they learned to recognize spoken language via their artificial ears. The artificial ear works with a microphone which picks up sounds, and sends the corresponding electrical signals to the skin of the patient. These are picked up by wires implanted in the patient's cochlea (inner ear). The electrical signals from the artificial implanted system are completely different to the signals the patients would have received from a biological ear. We already noted that the input sensory information is supplemented from the brain itself, and only about 20% of the total signal is the primary stimulus. The supplementary input from the brain (i.e., 'in-sound', analogous to 'in-sight' in the case of vision) would be based on information accumulated since infancy. The spatial and temporal extent, as well as the frequency range of the primary stimulus, generated from the artificial ear would hence be ridiculously mismatched with the input from the brain. In consequence, the brain would fail to 'understand' the input signals. However, after about a year of 'learning' with the artificial ears, the situation changed, and individuals could even carry out a telephone conversation! This clearly shows that very significant changes have occurred in the neural circuits in the brain. The cortex (i.e., the neural network of the cortex) has not only changed the weights at its nodes (i.e., changed the strengths of the synapses), but had also grown completely new neural connections (rewiring) to deal with the new structure of the inputs. This is neuroplasticity at work.

The *phantom limb* phenomenon has been recognized since 1871. It was reported by W. Michell, from clinical studies of patients having the subjective feeling that amputated body parts are 'still present'. Modern studies of this phenomenon have indicated that novel paths for cell signaling can develop, both in the cortical region, and also in the regions close to the sensory nerves themselves. For instance, the cortical region that usually deals with the signals from the hand ('hand region') would no longer receive those signals once the hand is amputated. If the left hand is missing, the left-hand region in the cortex has the corresponding left-face region, and the left-arm region on the two sides of the hand region. Ramanchandran (1992) and others were able to show that the neurons in the adjacent sides had formed new connections into the 'hand region' of the cortex which had become starved of inputs, i.e., the *deafferented region*'. Hence, Cajal's concept of plasticity applies not only at the level of synapses, but at the level of axons, dendrites, and perhaps at the level of new neurons as well.

12.6.1 *Types of memory and types of consciousness*

In this chapter we have examined how special adaptations of cells under evolutionary pressure led to the emergence of information-processing sensory organs like the eye and the brain. We have also indicated that as the number of neurons in an organism increases beyond a critical value, it becomes conscious of its environment in a pro-active way.

It is clear that the sensory capacity of prokaryotes and eukaryotes gave them the character of replicating cellular automata whose functions were related to the environment of the automata. We have identified these replicating cellular automata as living organisms. However, we do not normally say that such simple organisms have *consciousness*. The latter term is normally reserved for higher organisms having a nervous system and having the capacity to process information in a non-trivial way (i.e., in problem solving). This distinction may be challenged by pointing out that the special cells of the nervous system of simple organisms use chemical mechanisms closely similar to those of more complex organisms. We may also remark that the simpler organisms do with their cytoskeletons and other organelles, what the more complex organisms do with their specialized systems. That is, it may be argued that *consciousness* is a gradually emerging property found in all life. A more extended view of this, embracing even inanimate matter is known as *pan-consciousness*, or *panpsychism*. This view will be discussed in some detail in Sec. 13.1.1.

On the other hand, the existence of special emergent properties at each level of complexity, associated with different energy scales and length scales, is a characteristic of the physical world. Some critical number of elementary interacting objects are needed to show such emergent properties. How a many-body system shows various types of order (e.g., ferromagnetism), or specific properties, is well understood in physical theory. For instance, a critical number of gold atoms is needed in a gold crystal before it shows metallic behaviour. Interacting-neuron systems are no different. In the Darwinian model there is, in addition to *ordering*, another element, namely *organization*. Different types of information processing involve different levels, or types of modular organization. These arise from the evolution of corresponding specialized structures and modular processes in the brain.

In our discussions of memory, we have considered declarative memory (explicit memory), and non-declarative memory (implicit memory). We delineated the brain structures and molecular mechanisms associated with these two kinds of memory. It is natural to expect that we also have an implicit consciousness and explicit consciousness. The implicit consciousness deals with skills, heart-beat, emotions etc., which are in fact *unconsciously carried out*. So it is known by names like the subconscious', the 'unconscious', or the *adaptive unconscious* [107]. Since it is directly associated with implicit memory, there is good reason to call it the *implicit mind*.

William James, already in the 19th century was one of the earliest persons who recognized the importance and possible dominance of these implicit mental processes over the explicit consciousness. In this book we associate the explicit consciousness mainly with declarative memory, or explicit memory. However, William James pointed out that decisions made by the unconscious mind are posted onto the explicit consciousness *post hoc* and we become *aware* of them only then. According to James, 'we meet a bear, run, and *then* experience fear that played no role in our fleeing'. A driver who suddenly brakes his car in response to a cat which suddenly jumps into the road is also acting under the control of his adaptive unconscious. He becomes aware *post hoc*, that he has taken evasive action. Similarly, when you see your old friend Peter that you had not seen for a while, you decide to shake hands and extend you hand. But in fact, the adaptive unconscious has *already* made the decision, and the hand is already moving into the shaking posture by the time the explicit consciousness registers the volition 'I am going to shake hands with Peter'. Functional NMR and other techniques have largely established what William James proposed in the closing years of the 19th century!

Some 10 million (10×10^6) nerve signals are found to arrive in our brain in a second [72]. The conscious mind is able to process less than a 100 such signals ($\sim$.001%) in that time! It is evident that the adaptive unconscious is a secret back-room boss who is in charge most of the time. In reality, 'you', and your explicit consciousness are merely the spokespeople, having the 'power of attorney' for assuming responsibility, 0.001% of the time.

12.6.2 *Bit capacity of the brain and the genome*

In the early stages of evolution, organisms were simple and had only a few neurons. They could be used to hard-wire set behaviours, just as early electronics could do simple set jobs, but did not have adaptive feed-back loops or large banks of memory. The set behaviours were transmitted to the off-spring by the inherited DNA. That is, automatic protocols for carrying out a sequence of chemical reactions can be configured via the DNA into the organelles of the cells. These produce the inherited reflexes of an organism.

The sea slug *Aplysia*, used by Eric Kandel and collaborators [51] for pioneering work in neuroscience, contains some 20,000 neuron cells. When there are many neurons, the number of possible interconnections also increases rapidly. Thus, in a human with ten billion (10^{10}) neurons in the fore-brain, and typically a thousand connections per neuron to other neurons, there are some 10^{13} connections. Each such connection between two neurons could be off or on (a binary unit or *bit*, see Sec. 7.2.1), or have a graded response (multi-state representation). Thus, even if each connection were a single bit, the 10^{13} bits, i.e., 1.25×10^{12} bytes represent some 1,250 GB (Gigabytes) of memory, since each byte is 8 bits. This much information cannot be transferred via our genomes, with $\sim 3 \times 10^9$ base pairs in it. Since each base can be one of 4 (i.e., 2^2), i.e., two bits, the six billion bits of the human genome corresponds to (6/8) billions bytes or $\sim$750MB. Thus the fore-brain alone has a lot more information capacity than the genome. This shows that even if we inherit the genomic information into our brain, there is still room for *learning*, memory and other inter-neuronal activities which go well beyond reflex action. By exploiting the extra capacity for handling information, the higher vertebrates evolved the collective-mode behaviour of groups of neurons that manifest itself as 'consciousness'.

As we already noted, the above discussion of the bit capacity of the brain, and of the genome assumes information processing models based on each elementary unit, and ignores graded or multi-scale representations, as well as collective processes. However, a set of neurons may work together as a primitive neural net or perceptron (Sec. 2.2.5) and do much more than what we have estimated in the above analysis. That is, the bit capacities of the brain and the genome given above should be regarded as very conservative lower-bounds.

Chapter 13

Exotic, Quantum Explanations of Consciousness

We examine conjectures that attempt to exploit properties of quantum systems, quantum or classical coherent processes, or non-equilibrium, driven states to model 'consciousness'. These involve ideas on non-computability, quantum gravity, tubulin-based quantum computations, phase-correlated quantum states, collapse of wavefunctions, driven non-equilibrium condensates, quantum fluctuations, quantum entanglement and so on. These conjectures have insurmountable difficulties.

13.1 Do we need exotic explanations?

In previous chapters we have traced how the fundamental laws of physics apply as we change length scales and energy scales to bring us to the world of familiar everyday objects. The laws of physics, transforming through quantum chemistry, gradually become the laws of chemistry and kinetics of many particle systems. We showed that the usual 'paradoxes' of quantum mechanics disappear when many-particle effects, finite-temperature effects, and the boundary conditions of the actual physical systems are adequately included in the discussion (see Sec. 8).

We examined how the dissipation of energy from the sun, and from the locked-in energy of the earth drove pre-biotic matter into hydrothermal hatcheries, metastable micelles in Darwinian ponds, zillions of foam bubbles, lipid bilayers etc., that are like trial cells for increasingly complex replicator chemistry. The primordial cells became little containers which behaved like primitive compost pits. They evolve new pathways for the dissipation of the energy in wet, salty, mineral-catalyzed, methylated, ammoniated matter pushed out of equilibrium by the rays of the sun or the energy of the vents. Those trial cells that developed replicators grew in numbers. They became even better adapted to replicating and responding to the environment. This process may be thought of as Darwinian evolution, or the spontaneous working out of a Metropolis-Teller type algorithm (see Sec. 2.4.7) in an energy landscape maintained out of equilibrium by the energy-pumping action of the sun and the earth's core.

The lipid-bilayer-based cells evolved into complex multi-cellular organisms capable of monitoring and sensing their environment and also processing information. The sensory capabilities of cell membranes further evolved, generating specialized sensor organs like the eye and nose. Specialized electrical communication cells known as neurons became an important part of such organs. Some organisms evolved an organ (the brain) containing mostly of complex networks of neurons, with the capacity to holistically process the information brought in from the specialized sensor cells. The faculty of having a brain led to the emergence of a new property, i.e., consciousness. The *capacity to process information and carry out adaptive simulations for long-term strategy, using mainly the neural system of an organism is identified as consciousness*. The organism may not be 'aware' of all parts of its consciousness. In fact, an animal may be asleep, and much of its awareness is 'turned off'. Even when an animal is awake, it is not aware of millions of physiological processes automatically carried out in the cells of the body. Such processes are also carried out in organisms without brains. Thus consciousness is a phenomenon of the brain where a very small fraction of physiological and neural activity is presented together as a synthesized *feeling*, based on *selections* from the inputs of the five senses, as well as massive inputs ($> 80\%$) from stored memory and internal electrical impulses associated with the physiology of the body. In extreme emergencies, the brain by-passes its consciousness completely, and takes control to act autonomously. Thus *consciousness is a tool of the organism* that dwells on matters where no rapid action is needed. It is also a 'display board' for impending actions that the involuntary physiology may have already decided upon, but presented as 'contemplated action' by the 'individual'. More justification for the above summary will be presented in a later chapter.

This physico-chemical view of living organisms and consciousness (or some similar version) is the main-stream working model of physiology, clinical medicine, and indeed, science in general. This model has been the basis of the spectacular advances in our knowledge of the external world, as well as ourselves via physiology, neuroscience and molecular genetics.

However, this scientific model is viewed with great discomfort by a good majority of the non-scientific world, as well as by some noted scientists. Those who wish to 'go beyond the scientific view' sometimes invoke an undefined 'stream of consciousness', or 'soul' that is not subject to the laws of physics. They propose mechanisms having little or no experimental basis but invoke, say, meditative introspection. Such theories also have difficulty in establishing that there is a real external world. Only one's personal world is revealed by one's own consciousness. We limit ourselves mainly to the physio-chemical points of view in this book.

Some of the short-comings of this physico-chemical view are said to be:

(i) It is claimed that it gives no elucidation of the nature of individual inner feelings ('qualia'), personality, creativity, morality, freewill and volition.
(ii) It is claimed to reduce a human being to a predictable, automated machine.
(iii) It is claimed that subjective 'time-flow' is not explained.

This class of theories regards consciousness in the following manner.

(a) Consciousness is an unresolved, emergent property whose explanation is hidden in the unresolved parts of physics. That is, consciousness involves aspects of the quantum theory, quantum gravity, entanglement etc. This view is seen in the writings of Eugene Wigner, John Wheeler, and more recently in Roger Penrose, Henry Staap and others, as already discussed in Chapter 7. We label this class of theories the 'quantum-measurement based consciousness' (QMBC) theories.
(b) Consciousness is a highly correlated, exotic, emergent property of a condensate of dipoles, charges, polarizable units etc., in the brain, possibly involving non-equilibrium phase-correlated long-range order. This class of theories makes use of ideas similar to those put forward by Herbert Fröhlich in the 1960s, and now known as 'Fröhlich coherence'. We label this class of theories as the 'quantum coherent-state consciousness' (QCSC) theories.

In fact, many discussions use QMBC together with QCSC, as well as admixtures of ill-defined ideas. One such common theme is the existence of universal fields of consciousness, universal minds, or *pan-psychia*. This doctrine claims that even the most elementary of particles have some sort of consciousness (see Sec. 12.6.1). Stuart Hameroff [43], a collaborator of Penrose writes:

> "... greater computational complexity and ultra-reductionism ... features of consciousness, in particular the nature of conscious experience... Something more is required. If functional approaches and emergence are incomplete, perhaps the raw components of mental processes (qualia) are fundamental properties of nature (like mass, spin or charge). This view has long been held by pan-psychists throughout the ages, for example Buddhists and Eastern philosophers claim a "universal mind." Following the ancient Greeks, Spinoza argued in the 17th century that some form of consciousness existed in everything physical.

Although some Buddhist writers have expressed pan-pyschist views, the Buddha himself (circa 6th centure BCE) declared that the questions of mind (soul) and matter (body), the origin and end of the universe, and a number of other similar questions are indeterminate, and indeed a waste of time to debate about (see Sec. 2.1). Positivist philosophers might also take such a position, although we should note that today we have tools of investigation which can probe energy scales and length scales reaching deep into the possible origin of the knowable universe itself. However, Buddhist philosophers who did not wish to remain within the strict epistemology of early Buddhism continued their discussion of mind-body problems, in ways parallel to those of western philosophers. Thus the scholars of Nalanda (modern Afghanistan) in the 5th century CE, as well as Mahayanist and Chinese Buddhists developed the concept of the *Alaya-vijnana*, a form of pan-consciousness [12] that pervades all nature. The more well-known *Bagavad Gita* has several verses (e.g., 7.4, 7.5) codifying the common beliefs about universal consciousness prevalent in ancient Indian thought.

13.1.1 *Pan-consciousness or pan-psychism*

Spinoza's ideas on panpsychism or pan-consciousness (see also Sec. 12.6.1), as well as Hegel's discussions of *The Phenomenology of the Spirit* are well known. A popular evolutionary approach to pan-consciousness is also found in current writings and the Internet. Consciousness is regarded as a third property of matter, besides the duality of waves and particles. It is said to be present even in elementary particles, and that as the level of complexity increases through nuclei, atoms, molecules, crystals, cells and complex organisms, consciousness also evolves in complexity. It ultimately manifests as we find it in the higher primates. At first Max Delbrück, the influential theoretical physicist, biologist, and Nobel laureate, toyed with such ideas and considered the possibility of a quantized unit of consciousness, complementary to those found in inanimate-matter fields. In this he was influenced by Niels Bohr [25]. But Delbrück abandoned these ideas as he became more well entrenched in biology transformed by the DNA revolution. Bohr asked Delbrück to summarize his position (just before his death, 1962):

> '... it appeared for a long time that the regulatory functions in living organisms, disclosed especially by studies of cell physiology and embryology, exhibited a fineness so unfamiliar to ordinary physical and chemical experience as to point to the existence of fundamental biological laws without counterpart in the properties of inanimate matter studied under simple reproducible experimental conditions. Stressing the difficulties of keeping the organisms alive under conditions which aim at a full atomic account I therefore suggested that the very existence of life might be taken as a basic fact in biology in the same sense as the quantum of action has to be regarded in atomic physics as a fundamental element irreducible to classical physical concepts.'

Delbrük comments that

> 'cell physiologists had discovered innumerable ways in which cells responded 'intelligently' to influences from the environment, and embryologists had demonstrated such feats as each half of an embryo developing into a complete animal! Such findings were vaguely reminiscent of the 'wholeness' of the atom, of the stability of the stationary states. The stability of the gene and the algebra of genetics suggested something akin to quantum mechanics. The resistance of biologists to such ideas did not

surprise BOHR. He had met the same resistance to the complementarity view of atomic physics among his physics colleagues. . . . ' Indeed, we might say that the discovery of the Double Helix in 1952 did for biology what many physicists had longed for in atomic physics: a resolution of all the miracles in terms of classical mechanical models, not requiring an abdication of our customary intuitive expectations. The Double Helix, indeed! With one blow the mystery of gene replication was revealed as a ludicrously simple trick, making those who had expected a deep solution feel as silly as one might feel when shown the embarrassingly simple solution to a chess problem one may have struggled with in vain for a long time.

Nevertheless, the chimera and fascination of a deep solution beyond physics continue to make the abdication of commonsense not so uncommon. Hence a brief examination of panpsychism is justified.

The fact that a ray of light, or a projectile would follow the 'shortest path', obeying the principle of 'least action' is sometimes presented as an example of primordial pan-consciousness. It is claimed that the 'particle knows' the right path! However, in Sec. 3.6 we explained how this property is a result of the simplicity of the laws of physics, and arose from the relationship between the local picture (worm's-eye view of Euler-Lagrange) and the global picture (bird's-eye view of Path integrals).

The fact that a quantum particle is sensitive to its boundary conditions is sometimes presented as evidence for panpsychism. The specification of boundary conditions is needed to define the system (reduction) that is being studied. They set the initial conditions for solving the differential equations that state the relevant laws of physics — classical or quantal. Hence this use of the concept of pan-consciousness is, at least at this level, nothing but semantics.

We may claim that an eukaryote like an amoeba or paramecium that is relentlessly 'hunting' bacteria for food, is 'conscious'. Then do we also ascribe consciousness to a heat-seeking missile that relentlessly follows an aviator who is doing complex manoeuvres to escape the missile? What about the fly who knows to fly out of the window, but does not ever understand that beating against the glass pane could be avoided by simply shifting to the open window right next to the glass? The burying beetle *Nicrophorus* carries out a seemingly intelligent act of burying a dead mouse many times its size as a food source [106]. The beetles working as a pair of parents, roll the carcass into an under-ground crypt, coat it with preserving secretions, and lay their eggs. They then trim the family brood of young larvae to fit in with the food supply, i.e., in proportion to the size of the buried carcass! A bee may find that the optimal time to collect nectar is around 10 a.m., and it would return regularly to the flowering shrub at the appointed time! Such examples of seeming 'conscious' behaviour are found at every level of organic evolution. Unfortunately, as implied in the quotation from Max Delbrück, mechanisms involving pathways of protein syntheses, chemical feed-back loops etc., can be presented to give conventional physico-chemical explanations of all these, seemingly intelligent conscious acts.

In fact, these are various types of unconscious life skills, some of which are genetically programmed and already innate at birth, while others are learnt skills, similar to that of a performing seal or a juggler. In Sec. 12.6.1 we associated this class of skills with the *adaptive unconscious*, AU. The non-declarative memory, also called the 'implicit memory', as well as the autonomous nervous system are the actors in an invisible theater. Hence we may indeed assert that the burying beetle and the bee do have a type of AU mind. Unfortunately, writers on pan-psychism make no distinction regarding levels of consciousness reducible to those of sophisticated automata.

13.1.2 *Conventional quantum theory in living systems*

All living processes involve chemical reactions. They in turn involve electronic processes occurring around a set of nuclei which are, to a very good approximation stationary at the time scales of electronic rearrangements. A fundamental description of these electronic processes requires quantum mechanics. However, the quantum mechanics can generally be subsumed in the language of chemical bonds, radicals, conjugation, induction etc., developed by chemists. The limitations of that language can be further overcome by using the language of molecular orbitals, even at the level of Hückel theory, or its LCAO-elaborations (see Sec. 7.3.2) before *ab initio* quantum theory is called for.

Such *ab initio* theory may be necessary if, say, we are to understand the surprising efficiency of rhodopsin (see Sec. 12.3) in vision. It renders a $\sim 70\%$ quantum yield in a single-photon absorption process which occurs at sub-picosecond time scales! Similarly, how the same retinal molecule bound to different types of proteins (photopsins) could be sensitive to different types of photons, i.e., different colours of light, cannot be elucidated without quantum mechanics. The energy levels of the system of conjugated double bonds in retinal are perturbed by the specific electric charges in the photopsin, and quantum mechanics accounts for such shifts. Similarly, the sensitivity of the auditory system almost at the level of thermal noise, or the sensitivity of the olfactory cells to single molecules, requires explanations at the quantum-molecular level.

However, in the end, the underlying physico-chemical process cannot be claimed to be specifically characteristic of living systems or requiring new specifics special to conscious systems, i.e., a form of *vitalism*. Processes which were deemed to be a prerogative of the world of organisms have been successively demonstrated in laboratory systems, since the 19th century synthesis of acetic acid by Adolph W. Kolbe, to the 20th century syntheses of chlorophyll, hemoglobin, proteins, viruses, genes, and the 21st century work on building up whole genomes [40]. This is why molecular biologists and biochemists remain unconvinced. Nevertheless, various proposals have been suggested where macroscopic quantum phenomena are linked with the brain's known neural activity. I. N. Marshall in 1989, and possibly other writers at other times have suggested (without associated experimental work) that coherent quantum Bose-Einstein condensation occurs among neural protein units. As discussed in Sec. 7.7.6, preconscious-to-conscious transitions and volition were claimed by Stapp [202] as 'the collapse of a quantum wave function' in presynaptic axon terminals. In other suggestions, from Hameroff [45], and Penrose I-[160], microtubules are singled out as the seat of activity. Hameroff and Penrose have argued that the microtubules within the brain's neurons are self-organizing quantum computers (operating via 'orchestrated objective reduction' – Orch-OR) [44]. Let us examine in detail this popular 'quantum-consciousness' model.

13.1.3 *The Penrose-Hameroff 'Orch-OR' model*

Hameroff has claimed that the 'orchestrated objective reduction (Orch-OR)' model can provide an understanding of 'consciousness', qualia, free will, the 'coherent sense of self', and other issues in cognition [44]. Quantum states of tubulins found in the brain are claimed to act coherently like the instruments of an orchestra. This 'orchestral process' manifests as 'consciousness' in this theory.

The essential physical mechanism in Orch-OR is the 'orchestration' (by microtubules) of the objective-reduction mechanism proposed by Penrose (see Sec. 7.5.4.1). The OR mechanism was discussed in Sec. 7.5.4. The OR mechanism is claimed to provide for the self-collapse of entangled quantum states due to effects of quantum gravity working at the Planck scale of length and time (10^{-35} m, 10^{-43} s). Microtubules are part of the cytoskeleton of cells, and are made up (mainly) of arrays of two types of molecules, viz., α and β-tubulin molecules (see Sec. 11.2.2). The two tubulin molecules hold together to form a tubulin dimer which is the structural unit of the microtubules. These are hollow cylinders, 25 nm in diameter, and consisting of 13 columns of tubulin dimers arranged in a hexagonal lattice on the cylindrical surface (see Fig. 13.1).

The two possible conformations are actually *not* two discrete states. They represent two possible statistical distributions of an immense number of nuclear positions defining the nuclear framework of the molecule held in its hydration sheath and the 'double layer' of ions. The hydration sheath depends on the configurations of at least the first few neighbouring shells of water molecules around the tubulins. Because the protein conformation is determined by all these participating ions, water molecules etc., and in a dynamical state of flux (due to atomic movement), it is very difficult to actually *compute* the most probable conformation of a protein (see Sec. 8.6). However, under certain circumstances, a molecule or specific group held in place in a protein (e.g., retinal in rhodopsin) may exist in two molecular forms (e.g, its *cis*- and *trans*-forms, see Fig. 12.1) and define two 'states' of the whole protein. The location of hydrophobic pockets in folded proteins may also act as the source of conformation switching. Tubulins have such hydrophobic pockets.

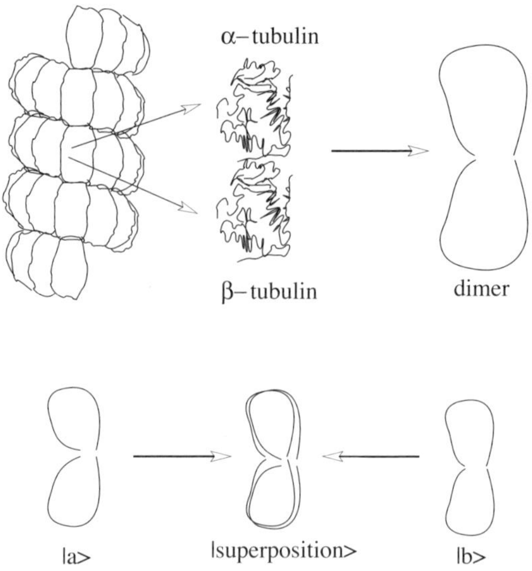

Fig. 13.1 The microtubule is a narrow cylindrical structure made up of dimer units of α- and β-tubulin. These are folded proteins whose conformation is delicately determined by intermolecular forces, electrostatics etc. Two possible conformations of the dimer are associated by Penrose and Hameroff with the electronic states $|a>$ and $|b>$. A quantum superposition state $|$superposition $>$ of the two conformations of the dimer is proposed. This is a qbit that, when coupled with other such qbits, is deemed to function as a quantum computer. The superposition applies to the probability amplitudes (wavefunctions) and not to the probabilities of occurrence of a given conformation (the superposition depicted is purely schematic).

These proteins that are capable of switching among at least two conformations may act as 'effective classical bits'. Pattern formation and replication similar to those of cellular automata (see Sec. 11.2.3) have been simulated by Hameroff and collaborators, with microtubules modeled using two-state elements (bits) arranged in hexagonal lattices on cylindrical surfaces, and interacting via dipolar forces (quoted in Ref. [44]). However, this is not at all surprising. The issue is whether the model used is in any way representative of the reality prevailing inside cells. Penrose and Hameroff go much further than the claim that arrays of tubulins could be used to process classical information. They claim that in spite of the extremely tightly-coupled environment of the cell, the temperature of the cell etc., electronic and nuclear positions of the switching conformations of tubulin molecules would form and persist as *quantum superposition states*. They also seem to suggest that these superposition states are not just confined to one dimer, but cover a macroscopic quantum-correlated state including many dimers. The main justification for this claim seems to rest on the validity of Herbert Fröhlich's model of long-range quantum coherence in biological systems [37] that we discuss in Sec. 13.2.

Penrose argued that valid results not accessible by algorithmic methods (cf., meta-sentences uncovered by Gödel) could be accessed by the brain, i.e., the mind can deal with non-computability I-[160]. A quantum computing process is presented as how the mind achieves this feat. The Orch-OR quantum computational process involving the objective collapse is described by Hameroff in his figs. 5-7 of Ref. [44]. Following him, consider three tubulins which are in one specific classical state (input state). However, the three tubulins have access to 2^3 classical states, which are all supposed to enter into a quantum superposition. The action of quantum gravity produces a collapse 'selecting' one of the eight classical states. In microtubules, the critical threshold for objective reduction is said to be achieved in coherence with other microtubule tubulins of a correlated system. 'Consciousness' is supposed to arise when the superposition collapses to a specific classical outcome. Thus there is a pre-conscious stage where classical computing gradually leads to a coherent quantum superposition, together with a stage of quantum computing (which requires isolation from the environment). The quantum states link to those in other neurons and *glia* by tunneling, and via coherent photons traversing cell membranes. Thus the coherent quantum state is said to spread throughout the brain volume.

The result of this process is a 'read-out' after about 25 ms (for a system of 2×10^{10} tubulins), when OR occurs. The physiological manifestation of this process is said to be a flash of consciousness. Hameroff describes the 'binding process' and 'time's arrow' as follows:

"Each instantaneous Orch-OR event binds superposed information encoded in microtubules whose net displacement reaches threshold at a particular moment: a variety of different modes of information is thus bound into a 'now' event. As quantum state reductions are irreversible in time, cascades of Orch-OR events present a forward flow of time and 'stream of consciousness.' ... The problem in understanding free will is that our actions seem neither totally deterministic nor random (probabilistic). What else is there in nature? As previously described, in OR (and Orch-OR) the reduction outcomes are neither deterministic nor probabilistic, but 'non-computable'. The microtubule quantum superposition evolves linearly (analogous to a quantum computer) but is influenced at the instant of collapse by hidden non-local variables (quantum-mathematical logic inherent in fundamental spacetime geometry). The possible outcomes are limited, or probabilities set ('orchestrated'), by neurobiological feedback ... The precise outcome – our free will actions – are chosen by effects of the hidden logic on the quantum system poised at the edge of objective reduction".

Thus the 'Orch-OR' model regards consciousness as a series of wavefunction collapses of coherent quantum states with significant macroscopic extension. The arrow of time is due to the irreversibility of quantum collapse. The many-molecular multi-organelle nature of the cell is replaced by a 'coherent quantum state' which binds sense data into a unified consciousness, qualia and feelings.

13.1.3.1 *Comments on the Orch-OR model*

The 'Orch-OR' model currently rests on a large number of inconclusively established physical and cognitive conjectures. We noted in Sec. 7.5.4.1 that the quantum gravity discussion of objective reduction is itself a conjecture for which we have no experimental evidence or unequivocal grounding in a clear theory. A number of authors [92, 82] have examined the theory critically, and have received responses from the criticized authors in Refs. [44, 94].

Penrose looked for quantum processes because of his belief that the mind can deduce non-algorithmic (non-computable) results [74]. However, invoking an experimentally unsupported quantum model is not the only way out. The brain is believed to function as a set of neural networks (Sec. 2.2.5). It may access results which are non-computable (via common algorithmic processes) but deducible with neural nets. Furthermore, no experimental evidence of quantum-correlated processes in microtubules have been found. The thermal de Broglie length (cf., Eq. (7.56)) of macroscopic systems at finite-temperatures is in the micro-femtometer range, and no quantum superpositions can be expected if $T > 0$. In our view, this is true even in regard to invoking a Fröhlich type condensate discussed later.

Hameroff's statements like *The precise outcome – our free will actions – are chosen by effects of the hidden logic on the quantum system poised at the edge of objective reduction leads to free will* offer no understanding of the conundrum of free will and determinism. In this book we have discussed in several places (cf., Ch. 9) how macroscopic systems shake themselves of quantum-unitary determinism as well as classical Laplacian determinism. Also, *who* is exercising the free will? If consciousness emerges fully baked from a quantum collapse, when is volition exercised? The experimental work on memory has shown the existence of declarative and non-declarative memory, and their seats as well as the molecular mechanisms in the cells of the brain have been located. It turns out that much of the processing is done in the non-declarative parts of the brain, involving the adaptive unconscious. Given the 11 million nerve signals that the brain processes in a second, only some 40 signals/sec are handled by the consciousness, while the remaining 99.99% is handled by the adaptive unconscious. There is very little room for volition, at least for actions which require only a few seconds.

The 'binding' of information (e.g., in vision, spatial location, auditory location etc.) occurs in modular neural nets which take the input signals, and add to it some 80% more internally manufactured frame material (as in 'Gestalt' explanations of duality hallucinations) to create the binding. The very tentative and conjectural proposals from Orch-OR do not reach anywhere near the currently available picture of binding that we have in standard physico-chemical models integrated within neural network theories (see Sec. 12.6). Pathological phenomena like 'the phantom limb syndrome' provide a clear indication of how signals are constructed and how binding takes place. These can be understood and even simulated using neural nets which require no quantum computations.

Another claim of the Orch-OR theory is that it establishes an arrow of time for consciousness via the collapse of the wavefunction which gives birth to a flash of consciousness. If so, can consciousness have a sense of the past given the discontinuity imposed by the collapse of the wavefunction?. Why should such collapse events give a psychological uniform 'flow' of time? Although answers to these could be invented, they would probably not be based on any established facts of physics or physiology. The problem of the arrow of time exists for text-book quantum mechanics (as well as classical mechanics) where the time evolution is unitary, and events flow equally well forwards or backwards. But living organisms are part of the diurnal cycle of a thermodynamically driven world. Cells have evolved

Circadian clocks to optimally use the cell biochemistry where some chemicals are best synthesized at night! The arrow of time is well encoded and seems to be understood by the organelles of the cell (Ref. I[164]). This seems to be so not only in multi-cellular organisms like humans, but even in sufficiently long-lived single-cell organisms. Furthermore, our discussion of the process of long-term potentiation (LTP), as well as Pavlovian conditioning showed how second-messenger modulation can build-in the concept of an unconditioned stimulus coming before the conditioning stimulus. Thus the arrow of time already exists within standard neuroscience, and the breakdown of quantum unitarity that exists in von Neumann QM is irrelevant to cell physiology.

The 'Orch-OR' model selects groups of microtubules as the seat of quantum-correlated condensates. Such tubules are found in abundance in plant cells, and even in in cilia, flagella and all sorts of other organelles as well. The account given by Penrose and Hameroff applies equally well to plant cells. Hence, if Orch-OR holds, one should expect plants to be fully consciousness organisms.

Even microscopic quantum systems show phase-coherent condensates only at very low temperatures, and when decoherence is minimized. However, Fröhlich proposed that such quantum correlated systems may exist in driven (i.e., non-equilibrium) systems, even at room temperature. As Fröhlich's model may be invoked by theories of quantum consciousness, we discuss it below.

13.2 Fröhlich's model of long-range coherence

Any dynamical system is characterized by its modes of excitation (see Sec. 2.3.3). If one or more of these modes is being continually excited (mechanically, electrically or otherwise), it is a 'driven system'. Driven systems can go into metastable stationary states where there are more particles in a higher energy state than in the usual lower-energy ground state. This is called population inversion. Driven systems can also go into chaotic dynamical states with intractable, unpredictable time evolution. These systems differ fundamentally from superconducting systems or Bose-Einstein condensates which are in their ground states. Systems with population inversion (e.g., in a laser) persist only as long as the inversion is sustained by the driving field which inputs energy into the system to balance off dissipation.

Such non-equilibrium stationary states are very common in classical systems. A system of eddies in controlled turbulent flow is a simple, common example. Enhanced plasmon modes, or plasma-beam instabilities in a driven plasma are other well-understood examples. Basically, when the energies and wavevectors of two modes match each other, they couple strongly, and exchange energy.

Biological systems are statistical mixtures of a vast number of electronic, configurational and conformational states at finite-T. A statistical sum over classical probabilities covers most situations. A density-matrix description is necessary if some quantum phenomenon involving bond-breaking and bond-formation is involved (as in photo-absorption). Enzymes play a major role in enabling chemical reactions that would otherwise not occur in cells which are at biological temperatures ($\sim$25 meV). Fröhlich proposed a mechanism of enzyme activity for concentrating energy in the low-lying modes of molecules, and thus enabling the rate of chemical reactions [37]. Since cells are energy-producing elaborate 'compost sacks', cell processes can be considered as a set of interacting modes driven by the energy of the cell. Such driven modes (modeled as oscillators) may, under suitable conditions generate a coherent collective state where nearly all of the energy is concentrated. This state is known as the *Fröhlich condensate*. The need for such *ad hoc* models has now disappeared as quantum chemistry has learnt how to address enzyme activity within its standard framework [42]. However, Fröhlich models continue to be used in some models of consciousness.

To be more specific, we may think of a large molecule, e.g., decanol, $C_{10}H_{21}OH$ as the object of interest. This has nine C-C bonds, 21 C-H bonds, one C-O bond and one O-H bond. We may regard this as a system of 32 coupled oscillators. A normal mode analysis would provide us with the vibration frequencies $\omega_i, i=1,\cdots,32$ of these oscillators. One would expect that the lowest energy mode ω_1 would contain a significant part of the motion of the whole carbon chain, as well as the C-OH bond, with the heavy O atom moving as little as possible. We may consider an ensemble of such molecules, arranged if necessary as a Langmuir film at a water interface, with the O-H group sticking into the water. The water would act as a 'heat bath' maintaining a temperature T, written in energy units (and not degrees, so that the Boltzmann constant k_B is unity). The decanol molecules interact with each other via van der Waals forces and weak intermolecular forces which are difficult to calculate. If there are M molecules, we have some M coupled oscillators for each of the 32 vibrational modes of the decanol molecule. Because of the normal-mode analysis, the 32 modes are essentially like non-interacting oscillators if anharmonic forces are ignored. But the mode analysis is compromised by the formation of the ordered Langmuir film. To be more accurate, one may need to do a normal-mode analysis of the whole film, showing that simple models lose their validity rapidly, when actual biophysical situations are envisaged.

At any temperature T, each of these oscillators occur in many vibrational excited states v. The occupation number n_{iv} of the v-th state of the i-th oscillator containing $v\hbar\omega_i$ of energy would be

$$n_{iv} = 1/\{e^{v\hbar\omega_i/T} - 1\}. \tag{13.1}$$

This is a simple Bose-Einstein distribution, and not a *Bose-Einstein condensate*. If $\hbar\omega_i$ are small compared to T, we can simply use the classical Maxwell distribution (high-T limit). In the low-T regime, Bose-Einstein condensation is possible. To make a Bose-Einstein condensate, we would need to have a very large number of molecules (in our Langmuir film), all in their lowest vibrational mode $i = 1, v = 0$, and with the phases of their wavefunctions all locked in step.

In Fröhlich's model, each of these oscillators, $i = 1,\ldots,32$ are 'driven' in the sense that energy is injected at a steady rate, irrespective of their vibrational state i, v. The interactions produce energy redistribution, and a steady state is reached, due to a balance between the energy-input and the energy lost into the heat bath. In this steady state, the vibrational energy is no longer distributed according to Eq. (13.1). Fröhlich shows that (i) the total number of vibrational quanta N (i.e., sum of n_{iv} over all i, v) increases far beyond the value N_T obtained from Eq. (13.1) for the system in thermal equilibrium. This is certainly not surprising, since energy is pumped into the system. (ii) most of the vibrational quanta are concentrated in the lowest-energy vibrational mode ω_1. (iii) this phenomenon does not require low-temperatures as in the traditional Bose-Einstein condensation. The last two results are not expected at first sight. The defining characteristics of the Fröhlich condensate are these three properties.

The Fröhlich condensate can be derived within a classical (high-temperature) approximation as well. However, crucial aspects of the model and its elaborations (e.g., that of Wu and Austin [110]) neglect important physics. The aqueous medium, upper vibrational states, and correctly constructed matrix elements for energy exchange (rather than simple coupling constants) need to be included. However, accepting these limitations, Reimers and colleagues have studied various aspects of the Fröhlich model using computer simulations, and concluded that persisting Fröhlich condensates would not form even within the model of Wu and Austin.

Although there is no experimental support for Fröhlich's proposal, it has become central to many exotic models of consciousness where the formation of the Fröhlich condensate is the essential physics. The 'Orch-OR' model of Penrose and Hameroff can invoke a Fröhlich condensate as the coherent state formed on arrays of tubulins in microtubules. As discussed in the previous section, a Fröhlich condensate intervenes in the classical-computing part of the 'pre-consciouness', as well as in the quantum computation part just before the claimed collapse of the superposition wavefunction of the condensate. The collapsed state appears as consciousness, and this is presumably a classical Fröhlich condensate or one of the quantum-correlated Fröhlich states contained in the superposition. Thus the existence of a Fröhlich condensate is relevant to the 'Orch-OR' theory as well.

The failure of Reimers *et al.* [82], to demonstrate the sort of Fröhlich condensate demanded by 'Orch-OR' may prompt one to review the Fröhlich model and consider some modifications. In defence of the 'Orch-OR' model, Hameroff has argued that even a very weak form of a condensed state (as seen in the simulations of Reimers) is sufficient for the Penrose-Hameroff model of consciousness. Reimers *et al.* used a linear chain of oscillators rather than a topologically richer structure like a 2-dimensional surface or a 3-D medium. One can envisage a model where the oscillators of the Fröhlich model are correctly embedded in an aqueous medium, with the energy exchange processes formulated using realistic matrix elements. Although no such calculations have been done, the common wisdom would be that the Fröhlich condensate would become even more unlikely.

The importance of the aqueous medium (H_2O) and its capacity to introduce new decoherence mechanisms are not fully appreciated by those who propose coherent quantum states in living organisms. Most amino acids and sugars have -O-H groups, N-H groups etc., which couple to the highly dynamic tetrahedral H-bonding network of the aqueous medium. Consider a molecule in a highly energized oscillator state of the sort proposed by Fröhlich. This energized mode (in a Fröhlich condensate) is the lowest-frequency oscillator mode. Since O (and N) groups are heavier than C, H, groups, the lowest oscillator would always include the oxygen atoms connected to the very light H atoms. The high energy of this mode would make the H atom also become energetic, and exchange into to the aqueous medium, when a 'cooler' H^+ ion comes in from the water lattice to recombine with the O^{--} oxygen. This process of proton exchange of energetic H^+ and the recombination of cool H^+ will produce a very efficient energy-relaxation mechanism which would rapidly quench off the Fröhlich state in low-energy oscillator states. Thus microtubules and other protein-containing materials in the cell medium would be unable to sustain any type of Fröhlich state.

Salari *et al.* [83] have commented on the simulations of Reimers *et al.*, and pointed out that the coupling constants etc., used do not reduce correctly when the $T \to 0$ limit and other limits are taken. On the other hand, the use of a constant phenomenological coupling constant would remain valid if its regime of validity is taken to be that of the simulation.

In our view, the nearly-zero de Broglie wavelength of macroscopic states at the brain temperature is fatal to any model using macroscopic superpositions of quantum states the brain. Since the full consequences of the aqueous medium and other features have not been taken into account in Reimers *et al.*, and given their model which is in effect designed to favour the Fröhlich picture, the negative conclusions of Reimers *et al.* are fatal to the 'Orch-OR' model and its variants.

13.3 Free will and attempts at 'quantum' explanations

Many writers get inveigled into 'quantum explanations' of consciousness because of the existence of 'uncertainty' and unpredictability in *individually observed microscopic* quantum phenomena. It is only the statistical average of observations of many events that must conform to the 'expectation value' given by the quantum theory. QM seems to give room for 'free will' and 'volition'. Science writers have spoken of the 'collapse' of the wavefunction ψ that describes any system as a source of irreversibility, breaking the rigid chain of events (causality) defined by physics. Both Stapp and Penrose reach for the collapse of ψ to escape from physical determinism. However, the 'collapse' of ψ is not even a feature in some re-statements of measurement in QM, e.g., in Bohm's reformulation.

As we discussed in Sec. 8.6, classical mechanics of complex systems already provides a very high level of practical indeterminacy. Given the initial conditions of a complex macromolecule immersed in salty water at room temperature at some time t, or a range of times Δt, we are still unable to predict the behaviour of the macromolecule in some future time segment. As long as we can make predictions to within a Δt which is faster than the rise-time of an action potential, that would seem adequate. But then, in dealing with a neuron, it is not just one protein that comes into play. We need to deal with many of the participating molecules in the cytoplasm as well as the lipid membrane, the axons, the dendrites, the synapses,..., and so on! Surely there is non-computability not just at one level, but at several levels. It is not as if we are asked to predict the fall of a single flipped coin, but that of a trillion flipped coins individually! So, just as we cannot make predictions of protein conformations, we cannot make useful retrodictions either. Hence there is no determinable causal chain! Thus it is clear that much of our worries about the dictatorial power of Laplace's demon arises from restricting our understanding and vision to very simple dynamical systems.

In spite of this, philosophers like Robert Kane have invoked a quantum uncertainty (in time) by postulating some 'quantum-fluctuations' in the neurons [52]. Attempts to use the Heisenberg uncertainty principle to generate 'elbow room' for some sort of leeway in physical law is doomed to failure. Most such attempts are based on the failure to appreciate the physical basis of the uncertainty used in quantum mechanics. The uncertainty principle says something about the impossibility of simultaneous precise observation of two non-commuting acts. So, if Kane wishes to have an uncertainty in a time specification, he has to *reduce* the uncertainty in specifying the energy of the corresponding quantum eigenstate. That is, he has to postulate the existence of sharpened quantum states in the brain, and look for exotic states like those of Fröhlich or Penrose and Hameroff.

But then, Kane is a philosopher and not a physicist, and so he prefers to specify the grand scheme and leaves the nuts and bolts to the scientists. These nuts and bolts are a challenge to even conceive. If the rule of physical law is broken down in a quantum fluctuation or what ever, during some time interval, Kane and others who are naturalist philosophers (i.e, those who do not wish to invoke miracles) are nevertheless saying that reality is riddled with moments where physical law does not hold. How does the 'self' who has become the unmoved mover of such moments come into the saddle? These details too are left open.

13.4 Conclusion

'Consciousness' has been a major enigma to every thinking person since antiquity. Heroically imaginative efforts that stretch the frame have been made to crack the enigma. We cannot do better than quote Marvin Minsky [68].

> But one can carry that quest too far by only seeking new basic principles instead of attacking the real detail. This is what I see in Penrose's quest for a new basic principle of physics that will account for consciousness. The trouble is that this approach does not work well for systems whose behavior has evolved through the accretion of many different mechanisms, over the course of countless years. For example, in physiology, the excretion of excess potassium in the urine occurs because our ancestors evolved an elaborate system of receptors and transport mechanisms, along with intricate machinery for controlling them. This is understood so well today, that no one feels that there's any need to postulate a separate, special principle for the Conservation of Potassium. Progress in this area is no longer news for biology because we have seen two hundred years of great success accrued from working out details. Since Harvey, Darwin and Pasteur, the idea of a Vital Force has nearly vanished from biology. Why is it still so much a part of present-day psychology?

The quest to comprehend consciousness with imaginative models will continue until neuroscience becomes of age.

Chapter 14

Addressing the Enigmatic Questions

In this chapter we examine the enigmatic questions about the sense of self, consciousness, free will, determinism, personality, self-interest, suffering and ethical behaviour, guided by the scientific perspectives of neuroscience, chemistry and physics.

14.1 The awareness of 'self', and of 'others'

In previous chapters we have traced how the fundamental laws of physics apply as we change length scales and energy scales to bring us to the world of familiar everyday objects. The laws of physics, transforming through quantum physics and quantum chemistry, gradually become the laws of ordinary chemistry and kinetics of many-particle systems. We showed that the usual paradoxes of quantum mechanics disappear when many-particle effects and boundary conditions of the physical systems are correctly included in the discussion (see Sec. 8). We also noted how the hard determinism of physics, when applied to sufficiently complex systems like macromolecules in aqueous solution, transform in most cases to a very weakly determined physics, or plain chaotic behaviour. The weakly determined systems have ground states having a large number of almost equally likely energy minima falling within a thermal width k_BT. The vast majority of living organisms, be they single-cell or multi-cellular organisms, use essentially the same macromolecules and chemical processes, played out within a narrow range of typical temperatures.

The evolutionary progression of living organisms began from single-celled organisms and today we have an incredible tapestry of multi-cellular organisms. We might think of an amoeba as a typical unicellular organism for our purpose. Such organisms were small engines or 'compost sacks' (where the 'sack' is the cell membrane) which could 'digest' other cells that they encountered as they float about at random. Darwinian evolution equipped them with the capacity to move around and scavenge other cells. Those cells that developed defence mechanisms to evade their own capture thrived. Thus, this process of being consumed or evading capture led to increasing sophistication in the sensory equipment and other equipment of the organisms.

As discussed in a previous chapter, many multicellular organisms developed advanced sensory organs, neurons and brains. The sensory equipment of such an animal is believed to receive some millions of electrical messages per second into the brain [72]. These messages are rapidly processed by the animal's brain and the signals are sent to the animal's motor muscles etc., ensuring that it flees away from danger, fights enemies, stalks prey for hunting, and defends territory. The early organisms did all this essentially automatically, with pre-wired behaviour. Once the organism evolved significant systems of neurons and synapses, it began to add to its pre-wired capacity by the learning of skills and the augmentation of its (non-declarative) memory. That is, they developed an 'adaptive unconscious' (AU). This is the brain facility which processes sensory information (both external and internal) rapidly and automatically, learns skills, and executes bodily functions and reacts to the external world. But it does not create explicit awareness. The 'adaptive unconscious' is an implicit information-management and execution system. As such we shall sometimes call it the 'implicit awareness'. Thus when a bee 'learns' that there is nectar in a certain flower at $\sim$10 a.m., it returns to that flower at $\sim$10 a.m. the next day as well, directed by its implicit awareness.

These highly sophisticated automata stalking each other needed new tools to get some evolutionary advantage. Some animals carried on specializing their vision, their auditory capacity or sense of smell, and so forth. Others developed their neurons and information processing capacity, and developed a new and powerful tool, in step with the growth of implicit awareness. The new tool that some creatures developed was a capacity to imagine (i.e., model) what it would be like 'if' a possible scenario were played out.

The space cells and other parts of the memory of these animals could be used for 'play acting' in their minds. They began to 'guess' *what could happen*! That is, these sophisticated automata developed an *explicit awareness* or *consciousness*. This consists of self awareness, as well as an awareness of others, enabling the consciousness to enact, say, future hunting plans in an imagined *mental world*. This awareness of oneself, the others and the setting, imagined out in the animal's mind is what Premark calls a 'theory of the mind' (see I-[172]). The conscious organism as well as others are complex adaptive systems. They are best thought of, at least in the first approximation, as *free* agents for time scales larger than those set by the adaptive unconscious. The capacity to use consciousness, and plan out strategies for conquest, survival or multiplication provided a novel tool in the evolutionary process.

What is the nature of this consciousness? Who or what is this construct called the 'self' that this imagination is aware of? How does it construct a mental world and how does the 'self' exercise its volition on the actual physical world? What is the relationship between the external world and the mental world?

While detailed answers are probably in the pipeline, studies of the brain and its neuronal nets are presently in the same state that atomic physics found itself in the 1900s, or that of particle physics in the 1950s when confronted with 'a zoo of elementary particles'. The clarifications came with the invention of new tools (spectroscopes, particle accelerators), as well as new theories. Similarly, the methods of molecular biology are now firmly entrenched in the cognitive sciences. Nuclear magnetic resonance, techniques based on positron annihilation and other tools familiar to physicists for many decades have now been adopted into neuroscience. Methods currently being fashioned in nanotechnology will soon have a place in the cognitive sciences as well. New insights to old questions can be expected before long. In this chapter we visit some of these questions using the background built up in the previous chapters.

14.2 Why consciousness?

An organism which has a brain equipped with an 'adaptive unconscious', i.e., an implicit awareness, receives and processes millions of bits of information (electrical signals) per second. The skills it has inherited and learned are established in the form of synaptic connections which form modules of interconnected neural networks. These generate suitable action potentials in response to the input stimuli. These action potentials travel along the nervous system of the organism and activate the animal's muscles appropriate for the stimulus. The time elapsed between the input stimulus and the response is usually short, leading to a suitably swift response. The adaptive unconscious is also capable of long-term programmed action. It has a Circadian clock, as well as a sense of the seasons, the arrival of summer and winter etc. The organism has developed Circadian chemical mutations over years of evolution of its species, and encoded them in its genetic heritage. Here we quote Marvin Minsky [68]

> It now appears that perhaps fully half of our entire genetic endowment is involved in constructing our nervous systems. This would suggest that the brain is nothing like a single large-scale neural net; instead, it would have even more parts than the skeletomuscular system — which can be seen to have hundreds of functional parts. If you examine the index of a book on neuroanatomy, you will find the names of several hundred different organs of the brain. A good fraction of those are already known

to have psychologically distinct functions. To pursue the analogy a little further, note that the skeletal anatomies of animals have been known for millennia, but only in rather recent years have scientists understood the mechanics of locomotion and its various gaits; that had to wait until scientists learned more about the mechanics of forces and materials. Similarly, mechanistic theories of psychology may have to wait even longer for adequate conceptual tools because the 'mechanics' of heuristic computation could turn out to be more complex than those of physics. Before these new ideas emerged, with the era of complex information-processing computer models, such models were not considered convincing — perhaps because there were no feasible experiments. I don't mean to say that there was no progress at all before computers, only that there was precious little. Freud himself was one of the first to conceive of 'neural-net-like' machines — only no one would listen to him except Fliess. Later came the astounding insights of Post, Gödel, and Turing, followed by those of Rashevky's group, McCulloch and Pitts, and Grey Walter's simple yet somewhat life-like mini-robots. But significant progress began only in the 1950s when more serious models could be conceived, tested, and discarded in days or weeks instead of years. Soon the researchers in Artificial Intelligence discovered a wide variety of ways to make machiness do pattern recognition, learning, problem solving, theorem proving, game-playing, induction and generalization, and language manipulation, to mention only a few. To be sure, no one of those programs seemed much like a mind, because each one was so specialized. But now we're beginning to understand that there may be no need to seek either any single magical 'unified theory' or any single and hitherto unknown 'fundamental principle'.

These many mechanisms associated with the nervous system enable the organism to deal with natural phenomena (e.g., falling rocks, seasons) of the physical world which follow an expected behaviour (i.e., a set of deterministic routines). Some of the longer term processes may be recently learnt knowledge, acquired by adaptive learning (involving LTP and the non-declarative memory), guiding the organism to local sources of food (e.g., edible fruits) and water.

It should be noted that the brain, and its associated faculty that we call the mind, carry out feats of computation that are mind-boggling to a computer scientist. Think of a person lying in bed, eyes closed, and just lifting his hand to grope at the night table to press the button on an alarm clock that s/he recognizes by touch. The computer coding necessary to correctly and effortlessly to do this seemingly simple job via a robotic arm, and the information processing needed are totally unimaginable. Yet, such tasks are everyday fare not only to humans, but even to circus animals.

Computer scientists have recognized that these actions involve incredible feats of computation and information processing that are effectively beyond the capacity of the computers that humans have so far manufactured. How are these computations organized? Marvin Minsky [67] presented in the 1980s, a point of view known as the 'society of mind', held to be valid to various extents by many scientists (as well as philosophers like Daniel Dennett [23]). Here we specifically mean computations in the 'adaptive unconscious', although Minsky and Dennett used it in the more general sense of 'mind'. According to this view, the computational organization involves a 'society of mind', populated by 'agents'. An agent is like a simple subroutine, a data structure or a basic neural net. And, as in modules of computer programs, agents can be connected and built into larger systems called 'societies of agents'. Together, societies of agents can perform functions more complex than any single agent could, and ultimately produce the many abilities we attribute to the amazing computational capacity of the mind. Dennett appears to suggest Darwinian competition among 'multiple drafts' of the modules as the evolutionary force optimizing these neural nets. It should however be noted that the molecular biology supporting the nature of memory etc. had not yet been established in the 1980s. The time scales and adaptive processes involved in this Darwinian picture have not been addressed. Since neural nets learn by an iterative self-consistent process, the learning process could be recast as a Darwinian process, which itself an optimization process.

However, these amazing, cognitive and computational skills of the adaptive unconscious turn out to be inadequate in *anticipating future events* generated by other organisms (complex adaptive systems) moving about in the external world. Organisms do not appear to obey the straight-forward deterministic routines of physical law, and instead behave as if they are subject to no laws — i.e., they seem to be indeterminate (see Sec. 8.6) or 'free' creatures. A swiftly operating automaton can hone in on a prey as long as it holds the prey in its field of sensory stimuli. But once a prey disappears from its sensory field, it has no capacity to anticipate or guess what 'these others' would do, in reference to oneself. It is here that the organism's capacity to deploy its consciousness, together with its adaptive unconscious, comes into play. *La raison d'être* for consciousness is that it provides an evolutionary tool for dealing with complex adaptive systems whose behaviour is non-deterministic — unpredictable in the usual sense. Consciousness is primarily a tool for guessing what the 'other' would do!

The adaptive unconscious is overwhelmingly the large component of the activity of the brain. The conscious part, or consciousness will be sometimes called *the conscious*, in opposition to the name 'adaptive unconscious'. The conscious is a very small component of brain activity while being *the* newest tantalizing tool unleashed by evolution.

If the practical reason for 'the conscious' is to guess and plan the evolution of other creatures (complex adaptive systems) that are interacting with the organism, then it has to have a conception of its *space*. The center of this space, i.e., the origin of reference, gives a meaning to the concept of *itself*. Thus, an ego-centric coordinate system is an essential component of this process of constructing the 'world' as sensed by the organism. The 'self' provides the origin of the spatial coordinate system. The 'person in the saddle' is just *a construct* serving as the referential point in our spatial consciousness. The spatial consciousness is itself a set of postings from the adaptive unconscious. Hence there is in fact no specific organ in the brain identifiable with the 'self'. As we shall see below, its locus is collectively given by the firing of a group of *place cells* in the hippocampus. That is, it is economical to associate the sense of self to the location of the animal as represented in its spatial consciousness.

The sensory data from the actual physical world, blended with four or fives times as much information drawn from the brain's internal store have been processed and they are available in the implicit awareness (adaptive unconscious). Neuroscientists have established that during the time that the implicit awareness processes some 2000-3000 messages, the explicit awareness (i.e., conscious) processes only just one or two pieces of information. That is, the conscious, much like a news anchor facing the cameras, works with a summary of the events given to it by the AU who has the the detailed dispatches. Of course, the brain does not work in pixels, but in images and patterns more appropriate to its modular structure based on neural networks. Clearly then, one of the tasks of the adaptive unconscious is to present a set of digests of the external and internal world that it has constructed, for its own actions which are largely out of control of the conscious mind. That is, the conscious is, to begin with, *a posting board* of the adaptive unconscious. Clearly, many postings are made by the adaptive unconscious, and only a particular posting captures our *attention*, depicting the most salient subjective construction. If the blood flow in some set of place cells in the hyppocampus were to suddenly increase due to a movement in the external world captured by the senses, then the attention of the conscious reverts to an updated posting from the AU specifically highlighting that movement.

14.2.1 *Constructing the subjective 'external world'*

The implicit awareness (i.e., the adaptive unconscious) is the brain facility that preceded the development of consciousness. This is the case even with a foetus. The latter is very much like a single-celled organism at the foot of the evolutionary tree. It has to undergo replication and become multi-cellular, and form organs in the safety of the mother's womb. It has to grow neurons, and its consciousness develops gradually, while the basic involuntary capabilities come first.

An organism foraging, hunting and surviving in the physical world soon acquires a working model of its world. This is, at least initially, a set of clues acquired and stored in synaptic adjustments (learning) and memory that we described in Chapter 12. The stored clues from the external world contain material held in long-term memory and short-term memory, including immediate memory. Inputs from place cells, vision, auditory-, olfactory- and tactile- sources, as well as from the information banks of the brain contribute to all the inputs that have to be processed and synthesized. This is known as the *binding problem*, to be discussed in Sec. 14.2.2. The adaptive unconscious pieces together (i.e., binds) all the clues, frames the information (if we are to borrow the language of the Gestalt psychologists) with internal inputs to make a synthesis. It uses this *mental construct* to activate the muscles of the organism in *involuntary* automatic reactions. However, if the animal has a consciousness, an additional step is also possible.

If the organism has a consciousness, then the AU posts a highly stripped version of the mental construct — *a cognitive map* — in the conscious. That is, the cooperative actions of many modules of neuron nets in the brain send action potentials which do not engage muscles, but *re-engage other neurons of the cerebral cortex*, forming images, internal dialogues, as well as conscious hearing of sounds, sensing of smell and feelings — the so called 'qualia'. This is the stuff of consciousness. What the conscious perceives is the mental construct delivered to it by the AU, That is, the conscious is aware of the constructed world of the adaptive unconscious, rather than the external world. The 'qualia' are the selectively presented fleeting components of this construction. They would correspond to definite biochemical processes — and sets of action potentials — in the brain. However, many people prefer to make a mystery (or a 'hard problem') out of these so called qualia, claiming that they cannot be explained in terms of anything in present-day science.

It is the use of a simplified cognitive map of the world (rather than the real world) which enables the brain to process information successfully, instead of getting bogged down in the overwhelming details and intractability of the computations that are needed to even recognize the face of a person. The molecular and cellular basis of this cognitive map is now well established, and in fact goes back to the work of Wade Marshal and collaborators who established in great detail the representation of the entire body surface in the brain of monkeys. Wilder Penfield, working in the same period (1930s) presented details of the human somatosensory cortex. Each part of the body is represented in a topographically connected way in the cortex, but the mapping is *not* a direct replica. Each part of the body is expanded or shrunk to reflect its importance in sensory perception. The mapping — known as the *homunculus*, reflects the density of nerve connections (innervation) servicing each sensory area.

For instance, the skin surface of the back, although large in surface area, has a much smaller representation than the finger tips or the mouth, since these latter areas are much more sensitive to touch. The work of Vernon Mountcastle and others showed, in the 1950s, how, for example, an individual cell in the left somatosensory cortex might be so specific as to represent the tip of the right thumb and nothing else. Another important result that came from such studies was the fact that sensory information is broken up and 'deconstructed' and associated with submodalities having their own pathways in the brain, and that this deconstruction is preserved at each relay in the brain stem and in the thalamus. These submodalities are combined, reconstructed and a 'binding' is done only in later stages of information processing, with generous admixtures of internally generated inputs.

The 'deconstruction' or breaking down of information into submodalities is found to hold even more dramatically in visual perception. Cells in the visual cortex do not respond strongly to excitation of the retina with spots of light. Instead, they respond to linear contours, contrasting lighter and darker edges, and shapes! Some cells respond to horizontal linear streaks (lines), while others respond to vertical lines. Evolution may have had reasons to develop such deconstructed vision. In fact, it is found that a frog's eye is sensitive only to movement, e.g., as when a fly moves across its field of vision, which is restricted to the near-field region within its striking distance. Thus the earliest eyes began by detecting the essential components of the visual field like linear motion of a fly that may become food. Others parts of the visual field were included at subsequent stages of evolution, as the need arose due to the pressure of competition. The initial step in visual encoding, i.e., deconstructing visual objects into line segments in different orientations may be simply how things evolved. Hence, all vertical line segments are associated with one set of cells grouped in a column, while other orientations have their specific cellular columns. The work of Mountcastle, David Hubel, Thorsten Wiesel and others shows that the brain not only stores and processes information at its connections, but also *transforms the information and pads it up* to knit a presentable story!

Other aspects of the visual field, e.g., depth, form, colour, and whether the objects are moving or not, are all separated out, deconstructed and transmitted via separate paths to the brain. In Fig. 12.4 we noted how the 'what' information and the 'where' information are separated out. Specific regions of the 'what' pathway do things like recognizing a face, or identifying a Matisse from a Renoir. The 'where' pathway conveys information about the movement of the object in the visual space. Our conception of visual space is directly associated with spatial memory. Spatial memory is, of course, an integral part of declarative memory (i.e., explicit memory). Although space is in some sense fundamental to our existence, there is no explicit sensory organ for space. This is not surprising because sensors depend on physical forces, i.e., changes in energy gradients. Thus spatial consciousness also has to be a construct; it has to be built up on a multi-sensory data base that defines 'space'.

In 1971 John O'Keefe began to unearth evidence to show that the hippocampus of rats contains a multisensory mapping of the space used by the rats. Research on the nature of declarative memory had shown that the hippocampus is a key player in memory formation. O'Keefe and others showed that when an animal (e.g., a rat) walks about in an enclosure, an internal map of the space develops within minutes. The brain deconstructs (breaks down) the multisensory information from the surroundings into many small overlapping areas, each represented by synaptic changes in specific cells in the hippocampus. These are the *place cells*. The moment the animal walks into a specific location that it has already got accustomed to (i.e., visited the location a few times before), the corresponding place cells fire their action potentials. When the animal moves to another location, other place cells come into play. Clearly, unlike vision, touch etc., which are directly linked to sense organs and involve processes inherited by the organism, the spatial map is based on a *combination* of external inputs which involves *learning* by exploration. The place cells are activated by a collective signaling process where the adaptive unconscious locates the animal's position in its internal spatial representation. Studies by Robert Muller, Eric Kandel and collaborators in the 1990s showed that the molecular mechanisms of the learning process involved in constructing the spatial map are the same as for short-term and long term memory, involving LTP, PKA, gene synthesis, protein expression and other familiar bits of biochemistry (see Table 12.1 etc.).

Thus, modifying the musings of the philosophers Immanuel Kant, spatial consciousness itself has to be learnt by an exploration of the environment. Perhaps Kant's ideas are partially correct since the ability to form a conception of space is innate. The neural pathways and connections are in place, as ordained by the genetic code, but they need to be nourished by actual experience — learning.

Given the deconstructed sensory information, spatial information etc., the adaptive unconscious is now ready to knit its own story in processing the external information by mixing it with large amounts of its own internal inputs. This story is further abstracted and posted on the billboard of our consciousness. However, this requires 'binding' all the dissected information into a coherent whole.

14.2.2 *The binding problem*

The binding problem is sometimes also known as the 'problem of integration'. The problem is presumed to occur at several levels. For example, if we consider visual information, information about motion, colour, form and depth have been dissected and conveyed into the brain by separate pathways. So they seem to need to be organized into a cohesive whole, ensuring the *unity of visual experience*. This is similar to the problem of taking a set of still pictures of a moving object in the three primary colours, at various random moments, and trying to edit and collate all of them to get a smoothly running color movie. This collation is done by some sort of 'master editor' (or 'virtual machine' in the sense of Dennett) working in the 'film lab'. The binding problem can also be resolved in principle by bringing together the separated out independent neural pathways in some master cortical area — the neural equivalent of the film lab. This is somewhat like the 'global workshop' idea of Herbert Simon and others proposed in the 1970s. However, we quote Semir Zeki [114], an authority on the primate visual cortex as well as on neuro-aesthetics.

"... here one comes across an important anatomical fact, which may be less grand but perhaps more illuminating in the end: there is no single cortical area to which all other cortical areas report exclusively, either in the visual or any other system. In sum, the cortex must be using a different strategy for generating the integrated visual images. Not only it is that there seems to be no such master area, but it is also the case that there is no master editor!"

The exact mechanism used by the brain to generate an integrated visual image, or any other sensory system, should await further experimental information. However, given that there is no master cortical area, a likely mechanism would involve a self-consistent adjustment of the deconstructed information (bottom-to-top flow) with internally supplied (top-to-bottom flow) information.

Just as there is no master master editor or single cortical area for 'binding', there does not seem to be a simultaneous convergence of information. Thus, according to Moutoussis and Zeki, we obtain colour knowledge 40 ms prior to knowledge of form, and 80ms before we obtain knowledge of motion of a visual object!

Perhaps the neural machinery used to deconstruct the bottom-to-top flow could be used in reverse, with top-to-bottom signals used to synthesize a coherent integrated visual image, generated from the visual cortex and not from the eye. It is likely that some such sanitized, simplified image, formulated using inner visual templates, decorated with the dissected information, is what is conveyed to the conscious by the adaptive unconscious.

However, it is easy to find problems where there are none. The medieval chemists looked for a principle that would integrate lead with a noble spirit to convert it to a noble metal. We too may be looking for nothing more than our own pre-judgments. The brain receives signals (action potentials) from the eye, ear, taste buds etc., and they are not distinguished in any way except by the pathway they arrive in the brain. There is no additional tagging principle. The brain, unlike our silicon-chip computers, uses the same neural nodes for storage and processing

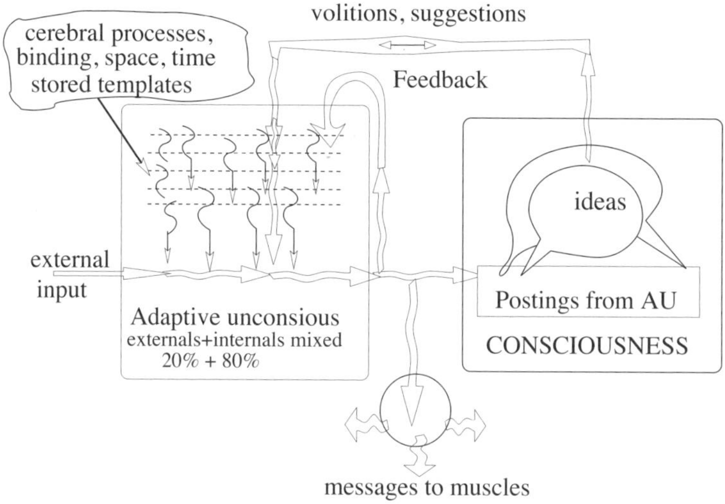

Fig. 14.1 Schema for the processing of external information by the adaptive unconscious (AU), and the conscious. The AU processes vast amounts of information, mixing in an order of magnitude more internal information. The AU acts without much reference to the conscious. A highly distilled summary of its actions is posted to the conscious, complete with spatial information and the location of the organism. The coordinate-origin of this 'ego-centered' space is the location of the *self*, a construct of the AU. Scenarios play out in this space, and are the 'ideas or thoughts' which compete for attention. They are re-engaged by the AU via feedback, and sent to muscles if suitable for action.

instead of separating memory and 'CPU'. There is no need to 'sort out' and tag the signals from vision, hearing, taste etc.

In the same way, the idea that we need to integrate the 'deconstructed' signals sent out via different paths in the visual cortex may also be our own creation of a problem. The information conveyed is heavily weighted by top-to-bottom inputs generated by the brain itself. Its treatment in the brain may well be *precisely the appropriate way* of handling it in a coherent manner, as judged by the framing process of the brain itself. That is, binding is in fact the essence of the deconstruction process itself! This may be clarified by an illustration. English readers are automatically used to reading a language from left to right, in a linear sequence. However, one can perfectly well have a language where the letters are written on a page (or even on a three-dimensional space) according to a completely different method of organization. This is shown in Fig 14.2 for a quotation from Einstein. It is simply a different way of organization, and one is no more consistent than the other. In the same way, the so-called deconstructed packing of sensory information probably *needs no re-binding*, as far as the adaptive unconscious is concerned.

The□most□incomprehensible □thing□ about□ the□ universe□
is□that□it□is□ comprehensible

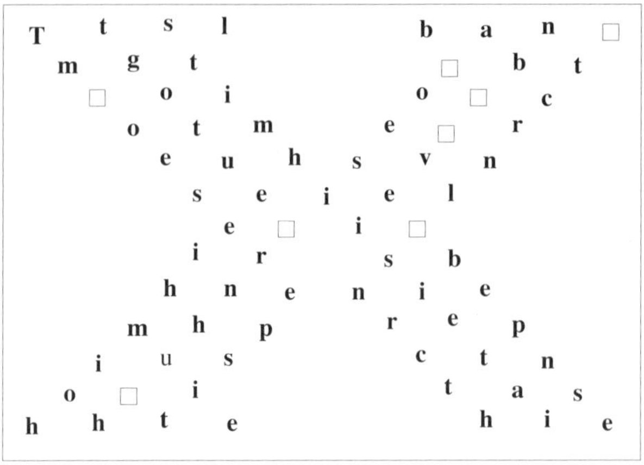

Fig. 14.2 A quotation from Einstein is paraphrased and deconstructed into a columnar sequence associated with an anti-clockwise rotation. This deconstruction requires no rebinding to make sense, unless we insist on a familiar left-to right linear script.

14.3 Causality and free will

The conscious mind is presented (see Fig. 14.1) as being able to exercise its *volition* and act back on the adaptive unconscious, there by activating the muscular system of the organism. The organism is exercising some sort of *free will*. Although, the issues of determinism and free will have been mentioned in several places in this book, we need to take fresh stock of the situation, fill in the gaps, and comprehend how the stark rigidity of physical law could spawn organisms having a capacity to defy those very laws.

Further, we could ask how 'cause and effect' or 'rigid causality' could co-habit with free will in this scheme of things. Many of these terms, like 'determinism, free will, cause, effect', etc., are so widely used in all sorts of ways that a clear discussion needs a preliminary re-charting of the arena of discussion. In fact, the exact sciences avoid these terms, and use mathematical relations among the variables, expressed as differential equations satisfying various boundary conditions. That is,

(1) the 'causal relations' are replaced by functional relations with are usually time-symmetric;
(2) the functional relations are differential equations, and hence their solution

requires specifying boundary conditions to make them applicable to a given situation;

(3) these boundary conditions specify initial or other conditions, coupling to the environment etc., and bring in irreversibility, decoherence and related effects.

(4) In sufficiently complex systems, e.g., a large protein molecule placed in an aqueous medium at a temperature T, we are in fact unable to specify the physical conformation of the molecule precisely. The 'ground state energy' surface is made up of many local minima coupled to the dynamically fluctuating locally tetrahedrally coordinated structure of water. Living organisms are made up of such proteins, other molecules, and dynamically hydrogen-bonded water, with *indeterminate* chemical structure if detailed knowledge is demanded.

The usual philosophical discussions of 'cause', 'origins' and 'ontology' do not carefully attempt to distinguish these three steps essential to the description used in physics. In particular, the 'ontological proofs' of the origin of the universe seem to attempt to infer boundary conditions from causal relations, instead of recognizing the need to specify the boundary conditions as a part of the physical description. Similarly, many philosophical discussions of quantum mechanics use 'states vectors' in abstract infinite-dimensional Hilbert spaces without reference to boundary conditions. This leads to many difficulties in the quantum theory of measurement, normalization of the photon vacuum, indefinite metrics etc. Bohm's reformulation of Schrödinger's quantum mechanics has the advantage that the effect of the boundary conditions are explicitly included in the dynamics via 'quantum potentials'. This avoids many of the pitfalls of the von Neumann formulations of quantum mechanics, and the need to 'collapse' a wavefunction to impose new boundary conditions arising from introducing a measuring apparatus (see Chapter 7).

Thus the concept of 'cause' does not sit easily within physics, while 'causality' is usually meant to imply the need for a correct time ordering, e.g., in products of field operators in causal Green's functions, in the S-matrix, or in Kramers-Krönig relations. Within such causal schemes, time can be legitimately reversed by converting particles to anti-particles which propagate backwards in time, as is conventionally done in computations using Feynman graphs.

In contrast, traditional philosophy, the social sciences, and everyday discussions tend to use the language of cause and effect, and with little mention of boundary conditions entering into the physical description. Hence some clarifications and recapitulations are needed.

The assignment of 'cause' in everyday discussions is best illustrated by a few examples. If we move a switch to the 'on' position, a light bulb lights up, and we may say that switching it 'on' caused the bulb to light up. At this point the discussion can be challenged or side tracked by stating that the cause was really the 'person' who moved the switch, or it was really his wife who asked him to 'put on the light', and so on. One could also argue that it is the 'electric current' which is the ultimate 'cause' of the light. These involve (a) different factorizations of the parameter space controlling the time evolution of the system containing the light bulb, electric wiring, and, (b) changing the boundaries of the problem by extending the number of people — causal agents — to be considered. The boundary conditions defining the system could be further extended to include the power-utility company and then many new 'causes' come into being. Hence discussions in terms of cause and effect can become meaningless. Only a formulation of a whole physical model, inclusive of all the relevant parameters within specified boundary conditions is satisfactory. That formulation ultimately becomes a description in terms of physics.

Let us consider a coarse-grained system, i.e., one where all the dynamical variables have been averaged over all the shorter length scales which are relevant to quantum mechanics (see I[83]). That is, we look at a part of our ordinary everyday world (e.g., the world of a civil engineer) where classical mechanics holds supreme. Then, if all the objects in such a system (abstracted from the external world by a suitable boundary) could be specified with sufficient precision, and in sufficient detail, then the subsequent evolution of the system is completely predictable, and hence there is complete determinism. For instance, let the state of the system at time t_1 be given by a function $S(p_1,\cdots,p_n;q_1,\cdots,q_n;t_1)$. Here p_i, q_i are parameters (e.g., positions and momenta of all particles) which are needed to describe the system at the given moment t_1. Then the state of every object in the system at some future moment t_2 is given by the new function $U(t_2,t_1)S(p_i,q_i;t_1)$, where $U(t_2,t_1)$ is the time-evolution propagator specified by the laws of physics for evolving the function from t_1 to t_2. If the system is sufficiently complex, the dynamics would be *chaotic* in the sense that the time evolution would be extremely sensitive to the specification of the parameters $\{p_i,q_i\}$. Nevertheless, the system is fully determined, and there seems to be no room for 'free will'.

Removing the coarse-graining, and looking at a system using a fully quantum mechanical picture does not help us to get away from this determinism. In the quantum picture, no single event at the quantum scale would be predictable. The specification of the state vector will no longer be simultaneously in terms of p_i, q_i, but in terms of momenta or positions and corresponding wavefunctions. A density matrix has to be constructed from the states and their populations. Only averages over the density matrix would be available, and they alone are of relevance to us living in the macroscopic world. Quantum fluctuations provide no escape from real-world determinism (see Sec. 13.3).

In such a world, causality seems to holds supreme. We use the world 'causality' in the pure physics sense that *the time evolution of the system is determined*. We do not say 'which cause' produced 'which effect', unless the dynamical parameters can be regrouped and reassigned as independent variables defining independent time evolutions. That is, if the dynamics of the system could be decomposed into a set of independent sub-dynamics, each controlled by a dominant parameter, that parameter may in some circumstances be used to define a 'cause'. Then, e.g., we may say that a given force caused an acceleration, or a specific classical collision led to a break up of a composite particle, etc.

Given the above discussion, there seems to be no hope for determinism to give away to indeterminism and free will. Although chaotic systems may be beyond computational predictability (due to the need to specify the initial conditions with impossible precision), they are nevertheless determined. One of the tacit assumptions that was made in the above discussion was the abstraction of the system under study from the external world. The time evolution of the objects in such isolated systems is indeed fully deterministic. However, if a realistic boundary which communicates with the external world is included, information is irrevocably lost to the external world and strict determinism is compromised.

For example, if a particle is enclosed in a large quantum system, e.g., a cube of linear dimension L, its behaviour is completely described by the eigenfunctions $\phi_n(\vec{r},t)$ and energies E_n, where n is a set of quantum numbers, and $\vec{r}$ is the position of the particle in the box (see Sec. 6.4). The energies are real, discrete quantities, with a finite separation $\Delta_{nn'} = E_n - E_{n'}$ between any two adjacent energy levels n, n'. The propagator (i.e., the Green's function) of a particle in the state n is given in the energy representation by

$$G(n,\omega) = 1/(\omega - E_n + i\varepsilon). \tag{14.1}$$

As already noted (Sec. 1.3), the imaginary part ε is an 'uncertainty' added to the energy due to loss of information at the boundary of the box. As the size of the box is increased, the discrete character of the spectrum, associated with the magnitude of $\Delta_{nn'}$ decreases, and the spectrum of energies becomes quasi-continuous. At this point, the analytic structure of the propagator changes drastically, and the set of discrete poles of the Green's function may be replaced by a *branch cut*, signifying the passage from a time reversible character to an irreversible character in the physics. Thus, damping introduced by the loss of information at the boundary of the system, if comparable to the discretization of the underlying spectrum, can completely change the character of the system [27].

Furthermore, living organisms, and the world we live in are best represented by a system in contact with a heat bath at some ambient temperature T. In our discussions we measure temperature in energy units, instead of in degrees, so that the Boltzmann constant k_B can be assigned the value of unity.

The low-energy dynamics of objects in such heat baths are easily seen to be quite different from those of isolated systems. Thus consider an inanimate particle, e.g., a minuscule grain of sand, having a mass m and slowly moving at the bottom of a pond, as it has some kinetic energy E (see Fig. 14.3). The grain of sand is buffeted by the random motion of the water molecules and moves erratically, displaying Brownian motion. Its kinetic energy E is, on the average, equal to T but hardly ever precisely at T. The bottom of the pond is irregular and has shallow pits, hills and valleys. These in effect define an energy landscape which is also of the order of the thermal energy T. The dynamics of the grains of sand are also best described as 'random motion'. Although the underlying physics is deterministic, the heat-bath, and the fluctuations in the average energy E create a situation different from the fully enclosed deterministic system.

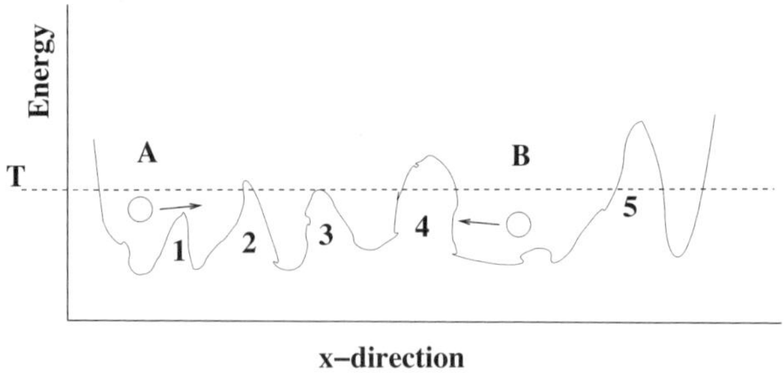

Fig. 14.3 A and B are two tiny inanimate objects, e.g., grains of sand, in a bath at a temperature T and moving in an energy landscape (e.g., due to the contour of the floor of the heat bath or 'pond') with peaks 1,2,. . . ,5. The grains move at random, and can occasionally acquire enough energy to jump over the topographical barriers.

We may also replace the grains of sand by a large protein molecule. The energy landscape of the heat bath is now convoluted with the energy landscape internal to the protein itself. As we have noted (see Sec. 8.6), it has a complex energy surface of hills and valleys with multiple minima due to the vast number of conformations possible for its atoms. The thermal buffeting would ensure that its active conformation is never the minimum-energy configuration, but any one of many within a width T of the thermal energy. That is, the actual conformation of the protein at any given moment is essentially unpredictable and indeed, *indeterminate*.

Hard determinism is beaten!

One could argue that if we take a sufficiently large number N of particles, then the effects of statistical fluctuations fall off as $\sqrt{N}$, and hence system fluctuations should be of no importance in collections of proteins etc. This argument is patently invalid for two related reasons. (i) The chemically active centers, NH_2 groups, -OH groups etc., function individually, and not as an averaged-out effect. That is, the *individual conformation* of a protein continues to hold the key to its activity. Many of these proteins are enzymes (catalysts) that have enabled many chemical reactions to happen within the range of the thermal energy T — these are reactions that would otherwise occur only at very high temperatures. (ii) The number N involved in a reaction step inside a cell is small and indefinite.

Now let us replace each of the inanimate grains by inanimate automata equipped with batteries and motors. The motors are only slightly more powerful than the thermal energy T. They are equipped with small propellers enabling them to move about in the pond, and go over the barriers 1,. . . ,5 using their own energy. We may also equip them with sensors, so that the automata detect the hills and valleys of its landscape. Now, given the slight excess energy of the automata over the thermal energy, their behaviour becomes more frantic ('hot'), but still erratic as there is *no 'purpose'* in their motion. They form a colony of Brownian particles, moving about and colliding with each other, as well as with the water molecules, and having an effective temperature T' which is slightly larger than T, as we have equipped them with extra energy. The effective temperature T' is established among the automata by their mutual collisions where kinetic energy is exchanged.

The next stage is to replace the sensors attached to these automata with better sensors that are sensitive to some chemical substance, e.g., sugar. The automata are now designed to be attracted to areas with higher concentrations of glucose. That is, the automata will now move around the pond 'hunting' for sugar-rich regions of the pond. If there are many such automata, they will jostle around and push, depending on the power of their motors, to get to the richer sugar pastures. Of course, these automata do not consume sugar as we have only built sensors into them and nothing more. They merely go to the sugar. They do not have any mechanism for generating energy from the sugar or for replication.

If these devices could plan out a scheme of action in advance, or take account of a developing 'ground situation' and modify their plan and enact another, then we begin to claim that such devices have 'free will'. Their capacity to select between alternative plans of action in an optimal way (e.g., to save energy or sugar) would be called 'intelligence'. A more complete discussion of 'intelligence' can be formulated along these lines. One well-known discussion of 'how we distinguish intelligent action' from 'optimized mind-less action' begins with Allen Turing's well-known 'test', where we pose the same problems to the automaton, and to an intelligent person. If the automaton's answers compare well with those of the person, then we should attribute a comparable intelligence to the automaton as well. Of course, this approach has been subject to much discussion. However, we believe that well-argued presentations of 'intelligence' as problem solving by machines capable of optimization has been given by many authors. For instance, Steven Pinker addresses such questions in detail [76].

Thus we see that the purely deterministic physics of many-particle systems, placed within an ambient spread of energies associated with the temperature and other state variables of the medium. and the boundaries, evolve into an effective physics of nominally free, i.e., non-deterministic objects. If these objects were complex systems with essentially chaotic dynamics, then the immersion in the heat bath makes it entirely impossible to specify the initial conditions of the dynamics, and hence the indeterminacy is essentially complete.

14.3.1 *Free will as a subdynamical concept*

In the previous section we examined the behaviour of a robotic device equipped with its own internal power, motile and sensory capacities. One aspect of the device was its capacity to move towards increasing concentrations of sugar, while negotiating the landscape of the environment it is positioned in. In discussing its motion, the effect of the 'random' fluctuations of kinetic energy arising from the molecular motion and solvation effects of the aqueous medium can be 'averaged over'. especially if the device has sufficient kinetic energy of its own. That is, a sub-dynamics emerges, where we only take account of the internal mechanics of the device. Now the question is, how deterministic are these internal mechanics? Its internal mechanics (empowerment of its muscles) is controlled, in the case of living organisms, by its adaptive unconscious. The mechanics of the automated bug is entirely controlled by its motors, sensors, and the information templates which optimize its motion between sugar gradients and the topography of the pond. The locus of motion arising from this internal mechanics is an effective macroscopic dynamics attached to the 'location' of the device in its spatial world. In our discussions of consciousness we noted that the location of an organism in space is mapped to a location in its spatial consciousness by the coordinated action of its place cells. In the case of the simple automaton, we have sensors playing the role of the place cells. Thus the dynamical 'location' of the organism is traced by the locus of the 'self' in the organism's spatial consciousness.

We begin to use the approximate language of 'cause' and 'effect', and say that the device (organism) is causing itself to move it. That is, we ignore the detailed electrical mechanisms of the battery and the solenoids, the motor and the propeller, the optimization algorithms etc., that makes the device move. We say that the device moves on its own, automatically. In the case of an organism we say that the organism moves by itself. An organism with consciousness would also 'feel' (through its place cells) that it is moving through space. That is, the adaptive unconscious posts into the consciousness, self-awareness of its motion.

In which direction does it move? Its sensors are equipped to detect gradients in sugar concentration. Thus, if its left sensor detects more sugar than the sensor in the right, it would begin to move to the left. If there are other devices which are similarly equipped in different parts of the pond, they would all begin to move towards the sugar-rich area and they would seem to 'compete' with each other. Thus *a new description of the devices, and their dynamics begin to emerge*. We begin to say that the devices behave as if they are *motivated by self-interest*.

These automata of the type A, equipped with sensors merely respond to the sugar gradients and topographical features of their environment. However, we may conceive of devices of a type B, equipped with sensors, as well as information processing devices which could plan out a strategy. For instance, the device would use stored topographical information (memory) to decide to go to the sugar-rich region in a more energy-efficient manner by taking an indirect, topologically optimal route R, instead of following the path P which defines increasing sugar concentration. The optimization algorithm may be like a neural network which can learn by itself, or a directly engineered AI-algorithm. Irrespective of the details of the algorithm, B is now using two sets of information (sugar gradients as well as topography) to optimally choose its path. As the capacity for optimization among more and more information templates is acquired by the device, its behaviour appears as being more and more 'intelligent', and it seems to act more efficiently to achieve its *self-interest*. Observing the behaviour of such devices of the type B, we would say that they are 'more intelligent' than the more primitive automata A that moved directly along the direct path P. (The automaton A is like the fly that beats against the glass of a closed window without moving to a nearby open window). In addition, we would observe that the type B devices seem to exercise their 'free will'. However, at any moment we can abandon the 'free-will' and 'self-interest' paradigm and revert to a more detailed discussion in terms of the detailed mechanics of the sensors, the information templates, and the optimization algorithms which use, say, some type of straight-forward Monte Carlo optimization routine. At that level we have abandoned the subdynamical factorization of the total dynamics, and the original full dynamics re-emerges. That is, the description in terms of fundamental molecular motions always exists, but the identification of higher-level attributes like 'free-will' and 'self-interest' cannot be done without the coarse graining and factorization of the sub-dynamics from the full dynamics.

14.4 Self interest

Is self interest expressed by the organism, or at a more elementary level? Richard Dawkins has argued that the evolutionary process itself is orchestrated by the self interest, i.e., *selfishness of the genes* rather than that of 'survival machines'. On page 33 of the 1989 edition of *the selfish gene*, Dawkins explains his views as follows:

> '... selfishness is to be expected in any entity that deserves the title of a basic unit of natural selection. ... some people regard the species as the unit of natural selection, others the population or group within the species, and yet others the individual. I ... think of the gene as the fundamental unit of natural selection, and therefore the fundamental unit of self interest. What I have done is to *define* the gene in such a way that I cannot really help being right.

Natural selection in its most general form means the differential survival of entities. Some entities live, and others die, but in order for the selective death to have any impact on the world, an additional condition must be met. Each entity must exist in the form of lots of copies, and at least some of the entities must be potentially capable of surviving — in the form of copies — for a significant period of evolutionary time. Small genetic units have these properties. ... What I have done is to define a gene as a unit which, to a high degree, *approaches* the ideal of indivisible particulateness. A gene is not indivisible, but it is seldom divided.

... The genes are the immortals, or rather, they are defined as genetic entities that come close to deserving the title. We, the individual survival machines in the world, can expect to live a few more decades.'

To understand the difficulties in Dawkins contrived definition, let us look at the subject of aromatic organic chemistry. We have vast sequences of organic reactions, where the benzene ring, and various structural groups G_i remains stable and unchanged. So are we to ascribe a concept of 'self-interest' to the benzene ring and other groups G_i, to interpret the kinetics of aromatic organic chemistry as a competition among G_i? However, one has no use for the concept of 'self-interest' to characterize the persistence of the benzene moiety in organic reactions. It has no predictive capacity here or in bio-chemical processes involving genes. If it is an 'emergent property' of genes (i.e., sets of base pairs with complementary molecules), what is the critical size, the order parameter etc, to formulate it in precise language?

Given that simple unicellular organisms have been unusually successful and able to survive even in astrophysical settings, genes can survive well in them. Thus evolution of 'higher' forms of life, sexuality etc. cannot be understood in terms of the self-interest of the gene.

Self interest is a higher-order emergent quality displayed by what Dawkins calls 'survival machines'. It is a manifestation of inertia, i.e., the quality of persistence in a given state. Newton's first law asserts that a 'free' particle would persist in its state of motion, but this is not interpreted as a form of self interest, but more as a result of the principle of least action. Genes do not make survival machines for them to persist for ever in a sequence of survival machines, any more than protons or electrons make various types of stars — their survival machines — to 'live' for ever in the evolutionary history of the universe. However, the the earliest survival machines (hydrothermal hatcheries, Darwinian ponds, froth bubbles etc.) happened to contain lipids, amino acids, phosphates, and sugars in water. Certain combinations of these are chemically stable, just as benzene rings are stable. These molecular combinations (genes) are no more, and no less, motivated by 'self interest' than benzene rings.

14.5 Personality

Even if each automaton in a group of 'intelligent' automata has self interest, the automata may not have *personality*. In a previous section we discussed automata of the type B which could respond to two types of information templates at the same time, viz., sugar gradients and the topography of the environment. If a large number of these devices were released into the pond, the jostling of the devices against each other would be similar to the interaction of hard-spheres (marbles) in a potential gradient, as no procedures have been built into the sensors of these devices to treat mutual device-device interactions, other than their capacity to deal with hard interactions in the topology. Further more, since the construction of the devices is essentially the same, only random engineering deviations would exist, and all the devices would behave virtually identically. An external observer would say that they seem to have no *personality* or *individuality* specific to them.

The growth of a multi-cellular organism from its embryonic stage is controlled by the genetic template as well as nutrients and ambient conditions in the mother's womb. The young infant comes to the world with a set of neurons and connections which are extremely plastic. The infant rapidly acquires new information and stores them in its memory. This storage is done by new anatomical changes at the synapses and axons. The visual, tactile, auditory and other information gathered by each organism is different from another. That is, even if we begin with two virtually identical organisms formed by a binary fission of a single cell (i.e., identical twins), the two daughter cells see the world differently. Also, the world treats them differently! These differences in the *experience* of the organisms produce different anatomical architectures in the synaptic connections and neuron networks of the organisms. Clinical work on infants born blind due to congenital cataracts, and who have had their vision restored at different stages of growth are very revealing in this regard. Although the genetically sanctioned pathways from the retina to the brain are ready and intact, it was found by Hubel and Weisel in the 1970s that early visual exposure is essential for the subsequent capability of the brain to acquire visual knowledge. That is, the different neural networks have to grow by learning. A large part of the 'innate contributions' of the brain too have to be learnt by nourishing the preset but pliable wiring of the brain by experience of the external world at a very early age.

These different neuron networks, preset but very malleable at birth, and modified by experience control the adaptive unconscious which is the main administrator of the muscles and organs of an organism. The adaptive unconscious largely determines the conscious part of the brain as well, and thereby reflects the anatomical differences in the architecture of the brain, via the biology of memory, emotions etc. That is, differences in brain-architecture provide the biological basis of *individuality* or *personality*.

It should how ever be noted that all organisms of a given species, e.g., humans, have a common brain structure and a common pattern of synaptic connections based on the shared genome of the species. That is, the genome has defined which regions of the brain are connected, and which groups of neurons within each region are connected to each other. In effect, a basic wiring diagram remains valid for all individuals of the same species, but the details vary from individual to individual. For instance, neuron connections of guitar players, and the representation of their playing fingers in the cerebral cortex would be much more extensive than that of someone who never plays such instruments, but uses the fingers to tightly hold a golf club. An individual like Mozart not only comes to the world with the 'right musical genes', but also to a family environment where he or she hears and participates in musical activity right from infancy.

If we view human societies as a new class of multi-unit organisms, where the units are human beings, interlinked by communication via language, culture and kinship, then we see that the common genome is supplemented by cultural differentiation at the memes level. The memes are themselves part of the brain architecture and reside in the synapses. This may be regarded as an added dimension in the make up of personality in that the 'person' is a part of the whole cultural group. That is, the cultural memes acquired by a person in a given social group, while having a common denominator – e.g., virtually everyone in the group has heard of the *Iliad* but not the *Mahavamsa* – would also have characteristics specific to his/her personality.

We noted that the information capacity of the genome ($\sim$750MB), and the available information storage in our neuron connections of the fore-brain (1,250 GB) seem to give much room for 'nature' as well as 'nurture' (see Sec. 12.2). That is, within generous limits, both the pattern and the extent of the synaptic connections can change, grow or weaken as a function of the individual's experience. This was vividly established in the experiments of Michael Merzenich and others [64]. Animal experiments show that this is indeed the basis of personality, not only in humans, but also in other living organisms.

14.6 Behaviour and guilt

If personality and behaviour are a matter of brain architecture, then is there a meaning to morality, guilt and criminality? Is there such a thing as ethical behaviour?

A counsel for an accused may plead that the accused acted in a particular way because of her/his brain structure, and assume no 'responsibility' for the actions or their consequences. Indeed, the legal system already recognizes that actions committed by some one under the conditions of a 'deranged mind', or 'temporary rage' cannot be judged under the same rules as for 'normal' individuals. The neurobiologist's job is to see how those rules have meaning within our new knowledge, and how such rules could be brought into line with improved knowledge of brain function. Pathologically clear situations like schizophrenia or bi-polar conditions are now fairly well understood at the neuro-biological level. This was formally recognized by the Nobel prize in physiology and medicine in 2000, awarded to Arvid Carlsson, Paul Greengard, and Eric Kandel. The problem is not with such easily definable situations like acute schizophrenia, or psychotic conditions, but more with those deviations from the norm that cannot be clearly characterized as pathology. Thus, what degree of deviations, i.e., eccentricity from the norm are acceptable to society, and what are not, are empirical matters of social acceptance and prejudice. However, that a better understanding of physiology can guide society away from prejudice is seen by the evolution of our attitudes towards same-sex relationships.

Societies have enacted rules of social behaviour and justice on the basis of collective empirical experience as well as the upholding of certain centers of power. That is, some aspects of 'justice' are not 'just, or fair' in any rational sense. However, they are promulgated to safeguard the property and privileges of those who are powerful — the domineering leaders of the pack. Such 'unfairness' is also part of the hierarchical organization of all societies. In the case of human societies, the very rich and powerful may have many acres of land, while hundreds of landless squatters may be crowding around the edges of the land. A squatter penetrating into the unused land of the rich man would be ejected and imprisoned for his unlawful act. That is 'justice', and the legal system originated to protect such rights of powerful individuals and groups! Quite often religion itself is recruited to justify such hierarchical justice, as in the *Manu dharma* doctrine of the Hindu Caste system, or in the doctrine of *the chosen people*. All social norms, including justice, religion etc., are the result of Darwinian evolution where individuals have to interact with each other, and with α-males, super-mums, and the 'Cleopatras' of each society.

In other words, the existing social norms, however imperfect, also have a microscopic neural representation in the brain, as a part of heredity, learning and acquisition of social skills. The adaptive unconscious of a normal 'well-bred' person would be implicitly aware of these norms. The conscious mind would also be aware of these norms according to the up-bringing of the person. We have concluded that complex adaptive systems (e.g., living organisms) that have a consciousness and the capacity to optimize among many plans of actions have a sub-dynamic where we can identify the property known as 'free will'. To the extent that this is true, there is a valid meaning for distinguishing between socially acceptable action (ethical behaviour) and unethical behaviour.

The inculcation of fear reactions, based on imbibing the concept of *sin* has been a means of control of human societies since antiquity. God, or other supreme being in different cultures tends to be an element of the social consciousness modeled with the attributes of Caesar — autocratic, powerful, patriarchal, and demanding your total and abject loyalty to Him. Similarly, Caesars, Kings and Pharaohs also claimed to rule by Divine Right. The Brahamins, i.e., the priestly caste of Hindus who exclusively officiate the religion, claim to have got their exalted position from God. The Church claims to get its exalted position by being the intermediary between God and sinner. Although modern heads of state, at least of non-Islamic nations, no longer justify their power using 'Divine Right', most religious organizations still do so. The history of religions, evangelism and imperialism, heresy and torture, even modern-day human-rights and attempts at 'protective intervention', all document the important role played by the belief in 'sin' — denial of 'God' — in human affairs [21].

Given that a person's genome is a significant indicator of the health, behaviour as well as the potentialities of the person, a fear reaction based on concerns of privacy has arisen against the acquisition of genomic data. In fact, a perusal of a catalogue of the human genome often reads like a list of diseases, e.g., a gene may be identified by its absence causing the Wolf-Hirschhorn syndrome. If there are over thirty five repetitions of that gene, it causes the deadly Huntington's chorea, and so on [113]. In countries like the USA where health insurance is largely a private business, some individuals may simply find that they cannot find an insurer. Often, the possibility of insurance is tied to being employed. Potential employers may, based on genomic data, openly or discretely refuse employment to vulnerable individuals. This is no different to the practice of an earlier age when women were refused employment because of possible interruption of work due to maternity. Clearly, these issues underline the fact that the right to health, as well as the right to be employed are matters that should involve a direct participation of the state. They can be left in the hands of the private sector once at least a basic safety net is in place. An educated and rational approach to this type of fear reaction has been spear-headed by the Harvard geneticist George Church. Church and colleagues launched the *personal genome project*, PGP, which aims to make personal genomes of consenting informed individuals publicly available, thus advancing the cause of biomedical research [4]. Otherwise, burying valuable information that could be used for the public good would be similar to the ban on dissections of the human body that pioneers of human anatomy had to face in the early days of the renaissance.

14.7 Biology, egalitarianism and democracy

Our genetic codes, although very similar, remain sufficiently different to create human inequalities in our individual physical and mental abilities. In the case of animals and plants, specific traits are selectively bred to create 'better' domestic pets, farm animals and crops. The early advocates of Eugenics argued that human beings may also be selectively bred to create super individuals, and also *eliminate* undesirable individuals. The logo of the second Eugenics conference in 1921 claimed that 'Eugenics is the self-direction of human evolution'.

Religious ideologies and social custom had dictated how societies should organize, bring up children and manage their affairs. Philosophers from Plato to Marx, and Mahavira to Krishnamurthy [32] had expressed their views on these matters. Eugenics was an old idea in a new garb, viz., coercive control of human reproduction 'for social good', but now using the newly minted ideas of biology — a biology that took at least another century to even find its own feet.

Sir Francis Galton, a first cousin of Charles Darwin, is often referred to as the 'father' of Eugenics, and he is credited with launching the name 'Eugenics' in 1885. However, the basic idea already exists in Plato. Plato's concept of communism of the family and society included the idea that the 'best sires would have the most children. Deformed children, and children of inferior parents will be put away in some mysterious unknown place, as they ought to be. Abortion and infanticide ought to be compulsory except for births sanctioned by the state'.

Plato also sanctioned the idea that the state has the right to deceive, in order to further the objectives of the state. In particular, Plato felt that the dogma that God has created three kinds of humans, viz., those with the ranks of gold, silver, and those of the common heard, of brass and iron, was a justifiable deception that the state should enforce. Those of golden rank were to be the guardians, while the soldiers were of silver, while the lower ranks should do manual labour. The children of each rank would generally inherit the rank of the parents.

It is not clear if Plato knew of the social structure that had developed in India. The Hindu caste doctrine promulgated that the *Brahamins* were created the highest caste, destined to be guardians of religion and ritual, while the *Kshatriyas* were the warrior caste. The mercantile classes were the *shudras*, and finally the *chandalas* — the untouchables — were at the bottom of the hierarchy. Each of these castes had their own pecking orders. Even today, 'low-caste' individuals may be treated as surfs, with no right to even draw water from a well; their very presence is considered to be polluting to a member of the upper caste.

The Eugenics movement found major advocates in the USA, in Nazi Germany and many other countries, among socialists as well as among staunch conservatives. England was a hot bed of Eugenics, where many intellectuals and politicians ranging from Bernard Shaw to Winston Churchill supported it. However, in the end no coercive legislation was passed due to the political efforts of strong liberals like Joshiah Wedgewood, a relation of the Darwins. Wedgewood and others successfully beat off a 1934 bill which claimed that it was 'desirable in the interests of the community that the feeble minded should be deprived of the opportunity of procreating children'. Also, although the Fabian society had supported Eugenics, the British labour Party had, by the late 1930s rejected it. Liberal-minded European countries like Holland also moved away from from it, while scientists like J. B. S. Haldane — a leading geneticist, biochemist and left-activist — began to attack Eugenics as bad science.

The Vatican, no supporter of individual liberty or free thought, opposed Eugenics just as it opposed abortions, family planning or the condom. The church had never faced any problems from the 'feeble minded' — in fact, its opponents had always been heretics and free thinkers who did not accept God's Plan as spelt out by the Church. Eugenics was an interference with God's plan. Thus, much of Europe (unless under the Nazis) also remained free of Eugenics legislation. However, in America, and in Canada in spite of its Catholic Quebec, legislation to forcibly sterilize people who were labeled as 'imbeciles, idiots' etc., was introduced in the 1930s. Such legislation continued in the USA and Canada well into the 1970s, in spite of the atrocities committed under the Third Reich.

When E. O. Wilson wrote his landmark work entitled *Sociobiology: The new syntheses* in 1975 [103], Ross, Kamin, Lewontin, Gould and others published a dogmatic riposte to it in the book *Not in your Genes* [81]. Today we know that Wilson's thesis stands justified, and modern research shows that genes matter very much, while the environment ('nurture') also matters very much. In fact, the idea of encouraging 'desirable' inherited traits and discouraging 'undesirable' traits seems reasonable at first sight. But then, the problems and pitfalls are equally evident to anyone but a zealot. Let us consider six rhetorical questions.

- Who gets to decide?
- What are desirable traits, and what are not?
- At what level (group, person, chromosome, gene, protein, synapse) do we look?
- How reliably can we predict outcomes of strategies?

- Do we know how one trait affects another?
- How is the program enforced?

The human species has its own cultural groups and national groups, Today science makes no use of such appellations because, given a specific characteristic or trait ascribed to, say some 'group', 'nation' or 'race', there is sufficient individual variability such that inter-group comparisons can be closer than intra-group comparisons, and hence the selection of traits based on 'race, nation', or group becomes scientifically meaningless. However, already in antiquity, concepts like 'chosen races', and 'elite castes' had become entrenched in human societies. Phrenology was a seemingly more 'scientific' attempt — but actually no better than palmistry — based on the physiological knowledge of the day, and based on the 'features of the skull'. This was proposed by the German physician Franz Joseph Gall in 1791. The Darwinian revolution sidelined many such pseudosciences. However, adolescent Darwinism spun out a few misconceptions of its own.

Biology has shown that multicellular organisms are tightly controlled systems in the sense that it requires that cells replicate correctly, 'according to plan'. Once a cell is formed (by division of a parent cell), it is managed to ensure that it operates correctly. Thus there are 'caretaker genes', and others which control 'cell death' wherein the cell is dis-assembled in a programmed manner (apoptosis). Cell division, where a new copy of the DNA is made, is a crucial operation which is accurately controlled. But given millions of base pairs, some errors arise — perhaps one in every billion or several millions — and these are enough to give rise to benign mutations as well as cancer cells. Most of these are immediately destroyed by the security forces (immune system) of the body. Thus the human organism is itself a tightly controlled highly totalitarian 'society' whose members are cells.

Cells forming multi-cellular organisms, and ants or bees forming insect colonies have great similarities. Cells signal each other using neurotransmitters and hormones, and control misbehaving cells using antibodies, bacteriophage and other weapons of the immune system. Insect colonies control members using pheromones in addition to hormones and neurotransmitters. The latter are spewed out into synaptic spaces, while pheromones are chemicals which are sent out into the air, or deposited as markers or as spatial clues. Dogs mark their territory with urine containing pheromones. Plants use the perfume of flowers in a similar manner, to attract insects. Insect colonies are multi-cellular organisms, dispersed in space, but tightly controlled by chemistry as well as signaling mechanisms that involve behaviour patterns which transmit information among members, as shown by Thomas Seeley and others. Such insect societies display collective unconscious adaptive behaviour (Sec. 12.6.1). A colony makes instinctive collective decisions, e.g., when it needs to found a new hive or search in a new area for honey. The 'mind', or consciousness of the colony is not found in the brain of any bee or in the queen's head, just as the mind of a human is not located in a particular neuron.

E. O. Wilson titled his famous book 'Insect Societies', rather than 'Insect colonies' [102]. Some human societies are also tightly controlled, while others are very weakly controlled but run smoothly on consensus. The latter are regulated by the public knowledge of every one's affairs ('gossip'), lack of a concept of 'individual privacy' or excessive privilege. There are remarkably egalitarian, peaceful societies, as Paul Gauguin found in Polynesia. In any case, it is an inescapable conclusion that humans are the 'cells' of a new class of 'organisms' that we call human groups, or human societies. However, human societies are knit together through verbal and visual communications rather than through pheromones. Although sexuality is very important, and led Sigmund Freud to give it the cardinal place in his psychoanalysis, we have to recognize that when the members of a species are no longer propelled by *estrus*, or 'heat', they have been liberated from a slavery to chemical control. Sex is not necessary for individual survival, while it is necessary for group survival (but genes can survive via asexual replication). Unlike animals, we have even lost our olfactory capacities, possibly since humans became bipeds and moved the nose away from the ground.

We have seen how humans, unlike, say insects, have a consciousness (in addition to the adaptive unconscious) which can visualize situations, use symbolic thinking, and have 'a theory of the mind' of others. Humans gesticulate and point to things. Many languages (e.g., French or Italian) are correctly spoken only with the appropriate hand gestures and with some eye contact. Infants come to the world ready to pick up these communication cues as well as languages. The interaction that knits human societies is not chemical, but communication using voice, vision, gesture and language. Technology has enhanced these immensely, and today we use a good part of the electromagnetic spectrum, copper wires, as well as the 1.55 μm fiber-optic communications to link us all together via the Internet, as well as other means. Would the supporters of the anthropic principle claim that the 1.55μm transmission window was created just for us?

Human societies are multi-unit organisms of a very different kind to insect colonies or multi-cellular animals. If we follow Dawkins, we may say that different groupings of human societies share common cultural beliefs or *memes* which are the components of the cultural genome inherited by an individual nurtured within a given group.

Does a society based on communication thrive best if we impose controls and create a highly authoritarian, controlled society as proposed by Plato or

St Augustine? Some societies, even today, impose detailed rules about how to bleed animals for slaughter, when to eat, and what days are good to bathe — or even assert that all bathing is a sign of sin. These are the cultural memes of those societies. If they are not voluntarily practiced, they become signposts of deep and pernicious control of opinion — i.e., suppression of 'heresy' — that characterize authoritarian societies. As societies advance, the need to have uncensored communication becomes even more important, as good collective decisions cannot be taken with manipulated information. Greek Drama and Elizabethan theater, carnivals and sports events, temple 'Mela' in India, were channels of public communication in those respective societies. Egalitarianism and democracy have advanced with the increasing spread of writing (as opposed to oral transmission), printing, journalism, broadcasting and the Internet. As communication and the exchange of ideas become more and more open, authoritarian memes which do not find justification in the collective consciousness of the group would be challenged and modified or discarded. Thus, the evolutionary processes involving the creating of mutant memes and their proliferation or extinction take place at the socio-cultural level.

Planning, especially at the level of controlling reproduction to emphasize selected characteristics requires the ability of detailed prediction of the outcomes of a chosen social strategy as applied to *individuals*. But we have already seen that we are unable to predict even the active conformation of a single protein immersed in water when it comes to cell function. At a more macroscopic level, we found that we have no way of predicting how an individual would blossom out. Isaac Newton was a sickly child who did not seem promising, and Einstein was not particularly outstanding when very young. Anatomists who have studied Einstein's brain have not come up with any credible clues to explain why he was a genius. On the other hand, we come across young children who seem to have much promise, and yet fizzle out before long. Then there are others who unmistakably develop their talent, nurtured and nourished by equally talented parents.

At an even more macroscopic level of socio-economic evolution, we are even less competent. The stock market is a phenomenon which is abundantly characterized by numerical data as a function of time, and constantly studied by technical experts in banks and other institutions, using the world's most powerful computers. But we have no predictive capacity. In effect, the judgment of a man on the street is almost as good as that of the expert in most such matters. As noted before (Sec. 3.10), the economist Friedrich von Heyak actually characterized economics as a 'pretense of knowledge' in his 1974 Nobel address. Economic and monetary dicta imposed by, say, the World Bank, are probably misguided as economies do not obey any deterministic laws, but follow chaotic dynamics. If the prognosis of the man in the street and the highly informed expert are equally unreliable, that is the best justification for universal franchise!

We are dealing with a complex adaptive system, possibly chaotic, but certainly indeterminate in the sense of being beyond our predictive capacity. Nevertheless, we are attempting to simulate it — govern it — with a number of 'representative movers', or 'participants' which sample the system. The bigger the number of participants in the sampling process, the better is the 'simulation'. Thus democracy and egalitarianism emerge by our need to assign essentially equal weights to all the participants who are equally unaware of their destiny. Due to our lack of knowledge of the time evolution of these participants, we simply have to let them play. Any other strategy — that is, a detailed plan which subsumes the individual dynamics via some model — will prove to be incorrect, as the envisaged time evolution of a chaotic system is guaranteed to diverge from the true trajectory very rapidly. The 'who' of the six rhetorical questions (e.g., who gets to decide?) is the collective mind of the society, and this works best if communication is not suppressed and if everyone's voice is heard loud and clear.

14.8 Happiness and suffering

Once an organism acquires a sensory system linked to modules of neuron nets, it begins to feel the environment outside it, as well as the conditions and dispositions of the organs inside it. The outside world is brought to the attention of the adaptive unconscious via axons connected to sensory neurons in the eye, ear, skin etc. The state of the internal organs are also transmitted as action potentials, and also by long-range chemical signaling via hormones. Pain and anxiety are reactions that the organism develops as a protective mechanism to ensure that it takes defensive or evasive action as soon as a dangerous situation is encountered. The conscious mind is put to the back burner; involuntary actions are controlled by the adaptive unconscious that leaps into action when danger is detected.

Reproduction is so fundamental to evolution that it would have been very surprising if the pleasure of sex was not a well encoded neuronal process. Even at the level of synaptic processes, we note the similarity of genital organs to a synaptic pair. Vacuoles containing neuro-transmitters being released by a synapse is similar to semen being released to fertilize an egg. Sexuality is a specialized form of cell signaling. Perhaps when action potentials reach pre-synaptic regions, the firing of the neurotransmitter chemical across the synaptic cleft leads to a successful message, and this leads to 'pleasure', while the inability to fire, and the consequent accumulation of unused neurotransmitter vacuoles in the presynaptic regions registers as stress detected by the AU by other subcortical neural networks.

Simpler pleasurable reactions, e.g., the taste of sweetness, reaction to pleasurable rhythms, are also involuntary processes controlled by the AU. Neurobiologists claim that the feeling of 'sweet taste' is linked with processes in the subcortical network and the brainstem [9]. Many pleasurable reaction are 'learned skills'. Thus, one has to learn to like the taste of beer or coffee, in much the same way that one has to learn to like the music of Arnold Schönberg or John Cage.

Pleasures directly controlled by the AU are linked with the non-declarative memory. However, the more interesting and probably more sophisticated pleasures are those linked with the declarative memory. Here we have pleasures of a very complex type harking back to old memories, nostalgia and also the anticipation of what may happen. We have the possibility that what we anticipate and dream of may not happen, and so we have the other side of the coin — mental pain, disappointment and acute suffering as well. Due to the strong control of the conscious mind by the adaptive unconscious, conditions of depression and other psychopathologies play a strong role in selecting the templates (neural-net modules) that the conscious mind has access to. The mood and the personality are largely set by the adaptive unconscious, and the conscious mind works with it.

Many writers have appreciated that the *normal state of the conscious mind* may be a state of discontent. Sigmund Freud called it the 'psychological misery of mankind'. The conscious mind has to grapple with formulating lines of action that satisfy the 'self interest' of the organism. Sigmund Freud in 1930 published *Das Unbehagen in der Kultur*, available in translation as 'Civilization and its discontents' [36]. Freud's early-manuscript title was in fact *Das Üngluck in der Kultur* and he was addressing the problem of *unhappiness* in society. Freud argues that the individual has to distance himself from the suffering that s/he faces in the world, due to the mismatch between his self interest and what is possible within the social and biological limits available to him in the external world:

> "..., what decides the purpose of life is simply the programme of the pleasure principle. This principle dominates the operation of the mental apparatus from the start. There can be no doubt about its efficiency, and yet its programme is at loggerheads with the whole world, with the macrocosm as much as with the microcosm. There is no possibility at all of its being carried through; all the regulations of the universe run counter to it. One feels inclined to say that the intention that man should be 'happy' is not included in the plan of 'Creation'. ... We are threatened with suffering from three directions: from our body, which is doomed to decay and dissolution and which cannot even do without pain and anxiety as warning signals; from the external world, which may rage against us with overwhelming and merciless forces of destruction; and finally from our relations to other men. The suffering which comes from this last source is perhaps more painful to us than any other."

Freud proposes 'universal love' as an antidote for *Üngluck*.

'far reaching mental changes in the function of love are necessary before this can happen.... What they bring about in themselves in this way is a state of evenly suspended, steadfast, affectionate feeling, which has little external resemblance any more to the stormy agitations of genital love, from which it is nevertheless derived'.

Freud regarded St. Francis as having developed a state of compassionate love. The greek *agape*, Christianized as 'God's love for mankind and the reciprocal human love of God, extended also to mankind', would have been known to Freud.

Similar views had been expressed in much greater detail than Freud by an ancient Indian sage. Some 26 centuries before Freud, Gautama Buddha, a contemporary of Heraclitus, had proposed an elaborate theory of suffering and emancipation from suffering that apparently escaped Freud's attention, although it was well known to 19th century Europe through the work of Arthur Schopenhauer. The first sermon of the Buddha, delivered in Benares announced four fundamental principles [80]. These were known as the 'four noble truths', or the four *ārya sacca*. The first ārya sacca was the principle that nothing has a permanent existence — everything is impermanent. However, all sentient beings are trying to maintain a state of permanent happiness, and this is impossible. Happiness must necessarily be followed by unhappiness, and unhappiness will be followed by another cycle of change. Hence *dukka*, i.e., 'suffering' or 'misery' — the *unglück* of Freud, is the principle character of the sentient world. Buddha's second ārya sacca, states that the cause of suffering is greed or *self interest* as given by the Pali word *thanha*, i.e., literally, 'thirst'. In the third ārya sacca the Buddha claims that there is indeed a way to abate suffering, namely, by subjugating 'thanha'. The fourth ārya sacca is, in effect, an exposition of a code for living, based on universal love or compassion ('metta'), methods of mind control and *jhāna*. It is this latter word, 'dhyana' in Sanskrit, which has morphed in to 'Zen' by the time it took the old silk route from India and arrived in the extreme orient.

The 'far reaching mental changes in the function of love' that Freud talked of in 1930 was also proposed by the Buddha. The word *arahant* in Pali refers to a person who has vanquished the 'foe' of greed (self-interest) and exudes compassion, not just for mankind, but for all *unglücklische* forms of *life*.

While the Buddha had argued cogently in the sixth century BCE that minimization of suffering and compassion are the ways to improve the human condition, the 19th century utilitarians like Jeremy Bentham and James Mill argued that *maximizing* happiness is the *summum bonum*. Bentham equated pleasure with happiness. The ethical doctrine is that those desires and those actions are good if they promote the general happiness. On the other hand, Byron and Nietzsche seem to hold that only a minority of the human race have any ethical importance. That is, the happiness or unhappiness of the rest should be ignored.

Since Byronic, Benthamite or Buddhist dispositions are themselves a result of neurobiology, some understanding of the molecular mechanisms that lead an individual to one or other ethical point of view can be achieved by suitable neuroscience experiments. That is, our moral dispositions are themselves a result of our genes and learning. Presumably, a person's moral disposition can be modified by the way he or she is brought up. Buddhist or some other practice of meditation directed to compassion, if shown to be effective, would be desirable in most civil societies, while a military education as in ancient Sparta may produce a very different moral attitude desirable in other circumstances. Hence our morality is also a matter of the synaptic connections in our brains. Hitler and Pol Pot were surely such morally sick individuals with pathological neuron wiring.

E. O. Wilson, in a 1998 book titled *consilience* has argued that morality is the distilled expression of our adaptive consciousness — i.e., a codified expression of our instincts. However, our synaptic connections are controlled by the $\sim$750MB of information coming via the human genome, and the 1,250 GB of information-space (memory) available to the fore-brain for learning and skill formation. Our instincts reside in our synaptic connections. How the adaptive unconsciousness arrives at moral decisions is unknown. What is the norm, and what is the pathological deviation, is a question where a value judgment is needed.

What organisms *ought to do* by way of moral behaviour cannot be settled by neuroscience (see Sec. 2.1.1). Perhaps the minimization of suffering, rather than the Benthamite maximizing of pleasure is a reasonable principle that could guide personal and social behaviour. Even if science is value neutral, the scientific *method* breeds its own values. It demands the need for extreme honesty with facts and the capacity to accept one's errors. Similarly, the need to respect the opinion of one's colleagues and peers is seen in the methods adopted by the learned societies, and even in the discussions at the famous Solvay conferences where Einstein, Bohr and other founding fathers of modern physics sorted out the essentials of the quantum theory. Unfortunately, the more egoistic, theatrical antics of some celebrities become the headline news.

Human society has moved out of direct Darwinian evolution and it is now capable of evolution based on the intellectual abilities of its collective consciousness — i.e., its culture and civilization. The sciences give us an understanding of nature that no living organism had ever possessed, instinctively or consciously. Human evolution itself is now in the hands of human beings [19].

Bibliography

[1] D. L. Adams and J. C. Horton, J. Neuroscience, **23**, 3771 (2003).

[2] Astrochemistry.net `http://www.astrochemistry.net/`

[3] Sidney Altman: The RNA World. Nobelprize.org (2010); `http://nobelprize.org/nobel_prizes/chemistry/laureates/1989/altman-article.html`

[4] Angrist M.: Personalized Medicine. **6**, 691 (2009), `http://www.medscape.com/viewarticle/712551` downloaded Dec. 2010.

[5] Bartels A. and Zeki S.: *The architecture of the colour centre in the human visual brain: new results and a review*, Eur. J. Neurosci. **12**, 172 (2000).

[6] Beers, T. C.: APS conference series **336**, 283 (2005).

[7] Bergquist, P. R.: Sponges, University of California Press, Berkeley (1978).

[8] Bernstein, H., Hopf, F. A. and Michod, R. E.: *The Evolution of Sex: DNA Repair Hypothesis, The Sociobiology of Sexual and Reproductive Strategies*, ed. C. Rasa and E. Voland, Chapman and Hall, London (1989).

[9] Berridge, K. C.: *Pleasures of the Brain*, Brain Cogn. **52**, 106 (2003).

[10] Bliss, T. V. P. and Collingridge, G. L.: Nature **361**, 31 (1993).

[11] White, J. G., Southgate, E., Thomson J. N. and Brenner, S.: *The Structure of the Nervous System of the Nematode Caenorhabditis elegans*, Phil. Trans. R. Soc. London, B 314, 1 (1986).

[12] Brian. E. Brown: The Buddha natrue, a study of the Thathagatagarbha and alayavijnana, Banarsidas Publishers, New Delhi (1991).

[13] Cairns-Smith, A. G.: Evolving the Mind. On the Nature of Matter and the Origin of Consciousness, Cambridge University Press, New York (1996).

[14] Thomas R. Cech: "Exploring the New RNA World", Nobelprize.org (2010); `http://nobelprize.org/nobel\_prizes/chemistry/laureates/1989/cech-article.html`

[15] Churchland, P.: The Engine of Reason; the seat of the soul, MIT Press, Cambridge, Mass. (1995).

[16] Chyba C. F., Sagan, C. in *Comets and the Origin and Evolution of Life*, p. 146, Springer, New York (1997).

[17] Cleaves II. , H.J. , Nelson , K. , Miller , S.L.: *The prebiotic synthesis of pyrimidines in frozen solution*, Naturwissenschaften **93**, 228 (2006).

[18] Corliss, J., Baross, J. and Hoffman, S.: *An hypothesis concerning the relationship between submarine hot springs and the origin of life on Earth*, Oceanol. Acta **4** (Suppl.), 59 (1981).
[19] Dawkins, Richard: The Greatest Show on Earth: The Evidence for Evolution, Simon and Schuster, New York (2009).
[20] Dawkins, Richard: The Selfish Gene, Oxford University Press, Oxford (1989).
[21] Dawkins, Richard: The God Delusion, Houghton Mifflin Harcourt, New York (2010).
[22] Deamer, D.W., Fleischaker, G.R.: *Origins of Life. The Central Concepts*, Jones and Bartlett Publishers, Sudbury, MA (1994).
[23] Dennett, Daniel C.: Freedom evolves, Viking, New York (2003).
[24] Delano, J. W.: Origins of Life and Evolution of Biospheres, **31**, 311 (2001).
[25] Max Delbrück: *Light and Life III*, Carlsberg Res. Commun. **41**, 299 (1973).
[26] `http://dictybase.org` downloaded 2010.
[27] Doniach S. and Sondheimer, E. H.: Green's Functions for Solid State Physicists, Benjamin Inc. Reading, Mass. (1974).
[28] Double-Helix: 50 years of DNA. `http://www.nature.com/nature/dna50/index.html`; Accessed on 28 Dec. 2010.
[29] Duve, Christian de: *Exploring cells with a centrifuge*, Science **189**, 186 (1975).
[30] Ehrenfreund, P. and Charnley, S.B. *Organic molecules in the ISM, comets and meteorite: a voyage from dark clouds to the early Earth*, Annu. Rev. Astron. Astrophys. **38**, 427 (2000).
[31] Ehrenfreund, P., Rasmusse S., Cleaves, J. and Chen Liaohai, Astrobiolgy **6**, 409 (2006).
[32] `http://www.strippingthegurus.com/stgsamplechapters/krishnamurti.asp` Downloaded 2010.
[33] Farquhar J., Peters M., Johnston D., Strauss H., Masterson A., Wiechert U. and Kaufmann A. J.: Nature **449**, 706 (2007).
[34] Feynmnn, R. P.: Feynman Lectures on Computation, Basic, New York (1999).
[35] Flores, J.J., Ponnamperuma, C.: *Polymerization of amino acids under primitive Earth conditions*, J. Molecular Evolution **2**, 1 (1965).
[36] Freud, Sigmund: Civilization and Its Discontents, Translated by: James Strachey, Notron & Co., New York (1962).
[37] H. Fröhlich: International Journal of Quantum Chemistry, **11**, 641 (1968).
[38] Gardner, M. *Mathematical Games; The fantastic combinations of John Conway's new solitaire game 'life'* Scientific American, **223**, 120 (1970).
[39] Garrison, W. M., Morrison, D. H., Hamilton, J. G., Benson, A. A., Calvin, M., *The reduction of carbon dioxide by ionizing radiation*, Science **114**, 416 (1951).
[40] Gibson, D. G., J. I. Glass, C. Lartigue, V. N. Noskov, R.-Y. Chuang, M. A. Algire, G. A. Benders, M. G. Montague, L. Ma, M. M. Moodie, C. Merryman, S. Vashee, R. Krishnakumar, N. Assad-Garcia, C. Andrews-Pfannkoch, E. A. Denisova, L. Young, Z.-Q. Qi, T. H. Segall-Shapiro, C. H. Calvey, P. P. Parmar, C. A. Hutchison III, H. O. Smith and J. C. Venter.: *Creation of a bacterial cell controlled by a chemically synthesized genome*, Science, Published online, May 20, 2010, Science, **329**, 52, (2010).
[41] Gilbert, W.: *The RNA World*, Nature **319**, 618 (1986).

[42] Gracia-Viloca M., Gao J., Karplus M. and Truhlar D. G., *How enzymes work*, Science **3003**, 186 (2004).

[43] Stuart Hameroff: `http://www.quantumconsciousness.org/presentations/whatisconsciousness.html` – downloaded 2010.

[44] Stuart Hameroff: Phil. Trans. R. Soc. Lond. A **356**, 1869 (1998).

[45] Hameroff, S. R. and Watt, R. C.: *Information processing in microtubules*, J. Theor. Biol. **98**, 549 (1982).

[46] Hargreaves, W. R., Mulvihill, S. J., Deamer, D. W.: *Synthesis of phospholipids and membranes in prebiotic conditions*, Nature, **266**, 78 (1977).

[47] Heinen, W. and Lauwers, A. M.: Orig. Life Evol. Biosphere **26**, 131 (1996).

[48] Hoyle, F. and Wickramsinghe, N.C.: *Astronomical Origins of Life; Steps Towards Panspermia*, Kluwer Academic Press, Amsterdam (2000).

[49] François Jacob: The Possible and the Actual, Pantheon Books, New York (1982).

[50] William James: *The Physical Basis of Emotion*, Psychological Review, **1**, 516 (1884).

[51] Kandel, Eric: In Search of Memory, Norton, New York (2006).

[52] Kane, Robert: The significance of Free will, Oxford University Press, Oxford (1996).

[53] Kelley, Deborah S.: *The lost city hydrothermal field*, Oceanography, **18**, 32 (2005).

[54] Kornack, D. R.: *Neurogenesis and the evolution of cortical diversity: mode, tempo, and partitioning during development and persistence in adulthood*, Brain Behav. Evol. **55**, 336 (2000).

[55] Lane, Nick: Life Ascending, Norton & Co., New York (2009).

[56] Lehninger, A.L.: Bioenergetics, W. A. Benjamin, New York (1965).

[57] Lohrmann, R.: *Formation of urea and guanidine by irradiation of ammonium cyanide*, J. Mol. Evol. **1**, 263 (1972).

[58] Macy, Joanna: Mutual Causality in Buddhism: The Dharma of Natural Systems. State University of New York Press, New York (1991).

[59] Margulis, L. and Schwartz, K. V.: Five Kingdoms, an illustrated guide to life on earth, Freeman, Ney York (1998).

[60] Martin, William, Baross, John, Kelley Deborah, and Russell, M. J.: *Hydrothermal vents and the origin of life*, Nature Reviews Microbiology **6**, 805 (2008).

[61] Martin, W. and Russell, M. J.: Phil. Trans. R. Soc. Lond. B **358**, 59 (2003).

[62] Maxwell, J. C.: *On colour vision*, Proc. Roy. Inst. Gt. Britain. **6**, 260 (1872).

[63] McClintock, B.: *The significance of responses of the Genome to challenge*, Nobel Lecture, Dec. 1983. `http://nobelprize.org/nobel\_prizes/medicine/laureates/1983/mcclintock-lecture.html`

[64] Merzenich, M. M. and Sameshine, K.: *Cortical plasticity and memory*, Current Opinion in Neurobiology, **3**, 187 (1993).

[65] Miller, S. L.: *A production of amino acids under possible primitive Earth conditions*, Science **117** , 528 (1953).

[66] Miller, S. L. and Urey, H. C. *Organic Compound Synthesis on the Primitive Earth*, Science, **130**, 245 (1959).

[67] Minsky, M.: The Society of Mind, Simon and Schuster, New York (1986).

[68] Minsky, Marvin: *Machinery of Consciousness*, 75th Anniversary symposium of the National Research Council of Canada, NRC Press, Ottawa (1991).

http://kuoi.com/~kamikaze/doc/minsky.html, downloaded Dec. 2010.
[69] Moody, J. B.: Lithos, **9**, 125 (1976).
[70] Nelson, K., Levy, M., Miller, S. L.: *Peptide nucleic acids rather than RNA may have been the first genetic molecule*, Proc. Natl. Acad. Sci. USA **97**, 3868 (2000).
[71] Nielsen, P. E.: *Peptide nucleic acid (PNA): a model structure for the primordial genetic material?*, Orig. Life Evol. Biosph. **23**, 323 (1993).
[72] Norretrandres, T.: The User Illusion, Viking, New York (1998).
[73] Orgel, L. E.: *Prebiotic chemistry and the origin of the RNA world*, Critical Reviews in Biochemistry and Molecular Biology, **39**, 99 (2004).
[74] Penrose, R.: Shadows of the mind, Oxford University Press, Oxford (1994).
[75] Pereto Juli, Bada, Jeffrey L. and Lazcano Antonio: Origins of Life and Evolution of Biospheres, **39**, 395 (2009).
[76] Pinker, Steven: How the Mind Works, W. W. Norton & Co., New York (1997).
[77] Ponnamperuma , C., Lemmon, R. M., Mariner, R., Calvin, M.: *Formation of adenine by electron irradiation of methane, ammonia and water*, Proc. Natl. Acad. Sci. USA **49**, 737 (1963).
[78] Ponnamperuma , C. , Mack , R.: *Nucleotide synthesis under possible primative Earth Conditions* , Science **148**, 1221 (1965).
[79] Ridley, Mark: The Cooperative Gene, The Free Press, new York (2001).
[80] Rahula, Walpola: What the Buddha Taught, Grove Press, New York (1974).
[81] Rose, S., Kamin, L. J. and Lewontin, R. C.: Not in your Genes, Pantheon, London (1984).
[82] Reimers, J. R., McKemmmish, L. K., McKenzie, R. J., Mark, A. E., and Hush, N. S.: *Weak, strong, and coherent regimes of Frólich condensation*, Proc. Natl. Acad. Sci. **106**, 4219 (2009).
[83] Salari, V., Tuszynski, J., Rahnama, M., Bernroider, G. arXiv:1012.3879v1 [physics.bio-ph] (2010).
[84] Sanchez , R. A., Ferris , J. P., Orgel, L. E.: *Conditions for purine synthesis: Did prebiotic synthesis occur at low temperatures?* Science **153**, 72 (1966).
[85] Schopf, W. J.: *Fossil evidence of Archaean life*, Phil. Trans. R. Soc. B **361**, 1470 869 (2006).
[86] Sephton, M. A.: *Organic compounds in carbonaceous meteorites*, Natural Product Reports, **19**, 292 (2002).
[87] Singer S. J. and Nicolson, Garth L.: *The Fluid Mosaic Model of the Structure of Cell Membranes*, Science **175**, 720 (1972).
[88] Manfred Spitzer: The Mind within the Net, The MIT Press, Cambridge, Mass. (1999).
[89] Squire, Larry R. and Kandel, Eric R.: Memory, from mind to molecules, Freeman & Co., New York (1999).
[90] Strickberger, Monroe: Evolution, Jones and Bartlett, Burlington, Mass (2000).
[91] Strother, P. K., Barghoorn, E. S.: *Microspheres from Swartkoppie formation, A review*. In Halvorson, H. O., van Holde, K. E. (Eds.), The Origins of Life and Evolution, 1–18, Alan R. Liss, Inc., New York (1980).
[92] Tegmark, M.: *Importance of quantum decoherence in brain processes*, Phys Rev. E **61** 4194 (2000).

[93] Tian, F., Toon, O. B, Pavlov, A. A. and De Sterck, H.: Science, **308**, 1014 (2005).

[94] Hagen, S., Hameroff S. R. and Tuszyński, J. A.: Phys. Rev. E **65**, 061901 (2002).

[95] UMIST database for Astrochemistry, Woodall, J., Agúndez2006, M., Markwick-Kemper, A. J. and Miller T. J.: `http://www.udfa.net/`

[96] Wächtershäuser, G.: *Before enzymes and templates: theory of surface metabolism*, Microbiol. Rev. **52**, 452 (1988).

[97] Wächtershäuser, G.: *On the Chemistry and Evolution of the Pioneer Organism*, Chemistry & Biodiversity, **4**, 584 (2007).

[98] Watson J. D.: Molecular Biology of the Gene, 2nd edn., Saunders, Philadelphia, PA (1972).

[99] R. Wayne: Plant Cell Biology. From Astronomy to Zoology, Academic Press, New York (2009).

[100] Wickramasinghe, N. C., Dharma-wardana, M. W. C. and Nandi, I.: Roy. Astr. Soc. (UK) Monthly Notice **134**, 25 (1966).

[101] Wickramasinghe, J. T., Wickramasinghe, N. C and Napier, W. M.: Comets and the Origin of Life, World Scientific Press, Hong Kong (2009).

[102] Wilson, E. O.: The Insect Societies. Belknap Press, Cambridge, Mass. (1971).

[103] Wilson E. O.: Sociobiology, The new synthesis, Harvard University Press, Cambridge, Mass (1975).

[104] Wilson, E. O.: Consilience, Little, Brown, New York (1998).

[105] Wilson, M. A. and McNaughton, B. L.: *Dynamics of the hippocampal ensemble code for space*, Science **261** 1055 (1993).

[106] D. S. Wilson: Evolution for Everyone, Random House, New York (2007).

[107] T. D. Wilson: Strangers to ourselves; Discovering the adaptive unconscious, Harvard University Press, Cambridge (2002).

[108] Wolfe, D. W.: *Tales from the Underground: A Natural History of Subterranean Life*, Basic Books, Cambridge, Ma. (2001).

[109] Wolfram, Stephen: A new kind of Life, Wolfram Media Inc. (2002). `http://www.wolframscience.com`

[110] Wu, T. M. and Austin, S. J.: J. Biol. Phys. **9**, 97 (1981).

[111] von Neumann, J.: The Computer and the Brain, Yale Univeristy Press, New Haven, Con. (1960).

[112] Von Neumann, J. and Burks, A. W.: Theory of Self-reproducing Automata, University of Illinois Press, Urbana (1962).

[113] Wexler, A.: Mapping Fate, University of California Press, Los Angeles (1995).

[114] Zeki, S. M.: A vision of the brain, Oxford University Press, Oxford (1993).

Index